# FISHERIES OF THE NORTH PACIFIC

Seattle Post-Intelligencer

# F THE NORTH PACIFIC

## History, Species, Gear & Processes

## by Robert J. Browning

WITH EDITORIAL ASSISTANCE AND REVIEW BY

DR. DAYTON L. ALVERSON, Director, Northwest Fisheries Center;

ARTHUR H. PRIDDY, Librarian, Northwest Fisheries Center;

JOHN A. DASSOW, Pacific Fishery Technology Center,
NOAA, National Marine Fisheries Service, Seattle, Washington;

HARRY L. RIETZE, Director, Alaska Region,
NOAA, National Marine Fisheries Service, Juneau, Alaska;

LAURENCE FREEBURN, Alaska Cannery Operator;

RICHARD H. PHILIPS, Pacific Editor, National Fisherman;

The editorial staff of ALASKA® magazine

## DEDICATION

*This book is dedicated to all the world's commercial fishermen, whatever their gear, whatever their sea, with special respect and affection for the Americans of the Northeastern Pacific.*

# CONTENTS

# ACKNOWLEDGMENTS

This book attempts to present something of the story of the major American and Canadian fisheries of the Northeastern Pacific Ocean. It is intended as a guide to the history of the fisheries, the biology of the species, the vessels of the fisheries, assembly of gear, fishing methods, the handling of the catch at sea and ashore, and processing of fishery products. There is an attempt at assessment of the future of these fisheries and there is included an appendix covering certain matters more completely than was possible in the text. The book does not pretend to be a scientific work.

The area of concern runs generally from the Mexican border north into the Bering Sea and westward into the Aleutians. Although there is a year-around presence of fishing fleets from Eastern Asia over much of that ocean area, they have been considered only as they impinge upon North American fisheries and fishermen. The Asians and the North Americans work at two entirely different levels of fisheries technology.

Necessarily, any book such as this must depend heavily on the work of people whose professional careers have been spent in fisheries and fisheries-related fields — scientists, engineers and technicians, fisheries company personnel, and working fishermen whose lives and livelihood depend upon the uncertainties of the sea and the quarry they seek from it.

Thus, I am indebted to many persons for their contributions to my efforts. Among them are five who, to borrow the military tribute, went beyond the call of duty to guide me in matters outside my own experience and observation. They are, alphabetically, Capt. Sig Jaeger, Division of Marine Resources, University of Washington, Seattle, Washington; Jerry E. Jurkovich, fisheries biologist and gear specialist, Northwest Fisheries Center, National Marine Fisheries Service, Seattle; Richard H. Philips, Pacific editor, *National Fisherman;* Arthur H. Priddy, librarian, Northwest Fisheries Center, and Don Reed, information officer, Washington Department of Fisheries.

In addition to some of the above, the publisher's board of review included Laurence Freeburn, Alaska cannery operator, Dr. Dayton L. Alverson, Director, Northwest Fisheries Center, NOAA, National Marine Fisheries Service, Seattle, and John A. Dassow, Pacific Fishery Technology Center, NOAA, National Marine Fisheries Service, Seattle. The proofs of the original manuscript were edited by Mrs. Ethel Dassow of Mercer Island, Washington, who gave generously of her knowledge of fisheries technical writing to bring order to syntax and style. A number of corrections and suggestions offered by this panel were incorporated into the final text. The author, however, remains solely responsible for any errors of fact or interpretation.

Others who had a part in the development of this book include:

Mario Puretic, inventor of the Power Block; Dr. William F. Royce, then associate dean, College of Fisheries, University of Washington; Peter G. Schmidt Jr., founder and president, Marine Construction and Design Co., (MARCO), Seattle; Robert Allen, vice-president and general manager, MARCO; Eldon Grimes, director, Fishery Development, MARCO; Merrily Manthey, public relations, MARCO, and her successor, Mary Alden; Walt Kisner, editor, *Fishermen's News,* Seattle; A. T. Pruter, deputy director, Northwest Fisheries Center, Richard L. McNeely and A. K. Larssen, NMFS; John Kitasako, Port of Seattle; Ward Turnbull, president, Nordby Supply Co., Seattle; Ormal Richardson, superintendent of the Nordby net loft, and his wife and assistant, Jan; John Wiese, Anchorage, Alaska; Jack Schott and Eric H. Knaggs, California Department of Fish and Game; Lanny Swerdlow, Oregon Fish Commission; Dan Stair, Luhr Jensen and Sons, Inc., Hood River, Oregon, and Eugene A. (Buzz) Johnson, commercial fisherman field agent, Washington Sea Grant, Clover Park Education Center, Lakewood Center, Washington.

To those names must be added those of the fishermen, skippers and crewmen alike, whose suggestions, opinions and methods have influenced me heavily — S. A. (Spuds) Johnson, Jerry Granberg, Larry Brandstrom, Robert Browning Jr., Don Granberg, Kenny Clayton, John Johanson, George Staton Sr., Harvey Moen, Robert Bassett and David Huswick. For any omissions I apologize.

This book was conceived by Robert A. Henning, editor and publisher of *ALASKA®* magazine and owner of Alaska Northwest Publishing Co. Mr. Henning underwrote the project in its entirety. I have tried to adhere to his concept as closely as possible.

*Robert Browning,*
*Edmonds, Washington,*
*1974*

# 1/ FOREWORD

## *This is the Eastern Ocean*

*T*he Northeastern Pacific Ocean rolls and tumbles across some 40 degrees of longitude and about 35 degrees of latitude, measuring out at about 3,000 miles from east to west and the same distance, more or less, from north to south. These distances are approximations from Vancouver Island to somewhere out in the Aleutians quite a bit to the west of Adak and from Baja California to the water wastes of Bering Strait. The Bering Sea must be considered a part of the Northeastern Pacific because its fisheries are an extension, an evergrowing one, of the fisheries of the Eastern Ocean although on the global projection, the Bering does not appear to be an integral part of the Pacific because the Aleutians fence it off. Geographers say the Pacific itself includes 63 or 64 million square miles of the surface of the planet and, of this total area, the Northeastern Pacific sprawls over 9 million square miles, a fair enough chunk of the parent sea and of the planet too. From this section of the great sea, American and Canadian fishermen alone take about 1.3 billion pounds of fish and shellfish each year. Foreign fishermen from East Asia take double and treble of the abundance of this shoulder of the Pacific, sometimes with apparently-calculated effrontery, leading thereby to problems of international scope that severely cloud the future of many of the major fisheries of the Northeastern Pacific.

This great geographic spread gives to the Northeastern Pacific and the lands that border it a diversity of climate and weather that reflects directly on its fisheries and their conduct. The fishermen of this area work under conditions ranging from the sub-tropics to the Arctic, from temperatures killingly high to temperatures killingly low. The hurricane-force winds off Baja California are easily matched or bested by the murderous storms, winter and summer, of the Aleutian Islands and the Bering Sea and between them, the south and the north have recorded an impressive list of lost men and lost fishing vessels. The Northeastern Pacific cannot log as many disasters as can the North Atlantic only because it has not been fished from a time in prehistory such as the Atlantic and the waters northerly and easterly have been fished.

1

## The Eastern Ocean Fisheries

The fisheries of the Northeast Pacific are more diverse even than its weather and its fish are equally perverse because fish are creatures of reflex and they (or most of them) are not as predictable through the day or through the season or over the years as some persons like to think they are. The Pacific salmon does cling to some kind of order through its life cycle but, often enough, the salmon confounds the prophets of its peregrinations. Some of the fish of these waters live and die in a usually narrowly-defined part of the sea, this being most true of those adults that live near or at the bottom of the sea. One fish, however, the albacore of temperate waters, (among others), is a true nomad, a pelagic wanderer that travels across most of the Pacific with the real extent of its movements still largely a secret. The five North American salmons manage to cover a good piece of the Pacific themselves and Americans and Canadians learned the hard way in the early 1950's that their homework on the travel habits of the salmon had been poorly conceived and poorly done, this after the signing of the United States-Canada-Japan treaty regulating certain fisheries of the Northeastern Pacific.

The fisheries of this Pacific area account for more than one-quarter of all Canadian and American fish landings and they are worth about $220 million a year to the 45,000 men who man some 20,000 fishing vessels of all sizes, ages and conditions from the Mexican border to Arctic Alaska. These fisheries begin on the northern rim of the tuna fishery below California and end in the high Bering where Americans crowd farther north and west each year looking for fresh king crab grounds. In between lie many fisheries, not all of them on salt water. The carp of Washington State contribute more than a million pounds a year to a healthy food and industrial market and some of the men who work this inland fishery may never have seen the sea. Some of these fisheries are as obscure as the Puget Sound fishery for sea cucumbers intended for a most select market; the Pacific salmon fishery may be the best documented fishery in the world. The Pacific halibut fishery, getting well along to the end of its first century with its bulwark of rules, may be the most fruitfully-regulated fishery anywhere. The harpoons of the small swordfish fleet strike home in the biggest of all fish, except for the occasional shark or sturgeon, commonly taken in the Eastern Pacific, north or south. And there are those fisheries that catch neither great big fish nor very small fish nor glamorous fish but, nevertheless, catch fish in quantity for food and for industry with their men plugging away by the season or the year at a hard and often poor-paying job.

## The Gear And The Vessels

The gear of all these fisheries is as diverse as their quarry. Some of it is ages old...truthfully, electronics and its applications are all that is really new because, in principle, the varied methods of catching fish were first used so long ago that they were old beyond calculation when the first history was set down in writing. The spear, the hook, the trap and the net...these are the only major ways to catch fish. The powder-charged harpoon gun may not much resemble the bone-tipped spear launched from a dugout canoe but the thought behind each is identical. Nor does the nylon seine powered by a Puretic block look close kin to a net weaved from lianas or papyrus or cedar fibers...but the nylon works just the way the lianas, the papyrus and the cedar fibers worked. Every fishery from California on north has its own gear and some part of it may be as strange to men from other fisheries as was the moon to Neil Armstrong and Edwin Aldrin. But almost all men who fish know the purse seine, the trawl and the gillnet because these take almost all the fish.

Of all American and Canadian industries, commercial fishing has been called the most disorganized, the least assembly-line oriented, one governed by impulse rather than rule. This is not really true, detractors to the contrary, because fishing has its own disciplines, its own ayes and nays, its own truths or mis-truths followed with devotion. All major gear is pretty much standard and so are most fishing vessels. Most boats of any size can go from one fishery to another as luck and economics order with only a minimum of re-gearing. Halibut vessels can become salmon seiners in season and draggers between seasons and a vessel goes where the money looks the best. Fishing vessels of the East Coast have their own jobs and they fall into several classes and sizes according to the work to be done. Similarly, fishing vessels of the West Coast have their own groupings and a general family resemblance within those groupings. One halibut "schooner" looks pretty much like any other halibut schooner while the ubiquitous western or northern seiner type looks as if it were built by the cable length and cut to size by the fathom. Albacore and salmon troll boats are all cousins, if not brothers, under their poles and no one who can see may possibly mistake them for anything other than trollers. (The big tuna seiners working out of California and Puerto Rico are in a class by themselves for good looks and efficiency at their own job but they have no part in this present work because theirs is largely a world fishery rather than a fishery of the Northeastern Pacific.) National origin plays something of a role in the looks of fishing vessels. Vessels built and used in the northern part of the Northeastern Pacific reflect the Scandinavian ancestry of the men who design and man them. To the south, from San Francisco Bay on down, some older vessels exhibit something of a Mediterranean influence. Italians and Portuguese were the first real marine fishermen on that coast and they are still dominant in certain

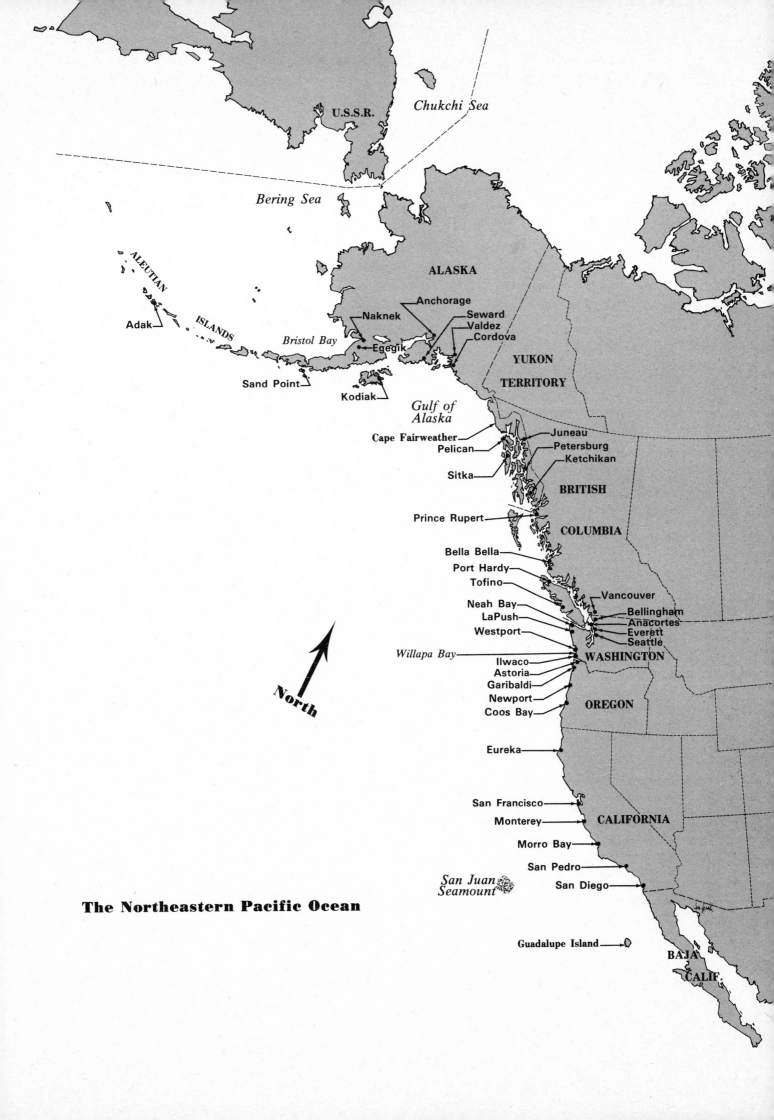

**The Northeastern Pacific Ocean**

fisheries of California. They have also a certain weight in the Bristol Bay red salmon fishery where Italians and Portuguese turn up in every season to show how gill-netting should be done. Of the distant water tuna fishery based out of San Diego and San Pedro, it is said that a strong strain of Portuguese in a man's ancestry is his best qualification for a job. Nor does it hurt to have a Norwegian name in the Pacific halibut fishery.

Dan Stair

## The Men Of The Fisheries

The men who work these fisheries of the Northeastern Pacific are as diverse as the fish they seek. They come in all races and ages with origins as disparate almost as the nations of the world. Some are first-generation fishermen, especially those men with names of Anglo-Saxon background. Many of those with names of Scandinavian, Latin or Slavic derivation are sons of long lines of fishermen, scions of fisher families perhaps generations old with, often, a family-owned boat as the focus of activity. Among American fishermen of the West Coast, the Scandinavians seem to gravitate toward the fisheries of the northern waters while the men of Mediterranean ancestry like better the fisheries of the south. Slavs once dominated the seine fishery of Puget Sound and Slavs, more than men of any other ethnic background, appear at home in all the fisheries of the Northeastern Pacific. Canadian fishermen of British Columbia are as ethnically-mixed as the Americans but because of geography, the Canadians do not range as widely north and south and there is not the tenuous separation of activity by nationality that is found among the Americans. Their tuna men fish from the Eastern Pacific to West Africa, of course, and their halibut fishermen go as far north and west as the Americans, but most Canadians are preoccupied with the close-to-home salmon, herring and drag fisheries and have no occasion to look much farther.

From Puget Sound north, a new face begins to appear in this Pacific fishery, that of the native of Washington State, British Columbia and Alaska, Indians mostly, with a scattering of Aleuts and Eskimos in Western Alaska. The natives fished these waters with effective enough gear long before the white men came poking along from the west and the south and their switch to white man gear has inevitably increased their efficiency. But, fisherman for fisherman, the natives are no more or no less effective than their white counterparts since a fisherman's efficiency depends, as much as anything, on a certain degree of pure luck and a willingness to work hard. Toughest competitors among the native fishermen are those of the seine fisheries of South-eastern, cutthroat affairs anyway, and hard words have been known to fly among white and Indian boats at work, while fists may follow if the respective crews encounter each other ashore, especially if the natives are Tlingit Indians. Almost two centuries of association with the white man has done little to inspire the hard-headed Tlingits of the upper Alexander Archipelago with respect for him and his ways, and this shows most heartily in the world of the purse seiners of Icy Strait. There, whites and Indians fish the Inian Islands to-gether. But with almost no exception, the Indians stay on grounds that have come to be theirs over the years while the whites stay on theirs. Middle Pass in the Inians is almost exclusively fished by Indians just as North Pass seems to belong to the whites. This usage is not invariable but few white skippers ever try Middle Pass...

The situation differs somewhat in Western Alaska where the Aleuts and Eskimos are fewer, milder and more disorganized and the fisheries are, with the king crab and shrimp fisheries excepted, mostly one- and two-man endeavors like the red salmon gillnet fishery of Bristol Bay. The rivalries that tend to spring from crew boat fishing seem not to start out to westward, partly perhaps because the short and intense salmon season keeps men too busy for anything but fishing and sleeping when the chance offers. In any event, white men rule all these fisheries of Western Alaska and the natives, as a group, have comparatively little to do with them except those of Bristol Bay and, to some extent, the growing shrimp fishery based on Kodiak Island. From Bristol Bay northward most native fishing tra-ditionally has been chiefly for subsistence although there have been small-scale commercial fisheries on the Kuskokwim and Yukon rivers since World War I. Native groups have become increasingly concerned with the expansion of these existing fisheries and the crea-tion of new ones. Their growth seems to be only a matter of time, training and money.

In the interest of perspective, it should be said that there is a division between white fishermen of Alaska and white fishermen from the states to the south although Alaska resident fishermen outnumber non-residents by about two to one during most seasons. There is a resentment, a silent one more often than not, borne by Alaska resident fishermen toward the outsiders, an attitude engendered understandably enough by economic rivalry, a rivalry that becomes more acute and more noticeable in years of shortened salmon runs. The State of Alaska has tried through various means to limit the entry of non-residents into her fisheries but no legal means has been devised to keep non-residents out of them. The most that has been achieved is the higher cost of licensing for out-of-state fishermen but this is a minor matter for most men from the South 49. There have been attempts at gear limitation aimed at both Alaskan and non-resident fishermen but they have been no more successful than similar embryonic attempts in Washington where at least twice too much

salmon gear works every season. Only in British Columbia, where a hard-headed and distantly-based federal government controls marine fisheries, has an effective gear limitation program been imposed.[1]

Here, then, is the Northeastern Pacific...from the following pages it is hoped that you may gain an insight into its fisheries, their gear, their men and their problems as well as an understanding of the complex procedures that send into the markets of the United States and Canada and much of the world a vast quantity of fisheries products of many kinds.

---

1. In 1973, the Alaska Legislature approved a limited-entry law. The act faced a future clouded by litigation.

Robert Browning

*Block seiner working on Puget Sound's Hansville ground begins early morning set. The seiner heads off the beach while her seine skiff steers into it as seine ripples across stern. Crewman morosely monitors purse lines to avert hangup.*

# 2/ The Fish and
# Fisheries of the Northeastern Pacific

The fish and shellfish taken in the commercial fisheries of the Northeastern Pacific run, by name, through most of the English alphabet—albacore to yellowtail—and they are taken across a great expanse of sea and sea bottom. For this study, about 75 species of fish and shellfish have been the focal point although note has been made of other, marginal species of fish and shellfish that are presently not favored as food by most Americans and Canadians or that exist or are caught in such small weights as to be negligible in relation to the total fishery.

The fin fishes have been classified in two groups, those found mostly at the surface or between the surface and mid-water (the albacore, again, is an example) and those species, many of them, that are taken almost entirely by the trawl. This second grouping obviously includes the flatfishes—with the exception of the Pacific halibut, a special species with a special story—and the numerous species of rockfish.

There are some 20 families of fin fishes encompassing about 50 species in the first grouping while the second grouping includes 55 or more species of rockfish alone (with only a half-dozen or so of commercial importance) in addition to at least a score of flatfish (only six or so important) as well as such other bottom dwellers as the members of the cod family.

The shellfish have been divided into the two usual groupings, crustaceans and mollusks. This classifica-
tion includes eight major varieties amounting to 25 or more species. Representative members respectively of the crustaceans and mollusks are the Dungeness crab and the Pacific oyster. In total tonnage taken each year, these shellfish comprise only a fraction of all West Coast American and Canadian landings although their overall value amounts to much more than landing or processed weights would indicate. One need only compare the price of a ton of shucked Dungeness crab meat with a ton of Pacific hake meal.

It should be added that only one of the tunas, again the albacore, is considered at length in this book because the other tunas are not properly creatures of the Northeastern Pacific although they account for the single largest processed weight because of the concentration of the canning industry in Southern California (much of its raw material comes from areas as far apart as West Africa and the far Western Pacific).

Throughout this book, the nomenclature of the National Marine Fisheries Service, the former Bureau of Commercial Fisheries, has been used although the final authority on any specific terminology must be the accepted name as it is rendered in Latin. This, in itself, becomes a question with some species of fishes because of a lack of international uniformity but it is not a subject to be considered here.

## ABALONE

Some of the finest and most expensive of seafood comes from the most unlikely of sources. The oyster, for example, does not look like much of a good thing to the unknowing nor do most persons conceive of the sharks, or some of them anyway, as food of high quality.

So it is with the abalone, a marine snail that might be described as the only shellfish that fights back and refuses to give up even long after he has been removed from the sea bed, bounced around in a boat and delivered finally to the shucking table. It takes a steel bar, considerable strength and equal skill to pry the abalone loose from its selected rock. Then, after the meat has been rudely extracted from the shell, the still-living creature must be left lying around for hours until it relaxes enough that it can be prepared for food.

The abalone, along with the mammalian dolphin, figures in some of the oldest writings of East and West. Aristotle admired the abalone in the Fourth Century B. C. And a journal of the ancient Japanese abalone fishery recounts, along about 425 A. D., the melancholy story of the too-deep dive and the quick demise of one Asaki, abalone fisherman, who sought to show his reverence for the emperor by bringing up from the sea for His Imperial Majesty the greatest of all abalone pearls. (Abalones and many other similar organisms produce pearls just as handily as do pearl oysters.)

The abalone, members of the genus *Haliotis*, are found through much of the world's temperate seas. They occur in many areas of the Pacific and Indian Oceans and in the Mediterranean where Aristotle encountered them in the Aegean, Homer's "wine-dark sea." They grow the largest off the coasts of California, Japan and South Africa and these three regions produce most of the world's commercial catch.

In keeping with its lowly status (zoologically, not gastronomically) as a snail, the abalone is a primitive life form with a simple anatomy. Most of the body consists of the shell muscle, the abalone's source of strength and the source of the edible flesh.

The abalone shell, with its tenant removed, often is a thing of beauty and some utility. The shell itself is moderately concave and moderately oval. It is characterized by a series of elliptic holes, three to nine of them according to species, that penetrate the shell near the ventral edge. When the overlay of other marine life is removed from the surface, the color is seen to range from brick red through shades of green, blue, brown and cream or white. The body of the abalone is of several colors too—black, olive green, brown, yellowish-green, none of them particularly pleasing to the eye of the potential human consumer. But beneath the color lies succulence. The shell interior with its mother of pearl facing displays a variety of pleasant shades of green, pink, pearly white, purplish pink. The largest and most colorful shells find a fair market after the abalone has been displaced and turned into steaks.

The abalones of the North American West Coast are spring and summer spawners. The spawning reflex appears to be triggered by a sustained water temperature of 68 degrees F. over most of its range. The abalones are single sexed and males and females discharge sperm and ova into the water in great quantities. The eggs are fertilized immediately. They hatch in less than a week and the larvae enjoy a free-swimming existence for one to two weeks. But when the shell begins to form, the minute creatures sink to the sea bottom under its weight where they remain fastened to rocks for the duration of their existence. Growth is fairly regular and depends chiefly on the availability of the favored food supply of several species of kelp. No kelp—no abalone.

The abalone is a dweller along the open coast, a lover of surf-racked reefs and rocky shores where the currents flow fast and the kelp grows thickly. They distribute themselves by species from the tide zone down to about 500 feet. In California, with its eight species, the big red abalone lies deepest while the smaller green lies highest. The other six species—pink, black, white, flat, pinto and threaded—scatter themselves between.

Almost everything, man, fish and beast, that dwells along the shores likes the abalone. California's sea otters use rocks to smash holes through the shells through which they can pull out pawsful of the flesh. Crabs, octopi and such strong-jawed fish as the lingcod crunch the whole abalone to bits. Boring clams, snails and sponges drill through the shell to get at the flesh. Sea urchins compete with the abalone for living space. Radiocarbon dating of materials found with abalone shells found lying in kitchen middens show that the coastal Indians used the abalone for food more than 7,000 years ago.

The abalone is scattered along the coast of North America from the Gulf of Panama into the Gulf of Alaska. Cold water limits spawning in most areas above Northern California just as it inhibits the spawning of oysters. Nevertheless, there are enough abalone in such places as Haro Strait between British Columbia's Vancouver Island and Washington's San Juan Island to present a potential for a small-scale market fishery. There are abalone along the West Coast of Vancouver Island, one of the world's most surf-swept shores, and along the west coasts of the islands of the Alexander Archipelago of Southeastern Alaska, enough of them to be worth limited exploitation at some time in the future.

But the Northeastern Pacific's only commercial fishery is that of California and it is one that does not much benefit persons not residents of California or not well-heeled enough to travel there. California law prohibits export of abalone and any person wishing to sample abalone necessarily must go where the abalone is.

Bootlegging of California abalone has carried the flesh across the state line on occasion and it has been known to appear on menus far beyond the confines of the Golden State. California's round weight landings run between four and five million pounds annually.

Of the eight species of abalone in California, only two, the red and the pink, figure to any extent in the state's commercial fishery. Sports fishermen take all abalone they can get their hands on. The red abalone, at 11 inches or so the biggest of the California species, is the mainstay of the fishery. The smaller pink is taken in good quantities but buyers sometimes put limits on their pink buys when the red is especially abundant. Point Conception, that line of water temperature demarcation for so many species of marine life along the California coast, does the same job for the red and the pink abalone. The red is found in greatest numbers north of the point, the pinks to the south. But the overlap is rather wide and fishermen find both on the same grounds.

The commercial abalone season opens in March and runs into January. Catch peaks follow weather patterns and over the years deliveries have been heaviest in May and in July-August. Some 1,000 men are licensed for the fishery each year with about one-third of them divers.

The fishery dates from the 1850's when some few of the Chinese brought into California as near-slave labor discovered there was at least one way better to make a living than toiling in the fields and mines of the white man. These abalone hunters, working out of San Diego, labored with iron-tipped rods from their small boats to find abalone, pry them loose from the rocks in shallow water and scoop them into the boat. The abalone immediately were popular in the market and the fishery prospered for almost 50 years with no advance in gear or technique.

The Japanese, innovative and aggressive with a long history of abalone fishing behind them in the home islands, moved into the fishery after 1900 when the county fathers of Southern California began to worry about the unrestricted abalone catch and, by ordinance, barred fishermen from shallow inshore areas. This effectively took the Chinese off the water. They were not fishermen essentially; rather they were farmers laboriously scooping a crop from the sea bottom. The Japanese were fishermen in the best meaning of that word. Moreover, they were expert hard hat divers and the new abalone grounds were in water far too deep for the Chinese and their patient probing. Although Caucasians trickled into the fishery off and on in the following decades, the Japanese dominated the industry until December 7, 1941, after which day patriotic concern and economic self-interest saw to it that the innocent Japanese were locked up in concentration camps far removed from the blue Pacific.

The fishery marked time pretty much during World War II as military service and other labor demands drew men out of it. Hard hat diving was one of the military skills taught during the war, however, and when peacetime came, there was a new supply of diving talent for the abalone fishery. Caucasians have been in the majority in the fishery in the years since the war but the methods are still very much as the Japanese established them.

All fishing, the one-man rig excepted, is a matter of teamwork and the abalone fishery as it is worked today is a supreme example of teamwork. More so than in any other fishery of the Northeastern Pacific, the diver on the bottom, very literally is in the hands of his shipmates. And the abalone fishery is the only fishery of the Eastern Pacific where the boat follows the fisherman.

The common boat in the California fishery runs from about 30 feet to 45 feet. In the past, its crew consisted usually of boat handler, line tender and one or more divers. But as semi-self-contained diving gear became practicable for use at the depths worked—from a legal minimum depth of 20 feet to the lowest workable depth —this system changed to accommodate the new gear. The old system was the rigidly-structured system devised and followed by the Japanese. Caucasian fishermen tended to be more individualistic in their approach to the fishery, a trend abetted by the switch to the more flexible gear. This consisted of free-diving outfits with a surface air supply rather than the back tanks of the pure SCUBA outfit.

Thus, husband and wife teams appeared in the fishery here and there, with the woman handling matters topside while her man did the hard work on the bottom. Two-man boats with each man a diver came into the fishery too. This allowed a more intensive working of a bottom area around the anchored boat but limited the area to the length of the air hoses, usually about 300 feet. Others followed the traditional system somewhat more closely, with one or two men on deck servicing one, two or three divers.

Any kind of diving is dangerous per se. It is hard work too. Abalone fishing is both dangerous for the man on the bottom and hard work because of the very medium he must work in. Depths are not excessive— up to 125 feet or so—but bottoms are rough, the kelp is entangling and sometimes the abalone are far between. On a good day on a good bed, the diver may move no more than a few hundred yards or a half-mile at most. Divers have been known to travel up to several miles, sometimes, on lean days, something that can be done only by the strong.

Diving periods run from 30 to 90 minutes and on good days a diver may spend from four to six hours in the water, a considerable time in the cold waters in which he works. This includes the necessary decompression time to escape the "bends," the decompression disease caused by nitrogen bubbles forming in the bloodstream under too-rapid decompression.

As in all other fisheries, weather is a governing factor in the abalone fishery, both for the day and for the season. In ordinary weather, the sea tends to be flattest around daylight and for a few hours thereafter. Thus, abalone fishermen, like almost all other fishermen, are early risers with work starting just at daylight. On too many days, rising winds and seas force an end to the operation by late morning or early afternoon.

Each man on the abalone crew has an exacting job and a mishap may be fatal. The boat handler must keep his craft positioned properly over the diver or divers and, obviously, clear of rocks and other obstructions. On much work, this calls for boathandling of the highest order with sea and wind and reef to be reckoned with every moment.

It is up to the line tender to see that his diver(s) stays alive and healthy by maintaining for him a continuing supply of clean air and by keeping his air hose from fouling at the surface. The line tender has another job too, that of clearing away the ever-present kelp that comes to the surface, following the air hose, as the diver slashes through it. Kelp is always a problem to an abalone crew although without the kelp there would be no abalone to harvest. When kelp vanishes because of disease, pollution or wind and water scouring, the abalone vanishes too.

Any man who has put a small boat into a kelp bed, accidentally or otherwise, knows what a pest it is on his screw, skeg and rudder or outboard engine underwater unit. On the bottom, it is a problem multiplied many times over and its entangling strands have taken lives. The diver working his way through the kelp is like a man trying to force his way through the gloom of a dense bamboo thicket while, at the same time, he is trying to find something he lost on the ground. It takes muscle, a sharp knife and good vision on the poorly lighted sea bottom.

The abalone diver carries his knife on a lanyard at his wrist, handy to his every move in this marine jungle. The knife is necessary to almost each movement but as soon as he cuts away the kelp before him, it ceases being his problem, one then bucked topside to the man in the boat.

It was noted a few paragraphs ago that the abalone is the only shellfish that fights back. This is not to be taken literally because the abalone has nothing to fight back with. But getting the abalone off its rock is a job for a strong man unless the abalone is taken by surprise. In this case it can be pried loose rather easily. But if the creature has a chance to pull his shell tightly down around him against the rock, it takes strength to pry him loose with the short steel tool with a dull, flattened tip used for the job. By this stubborn refusal to be moved, the abalone does, at least, put up passive resistance. The tip of the pry bar must be dull because one sharpened would kill the abalone or, if he were undersize, hence illegal, he might die from his wound or be vulnerable to predators. Legal length, seven and three-quarters inches for red abalone, for example, is determined by two spurs welded onto the bar.

Once the abalone is off its rock, the diver crams it into a narrow-throated small-mesh basket, resembling a miniature fish brailer. When the basket is full, the diver tugs on its line, the tender hauls it to the surface, and sends down an empty on the weighted line.

Processing of the abalone after delivery to the buyer is not done quickly or easily. The abalone gives up hard and after the creature is brought to the shucking table, it is allowed to lie quietly for a time so that the outraged muscle, eventually to be turned into steaks, may relax. Shucking the abalone is somewhat more complicated than shucking the unresisting oyster, a hard-learned trick in itself, and the shuckers must be fast and skilled. Skilled and quick people save money for the company because almost all abalone processing is hand work and there is no place for the blunderer. Almost all shuckers are men because they have the arm strength needed to force the shucking tool, a flat, slightly-curved steel piece, between the shell and the flesh to pop the abalone out neatly in a single, quick movement.

When the abalone and its shell are two, the viscera is cut away and the remaining flesh is washed. This may be done by hand or in a washer somewhat resembling a light-weight concrete mixer. After the washing, the still-living abalones are spread in a single layer on a table to let the muscle relax further after the unceremonious matter of shucking, butchering and washing. Relaxation is complete when the muscle mass stops curling at the cut edges. Now, the foot and the dark tissues are trimmed away, again by hand. When this has been completed, the abalone or what is left of the body mass weighs only about a third of the original weight. The usable remainder is a flattened oval of muscle to be machine-sliced across the grain into steaks about one-half inch thick. After the slicing, the highest art of the abalone processing line is brought into action.

The abalone, as it is now, is no more tender than the finest of shoe leather. To be eaten by any human, it must be tenderized. This must be done by the human hand because no machine substitute has yet been devised for the trained hand wielding a wooden mallet with just the right degree of weight on the tough abalone muscle. The steak must not be pounded too lustily lest the fibers be broken down too much. Nor must the steaks be left too tough for the teeth of the consumer. Somewhere between. . .gently.

After tenderizing, the steaks are sorted and graded by size and color with the largest and lightest leading the rest. The steaks are packed gently in institutional containers of several sizes. All California abalone goes into the fresh or fresh-frozen market. None is canned although a small amount of dark flesh and white trimmings is ground to be used in patty form.

## ALBACORE

The albacore, *Thunnus alalunga*, occurs through most of the world's temperate seas. It is an abundant sport and commercial fish off Florida and through most of the Caribbean Sea and the waters along the east coast of South America. It is equally at home in the Mediterranean Sea, along both coasts of Africa, the west coast of South America, and around Australia and New Zealand.

It roams through much of the North Pacific from Japan to Baja California and northward off the Pacific Coast, in some years, almost as far as Kodiak Island. Over the entire length of its coastal migration, the albacore contributes to an often-prosperous fishery, one that is among the fastest-paced of all fisheries.

The albacore is fastest of all tunas and is particularly distinguished by the long pectoral fins extending back to the anal fin. The body is completely scaled and the flesh is the lightest of the tunas and the only one that, in the United States, may be labeled "white meat."

The living fish is among the most colorful of major commercial species. The back is a steely blue, shading into a silvery cast toward the belly. In sunlight, the freshly-caught fish flashes with touches of gold and bronze and green as it writhes on deck. But the iridescence fades as quickly as the fish dies.

The complete life history of the tunas is still somewhat uncertain and the albacore is no exception. Like all tunas, it is completely pelagic at all stages of the life cycle from egg to adult. It prefers to remain well offshore over its chosen range and albacore taken in the American-Canadian coastal fishery are rarely caught less than 30 to 50 miles from land. The bulk of the catch usually comes at distances of from 50 to 150 miles off the land and at times large concentrations of albacore have been worked up to 300 miles off the coast. The albacore taken relatively close to the coastline are older and bigger fish than those caught farther at sea, some fishermen believe, although there is no firm scientific evidence to support that theory. In any case, all albacore caught in the Pacific Coast fishery are immatures.

The albacore apparently spawns in many areas of the high seas in water above 57 degrees F. Females with ripening eggs have been found in the Hawaian commercial fishery. In the Northeastern Pacific, the albacore feeds chiefly on juvenile rockfish, saury, anchovies, euphasids and squid. The largest known specimen weighed 66 pounds, four ounces. The average weight of those taken along the Pacific Coast runs from 10 to 20 pounds although larger ones show fairly commonly. Characteristically the albacore is a school fish although the schools seem to scatter widely when the fish are moving rapidly. When the migrants run into the 57-degree water barrier, however, they tend to school up and mill around as they seek to get away from the cooler water. Despite their dislike for temperatures below 57 degrees, feeding fish often move into colder zones in their pursuit of feed. Their favored temperature span runs from 58 to 68 or 70 degrees. (Another tuna figures, to a greater or lesser extent, in the coastal fishery. This is the bluefin tuna, *Thunnus saliens,* the "horse mackerel" of the East Coast, a much larger fish than the albacore and one just as unpredictable in its move-

ments. It runs in sizes from 60 to 200 pounds although fish up to 400 pounds or so are not uncommon. The largest measured specimen weighed 977 pounds. The fishery for the bluefin centers around the offshore islands of Southern California and Baja California. It is taken by seines, by troll gear and by bait boats. Annual catches have been as little as several hundred tons while the biggest catch, recorded in 1966, was 17,410 tons. Its flesh is darker than that of other tunas and because of this, ex-vessel prices are lower than those for the albacore, yellowfin and skipjack tunas.)

An unknown number of albacore, some millions of them, come into the West Coast fishery in "normal" years and despite great pressure, the stocks fished off North America appear to be holding up well. It has been estimated that some 4,000 vessels enter the albacore fishery in season when a fair return seems indicated or in those seasons that the salmon troll fishery, the alternate fishery for many of the vessels, turns out poorly. There is no reason, however, to believe that the numbers of albacore will continue at present favorable levels because it seems apparent that fishing pressure will become heavier as other fisheries continue to decline.

The time and place of appearance of albacore in the American-Canadian fishery is determined solely by water temperature. During years of exceptional water warmth such as occurred along the Pacific coast in the late 1950's, albacore can be found the year around off Southern and Baja California. In 1959, they were taken by vessels fishing out of Kodiak.

The northward movement of albacore from their usual first appearance below Guadalupe Island off the coast of Central Baja California hinges on the warming of surface waters to a point somewhat above 57 degrees F. The earlier this warming influence moves north, the earlier the albacore move with it. The farther north this warming moves, so move the albacore.

Under most conditions, the albacore enter the southern fishery in early June, then edge northward with the warming water until they appear off Oregon and Washington in early July. As the first storms of September begin to cool the surface of the sea, the schools swing westward or southerly toward the Central Pacific and eventually appear in the Hawaian fishery or even in the Japanese fishery in the Western Pacific.

It is difficult to define accurately what should be called normal in the albacore migration timing and routing. Dislocations extending over 10 years are recorded. The fish almost vanished from their usual Pacific Coast grounds in the 1926-1934 era. The late 1950's saw five years of major change from the "normal" pattern of movements.

In the 1951-1955 period, the albacore first came into the fishery well south of Guadalupe Island. But in 1956, water surface temperatures began to rise and the migration passed right through the Guadalupe area. In 1957 with still warmer water, the fish began their swing northward in an area to the north of Guadalupe. And in 1958, with the warming trend still on the rise, the first fishing began near San Juan Seamount off Southern California, almost 300 miles north of Guadalupe and some 500 miles north of their "normal" area of first appearance below the island.

In 1959, the cycle of warmth was at its peak and the albacore showed themselves first well north of the seamount area and went farther into the Gulf of Alaska than they had previously been found. But in 1960, water surface temperatures began to go down and the albacore made their first appearance south of the seamount. By 1961, they showed just north of Guadalupe again. In 1968, a year of exceptional albacore catches off Oregon, Washington and British Columbia, the fish first were encountered just south of San Juan Seamount early in June. This season may be regarded as an unusually fruitful year and an account of it appears in order.

No substantial fishery took place off California that season, however, because the economics of fishing interfered. Fishermen and their organizations held out firmly for a favorable price and on July 10 settled for $425 a ton. But by then, the offshore waters had warmed quickly and the bulk of the albacore had moved northward to the Oregon and Washington coast where California fishermen were forced to follow them.

First sightings were made off Southern Oregon during the first week of July and after July 10, fishing areas ranged from Cape Blanco, Oregon, to Grays Harbor, Washington, and up to 100 miles offshore. Fishing continued good through August but fell off rapidly in September for trollers. But during that month, bait boat catches picked up sharply. The trollers found their best fishing from 60 to 100 miles out for albacore averaging 13 pounds. Bait boats did best in the 20- to 40-mile range for fish running from 18 to 25 pounds. The fishery was over for most vessels by October 10. Landings by all vessels were chiefly in Oregon because of its proximity to the main fishery areas and because of the refusal of most Washington processors to accept albacore. Total Oregon landings have been placed at from 34.7 million pounds to 41.6 million.

For the fisherman, the albacore is the most valuable of all the tunas. This modern-era appreciation of the albacore was not always true. When commercial fishing was in its infancy off Southern California in the 1860's and '70's, albacore scarcely received its just due. The beautiful albacore was regarded as a trash fish, one to have its head smashed and tossed back into the water along with the likes of sharks, skates and barracuda.

It was after 1885 before the true value of the albacore and its adaptability to processing began to be known. Even then, it was 1906 before the first albacore went into the can in an experimental venture into a new field. By 1911, two plants, one in San Diego, the other in San Pedro, were working tuna in volume and in 1914, 11 canneries were using albacore because none other of the tunas was being canned. It was to take another 20 years or so before fishermen, processors and consumers began to fully explore the potential of what is now one of the world's great fisheries.

While the change of attitude toward albacore was taking place, there was another change underway at sea. In the 19th Century, albacore fishing had been a hit-or-miss affair with albacore being taken as part of a general white seabass-bonito fishery conducted by early-day bait boats. There is scarcely any mention of trolling in the fishery but there seems to have been a minor amount of it taking place, although neither boats nor gear much resembled trolling as it is known today.

It is clear that by 1900 the bait boat using the so-called "Jap method" of hook and line fishing after fish were attracted by chumming them up dominated the fishery. But the troller began to appear on the record as a fairly profitable and inexpensive method of taking albacore. This virtue of the two-man or three-man (or even one-man) troller attracted more and more of them into the fishery as engines became safe and reliable for small boats.

Movement of trollers into the albacore fishery picked up sharply between 1920 and 1925 as the full value of the other tunas became apparent. The bait boat fishermen turned to the other species and began to build bigger boats to follow such species as the yellowfin ever farther from home. Thus the tuna clipper was born and reigned queen of the tuna fleets until the "purse seine revolution" of the 1960's.

Matters were shaping up in great style, then, for the albacore trollers as they came to make up the majority of the albacore fleet. But the unpredictable albacore with their sensitivity to water temperatures vanished or almost so in 1926 and it was 1934 or 1935 before the fishery began a revival. Even so, there were small deliveries of albacore every season although most trollers stayed with salmon during the hard-luck decade.

With the return of the tuna to their accustomed routes, the albacore fishery began to move northward with the fish. In 1936, commercial deliveries began in Oregon and the next year the first commercial landings were made in Washington. In the years since, a vigorous fishery has existed in the Pacific Northwest, subject, of course, to the vagaries of fish, weather and price.

Owners of larger salmon trollers, boats in the above-40 feet classes, found themselves particularly attracted by the albacore fishery because of their ability to follow the albacore far offshore. This situation still exists and many boatowners plan each year on "making" their season with the albacore.

Almost all albacore caught in the West Coast commercial fishery are taken by trollers, also called jig boats, and by bait boats. Seiners sometimes account for a percentage of the landings. These seiners mostly are vessels from the northern fleet discouraged by a poor salmon season or draggers re-rigged for a fling at the unpredictable tuna fishery. Most of these seining ventures do not pay off well because shallow, salmon seine gear is not suited for albacore. Big, far-ranging tuna seiners with proper gear are concerned with species better suited to them than the albacore.

Tuna trollers are drawn into the fishery from the Pacific Coast states, British Columbia and Alaska. Most bait boats are from the historic tuna ports of Southern California. Trollers greatly outnumber bait boats. But the catch is more evenly distributed between them than numbers would indicate because the troller can seldom handle more than 30 tons of iced fish while the bait boat may have room for several times that amount.

Single-manned trollers are common in the fishery although the larger vessels may carry three men. Bait boat crews range from four to 10 men. No matter the method, albacore fishing becomes hard work during periods of intensive fishing because of the hand labor involved. Many trollers still haul their lines by hand although fast and efficient hydraulic line haulers are available at a reasonable price. They are seen increasingly in the albacore fleet.

## ANCHOVY

Most Americans and Canadians are familiar with the anchovy as a small, pungent fish fillet, sometimes rolled around a caper, a standby of cocktail parties where its chief function seems to be to send the consumer in search of another drink.

Most of those who know the anchovy solely as a party food do not realize that they are seeing, so to speak, merely the tip of the iceberg. For every anchovy fillet finding its way into the small flat can customarily used for them, a ton or more of anchovy goes into the world's reduction plants to be converted into meal and oil.

The anchovy, along with the herring and all those fishes called sardines, must be counted among the most abundant of the world's fishes. The center of this abundance appears to be off the coast of Peru where the upwelling of the cold, rich waters of the Humboldt Current supports a standing anchovy population that must be 20 to 25 million tons and may easily be double this amount. The Peruvian anchovy fishery, originally financed by American money and operated with American technology (including vessels and fishermen), is one of the world's richest fisheries.

A secondary center of this anchovy abundance lies along the Pacific Coast of North America from Baja California at least as far to the north as Vancouver Island. This anchovy stock has been estimated by the National Marine Fisheries Service at from 4.5 million to 5.6 million tons. It may be substantially greater.

These Pacific anchovies are almost untouched. They have been characterized as:

"...an enormous, increasing and unutilized resource with an amazing economic potential for human food as well as fish meal and fish oil...herein lies one of the untouched reserves of the United States fishing industry, unused because of a curious blend of selfishness in the name of sport, bureaucracy, politics, industrial caution and financial reluctance toward innovation."[1]

Large schools of anchovies appear in the summer and fall off Washington, Oregon and British Columbia but they are largely ignored. There have been no commercial landings in British Columbia in recent years. Washington and Oregon landings have averaged about 150,000 pounds a year since the early 1950's.

In 1970, 192.3 million pounds were landed in California's southern fishing districts with little industry enthusiasm visible for this fraction of the total anchovy stock.

Nor was there any interest in packing anchovy for human consumption. California law dictates that 13.5 cases, 48 pounds each, be packed from every ton of anchovy purchased for human use with the remainder, almost two thirds of a ton, allowed to go for reduction with no strings attached. Most of the industry simply wanted no part of human-use packing under any circumstances.

There was, however, a segment of the industry that did argue that post-World War II processing advances and changing market demands could make anchovy packing economically feasible, especially with the resulting by-products as a fringe benefit.

California's state government, ever mindful of the ballot box power of its sportsmen, for years over-ruled recommendations of its professional fisheries people that a large-scale reduction fishery be permitted on an experimental basis. In 1965, however, the California Fish and Game Commission unbent enough to set a 75,000-ton quota for reduction on a year-to-year basis. Late in 1969, after considerable soul-searching by the commission, the quota was raised to 140,000 tons a year. Even this generosity did not provoke a rush to take advantage of it although more gear did go into the fishery.

The fishing industry, with a caution not usual in what often has been a free-wheeling enterprise in the West, looked down its nose at the 75,000-ton quota and did not lift its eyes perceptibly when the ante was raised to 140,000. Its spokesmen continued to claim that the bigger quota still did not justify the necessary plant investment, especially in a time of tight money and an apparent slump in demand for fish meal and oil.

The reason is apparent and is contained in the last sentence of the quoted matter—"...a curious blend of selfishness in the name of sport, bureaucracy, politics, industrial caution and financial reluctance toward innovation."

Here the key words are "sport," "politics," and "industrial caution."

California's politically-potent horde of sports fishermen oppose the use of anchovies for open reduction, that is, reduction with no qualifying conditions. They wish to see these immense schools preserved as feed for

---

1. Unsigned article, "New Fisheries," *National Fisherman Yearbook*, 1969.

sport fish although it is improbable that all the sport fish in the Eastern Pacific could make much of a dent in the anchovy schools.

This industry faction, too, points to a distinct and unwelcome possibility, that of encroachment by foreign fleets on the Pacific anchovy stocks. In this event, the United States and Canada would have little cause to claim that those two nations have been properly exploiting the anchovy. It is this provable claim that has preserved to some extent other fisheries along the North American East and West coasts.

California's marine fisheries professionals view the anchovy and the fuss surrounding it with clear eyes. One asks:[1]

"What of the future? Will the controversy over the little anchovy continue?

"Top...biologists look to the future with confidence. While the ocean is not a bottomless cornucopia, scientific findings point toward the existence of anchovy stocks that could support a commercial fishery greater than has ever been known in this state.

"And they have no reason to believe that their harvest would impinge on the legitimate requirements of sport fishermen, *given only realistic controls by realistic men.* (Italics by the author.)

"The anchovy will not go the way of the sardine. On the contrary, judicious harvesting of the anchovy could some day lead the sardine back to a position of prominence in the coastal waters of California."

And Oregon, Washington and British Columbia too, hopefully.

(The reference in quotes above that "judicious harvesting" of the anchovy could lead to the commercial reappearance of the sardine is based on the belief among many fisheries scientists that the mushrooming anchovy stocks, filling the vacuum left by the sardine, may act as a "cover" or "umbrella" that the sardine now cannot break through to regain its former abundance.)

What is the little fish that is the cause of all this fuss and bother?

It is the Pacific anchovy, *Engraulis mordax,* one species of many scattered around the world, an unassuming creature four to six inches long with an olive back, lighter sides and belly and a silver lateral line running from the eye to the tail. There are many dark spots sprinkled on the small body and fins. Distinguishing features are the overhanging snout, the large eyes and the streamer-like anal fin.

The anchovy spawns during most of the year, shedding almost transparent eggs in inshore surface waters where they hatch in about 24 hours under warm temperatures. For its size, the anchovy is a rather fearsome carnivore, feeding voraciously on planktonic shrimps

---

1. Messersmith, James D., "Anchovy—Small Fish, Big Problem," *Commercial Fisheries Review,* January, 1970.

and other life small enough to swallow, including its own young and the young of all other fishes that come within its reach.

The anchovy is fished by purse seines under ordinary conditions although the lampara net is used when live anchovy are desired for bait and chum purposes. The seines are of fine mesh and are deeper, longer and more heavily corked than northern salmon seines. Suction pumps with wide-mouthed, heavy hoses are used to transfer the catch from seine to fish hold rather than the slow and cumbersome brail.

## BARRACUDA

The "man-eater" sharks, an imprecise term at best, have acquired a bad name and one sometimes undeserved through all the centuries that man has known them and feared them.

The real terror of some coastal waters of the world's tropical and sub-tropical seas are the barracudas, some 20 species of the genus *Sphyraena,* fierce predators that a happy quirk of nature has kept down to four or five feet of length for most big specimens although a few up to 10 feet have been recorded.

It is certain that many attacks on humans that have been blamed unfairly on sharks instead have been the slashing attack of the barracuda, unafraid of anything that moves in the water, including the toughest of the sharks.

E. C. Raney, the great fishing writer, once described the barracuda and its habits thus:

"...a pugnacious, five-foot barracuda, its long jaw lined with razor-sharp teeth, endangers man or any other large animal. Most fishermen in the tropics dread the barracuda more than any shark...it displays no fastidiousness in choice of food, lunging for any size or shape of prey."

Like the shark, the barracuda rallies fast to blood in the water and any creature, human, animal or fish, bleeding for any reason in tropical or sub-tropical waters should be regarded as a high-risk case and one quite likely to suffer further severe or fatal injury if not removed quickly.

No man who respects his fingers makes the mistake twice of bringing a barracuda aboard by line or in net unless the means to dispose of him is right handy.

Only one species of this predator is found along the Pacific Coast from California to Alaska. This roughneck is the Pacific or California barracuda, *S. argentea,* a species with a maximum weight of 12 pounds and a length of about four and a half feet. Despite his lack of heft, this barracuda is fully as nasty as any other of his kind.

People who use northern waters along the Pacific Coast have no need to worry about the barracuda nor the sharks, for that matter. Only in periods of extremely warm water is the barracuda found north of Point

Conception, California, well down in the lower third of that state. Such a time of warm water came during the late 1950's when albacore were caught out of Kodiak and there were confirmed reports of barracuda taken off Southeastern Alaska. It is certain that at least three barracuda were taken in purse seines on the West Beach ground off Puget Sound's Whidbey Island in 1957. The biggest of these was about 30 inches long a skinny fish with underslung jaw, too many teeth and all the sweetness of a wounded grizzly bear.

This extreme length of body in relation to body depth makes the barracuda hard to overlook among all the fishes of the Pacific. Its back is gray or olive, its belly a silver. The tail is yellow. The fish has the ability to change its shading somewhat to accommodate itself to differing bottoms.

All barracuda of the Pacific species appear to mature sexually at three years. They spawn principally from May through July on their customary habitat. They are mass pelagic spawners with large quantities of milt and eggs shed simultaneously. Females produce from 45,000 to something like 500,000 eggs according to age and size. The eggs hatch in four to five days and the young begin to feed as voraciously as their elders within four days thereafter.

The flesh of the barracuda appears to spoil more rapidly than that of most other tropical fish. In some regions where it is taken, it is suspect as a poisonous fish although the flesh often is sold under other names. This allegation of poisonous flesh may be based solely on the poor keeping quality of the flesh in many instances. But the flesh of some species of tropical fish in certain areas may contain a poison and this sometimes is true of the barracuda. The poisonous element is most apt to occur among older fish.

The only American commercial fishery for barracuda is conducted in a small way out of Southern California where gillnets and troll gear, the only permissible gear, take up to 300,000 pounds a year. Barracuda are mostly an incidental catch in fisheries intended for other species.

The commercial fishery began in the early 1920's when the public and the industry began to realize respectively that the barracuda was fair eating and could be reduced to meal as handily as any other fish. Landings through the 1920's and the 1930's sometimes ran from 6 million to 8 million pounds a year. The rise in catch figures was concurrent with the development of the purse seine in Southern California.

After World War II, it was evident to California fisheries people that the barracuda was being sadly overfished and that it would soon disappear from either sports or commercial fisheries if the depletion were not checked. Despite its unsavory reputation, no one had any desire to add the name of the Pacific barracuda to the list of the world's creatures already extinct or approaching that sorry stage. Strict control measures were imposed on all fisheries and the Pacific barracuda

native to California waters, at least, now appears safe. There exists a barracuda commercial fishery conducted by Mexican nationals in that nation's waters that threatens the stability of the stocks there.

## CARP

A most unlikely fish to appear in the pages of a book called the *Fisheries of the North Pacific* is the carp, a stolid, widely-occurring fresh-water fish closely related to the common goldfish and other members of the minnow family.

But there is something of a commercial fishery for carp along the Pacific Coast from California to Washington. The latter state outdoes its sisters to the south in carp catching with total annual landings of more than one million pounds.

Some of the catch goes for human consumption, especially in communities with substantial populations of southern origin, both black and white. Carp may be found most of the year in the famed Pike Place Market in Seattle, Washington, where its sale is determined by its comparatively modest price as much as by regional heritage. The carp also is used for mink feed and for reduction.

The several varieties of carp, all members of the genus *Cyprinus*, are native to Asia. The carp appeared in Europe about 1227 where it became and still is a skillfully-cultivated food fish. In 1877, it was imported into the United States in another of those ill-conceived transplants such as those that sent the rabbit to Australia, the starling to North America and the gray squirrel to England.

The carp probably would have spread across the continent on its own but the process was accelerated by still another poorly-planned scheme, that of stocking every farm pond in the country with a fish likely to thrive under rough conditions and supply food to hungry farmers.

That the carp thrived is an understatement of the first order. It has thrived so well in the near-century since its first tail wiggle on the American shore that it now is found in every state except Alaska and Hawaii and in most of the provinces of Canada. Despite its high standing in Europe and Asia, the carp on this continent is regarded as a nuisance by everyone, including the descendants of those gullible farmers who listened to the blandishments from Washington, D.C., and accepted the carp plantings in 1879 and the years after. Those descendants are still trying to rid their ancestral waters of the myriad descendants of the first carp. In one way or another, the carp has come to occupy water bodies that might be better used by such as the basses and the sunfish. Where the carp moves in, the catfish is about the only fish not crowded out.

The carp is a chunky, humpbacked fish with large scales and two barbels on the sides of its mouth. The largest ever recorded weighed 83 pounds. Twenty pounds or so is not uncommon and carp from five to 10 pounds are to be found in almost every place where the fish occurs. The carp is a prolific spring spawner with large females capable of shedding up to two million eggs a season. The eggs are adhesive and cling to vegetation or whatever happens to be handy. They hatch in from six to 12 days and the young grow up to nine or 10 inches in length in a year under favorable conditions.

Carp feed on the bottom, often in water so shallow that their arched backs show above the surface as the fish work their way through the roily water. They are not particular in matters of diet, impartially devouring vegetation, insect larvae, crayfish, snails and other fish, including their own young. A characteristic of the carp is its night-time habit of leaping from the water to fall back splashingly, a fuss that can be heard at some distance from the lakes and sloughs the fish prefers. It is a sound in the night remembered always by those who have heard it as children. The carp has a singular virtue in connection with children, it must be added. It is one of the easiest of all fishes to catch and a very young fisherman with the most primitive of cane pole, twine line, cork bobber and a few worms or a handful of dough balls can catch carp as long as he wishes.

For all the hard things said about carp, it is as good eating as many other fishes more highly regarded. The flesh is firm and flaky although it sometimes is grayish more than white. The flesh of older fish may have a distinct fishy taste or even a taste of the mud it lives in but the flesh of young carp is no fishier than the flesh of the lingcod or the rockfishes. Most of the fishy flavor may be removed from carp by soaking fillets or chunks in cold salt water, fortified by the addition of sliced onions, for several hours or overnight if one cares to go to that much trouble.

Much of the carp caught commercially in Washington come from the sloughs of the lower Columbia River. The pools behind the Army Engineers' carelessly-planned Columbia River dams offer ideal conditions for the propagation of carp while, at the same time, their warm and stagnant waters kill off the fine anadromous species that once used the free-running river in such vast numbers.

Almost all commercially-caught carp are taken by hand-hauled beach seines in one of the simplest of commercial fisheries. Set lines are sometimes used also.

## CLAMS

During pioneer days on Washington's Puget Sound, there reputedly sometimes was heard the wry comment:

"When the tide's out, the table is set."

This bit of folklore, if it ever existed beyond the fancies of versifiers, was more apt than most attempts to make the best out of what sometimes was a bad situation. In the times the quip refers to, a living often was hard to come by along the beaches of Puget Sound and many early-day families found themselves, of necessity, waiting until the tide was out to set the table.

When the tide was on the ebb, breakfast, lunch and dinner—clams, oysters, crabs and a host of other shelled sea life—could be rustled from the beaches, an infinitesimal part of what then seemed to be perpetual supplies of seafood. These settlers did eat a lot of it and there are folk tales of children developing embryo shells along their shoulder blades because of this high proportion of such food in their diet. Another story alleges that descendants unto the fourth generation of these pioneers found themselves experiencing a certain rise and fall around their midriffs in tune with the ebb and flood of the tide because of the great quantities of clams consumed by their forebears before they all got rich by stealing land from the Indians and one another.

In those days, Puget Sound on the ebb tide truly was a land with the table set for those who liked shellfish or were forced by hunger to seek it out. There was a lot of it and it was easy to find. There were fat crabs by hundreds and by thousands and oysters and cockles and mussels and geoducks, and scallops and abalone and clams by the bucketful or the wagon load, depending on the particular need. From the beaches and the waters of Puget Sound came in the old days the major part of the diet of all the people, Indian or white, who dwelt near the water before husbandry made available sufficient meat and grain for those who could afford them.

Where the white man has put his foot on this continent, aboriginal abundance has vanished forever. Puget Sound is no exception. But there is a remainder that does make Puget Sound and coastal Willapa Bay and Grays Harbor the West Coast's leading purveyor of clams and oysters. California's commercial clam production is infinitesimal. Oregon does not have extensive inland waters where most types of clams flourish best. British Columbia and Alaska have great stocks of clams but the former has felt no call to supply other than a small local demand and Alaska, although it has had substantial razor clam packs in the past, has felt the pinch of geography and increasingly-expensive labor since the 1950's. Clam processing is mostly a hand operation and Alaska packers cannot compete with processors from elsewhere in the United States and Japan.

Clams (and oysters where they occur together) are subject to toxicity on some beaches of Alaska and British Columbia. This toxicity produces paralytic poisoning in humans. Alaska's health authorities contend that "all beaches in Alaska are suspect and may have toxic clams." This blanket claim has been disputed by some elements in medicine and the industry, such as it is, but the pertinent processing regulations have never been effectively challenged. The University of Alaska notes, in connection with razor clam processing:

"The crux is that razor clams are 'suspected' of being capable at times of containing a deadly alkaloid residue. This is a type of poison that differs from a bacterial contaminant which can be eliminated by heat sterilization. It (the alkaloid element) is however, effectively avoided by complete evisceration, a process that always is applied to razor clams. That probably accounts for the poison never having been detected through the 50-odd years that razor clams were marketed in the past..."

That such a poison element may exist on some clam beaches during some part of the year cannot be disputed because illness directly traceable to clams turns up every year.

Despite Washington's pre-eminence in production of razor and hardshell clams, the commercial fishery still does not amount to much in comparison with other money fisheries. The round-weight landings of all presently-used species of clams amounts to about 2 million pounds a year. In contrast, Washington's production of shucked oyster meats, another minor fishery in terms of dollars, always runs close to 7 million pounds, something like 80 or 85 percent of the Pacific Coast's entire output.

Three major species dominate the Washington clam fishery. These are the razor clam and two hardshells, the little neck and the butter clam. Other minor species are taken, however.

The Pacific razor clam, *Siliqua patula*, is found in varying abundance from Pismo Beach, California, to the Aleutian Islands on exposed, surf-pounded sand beaches where currents induce quick and continual change of water over the beds. The razor clam occurs scantily in comparison with its northern numbers on beaches south of the Columbia River except for a short strip of Oregon's north coast. In Washington, it exists in great numbers from Cape Disappointment northward to the Queets River although it may be found in smaller stocks on all sandy ocean beaches, few as they may be, as far northwesterly as Unimak Island in the Aleutians.

The razor clam is exceptionally rich in meat and is a favorite target of personal-use diggers. These sports users take far more razor clams each year over the clam's habitat than do commercial diggers. In British Columbia and Alaska this is no threat to clam stocks, especially when most of the clam beaches are not visited from one decade to the next by any life other than sea lions and birds.

That happy situation is distinctly modified south of the Canadian border. Western Washington has experienced rapid population growth since World War II with consequent pressure on all forms of outdoor recreation. Many Washington residents as well as a great number of Oregonians consider razor clam digging Number One among outdoor activities. As more and more people seek more and more clams on the state's ocean beaches, actually with only about 60 miles of first-rate clamming grounds, state fisheries biologists have had to cut bag limits and digging days to keep the clam fishery in a reasonable state of balance. Every favorable tide finds thousands of persons, armed with every conceivable type of weapon against the inoffensive clam, hastening to the seashore.

(The popular, easy-to-reach Washington beaches used by sports diggers were closed to commercial digging in the early 1960's. Commercials may work on the rather dangerous "detached" spit areas of Willapa Bay exposed by low tides, the only commercial fishery on state-controlled lands. The only substantial source of commercial razor clams are the beaches of the Quinault Indian Reservation. These beaches were closed to all use by whites by order of the Quinault Tribal Council in 1969.)

The enthusiasm of these personal-use clammers can be excused. Their apparently-universal habit of throwing away clams not big enough to suit them cannot. State regulations say that all clams taken must be retained until the permissible bag limit is reached. That rule means that one-inch and two-inch and three-inch clams must not be discarded in the hope the next hole will turn up a five-incher. But the human tendency toward a small amount of larceny manifests itself here among persons of all ages and all states of society who otherwise might not tamper with an unguarded postage stamp. One person throwing away one small clam doesn't amount to much on one day. But 10,000 persons throwing away a clam apiece that same day amounts to a considerable amount of larceny and an irredeemable mortality of young clams. It happens on every good clam beach every day clams can be dug.

The commercial season is short and closely-watched. Here, a substantial portion of the catch finds a fate as deplorable, almost, as that of the undersize clams taken by the personal-use diggers. This part of the take goes to induce Dungeness crab to find their way into crab pots from which they cannot find their way out. It has taken four or five years to grow those clams and their use as crab bait seems rather a sorry end after that comparatively long time. Razor clams would seem to merit at least the frying pan or the chowder pot rather than the crab pot. In view of the deterioration of the state's razor clam stocks, it appears that this waste of a valuable food resource should be prohibited until the time, if it ever comes, that the razor clams can stand up to the growing pressure on them. Crabs can be taken on many baits. The commercial fishery should be made to limit itself to what the human use market can handle.

The razor clam comes by its name from its long, rectangular, slightly-curved shell. Shell color runs across several shades of ivory or yellowish brown with narrow streaks of black. The clam reaches sexual maturity at from two to three years and spawns in late spring when water temperature gets to 55 degrees F. for a few days. Milt and ova are discharged through the siphon of the clam and fertilization takes place as the clouds are diffused in the water. The larvae drift around with the currents for about six weeks after the hatch. When the shell begins to form, the juveniles settle to the bottom to dig into the sand and adopt the adult style of life. There normally is high mortality during this early adult stage because of bottom scouring set off by wind and wave action during the winter following the set. This destruction is offset by heavy egg production with each female shedding from 6 to 10 million eggs each spawning season.

The razor clam has a rather good life span in the northern part of its range and Alaska has yielded specimens up to 19 years old although the average is from 11 to 15 years. Washington razor clams live about eight years while clams from California's Pismo Beach average only four years. The clams grow slowly and a four-year-old is only about five inches long.

Actually, a razor clam eight years old is something of a rarity on heavily-used beaches in Washington because the intensive "fishery" precludes those advanced years for most of the race. Razor clams four years old on those lands are considered elderly by shellfish biologists because most are harvested by the end of their second year.

Novice razor clam diggers, big strong men though they may be, have been known to lay down their tools in something like teary frustration to watch veterans of the art, little old ladies in tennis shoes even, produce clams as rapidly as they can wield a shovel. The razor clam, like the scallop, can use a measure of evasive action to avoid confrontation with an enemy. The clam can move vertically for two or three feet with what is speed for a mollusk, something on the order of nine inches a minute in soft sand. It takes practice to track the clam downward from the dimple it leaves on the sand when it withdraws its neck as alarming vibrations reach it. The most usual tool for the razor clam is the "clam gun," a shovel with a narrow blade set at a 45-degree angle to the haft. It is this big angle that throws the novice digger off as he strives for his first clams. In most cases, the unaccustomed digger cannot drive a truly vertical shaft into the sand and too many times his blade slices through the quarry instead of lifting it easily from the sand.

The razor clam is the glamour clam of the Washington commercial fishery but the razor clam catch accounts for only a fraction of the state's production. The many hardshell clams, more pedestrian and usually easier to harvest, compose the rest of the fishery. This part of the industry is far removed from the open ocean beaches. Almost all hardshell clam production comes from the state's Puget Sound fisheries district. This area includes Puget Sound proper, Hood Canal, the Strait of Juan de Fuca west to Sequim Bay and all state waters north and east of Point Wilson at Port Townsend.

The hardshell clams inhabit several types of beaches from near the top of the tidal zone to 30 fathoms or more below the low water mark. According to species, they live in sand, gravel, mud, clay or mixtures of these.

The butter clam, *Saxidomus giganteus,* and two species of little necks compose two-thirds of the hardshell catch. The little necks are the native little neck or rock clam, *Venerupis staminea,* and the Japanese little neck or Manila clam, *Venerupis japonica.* The little necks are the smallest of the major species with shell sizes usually not exceeding 2-1/2 inches and are the clams most favored for steaming. The butter clam is similar to the quahog of the East Coast with shell sizes of from three to five inches, and is most frequently seen as the minced or chopped major component of clam chowder. Other hardshell clams appearing in the fishery are the horse clam, the Eastern softshell or mud clam, the cockle and two or three species of piddock or boring clam. They too are usually minced or chopped for canning.

Several styles of mechanical diggers or dredges have been approved for use on Washington's privately-owned clam farms but much digging is done by hand according to species and depth of habitat. The abundant Manila clam can usually be taken with an ordinary garden rake because of its liking for a near-surface habitat in soft sand. The old-fashioned potato fork with its tines set at right angles to the haft is a most efficient harvester of clams and, while digging, it aerates and loosens the beach material to the benefit of the remaining clam population.

A new face, so to speak, appeared hopefully in the commercial clam fishery of Washington State late in 1969, an appearance that has yet to be evaluated but one, nevertheless, that may have a good potential. This late starter is the geoduck, *Panope generosa,* a one-clam meal for a small family in the pioneer days along Puget Sound and adjoining waters. The geoduck is the largest of the hardshell clams of the Northeastern Pacific with a weight of three pounds or so, run-of-the-bed. The geoduck was well- and favorably-known to the aboriginal inhabitants of the Pacific Northwest and the first whites quickly acquired a similar appreciation of it. So well-liked was the geoduck that its one-time abundance along the tide line had been dissipated before World War I and the geoduck could be taken only at extreme low water mark on some 20 tides or so a year from then on.

Until 1969, the geoduck had been reserved for personal use. But the Washington State Department of Fisheries had been aware for many years that great numbers of geoducks existed well below the low water mark, far out of the reach of all personal-use fishermen except those with diving gear. Occurrence on the widespread beds below the low water mark has been proved to run up to about 25,000 pounds per acre, with a commercially-usable area of 32,925 acres with 64 million geoducks on Puget Sound, certainly a wealth of geoducks. This knowledge lay around uselessly for a long time before the department decided to do something with it. Department personnel had (and have) a well-founded suspicion of the reaction of pseudo-conservationist and sports fishermen groups toward any commercial fishery and most especially toward any suggestion that a new commercial fishery of any kind be established anywhere.

Thus, it took several years of "study," laced with a certain uneasiness, before the department came forth with its perfectly logical suggestion to the Washington Legislature that some use be made of this great abundance of geoducks lying peacefully in their beds far beyond the touch of most sports takers. The tentative suggestion drew hardly any fire, however, perhaps because most sports fishermen are accustomed to conceive of the quarry as finned, scaled and mobile, not something to be rooted out of the sand with a tool. A subsequent public hearing suggested rules for the fishery and the whole affair was made official on December 9, 1969.

The success of the new fishery depends, of course, on the existence of a market for the product. No such market existed specifically for the geoduck at the time the fishery was established, but clams and clam products had been produced commercially in Washington for 75 years or more when the geoduck first entered the marketplace. The geoduck is big and it is almost entirely edible. Some of it can be quick-fried as the razor clam is fried and served in restaurants at the same handsome price or at a better one. The rest of the geoduck is quite suitable for mincing or chopping for canning in that form or as the name ingredient of clam chowder. The only certainty of the geoduck market is that geoducks canned whole will not be offered.

The geoduck (known and pronounced variously as guiduck, giuduck, goduck, gweduck and gooeyduck) is found from Elkhorn Slough in California as far north as the Queen Charlotte Islands of British Columbia, with the apparent center of abundance in Puget Sound. The creature is another of those mollusks too big for its shell. Even when it is disturbed, it cannot retract its neck and its body into the protection of that chalky-white structure. The geoduck usually is found in muddy or sandy ground, lying from 18 inches to four or five feet deep. It can be found and identified by the tip of the neck or siphon as it lies at ground level to feed. The siphon is smooth over its entire length and the tip does not collect seaweed or barnacles as do the siphons of the piddock and the horse clam. The geoduck cannot dig itself away from its pursuer as the razor clam can, although the rapid retraction of the siphon gives that impression.

## DUNGENESS CRAB

From the earliest times, shore dwellers the world around have been aware of and equally appreciative of the merits of the edible crustaceans. Kitchen middens from most of the world's seasides are heavy with the remains of the invertebrates that, in their day, contributed so much to the gastronomic well-being of the sea people. (If the world's bravest man was he who first dared the oyster, only slightly less heroic was he who cracked the first crab for dinner because, under heaven and on earth, there are few living things with less eye appeal to the uninitiated than the crabs.)

The *crustacea*, the jointed, shelled denizens of the bottom lands of the sea, embrace thousands of species, all the way from the lobster through the crab and the shrimps to the barnacles and sea lice. Comparatively few of this number are used as food, partly because people have not been trained to eat them and partly because of their unavailability to the usual gear of the fisheries.

The crab caught in greatest volume in the Northeastern Pacific is the king crab of Alaska. The oldest crab fishery of the area centers around the Dungeness crab, *Cancer magister*, taken from mid-California to the Aleutians almost the year around. The lower Pacific coast Dungeness fishery is a winter fishery; that of Alaska is a summer fishery.

The Dungeness is a toothsome, medium-sized crab as crabs of the world are measured. The only larger ones found in the Northeastern Pacific are the Alaskan king crab and the tanner crab, a native of Alaska too but found also as far south as the California line.

Most Dungeness females never get much larger than six inches across the carapace while males up to 10 inches in width have been recorded. Commercial and personal use regulations along the entire West Coast range of the Dungeness crab require a minimum shell width of from six to seven inches for males. Females may not be taken at any time. There is no valid reason for taking them anyway since the female is a scrawny creature with little usable meat, one far better suited for reproduction than for market purposes.

Dungeness crabs usually breed in May and June in shallow coastal waters with one male capable of servicing five or six females. The female carries the fertilized eggs attached to her abdomen until the larvae are ready for release during the following January-March period. The shrimp-like young spend up to 12 weeks in their free-swimming form. During this time, they are greatly attracted by light and often can be seen in dense swarms around piers, floats and vessels with lights shining into the water.

By June, the young crabs assume the adult form and settle to the bottom, roughly a year after breeding. The Dungeness prefers a sand bottom although it can be found in light numbers on rock and gravel. After settling to the bottom, the juveniles go through a succession of molts with growth increasing with each shedding of the shell.

Within a year of assuming the adult form, the young crab are about 1.5 inches across the back. At two years, they average four inches. They achieve sexual maturity at the end of the third year. Most males have gone through the breeding cycle at least once before becoming subject to the fishery.

The crab, in common with certain other cold-blooded creatures, has the ability to regenerate missing appendages. The loss of a leg or legs may reduce the chances of survival of older crabs but apparently is a matter of little concern among juveniles. Two or three molts may be necessary to restore the missing member to its full size.

Because the male crab does not grow appreciably after the fourth year, it is considered uneconomical to leave males on the grounds after that time. It is estimated that about 90 percent of all legal males are taken in the coastal fishery of the three contiguous Pacific Coast states. The catch is less intense over most of the range of the Alaska and British Columbia Dungeness fisheries.

It has been established that the reproductive capacity of the crab stocks is so high that the general level of abundance is not disturbed by the catch of the males. Because of the high growth rate of the Dungeness, the marketable population is replaced year by year. Thus, it has been concluded that no danger of overfishing exists as long as reasonable season, sex and size regulations are kept in force.

This conclusion, of course, does not pretend to take into account the fluctuations in population due to natural factors such as disease, temperature extremes, salinity changes and predation.

The first commercial crab landings on the Pacific Coast date back to the 1880's with the market confined to the immediate area of the landing. The crab is a highly perishable product and a large-scale commercial fishery was not economically possible until developing technology showed processors how to keep crab edible by one means or another after its removal from the water.

A "significant" fishery began to develop along the Pacific Coast in the mid-1920's, coincidental with advances in refrigeration techniques. Much of the Dungeness catch from the lower Pacific Coast states is cooked, cleaned, sectioned and frozen while a considerable amount of meat is shucked and frozen for off-season use. Almost all of the Alaska catch is so handled or is canned because all but a fraction of the state's catch is used elsewhere. In the four states and British Columbia, whole, cooked fresh crab may be found in retail markets up to several hundred miles from the port of origin because of the ability of processors to hold them safely for several days after cooking.

The Dungeness fishery of Washington, Oregon and California appears to have reached its maximum yield with a catch of something like 35 or 40 million pounds available during big seasons. Alaska and British Columbia must be the source of further supplies and production from these areas will be regulated, for the near future anyway, solely by the effort put into the fishery.

British Columbia, with its thousands of miles of coastline, possesses crab stocks that have not been touched or even assessed. The province averages from 5 to 7 million pounds of Dungeness crab landings every year. The crab fishery should continue to grow as other fisheries reach their peaks. More effort in the crab fishery must be expected as the Canadians weed out their salmon fleet in a well-directed move to cut it by at least half.

The Alaska Dungeness fishery, just beginning to reach substantial proportions after World War II, was almost obliterated by the better-paying king crab fishery when that fishery began its remarkable growth in the post-war years. In 1954, Alaska fishermen took only 500,000 pounds of Dungeness crab. But an increasing United States demand for the Dungeness, coupled with a two- or three-year slump in Pacific Northwest and California landings, gave the Alaska fishery new life. Alaska has, like British Columbia, unknown and untouched stocks of Dungeness crab scattered along its coasts. Like British Columbia too, Alaska's production will depend on incentive.

## KING CRAB

The Alaska king crab fishery is one of the newest and one of the most hazardous of the world's major fisheries. It is conducted from Southeastern Alaska around the curve of the Gulf of Alaska as far west as Adak in the Aleutians and northward into the Bering Sea.

But it is a rich fishery, one that has had no problem in recruiting men despite the inevitable loss of boats and crews in winds that often top 100 knots.

It was not until after World War II that the king crab first was eyed seriously by fisheries company men seeking a new field, one that was to demand much in the way of experimentation and improvisation during its early years. By 1953, the king crab was being recognized as a significant market factor and production rose quickly during the next 15 years. The 1953 catch was 4.5 million pounds in round weight. In 1964, the catch figure came to 86.7 million pounds.

The round weight figure for 1966 was 159 million pounds worth $15.6 million to fishermen. But during the next year, the catch began a decline to 127 million pounds as heavily fished crab grounds began to yield fewer crab per gear unit. In 1968, the catch was only 81 million pounds and in 1969 it fell to 56 million.

As a result, the Alaska Department of Fish and Game ordered the first general season closures in the brief history of the fishery. Essentially, the first closures ran from March 1 through July 31 for the areas from Prince William Sound westward. The Bristol Bay area north of the Alaska Peninsula was kept open the year around. The minimum size of male crab was raised from 5-3/4 inches across the carapace to seven inches for the Bristol Bay area for a brief period also. The regulations have been changed to some degree each year since 1969 to reflect conditions in specific areas although it must be said that king crab regulation also has acquired a political tinge in the highly-charged atmosphere surrounding all Alaska commercial fisheries. The future direction of regulation of the king crab fishery obviously will be concerned with finding the "maximum sustainable yield."

The creature, scientifically labeled *Paralithodes camtschatica,* is a widely occurring crustacean of the continental shelf found over great areas of the North Pacific from its Alaska haunts west to the Kamchatka Peninsula and the Sea of Okhotsk. The full extent of the king crab range has not been determined although exploratory fishing has been carried out by Americans, Japanese and Russians over much of the North Pacific and the Bering Sea and adjacent waters. During most of the history of the American fishery, the heaviest concentration of gear has been from Cape St. Elias to Unimak Island. Increasingly, however, fishermen are pushing westward toward the end of the Aleutians and farther north into the Bering Sea in their search for unexploited stocks. Fishing intensity in Southeastern Alaska has always been relatively light.

## King Crab Biology

King crab catch figures from Alaska list three types of kings—the red king, the blue king or blue crab and the golden king or deepwater crab. The red and the blue are rather closely related while the golden is not even a member of the genus *Paralithodes*. Nor are the king crabs true crabs. They belong to the stone crab—*lithode* —family. They are *anomurans* properly, their chief distinction being that their fifth pair of legs is reduced in size and sometimes hidden. Thus, the king crab seems to possess four pair of legs, including the big frontal set with their out-size claws, instead of the five pair so apparent in the true crab. It appears doubtful in any event that any except the most scientific-minded of fishermen are truly interested in any difference other than ex-vessel price between *Paralithodes camtschatica* and *Cancer magister*, the Dungeness crab, for example.

Breeding is a somewhat lengthy process for the crab since it begins with a slow migration by sexually-mature male king crabs into shallower water from the depths around the 100-fathom curve where they have spent summer and autumn. This migration begins in late December or early in the new year and may take a considerable time. The crab is not a free swimming creature and it may begin its journey from far offshore over much of its habitat.

Despite the distances that must be covered, there seems to be little interchange among regional populations and adult crabs appear to return year after year to approximately the same breeding grounds. They seem also to live in the same general area over their life span.

As the crab labors through its journey into the breeding grounds, molting takes place with the time of the molt varying according to race. It extends generally from early in the year into April.

The migrating males are followed into the shallow water by the females pursuing their own migration. The young adults precede the older adult females. The favored breeding areas may be along the shore itself or on offshore shallows in depths running from three to 10 fathoms. Kelp-covered reefs are preferred. Younger females breed mostly in April, their elders in May.

The female must molt to breed and here the male instinctively offers a greater degree of aid than is customary among the mindless creatures of the oceans.

The male helps the female to rid herself of her outgrown shell by grasping her to him and removing the shell with his claws. The eggs, then free for release with the old shell gone, are quickly fertilized and deposited in the flexed-open abdomen of the female where they are carried for about 11 months before hatching. One male may service up to a half-dozen females with sexually-mature males not of commercial size mating freely with adult females of any size. It is estimated the average legal-size male taken by the commercial fishery may have spawned three or more times. When breeding is completed, both sexes return to deeper water, sometimes in depths up to 150 fathoms.

After the hatch in the spring following breeding, the young crabs spend from 10 to 12 weeks as free-swimming larvae. During this time, they are subject to heavy predation by every form of sea life that feeds on plankton.

After about three months in the larval stage, the young quickly assume the shape of the adult and settle to the bottom where, during the first year of life, the juveniles abound in shallow, rocky areas where food and comparative safety from predators may be found.

About 21 days after attaining the juvenile form, the tiny crab molts for the first time. Growth is rapid under favorable conditions and the small creature may molt from six to nine times during the first year. Its size increases by about one-quarter after each molt.

The juvenile continues to molt at a slower rate after the first year, perhaps as few as three or four times during the second year, two times the third year, once usually during the fourth year. Molting occurs once during each succeeding year until about the seventh when males often pass one or more years without a molt. The females, of course, must molt annually to breed.

A not-completely explained phenomenon called "podding" occurs among the juveniles from the second through the fourth or fifth year of life. Here, the small crabs mass themselves into piles often containing several thousand individuals. Up to 6,000 have been counted in a single pod. All members of the pod face outward and the pod itself may be a mass protective device. The pod consists mainly of members of the same age class with all in various stages of molt. Pod depths range generally from one to six fathoms or deeper.

Sexual maturity comes at five or six years. The male usually reaches legal commercial size at seven years. Females may not be taken. The king crab is comparatively long-lived and individuals 10 years old and even older have appeared in the commercial fishery. The largest recorded specimen, caught off the northeast shore of Kodiak Island, weighed 25 pounds with a 74-inch span leg tip to leg tip.

## TANNER CRAB (And Other Species)

There is another crab, or several similar species of crab lumped under one name, that must be considered in connection with the king crab fishery since it is taken in the same areas by the same methods and has become of commercial importance in its own right and as a complement to the king crab industry.

This is the so-called tanner crab, *Chionocetes opilio, bairdi* and others, a crab somewhat smaller than the average king crab but much larger than the Dungeness. A husky tanner crab measures about two and a half feet leg tip to leg tip and weighs about five pounds. Meat recovery from both species comes to about 20 percent of body weight, two pounds net from a 10-pound king crab, for example. *C. tanneri,* from which the tanner crab gets its name, has never been identified in Alaska although it is common in deep water from California to mid-British Columbia.

The difficulty of extracting the meat from the tanner crab has been one of the reasons for its neglect by Alaska crab fishermen although it is highly palatable and may be, the Alaska Department of Fish and Game has reported, the most abundant commercial crab in Alaska. It is found, mainly in shallow water, from Southeastern Alaska clear around the coast of the Gulf, out into the Aleutians, into the Bering Sea and north past Bering Strait.

Until 1967, the tanner crab got little attention from crab processors because of the dominance of king crab. But when the king crab began its decline, crab fishermen and processors began to take another look at the tanner. The Bureau of Commercial Fisheries worked out processing techniques that somewhat eased meat extraction, while Japanese processors who had been handling tanner crab for years passed along some of what they had learned. Kodiak processors took their first tanner deliveries that same year although only 118,000 pounds were processed. Some 30 million pounds were recorded for the Kodiak area in the 1972-73 season.

For market purposes, the tanner crab is known as the snow crab. Packers proposed, when the tanner first became a substantial item in their inventories, to call it "queen" crab, an obvious parallel to king crab. But the federal Food and Drug Administration ruled that the tanner should be sold as snow crab because of the long use of that name by the Japanese for the tanner. This was the name given the product when it was imported into the United States from Japan and upon this nebulous factor, the FDA shot down a valid promotion balloon.

## History Of The Fishery

The Alaska king crab fishery did not begin to come of age until the end of World War II and the lifting of the restrictions that largely had channeled men and their efforts into other fields.

Men whose business backgrounds were essentially oriented toward Alaskan fisheries began to seek other resources than salmon and halibut, then as now being worked at about their maximum sustainable yield.

These men had little but great expectations to base their efforts on, and more of them lost money during the first decade of the fishery than made money on it.

King crab was not really new even though Americans knew little about it. The Japanese had been canning it successfully since 1892. In 1933, the United States imported 7 million pounds of canned king crab from Japan and 2.7 million pounds from the Soviet Union. The trade continued on about that basis each year until the beginning of World War II abruptly ended it.

The first recorded processing of king crab in the United States was in 1920 when a small amount was packed by the Arctic Packing Co. in Seldovia. The next year, the Alaska Year Round Cannery, also in Seldovia, canned 60 cases. But after a series of on-again, off-again efforts, both companies stopped bothering with it.

Other pioneer, small-scale operators included a floating cannery in Kachemak Bay backed in 1923 by Jack Salmon and Bert Butler, salmon broker and attorney respectively; a Skinner and Eddy operation at Sand Point through the late 1920's, and a pack by Kinky Alexander, a Pacific American Fisheries Co. cannery superintendent during the salmon season, who had king crab operations variously at Olga Bay, Trap Point and Seldovia.

It was not until 1938 that another company, Pacific Fishing and Trading, made the first sizeable effort to catch, process and market king crab. The company sent Kinky Alexander in command of the 113-foot factory ship *Tondeleyo* to move from Cook Inlet westward to purchase and process king crab wherever it could be gotten. Alexander put up a good pack but the operation lost money, standard procedure for these early-day processors.

The first semi-scientific survey of the potential of the king crab fishery came in 1940 when Congress authorized a special one-year study "to locate the areas of abundance and to develop satisfactory methods for taking and canning king crab."

That August, the *Tondeleyo* with Arthur V. Nelson her skipper, in company with the 93-foot schooner *Dorothy,* Ellsworth F. Trafton skipper, worked over much of the area from Kodiak to False Pass with particular attention to Pavlof and Canoe bays on the south side of the Alaska Peninsula. The *Dorothy* was to drag for crab; the *Tondeleyo* was to can her catch.

But the mobility of the catcher boat was hampered by the slower-moving *Tondeleyo,* forced as she was to find shelter to do her work, and when the second half of the exploration was begun in the spring of 1941, the *Tondeleyo* was left behind. In her place went two seiner-type vessels, the *Champion* and the *Locks.*

The trio worked their way from Frederick Sound to Unimak Pass and north as far as St. Lawrence Island. They used pots, trawls and tangle nets and over much of their range, the catch was poor. But the expedition's report did establish "the presence of a large crab population in the Bering Sea and smaller but still commercially-important populations in Pavlof and Canoe Bays...around Kodiak Island and in lower Cook Inlet."

The report, in glittering understatement, concluded:

"This fishery may well supplement the established salmon industry..."

This was written during that uneasy autumn of 1941.

World War II brought most Alaskan fisheries exploration to a standstill. The old-line fisheries—salmon and halibut—had all they could do to scrounge men and materials to handle their own badly-needed products. The armed forces were commandeering almost everything that could float; the young men to man vessels were mostly wearing one uniform or another—blue, olive drab, field green. There were neither men nor boats to pursue will o' the wisp adventures out in the Aleutians or far into the Bering Sea. Besides, there were Japanese on the outer islands.

But over at Port Wakefield on Raspberry Island northeast of Kodiak, the people whose name has become synonymous with development of the king crab fishery were beginning to experiment with king crab in a small way.

These were the Wakefields and chief among them was Lowell Wakefield of whom the magazine *Pacific Fisherman,* said in 1965:

"Lowell Wakefield has contributed more to the growth and development of the United States king crab fishery than any other individual..."

From 1942 to 1945, the Wakefield family, whose chief interest then was their Raspberry Island herring operation, canned from 200 to 400 cases of king crab a year. And at the other end of Kodiak, another family, the Suryans, were putting up a few hundred cases of crab between seasons at their salmon cannery at Moser Bay.

The Wakefields and the Suryans were still going strong in the world of king crab when all their early-day competitors and many of their latter-day imitators had given up.

Lowell Wakefield not only had the vision some of his contemporaries lacked; he had also the capital to do what he envisioned. This he demonstrated in 1946 when he sent his combination trawler/processor *Bering Sea,* a converted East Coast dragger, into the Aleutians and the Bering Sea itself in the first of his highly-organized king crab exploration/processing cruises.

The 1946 effort was basically a trial run because Wakefield had bigger things in mind. What he and his men learned the hard way about crab catching and crab processing aboard the *Bering Sea* was incorporated in the then-building *Deep Sea,* the most famous vessel ever to enter the fishery, the first American ship built solely as a crab catcher and processor.

The Wakefield company learned well and fast. In 1955, Wakefield produced 85 percent of the United States king crab catch.

There were other notable figures in the crab fishery in these years just after World War II. The late Nick Bez, just beginning to become a major figure in Alaskan fisheries, took a fling at it with his *Pacific Explorer,* operated by the Pacific Exploration Co., a firm heavily-subsidized by the Reconstruction Finance Corp. *Pacific Explorer* was the result of a project conceived during the war to supply protein to the war effort by exploiting the fisheries of the North Pacific and the Bering Sea—after the Japanese had been ousted from these areas.

But the war was over before *Pacific Explorer* could go to work on her original mission and she was diverted to the Bering Sea and the Shumagin Islands to work with 12 catcher boats in the king crab fishery.

The catcher boat fishermen were guaranteed $1,000 a month and a production bonus and production was described as "fantastic." The boats delivered crab faster than the ship's processing gear could handle them. There were more than 17,000 cases of crab canned aboard that season but the ship lost money nevertheless. Too, the private processors began to complain about the competition from a federally-subsidized ship and *Pacific Explorer* was withdrawn from the crab fishery and worked over for tuna processing at sea.

Capt. Ed Shields, son of the noted codfish schooner captain, J. E. Shields, took his *Nordic Maid* to the Bering in 1949 and became the second operator to can part of his catch rather than follow the then-prevalent practice of freezing all of it. He worked the fishery for five years.

Bill Suryan, whose father, John, had done the early-day packing at Moser Bay, went into the Bering Sea in 1950 with the *Reefer King* and a fleet of catcher boats. *Reefer King,* following the lead of Bez's *Pacific Explorer,* became the second United States vessel to work as a processor only, depending on the catchers to supply her with raw material. Lowell Wakefield used his *Deep Sea* as both catcher and processor until 1956.

Of the *Reefer King, Pacific Fisherman* remarked wryly in 1965:

"During the first few seasons, *Reefer King* operated on the open sea and loading crab from the catcher vessels presented quite a problem when the sea was running...a system was devised whereby the catcher vessel tied up to a line behind the *Reefer King,* butchered the crab, then transferred the shoulder and leg sections to the processor in a cargo net dragged through the water from one vessel to the other.

"Although the system did not please everybody in that the weights recorded on the fishing vessel and those found on the processor were sometimes at a variance, the system did have the advantage of thoroughly draining and rinsing each section of crab during its journey through the water."

By 1955 or 1956, the "pioneer days" of the king crab fishery were over. The giants of the Alaska fisheries industry were deep in it or getting there. Companies or combinations of companies were formed, shifted, died and new ones came to take their places.

Gear and methods became more efficient, new grounds were opened up, new vessels designed for the fishery began to come along, gradually taking over from the somewhat rag-tag and bobtail fleet of the early years.

It is doubtful that anyone really missed the old days.

## The Toughest Fishery

Because of its location high in the north latitudes in a region stormy through much or most of every year, the king crab fishery presents more hazards to the men who work it than do most of the world's fisheries. Winds well above 100 knots are common through the winter months and even in the deep of summer, sudden blows of great strength can be expected.

The winter months are the peak months of the fishery, the months when most boats and most men are at work in it. Some of the boats and some of the men were hardly qualified for this endeavor in the early days. Until 1965, the Wakefield company's *Deep Sea* was the only vessel in the fishery that had been built to its exacting specifications. Most of the fleet was made up of Alaska-limit boats, up to 58 feet long at most. There were some under-powered self-propelled scows from the salmon industry and there was a clutch of one-time sardine seiners too big to fish salmon in Alaska and out of work otherwise because of the disappearance of the Pacific sardine.

A compilation of lengths of the 190 catcher boats licensed for the Kodiak area for the 1964-65 season was published in the June, 1965, issue of *Pacific Fisherman*. Here the inadequacy of the fleet of that day was clearly underlined.

Of the 190 vessels, 120 of them were in the 20- to 59-foot class. There were 42 in the 60- to 79-foot range. Another 14 ran from 80 to 99 feet. There were eight more in the 100- to 169-foot size. The 169-footer was a converted minesweeper.

The peculiarities of the king crab fishery impose several demands upon the vessels that work it.

A dead crab, theoretically, will not be knowingly accepted by any processor. Thus, catcher boats must have live tanks capable of complete changes of sea water every 20 to 30 minutes, preferably every 20 minutes. These tanks, filled with water, and, often, a shifting mass of crab, strongly and unfavorably influence stability of the vessel—particularly the stability of vessels not designed originally for tanking.

Too often for comfort, the catcher vessel must move its heavy pots from a barren ground to one with more promise. The pots stacked high on deck shift the center of gravity of the vessel upward and matters bad enough already because of the live tanks quickly become worse. Add to this minimum freeboard aft, freeboard often no more than 12 or 18 inches because of the ever-heavy load of sea water, crabs and pots.

Multiply these factors by hurricane-force winds, seas masthead high or higher, temperatures 25 degrees below zero with consequent heavy icing, and water so cold that the strongest man can live in it five minutes at the most.

The wonder then is not that so many men and boats have been lost during the brief history of the fishery; the wonder is, considering the makeup of the early fleet, that the loss ratio has been so low when the chances for disaster have been so great.

Since 1965, however, the complexion of the king crab catcher fleet has changed to a degree with the introduction of strong new vessels built for it. All of the new construction has been of steel, vessels in the 80- to 110-foot range, properly designed to allow for the dangers posed by the live tanks, deckloads, icing and demon winds.

Insurance rates for fishing vessels always are high. They are higher than most in the king crab fishery. The concern of insurance men about growing vessel losses resulted in imposition of stringent insurance requirements after 1965-66, a season when vessel losses were high. Almost every vessel in the fishery was subjected at one time or another to stability tests.

These tests resulted in removal from the fishery of some vessels ill-adapted for it and it also caused the rebuilding of others to qualify for insurance.

The knowledge gained over two decades of the fishery has tended also to cut the number of losses as skippers learned the hard way that one bad guess in this fishery is one bad guess too many.

Nevertheless, every year still, several king crabbers are lost, simply overwhelmed by weather that no vessel in the world of their size possibly could withstand.

Fishing for king crab does not seem to require the high art that fishing for some other marine species demands. Troll fishing for salmon or albacore is a fair example of fisheries where individual skill is the prerequisite, fisheries where the specific shaping of a lure, the selection of depth and/or speed, an almost-intuitive feeling for the habits of highly-mobile fishes make the difference between the good fisherman and the average fisherman, the difference between a good season and a poor one.

In the latitudes where the king crab is fished, seamanship, the ability to handle a cranky, heavily-laden vessel in rough water, would appear to be the first skill a catcher boat skipper must have. Even so, there are men who are masters of vessel handling who are not more than run-of-the-fleet crab fishermen. And there are men equally as good seamen who are good king crabbers too, men who consistently are highliners, even in the bad years.

If someone ever were to define this difference among men and arrive at its specifications, all fishermen would be highliners.

Most king crabbers fish grounds well-known through personal experience and the accumulated experience of others. As in most developing fisheries, there is a quite human tendency to let the other guy take the risk in exploring for new grounds. Probably fewer than 10 percent of the king crab fishermen have pushed past their fellows to seek out the unknown king crab areas farther west along the Aleutians or north through the island passes higher and higher into the Bering Sea; the other 90 percent just followed...

The bulk of king crabbers—and this appears to be the reason for the 1967-68 decline of the Kodiak Island area take—today largely are fishing the grounds first explored in the late 1940's and early 1950's by the Wakefields and the Suryans and their like, men who dared and lost money for most of the decade by their daring but who brought into being a fishery that still might lie dormant had it not been for them.

King crab fishing and all pot crab fishing is akin to the seining or gillnetting of salmon. In the salmon net fishery, catching fish often is a mere matter of setting a net along routes salmon customarily use and waiting for fish to blunder into it. This is a simplification, of course, but it is essentially the way many salmon fishermen do it.

The catching of king crab is just about the same process. The fisherman lays out his strings of pots where he has caught crab before or where he knows other men have caught crab and he waits for crab to find their way into his pots. The average or mediocre crab fisherman does, anyway. The highliner does not follow other men's patterns. He makes the patterns other men follow. He looks for the crab rather than wait for the crab to come to him.

## Catcher Gear

American king crab fishermen are forbidden to use tangle or trawl nets in the crab fishery because the nets, the tangle net especially, make it difficult to return females and sub-legal males to the sea without injury.

The early fishery saw both types of nets in use and they took most of the crab in those days. The tangle net was not outlawed until 1954 and the trawl not until the mid-1960's. But by that time, both types had been minor factors in the fishery for years.

The tangle nets, leftovers from the old shark fishery, were expensive, hard to maintain and hard to work and long before action was taken against them, fishermen had begun to look for something to take their place. The Dungeness crab fishery with its small round pots fished in usually-shallow water was generations old by then and it was the natural place to look for ideas for king crab pots.

But the light pots suitable for the small Dungeness crab obviously were not quite the pot for the large king crab and the depths at which most king crab are fished. Many men tried their skill and their experience at re-designing the Dungeness pot for king crab or designing an entirely new type. The evolution progressed satisfactorily enough and by 1950 John and Sam Selvog had developed the first king crab pots along the lines of those used today. The Selvog pots were six feet square by 36 inches deep, markedly more effective than the round Dungeness pot because they were less subject to water action on the bottom than the old pot.

Modern crab pots, fished from bigger, more seaworthy boats than most of those used in those first years of the postwar fishery, range from seven to nine feet square and from 30 to 36 inches deep. They weigh from 300 to 400 pounds.

Pot frames usually are constructed of welded steel rods with a nylon mesh laced to an interior frame. The mesh is kept away from the exterior structure to minimize damage caused by abrasion through water action on the bottom or from sides and deck of the catcher boat as the pot is handled or deck-loaded.

A variety of folding and nesting pots has been developed as fishermen sought ways to cut down on their top-heavy deck load when shifting pots from one ground to another. But none has proven entirely satisfactory and the standard pot of the fishery still is the heavy, standing type.

The drawback to most collapsing pots has been their comparative fragility. King crab pots must be as rugged as any piece of catcher gear anywhere in the world to stand up against the stresses of the conditions under which they are used. The working gear aboard the crabber must be just as rugged. The men working the deck must have skill and exquisite timing to take aboard a pitching, rolling boat a 400-pound pot with perhaps 1,000 pounds of crab in it in order that neither they, their gear nor the pot itself suffer injury or damage.

With the rise of the pot fishery the then-Bureau of Commercial Fisheries and, later, the Alaska Department of Fish and Game put strict limits on the number of pots a boat could fish. During the late 1950's, a boat working the Kachemak Bay area of Cook Inlet was allowed only 15 pots. Boats in other areas such as Kodiak could work a maximum of 30.

To beat these restrictions, fishermen began to use bigger and bigger pots, in many cases pots too big for the size of their vessels when deckloaded. This dodge on the part of the fishermen undoubtedly contributed to loss of life in the fishery.

When it became accepted by 1964 that the pot limit did little as a conservation measure, it was abandoned until 1970. Some boats fished strings of up to 120 pots during the intervening years. But in 1970, the State of Alaska imposed a pot limit of 60 to a boat.

Richard Philips, Pacific editor of *National Fisherman,* wrote then:

"Pot limits do not serve the purpose of managing (the fishery) for conservation purposes. They were imposed in an effort to force the big out-of-state boats out of Alaska so the smaller, more inefficient local boats could operate.

"Fifty Alaska-limit boats (58 feet overall), each with 60 pots, can fish a total of 3,000. Twenty-five good king crabbers fishing 120 pots each can also fish 3,000 pots.

"The difference is that the 50 limit boats will fish bays and protected waters, overfishing those waters and under-fishing exposed distant waters. The larger boats will fish where the fishing is best and...equalize fishing effort..."

Loran and an adaptation of the hydraulic line haulers used in other fisheries have made life easier for all hands aboard a modern king crabber.

Loran guides the skipper through the usually-poor visibility of the northern seas back to the area where he laid out his strings of pots. Loran A, properly calibrated and understood, should return him to within 1,000 yards of the beginning of his first string. Loran C, working at top efficiency, may bring him to around 100 or 200 yards from his target. Under optimum conditions, these systems may perform up to their maximum capability. But seldom do either men or electronics function at their optimum capacity, especially not in the high latitudes and bad weather of the king crab fishery.

The United States government has built loran stations in the Pribilofs and the Aleutians and, over most of the vast area where American fishermen work, the necessary master-slave station combinations can be acquired. But electronics are much afflicted, in this upper part of the world, by the disturbances set up by the Aurora Borealis and related phenomena and loran and radio often cannot do their job properly.

King crabbers, in the early days of the fishery, got by the hard way with the power from a seine winch and a combination of blocks and boom tackle. This was slow, hard work, rough on men and rough on the gear.

Several manufacturers have developed hydraulic blocks to pull king crab pots. Although they differ in detail, the principle is the same. The blocks weigh about 275 pounds and are hung from a starboard davit at the waist. Hauling a crab pot still is not child's play but men and gear fare better with the big block, the work goes faster and there is less chance of accident. The block automatically compensates for sudden strain on the warp, caused by a surge of the vessel, by way of a relief valve which lets the block pay out line when it senses the added stress.

Adoption of the crab block has a fringe benefit. Boats that once used five or six men now can fish with three or four. Net man shares are more and for most men, money is the only attraction the king crab fishery has or ever conceivably could have. This may not apply especially to residents of the crab fishing country, the men of Kodiak, for instance, men who have grown up knowing little other work than one fishery or another. But many of those in the king crab fishery are men from out of the state who are recruited into the fishery by the vision of financial reward. Not for them for any other reason are the barren and lonely sea and land of westward Alaska to be endured. More men than one have walked off a king crabber after a few weeks or months of it, muttering to all who will listen:

"Life is too damn short to spend it up here."

## HERRING

The herring is one of the world's most numerous and most famous fishes, a species caught by several million tons each year from the seas of the Northern Hemisphere. In Northern Europe, its use has been traced beyond 3000 B.C. and undoubtedly it has been a much-used food fish since man first began to explore the sea.

The herring—its name derived from an archaic Germanic term meaning "army"—has caused the construction of fishing and merchant fleets, determined the location of cities and towns from the Baltic to the coasts of Iceland and even, such was its worth, touched off a handful of wars through the countries ringing the Baltic and the North Sea.

This herring of history is the Atlantic herring, *Clupea harengus,* a somewhat larger cousin of the Pacific herring, *C. pallasii.* The Pacific herring has no such place in history as that earned by the herring of Europe but it too is an army, caught in great numbers and great weights.

For many years, British Columbia was the major contributor to the West Coast segment of this fishery with an annual production often exceeding 200,000 tons. Peak production was achieved during the 1962-63 season with 265,647 tons landed. From that time, landings trended downward until 1967 when the catch was only 58,370 tons. With such a drastic decline in stocks becoming apparent, the Canadians, fully remembering the defunct Pacific sardine fishery, clamped a complete ban on commercial fishing for herring.

The closure of the British Columbia fishery began to show results in 1970 when fisheries biologists traced down some 290 miles of spawn deposition, more than the annual average of the preceding 25 years. A limited food fishery was re-established that year but resumption of the large-scale reduction fishery, the real money fishery of the herring industry, was held in abeyance until it could be determined that the return of the herring was on a permanent basis or, at least, a basis as permanent as any fish stock, with its myriad reasons for fluctuation, can be.

United States Pacific Coast fishermen never went after herring with the same enthusiasm showed by the Canadians except in Alaska. Total annual landings in the territory never exceeded the B.C. figures although the fishery had a colorful, financially unstable history from the 1890's into the mid-1920's when fortunes were risked and lost on herring. Much of this effort went toward packing of herring in various ways for food. But, except for the World War I years, the Alaska pack could not compete economically or gustatorily with the sophisticated efforts of the North Europeans who had been experimenting with herring since before the written memory of man in that part of the world. The total Pacific Coast herring catch by Americans averages from 12 to 25 million pounds with Washington accounting for from 6 to 8 million pounds and Alaska for almost all the remainder.

Despite its prominence in the fisheries of the Pacific, the herring is, as an individual, an insignificant creature no more than eight or 10 inches long when it reaches maturity at three years. Its oily flesh is bony but tasty and as fresh fish, smoked, pickled, dry-salted or brine-cured under a variety of names, it is a favorite food of many ethnic groups in the United States and Canada where it is particularly favored by those of North European extraction. The herring is sold as a food fish everywhere in the two nations but almost all this food herring comes from Europe or the East Coast of the U.S. and Canada. Almost all Western-caught herring now goes for bait with demise of the British Columbia reduction fishery.

The herring is a pretty fish although not a spectacular one. Its lower jaw protrudes somewhat but no more than that of most round fishes. The tail is deeply forked and the dorsal fin is placed squarely in the middle of the line from snout to tip of the tail. The living fish has the blue-green back characteristic of so many marine fishes with the darker color fading away to a silvery belly. The herring is a school fish of the first order. Solitary specimens are almost never seen.

Although there are rather small resident "inside" populations along the Pacific coast from Puget Sound northward to Prince William Sound, most Pacific herring spend about eight months each year—April through November roughly—feeding and growing in the open ocean beyond the coastal islands. Herring stocks of Washington and lower British Columbia, for example, spend this pelagic period in the waters from Grays Harbor on the Washington coast north along the West Coast of Vancouver Island.

The spawning migration for these particular herring begins early in the winter as the schools move eastward through the Strait of Juan de Fuca. Their spawning grounds are among the San Juans of Washington and the Gulf Islands of the Strait of Georgia in British Columbia. Spawning is not an immediate process however, since the schools spend the winter massed sluggishly together above the sea floor (and under the ice in the northern part of their range as in the Bering Sea) awaiting sexual maturity. Most herring fishing takes place during these months. Herring for home consumption is taken in the summer months after the spawned-out fish have fed long enough to restore themselves to prime condition.

Spawning begins, according to race, in mid-February and runs through mid-July in the upper Bering Sea with the spawners in their great schools moving onto selected beaches. Not every beach is acceptable to the herring. A British Columbia survey of 1962 showed only 198 miles of beaches used by the spawners among all the thousands of miles of beach in the province. But it was estimated that those 198 miles contained more than 2 trillion herring eggs.

Although there are about equal numbers of males and females among the spawners, there is no actual pairing as with the Pacific salmon. The females discharge their streams of adhesive eggs—about 20,000 each—onto vegetation or other objects in the inter-tidal zone. Simultaneously, the males shed their milt and this fertilizing medium, entering the water in great quantities, forms bands of milky water visible for long distances—especially to seagulls and other aerial predators. These swarm by the thousands to the spawning beaches, destroying uncountable numbers of the herring eggs.

The eggs, clinging to eelgrass, kelp and other sea vegetation, piling, oysters, water-logged driftwood and boughs, are subject to widespread depredation and natural hazards as well as a unique fishery. The herring spawn is esteemed as a food delicacy by native populations of British Columbia and Alaska. In Alaska, where much of the spawn goes to the Japanese market, the short and sharply controlled commercial fishery nets 400,000 pounds or more of herring eggs on kelp with a value to the fishermen of $500,000 and up. This is distinct from the pack of herring roe, taken from the fish for a prosperous trade with Japan.

Those millions upon millions of eggs that do escape destruction hatch in from 10 to 20 days according to water temperature. The larvae, less than a quarter-inch long, are transparent and completely unable to swim. Larvae die in large numbers when they are ready to feed simply because they cannot move the few inches or feet from a sterile area into one rich in the planktonic material they need. It is estimated that 98 or 99 percent of all herring spawned die during the eight-week larval period.

But the survivors grow rapidly and after two months, they have become miniature adults already displaying an urge to school up. By October they are about four inches long, juveniles beginning to move in their schools toward the sea where they will await maturity.

Despite larval and juvenile mortality, there are lots of herring left. The Canadian Department of Fisheries, once estimating British Columbia herring resources, judged that only one in each 10,000 herring lives to spawn. The report concluded:

"On the average, 200,000 tons are taken annually. At 10,000 fish to the ton, this represents a catch of at least 2 billion herring a year. The fishermen take only 50 percent of the fish...there must have been between 3 and 4 billion herring at the start of the fishing season. Furthermore, if 50 percent of the stock dies naturally in a year, then altogether there must have been between 6 to 8 billion fish at the start of the year."

These figures, almost incomprehensible, stem from "the good old days."

Alaska too has an abundance of herring and when the Americans acquired the land in 1867, they began immediately to exploit its furs and fisheries. In the sea, the salmon were their first love and the saltery was their processing plant. Late in the century, a few enterprising men, most with more courage than money, began to experiment with winter herring at salteries scattered about Southeastern Alaska. This work was done mostly in Juneau and Ketchikan although from the beginning there were plants in the outports such as Big Port Walter and Little Port Walter on Baranof Island and in Taku Inlet. By 1910 the trade had spread to Prince William Sound and from there it went to Kodiak and its islands.

This budding herring industry was not aimed at an Alaska trade because few Alaskans wanted herring when there were other fish around. But the eastern part of the United States, heavily settled by European immigrants, liked herring and from the '90's well into the 1920's herring salting did well enough, by some packers at least, to keep them interested if not prosperous, and thousands of barrels went "Outside" every season. New York and Chicago were the market cities of the herring trade and here Alaska's herring had to compete with the herring of Scotland and Norway and sometimes Alaska herring fared poorly. Even at its best, Alaska herring sold under the import price, even after the Alaska packers sent for Scotsmen and Norwegians to teach them the art of curing herring properly. When World War I ended the North Atlantic traffic in herring, Alaska herring moved valiantly into the breach and the more efficient packers broke even for several years. It took only a short while after the end of the war for the European herring to make second-class herring again out of Alaska herring and the human-use packing of herring gradually dwindled away. The Alaskans were handicapped by expensive (for those days) labor and high transportation costs. The Scots and Norwegians had to ship their product across the Atlantic, of course, but water transportation costs considerably less than rail transportation. The Alaskans shipped by water to Seattle and then had to trans-ship by rail across the continent. The Eastern brokers inspected and graded in Seattle and much Alaska herring never got off the pier head for any useful purpose.

The best Alaska herring, the only herring that could compete on anything like even terms with the European herring, was packed as "Scotch cure," a process requiring too much hand labor in view of Alaska's unfavorable geography. The packers imported girls from Seattle and Vancouver to man the packing lines. Until the beginning of the herring season, the saltery locations were usually predominantly masculine and the young ladies from Outside performed at least one other welcome function beside "gibbing" herring. Some few of the girls supplemented their herring piece work money with other piece work money earned generally at night and at the end of every season, the more industrious and willing at both trades went back home with a considerable stake. Others—and some of the first—stayed on at the salteries or married and moved to town with their new husbands and became quite respectable, mothers to the next generation of Alaskans and now are enshrined in the civic annals of their new homes. So are some of the dance hall girls of the Klondike, Juneau, Nome and Fairbanks gold fields.

The Scotch cure as practiced in Alaska called for herring as fresh as possible, in good condition, free of feed, unbruised, still moist from the sea but held without sea water or ice for the brief time before the girl gibbers got to them. Their tool, a gibber too, was a small knife much like the familiar kitchen paring knife but with a blade only two inches long, a handle about four inches. The trick was to use blade and thumb between the gill cover and the pectoral fins to remove throat tissues, gills, pectorals, most of the gut material and the heart with an outward rather than upward pull. A girl working at this job obviously had to develop an almost superhuman dexterity to gibb her herring in quantities large enough to make her effort profitable.

The fish, after gutting, went into a tub from which the gibber packed her barrels when she had enough to make the job worthwhile. The standard Scotch barrel was of three-quarter inch staves with a 17-inch head and stood 30 inches high. Its capacity was 32 United States gallons. Barrels holding only 125 pounds net, however, were reported to be most common. The fish were roused in salt and packed bellies up with no salt between the first layer and the bottom of the barrel. The usual process called for one barrel of salt to three barrels of herring. A certain amount of shrinkage followed this and other methods of cure and for the Scotch cure, it usually required one extra barrel to repack five barrels to bring them up to advertised net weight. No water was added to the barrel and the fish cured in their own pickle, colored and flavored by the traces of blood and viscera that remained with them. This is the "blood pickle" of the traditional European herring pack, dependent for success on exclusion of air.

Some of the Alaska plants tried canning herring sardine-style in the ubiquitous flat can while one or two operators tried dry salting for the China trade. This latter didn't pan out at all because herring fat, of the type called "unsaturated," does not preserve well by such a method. Besides, the few herring that did get to China in the early 1900's were promptly confiscated because the Chinese government of that day owned the salt import monopoly and the Alaska herring always had at least a few grains of salt adhering to their by-then-malodorous carcasses. Other herring was lightly salted and vinegar-pickled but this could not compete with the German work in that field. Another product was the bloater, a large herring lightly brined, then smoked to a stage between the softer herring kipper and a hard smoke. This didn't prove any more profitable than most others of the herring food product enterprises for most packers because the fish smokers of Great Yarmouth in Norfolk on the North Sea seemed to own the process in that endeavor too.

## MACKERELS AND MACKEREL-LIKE FISHES

All around the world, there is more confusion than conformity when it comes to pinning names on fishes. Sometimes, even, there is a discouraging lack of unanimity among scientists concerning the Greek or Latin tags of some fishes. This confusion is more apparent perhaps among the many mackerels or mackerel-like fishes than almost anywhere else in the realm of the fishes.

Many species of fish the world over are called mackerels of one kind or another. Sometimes one small fish may carry around a half-dozen names with none of them having any bearing on the true identity of that fish. Those called mackerels are as diverse in name and appearance as the Atka mackerel of the Aleutian Islands, which is not a mackerel but a greenling, and the common crevalle which does not resemble a mackerel—although that name is applied indiscriminantly to two or three or more other fishes that do look like mackerels.

There is, however, a look-alike relationship among most of the fish rightly or wrongly called mackerel, a similarity of appearance that hopelessly compounds the mixup in nomenclature, a mess that probably never will be straightened out to the satisfaction of ichthyological purists.

These fishes possess in common a symmetrical streamlined body and are among the swiftest of the fishes. All have double dorsals. The mackerels and their relatives have an array of finlets between the rear dorsal and the tail while the false mackerels do not. Back coloration ranges through many shades of green into blue or blue-black and metallic silver while the sides and bellies fade away to dull silver or an off-white. None of these fish is of any great size and large specimens of most of them would weigh about 25 pounds. Greater weights have been recorded for most of them, up to about 125 pounds in a couple of cases.

Numbered among these related species, members of the sub-order *Scombridae,* are the Spanish mackerels, the true mackerels, the bonitos, the skipjacks and the tunas. Look-alikes include the jacks and members of several other similar-looking but unrelated families. Small members of the tuna families such as immature albacore sometimes are called mackerels by persons who overlook the long pectoral fin of the albacore, the other tunas and their cousins, the mackerels.

The tunas can stand by themselves as fish of great importance economically wherever they are found. Three only of the "mackerels" are presently of commercial value on the Pacific Coast. They are the jack mackerel, *Trachurus symmetricus,* one of the world's most palatable food fishes, also called Monterey mackerel, Spanish mackerel and mackerel jack; the Pacific or true mackerel, *Pneumatophorus japonicus,* called greenback, greenjack, zebra fish, striped mackerel and horse mackerel, and the California bonito, *Sarda lineolata,* known too as ocean bonito, striped tuna, skipjack, Chilean bonito, striped bonito and chanchilla. (Each of these has another handful of names of regional origin and in several languages. It is just as possible to misname a fish in Spanish or Portuguese or French or Italian as it is in English. The fish best known as the Spanish mackerel, for example, is not one of the fish named above but is of another species, *Scomberomorus sierra,* and is called sierra by the people of the shores of the Gulf of California. In English it is also called the cavala, the king fish and the cero. The true cero, *S. regalis,* is also called painted mackerel, king mackerel, tazza and carite. The true king mackerel, however, is known also as...ad infinitum.)

Each of these commercial species ranges in general from the tropics to Southeastern Alaska with the Pacific mackerel occurring often in large schools off the Oregon-Washington coast and as far north as mid-Vancouver Island. It is taken fairly often on sports gear fishing well offshore but there is no record of commercial landings in the Pacific Northwest in recent years. The fishery for them is concentrated from Santa Barbara Channel southward to the waters off the tip of Baja California.

The jack mackerel (not a true mackerel, remember) is taken in far greater numbers than either of the other two species. On the average, about 50 million pounds of jack mackerel are landed in California each year. Pacific mackerel landings used to come to about 2.5 million pounds while bonito totals about 15 million pounds. The Pacific mackerel might be regarded as an "endangered species" because of its increasing scarcity.

These fish are taken almost exclusively by purse seines. Small tuna seiners often turn to this fishery with their tuna gear when tuna are hard to find, quotas are filled or the market is stagnant for various reasons.

Some of these mackerel and much of the bonito are canned but the bulk of the mackerel catch goes to market as fresh, fresh-frozen or smoked fish.

These species and other species of the mackerel look-alikes sometimes occur in mixed schools and a catch landed and reported as jack mackerel, for example, may contain substantial numbers of other species.

The members of this group come to sexual maturity at two to three years. They are prolific pelagic mass spawners with the free-floating eggs hatching in about three days. Most spawning appears to take place in the spring in areas well offshore from the Southern California and northern Baja California coasts. The mackerels are comparatively long-lived and big specimens may be 20 years old.

## OYSTERS

It has been written that the bravest man in all of human history was he who first shucked and gulped down a raw oyster. Certainly, the oyster, uncooked and naked on its half-shell, is an object of intimidation to the timid or weak of stomach. But the oyster is sublime to him who steels himself, plucks out the oyster boldly, laces it with pungent sauce or the juice of the lemon and downs it heartily. For the faint of heart who cannot thus confront the oyster, there is always the oyster fried or the oyster stewed, the oyster with its true flavor driven out beyond recall.

The various species of oyster occur through all the world's warm and temperate waters. Seashore peoples discovered their usefulness before true man, *Homo sapiens,* made his first appearance about 50,000 years ago. The fossilized shell of the oyster (and those of clams and cockles and mussels too) have been found with sub-human remains more than 100,000 years old. Oyster eaters in those days were the locals only. The oyster is a fragile creature, one that spoils fast once removed from the water. It could not be taken too far from the sea in a world without refrigeration even though the nobles of old Rome discovered early how to use the ice and snow of the Appenninis to keep fresh the oysters and other shellfish from the seas around them on their slow journey from beach to banquet table. During the last century, foolproof canning techniques have made the oyster available wherever seafood is in demand.

Not only does the oyster taste good; it is a valuable addition to the human diet because of its high protein content and its content of trace minerals needed by the body. These include iron, iodine, copper, phosphorus, cobalt and manganese. Few foods as palatable as the oyster do as much good for the consumer.

The commercial oyster industry of the Pacific Coast from California through Oregon, Washington, British Columbia and into Alaska produces about nine to 13 million pounds of shucked meat every year, not enough to satisfy all the market. Washington, with its inland salt-water areas of relative warmth and high nutrient levels, accounts usually for about 85 percent of the total. California runs a poor second while production from Oregon and Alaska is negligible because of their cold coastal waters and lack of suitable area in Oregon. British Columbia produces about 1 million pounds of oyster meat each year. The province's Pendrell Sound has become important to Canadian and Washington oystermen as a reliable source of seed because natural reproduction through the setting stage of the oyster's life occurs there nearly every year.

Washington's Dabob Bay and Willapa Bay are dependable in many years as seed sources but their waters sometimes do not reach the required, sustained temperature of 65 degrees long enough to permit the free-drifting oyster larvae to set. Thus, despite its advantages, Washington still must depend on Japanese seed some years for a substantial quantity of oyster seed. (In addition, Dabob Bay setting is hampered by a water exchange mechanism which often sees warm water replaced by water too cold for the larvae to thrive.)

The Washington oyster industry presently utilizes three species of oyster. One is the tiny, native Olympia, *Ostrea lurida,* once a flourishing variety on Willapa Bay whence their fame and the oysters themselves reached as far as San Francisco. There, a plate of Olympias on the half-shell sold for the equivalent of $50 in gold. The rape of the Willapa Bay beds began about 1850 and lasted until about 1890. Now the Olympia grows only on a few secluded beds in southern Puget Sound. Pulp mill pollution in those waters seemed, during the years right after World War II, about to drive the Olympia to extinction even there. But the closure of the major offending mill for economic reasons, not because of any

concern about the Olympia oysters, started the Olympia on a comeback in the middle 1960's. This, along with adoption by the State of Washington of some fairly-stringent water pollution regulations, seems to offer the Olympia a chance to thrive in something like its former numbers. Olympia production does not begin to approach demand. The entire Washington production in 1967 was only 42,336 pounds. United States Bureau of Commercial Fisheries production reports commonly listed only two or three gallons a day available for the Seattle market during the early 1970's. Between February, 1969, and December, 1972, the Olympias rose from a price of $37 a gallon to $50. During this same period, the Pacific oyster, the major production oyster of the West Coast, remained stable at from $8 to $10 a gallon. A gallon of oyster meats weighs about 8.5 pounds.

Another oyster of limited production is the potentially-valuable Japanese oyster called Kumamoto. Its full use has been hampered by difficulty in getting seed supplies from Japan. During the 1960's Washington produced only 7,000 pounds of Kumamotos a year.

Because of over-exploitation and by default then, the giant of the West Coast oyster industry has become the Pacific oyster, *Crassostrea gigas,* introduced into the state in 1902 from Japan where it has been artfully cultivated for centuries. It is the seed of this oyster that West Coast oystermen turn to when their own waters do not allow adequate replenishment of the stocks. Seed imports range from a 1956 high of 74,000 cases to a low of 15,000 cases in 1965. Each case of seed must be inspected for predators by biologists of the Washington State Department of Fisheries before shipment from Japan. Some Japanese have been detected replacing inspected clean seed with inferior seed infested with various predators. The predator list includes the Japanese drill, a sea-going snail with a strong proboscis and a great liking for oysters. The drill sneaked into the United States with early seed imports and no control measures have been entirely successful.

## Oysters By Billions

The successful harvest of a Pacific or Olympia oyster crop is the culmination of a reproductive cycle that may aptly be described as a bit topsy-turvy, at least from the viewpoint of the human mammal. If the oyster were capable of conscious thought, it might at times quite earnestly plead for the services of a psychiatrist. Sexually, these species of oysters are ambivalent. They begin life as members of one sex and at some time during its span, they change sex. Then they may change sex again. And this may be repeated several times. This biological curiosity is of no importance economically since the sex switches take place in equal proportion among each sex so that there is no sexual imbalance at any time.

The Pacific oyster is about equally male and female for the first two years of life. At this time, however, the reversal process begins with balancing numbers of females becoming males and similar numbers of males becoming females. The Olympia is even more of a changeling. During spawning periods, most Olympias start out as males, turn into females for a period, then revert once more to the male phase.

The female of the Pacific species sheds about 200 million eggs during her spawning time. This horde of eggs is fertilized, at least in part, by the drifting sperm of the male as they hang in the water. They develop quickly into minute shelled larvae and spend three to four weeks moving freely before they seek, at pinpoint size, a clean surface to attach themselves to for the rest of their existence. It is this instinctive drive for stability by the spat oyster that makes possible the cultivation of seed oysters as practiced by the Japanese with their handy strings of shells. The Olympia is not as fruitful as the Pacific since it produces only about 250,000 eggs during a spawning season. Despite the great number of eggs produced and fertilized by either Pacific or Olympia oyster, only the smallest fraction survive to the setting stage. Otherwise, the world might well be up to its chin in oysters.

No matter what the source of his Pacific oyster seed may be, Japanese, Canadian or American, the oyster farmer finds himself with a supply of whole or broken oyster shell, each whole shell or segment the home of a score or hundreds or thousands of miniscule oyster spat. This shell material is called cultch or culch and the origin of the word is uncertain although Webster says it may stem from the English word "clutch" because of the spat's hold on its home or from the Old French noun culche, meaning couch. In any event, the term also includes rock and any other material on which the spat may locate itself. Oystermen who grow seed for sale make a few yen or pennies extra on their shell culch because the original tenants have already supplied one profit by virtue of being harvested and sent off to market before the shells are set out to gather a new crop.

Seed oysters are planted at any desired rate on nursery beds for first growth. Water, bottom and prevailing winds dictate location of the nursery beds and the planting rate. This may run from 20 to 100 cases of Japanese seed to an acre, for example, with each case containing from 12,000 to 15,000 spat.

The Pacific oyster gets toward market size in from two or three years depending upon food supply, purity of water and temperature and tide level. Even though water pollution may not kill, it can seriously slow the growth rate of the oyster and limit its size and fatness. Sometime during these nursery years, the oysterman normally moves the juvenile oysters to give them room for proper growth. It is here that the use of broken shell cultch shows best. Large oyster clusters must be broken up so that maximum, well-shaped growth may be attained. Separation of clusters sometimes is done by hand. It is expensive and causes considerable oyster mortality. Because of this, broken shell is preferred since it eliminates most of the hand labor of separation.

The oysters may then be scattered evenly over the nursery ground or moved considerable distances to other grounds suitable for oysters coming to market size but not satisfactory as nursery beds. Washington's oldest oyster farm company uses the comparatively warm and shallow waters of Dabob Bay off Hood Canal, an arm of Puget Sound, as a nursery for both imported and domestic seed. Many of the growing oysters are transplanted some 60 miles to the northeast to Samish Bay across Puget Sound where they are brought to maturity in water colder and more subject to pollution.

Pacific oysters are grown on tide flat beds exposed at low tide and on some grounds below the low tide line with a depth of no more than 20 feet. Olympias are raised in diked intertidal zone beds where they are never exposed to the weather. Fairly firm bottom is desired for nursery beds so that the young oysters may not be smothered by mud or displaced by water action.

Oysters are harvested during most of the year although commercial production declines sharply during the summer spawning months when the oysters, male and/or female, become skinny and milky in appearance. (The old wives' belief that oysters should be eaten only during the months with the letter R in the names does have some basis since those are the normally colder months. The story goes back to the days when there was no available ice or refrigeration to hold oysters until they could be eaten and it seems apparent that a lot of people during all those centuries got awfully sick from eating oysters too long out of water.)

Oyster farming on the West Coast is a fair-sized industry with its product having a wholesale value of $8 or $9 million a year. But the average oyster farm employs as much hand labor or more than the Appalachian hill farmer working his few acres of corn and beans. Olympia oysters are entirely hand-picked from their small beds and hand-shucked, a laborious and expensive job.

## Oysters One By One

The oysterman raising Pacifics comes out somewhat better because most of his beds are large enough that most of his crop can be harvested by dredge rather than by hand. Those missed by the dredge must be hand-picked at low tide and much of the catch must be hand-shucked. It is this necessity for patient hand work that has made Japan, with its abundant and docile supply of women workers, pre-eminent in the world's oyster industry.

The oyster dredge vessel commonly is of the power-scow type with a house and engine aft and deck working space forward. The dredge itself is smaller and lighter than the heavy-duty scallop dredge. The dredge employs four to six angle-iron teeth, independently mounted on a frame forward of the chain webbing, to rake the oysters from the bottom so that flat lip of the sack can sweep them up.

At dockside, the oysters are moved by conveyor belts to storage bins usually placed above the opening tables. As the oysters move along the conveyors, water jets give the shells a rough wash to remove the bottom material inevitably clinging to them.

From the storage bins, the oyster may travel one of several paths on its way to the consumer. It may be sold fresh or fresh frozen; it may be breaded and frozen; it may go to market canned in one of several styles and it may be smoked, whole or in pieces. Smoking is an especially flavorful way of preparing the oyster and it is a method that often appeals to the consumer who does not appreciate the delicate flavor of the fresh, raw oyster.

Three methods of preparing the oyster call for hand-shucking, an ancient art and one practiced in the West Coast oyster industry chiefly by women. Women's wages are less than those of men; their dexterity of hand is superior to that of most men and women have patience more than do men to accept a monotonous working day opening oysters that do not want to be opened.

The shucking of oysters and other bivalves need not necessarily follow the ancient practice of hand-shucking with the knife.

Steaming has been used for years to open oysters to be canned or otherwise processed, but, although the meats may be easily removed and are prime for canning, they are of no use on the fresh market. More oysters are sold raw in the United States than oysters cooked.

In 1965, the Bureau of Commercial Fisheries experimented with the use of microwave energy to open or "gape" oysters enough to make shucking faster, easier and less expensive. The process is not complex and the experimenters reported:

"Oysters and other bivalves that are hard to open have been successfully opened with microwave energy. The method can be adapted both to a batch-type and to a continuous-type commercial process without noticeably changing the organoleptic (taste, sight and odor) assessment of the shellfish. Hence, after being processed by this method, oysters, for example, can still be sold raw on the half shell. A comparison of costs with the hand-shucking method shows that a 33 percent saving can be realized by using microwave energy.

"The method also has other advantages such as increasing the productivity of a plant, enabling labor to be recruited more easily, lowering accident rates and resulting in a product free from shell fragments."

Actually, this method of opening oysters for use on the half-shell was anticipated some years before the BCF experiments by a handful of gourmet restaurants whose owners gained the benefits of the service of raw oysters while doing away with the expense of the hand labor formerly associated with the process.

Hand-shucking is used for oysters to be sold fresh, fresh-frozen or fresh-opened, water-blanched and canned. From the opening tables, the oysters go to a vat of cool, fresh water where they are tumbled and rolled by compressed air blasts to remove any foreign matter still remaining in the meats. The flesh of the oyster absorbs water during this process and gains size and weight so blower time is limited as closely as may be consistent with proper washing.

Hand-shucking preserves seed on the oyster shell and much of the empty shell finds further use as cultch for a future generation of oysters. Some of the shell opened by steam is finely ground to be used as a lime supplement for poultry. But every oyster processing plant is marked by a growing midden of shell, most of which will, in time, be used for cultch. Ancient men who used oysters thus unwittingly marked their own campsites.

To sell an oyster at its proper price it is necessary that it be graded and oyster grading, like so much in that industry, is an art more than it is a science. No instrument measures the oyster for size because the trained human eye does better than any mechanical device. The oysters are flooded from the blower vat onto the grading tables where residual foam is washed from them and deft hands group them by size for packing.

Size and quality standards for the West Coast oyster industry, including British Columbia, are suggested by the Pacific Coast Oyster Growers Association, with headquarters in Olympia, Washington. Actual enforcement of standards is a duty of federal and local agencies. Size standards call for four grades of Pacific oysters. These are large meats, less than eight to the pint; medium, nine to 12 to the pint; small, 13 to 18 to the pint, and extra-small, 19 or over to the pint. Retail prices for oysters go up as size goes down. The tiny Olympia runs from 1,700 to 3,000 meats to the gallon and is not size-graded.

Pacific oysters to be sold fresh are packed in containers of several sizes. The largest is the No. 10 can, seven inches tall and 6-3/16 inches in diameter. This container goes to military, restaurant and institutional use. The smallest is the half-pint jar found in retail markets. Smaller containers are sometimes used for Olympias.

Containers for fresh oysters are hermetically sealed with crimped covers. The cover carries the oyster grade, the net weight and the name and certificate number of the packer. It would seem that the cover might well carry the date of packing because of the short shelf life of the oyster even under refrigeration. The industry recommends that the fresh oyster be kept at from 33 to 35 degrees F. with the lower figure preferred. Most retail counters register higher temperatures. The National Marine Fisheries Service reports that fresh oysters kept at 33-36 degrees lose most of their flavor after a week. Bacterial action takes place easily in that temperature range and spoilage becomes apparent in about 14 days. Many potential oyster consumers have been disenchanted by oysters on the edge of spoilage because greed or neglect have kept them for sale too long. An easily-seen packing date and, perhaps, a warning concerning the brief shelf life of the oyster might prevent such situations so damaging to the industry.

The bulk of American oyster farming is as dated as Henry Ford's Model T. Most American farms got out of the one-horse stage a generation ago but most American oyster farms have barely gotten into that stage. Americans are still farming oysters much as the people of the north rim of the Mediterranean farmed them 2,000 years ago. American oyster culture is inefficient and unnecessarily costly on the oyster bed and in the processing plant. Matters will not improve much until the oyster farmer can achieve almost complete control over his crop. As it is now, he is attempting largely to ride herd on a "wild" shellfish. He is almost completely helpless to deal with any of the natural forces that affect his oysters for good or for bad. The only real authority he has is his ability to determine the rate of spreading of seed stock. After seed has been spread, things are mostly beyond the oyster farmer's ability to do anything until it comes time to thin whatever oysters may have survived from his planting. Pollution and predation are subjects he usually can do nothing about other than to complain.

The comparative inefficiency of a major segment of the American oyster industry was spotlighted by a Washington State Department of Fisheries study released in 1969 on oyster culture in that state. The report puts the annual yield of oysters on Puget Sound under "conventional bottom oyster culture" at about 800 pounds to the acre per year. Japanese oystermen using raft culture harvest from 16,000 to 64,000 pounds of oysters per acre per year. Obviously, the Puget Sound oystermen have a lot to learn from their Japanese contemporaries.

(Sharp disagreement with the preceding assertion is expressed by Cedric Lindsay, assistant chief, Management and Research, for the Washington Department of Fisheries. He said:

("Oyster growers are not stupid as this implies. Economics enters the picture here. The West Coast oyster grower gets $5 to $6.50 (then) a gallon while the Japanese grower gets the equivalent of $15-$18 a gallon.

("Actually, the competent and observant oyster grower learns the peculiarities of his ground and learns to farm it efficiently within its capabilities. Even the raft culturist has problems such as ice and fouling organisms.

("...you have to find people willing to buy oysters instead of hamburger or steak.")

Puget Sound is described by oyster biologists as one of the world's richest food areas for oysters. There are, on Puget Sound, they estimate, about 442 square miles of water area suitable for raft culture. One-half of this total area alone, some 187,000 acres, could produce up to 6 billion pounds of oysters every year, they estimate further. That is a lot of oysters. Six billion pounds equals the current total production of all United States fisheries. The chance of profitably disposing of 6 billion pounds of oysters under present conditions is exceedingly small. The entire world demand does not approach 6 billion pounds. But there is a promise for the future, the kind of promise that should interest those people who some day must face up to feeding more and more people from less and less land. Six billion pounds of oysters every year from half of Puget Sound would supply needed protein for a fair portion of all the world's people.

Raft culture has been tried experimentally on Puget Sound but it may not be even part of the answer to increased economical oyster culture on Puget Sound and elsewhere, whether that increase is considered in terms of six billion or 60 million pounds. Raft culture has become a high art in Japan for several reasons, among them the existence of an inexpensive and highly-skilled labor supply. Raft culture in America probably could be done economically only if some of the high incidence of hand labor could be eliminated. The probability of development of machines to undertake all or part of this work is small just as are the chances of development of a device to take over much of the hand labor in tuna canneries. The labor problem might be offset in some measure by high-volume production in something of an assembly line technique. Such a hypothetical system, once perfected, might just make raft culture a practicality in the United States. First cost might be high but it costs just as much for an auto plant to tool up to produce one car as it does for a million cars. Just so with oysters—under any system like the one just postulated, the first 100 bushels of oysters might be prohibitively expensive. But not the first million bushels and all the other bushels that might follow.

(In relation to this hypothesis, Cedric Lindsay remarked.

("There is a small matter of capitalization and interest on your loans.")

There is at least one other advance in oyster culture that may be of future importance in large scale oyster production, production at a rate more dependable and quicker than the methods now in use. This is the use of "free" or cultchless spat to get a marketable crop underway with more certainty than the prevalent cumbersome cultch method. Oyster hatcheries, public and private over much of the United States, spent most of the latter 1960's devising ways to do this. When shell cultch, either whole or broken, is used, the spat set thickly on the shell and the clusters must be hand-separated. This takes time and money and is one of the limiting factors in oyster production. There now exist a variety of patented processes for growing free spat although none has yet been used on a big commercial scale. One process, the work of the Virginia Institute of Marine Sciences, offers a fine netting to oyster larvae as a place to set when they come to the end of the free-floating stage of their existence. The tiny creatures set readily on the threads and can easily be washed off when they reach a proper stage of development. The oysters then can be grown on trays suspended from floats or rafts or on satisfactory natural bottom. These individual oysters grow faster and can be marketed in two years, the VIMS work has shown. A combination of raft culture and free spat appears to be possible with the use of trays suspended in series rather than the vertical lines loaded with oysters customarily employed by the Japanese. A raft of given flotation probably would not support as many free oysters as the older method but any combination of raft and free spat would be many times more productive than the present bottom culture, at least as it is practiced on Puget Sound.

Raft culture would eliminate the tedious and slow hand methods of harvesting employed at the present time. Tonging and raking oysters belongs with the harvest of grain by the scythe. The oyster dredge is somewhat better but not much and it often must be abetted by hand picking at low tide. Several hydraulic dredges exist and have been used successfully in clam farming. But they are expensive and beyond the reach of most West Coast oyster farmers. Perhaps some pool or co-op method such as that practiced a generation or two ago for community use of a threshing machine might be set up. The oyster harvest runs nearly the year around and such a dredge could move from one farm to the next without its owners and operators having to worry about a harvest peak such as that of the wheat lands.

All the efficiency in the world on the oyster bed, raft culture or no raft culture, is of little economic value if everything must be bottlenecked in the processing plant. Again, hand labor can be blamed for the slowdown. Too many oysters still are opened by hand. Such hand-shucking gets more expensive every year. Machine shuckers that work precisely enough for most purposes have been operated on the East Coast on something of an experimental basis. Their cost ranges around $50,000 and that makes them a major investment for the average West Coast oyster farmer, too much of an investment for most of them. These machines can shuck about 100 oysters a minute, something like 40 or 50 times the rate of the average hand-shucker. Grading is another squeeze. It takes a half-dozen people to keep oysters flowing across the grading table in an operation of any volume. A machine to do this job, one employing a photo-electric cell coupled, conceivably, with an electronically-activated grading table to distinguish among sizes and channel them accordingly, would trim labor costs enough perhaps to make American oysters competitive all over the world.

## PACIFIC HALIBUT

The Atlantic halibut, *Hippoglossus hippoglossus,* has been a prized product of the North Atlantic fisheries since the Middle Ages. Its Pacific cousin, *H. stenolepis,* has been the object of a commercial fishery for just about 85 years but it took this halibut less than 10 years to establish itself as a premium market fish. This took place as soon as the industry learned how to ship halibut from Seattle to East Coast buyers with a guarantee of arrival in good condition.

The halibuts are the largest of the many species of flatfish and are the ones of most value to the fishermen.

The Pacific halibut may be distinguished from its Atlantic relative—by the trained eye at least—by the somewhat narrower body of the mature fish. In habits they are similar, both favoring deep, cold water in the range from 37 to 46 degrees F.

The halibut normally displays a brownish or olive cast to the eyed side with overlying paler blotches. The blind side is white although a grayish tint appears among some specimens. These "grays" bring a lower price from the buyer.

Although its comparatively great size differentiates it from others of the flatfishes, the shape of its fins does so too. Both dorsal and anal fins run up to a point at mid-body and recede with a consequent diamond shape to the halibut. All other flatfish show rounded fins with no suggestion of the triangular shape.

The mouth is big and deep and equipped with sharp, heavy teeth. It feeds on other flatfishes, rockfish, hake, cod, crabs and clams.

The halibut occurs in the Northeastern Pacific from Point Arena, California, to the Bering Sea. The northern limit of its range is placed at Norton Sound on this side of the Pacific at about 64 degrees north latitude. However, individuals have been taken several hundred miles farther north. In the Western Pacific, the halibut is found as far north as the Anadyr Gulf. In North America, the halibut is more abundant in the northern reaches of its range and the commercial fishery is so concentrated.

The halibut is a deep water spawner. The species is prolific and one 200-pound female has been shown to hold more than 2 million eggs. Eggs and larvae are buoyant and subject to drift before wind and current, a fact that plays a major part in the eventual distribution of the halibut.

When the drifting larvae are less than an inch long, the blind-side eye begins its migration to the eyed side. At from four to six inches in length the juveniles settle to the bottom. This location may be far distant from the spawning areas because of the drift tendency. Eggs and larvae from the Gulf of Alaska spawning grounds are carried westward along the Aleutians and north through the Aleutian passes into the Bering Sea.

The halibut grows slowly. It takes about five years for its length to approach 20 inches. The mature male weighs from 50 to 100 pounds while a prime female, given a chance to achieve full maturity, may be more than eight feet long and weigh 500 pounds or more at 15 years or so.

Ideally, halibut should not enter the fishery until they are 10 to 11 years old, both so they may spawn several times and so they may gain something of their full growth. Areas of heavy fishing show both sexes at a much smaller average weight than the optimum.

The Pacific halibut commercial fishery is a text-book case of massive over-exploitation of a seemingly-inexhaustible resource. But the halibut supply was not inexhaustible. Fortunately, it was and is a renewable resource. It has become, perhaps, the world's outstanding example of restoration of a fishery by intelligent cooperation between nations and among fishermen and the industry.

The major threat to the fishery now is that presented by the big trawl fleets of the Soviet Union and Japan as they work halibut grounds where American and Canadian fishermen have practiced conservation or total abstention for almost a half-century. Ostensibly, the foreign fleets are not seeking halibut. But bottom trawls do not distinguish among species and undeniably there is a vast waste of young halibut totally useless for any purpose in their juvenile stage. American and Canadian drag fishermen are not allowed to keep halibut caught incidentally while fishing for other species. No such rules are observed by the foreign fleets.

The Indians of Washington and British Columbia and the Indian and Aleut populations of Alaska must have quickly discovered the halibut when they began exploring the seas around them in their distant past. The Englishman, James Cook, and the Frenchman, La Perouse, remarked admiringly in the 18th Century of the skill of the native Alaskans in their pursuit of the halibut. And La Perouse had nice things to say about the eating qualities of the big fish, truths that are just as valid today as they were 200 years ago.

The Makah Indians of Washington's Neah Bay seem to have been especially adept in the art of taking halibut and the first white men in the country copied the principles of some of their gear, substituting metal for bone and manila for twisted cedar fibers.

Not only did the whites copy the gear and methods of the Makahs—they also started the first commercial fishery for halibut off Cape Flattery in 1888. The halibut grounds there are small and what had been sufficient for the Makahs for centuries lasted the white man for only a few years.

It was just about this time, the early 90's, that workable methods of icing halibut for the long railroad haul east began to be used. The demand for this excellent fish grew, experience grew, gear improved and the coastal halibut began to disappear.

Fishermen began to explore more distant grounds and of necessity they had to look to the north. After the first decade, the halibut fishery southward from Washington's Willapa Bay was never much but marginal. The great banks that characterize the North Atlantic fisheries do not exist along the Oregon-Washington coast and since about 1900 the southerly fishery has supported only a handful of vessels.

But northward, inside and outside, lay large virgin areas, untouched by any but a few Indians fishing just off the beaches. So bank by bank, the fishermen fished themselves north. By 1911, they were at Cape Ommaney on the south end of Baranof Island in Southeastern Alaska. Two years later they were at Yakutat Bay and Cape St. Elias at the top of the Gulf of Alaska, beginning to look out to westward. They promptly moved on west.

By this time too, it was realized that inside fishing for halibut, the fishery in the sheltered waters of coastal British Columbia and Southeastern Alaska, would no longer support a commercial effort of any consequence. The only place left was the deep sea where most of the halibut were anyway, and to the deep sea the fishery moved and there it has stayed.

To work the high seas, the fishermen had one outstanding tool—gasoline power. They now had engines reliable enough and strong enough to take them in comparative safety from Seattle deep into the Gulf at speeds no sailing vessel could match. By 1910, all but three vessels of the big Puget Sound fleet had power in them. Soon, sail was left to the big dory schooners fishing the far-out banks, each with its nests of dories and its couple of dozen fishermen following one of the most hazardous of all methods of fishing. The dorymen did not have too much of a future, either as individuals or as exponents of a fishery style. In 1935, international agreement chased the dory vessels out of the halibut fishery.

Before the beginning of World War I, some men with a bit of foresight—fisheries officers of Canada and the United States, fishermen of both nations, and halibut dealers on both sides of the border—began to concern themselves that more and more gear was catching less and less halibut per unit of gear.

Since the fishery was solely an affair of the United States and Canada (the Soviet and Japanese intrusions against bottom fish were almost 40 years away), the interested parties began to work for some kind of arrangement to regulate the halibut fishery. There were only the two nations and there was no real opposition. But World War I intervened.

Where it took almost a quarter-century to get a US-Canada treaty to govern the Fraser River sockeye and pink salmon fishery, the two nations ratified a halibut conservation treaty on March 2, 1923, just about 10 years after real thought first had been given to it. If it had not been for the war, the treaty probably would have been signed much sooner.

That 1923 agreement set up a four-man body to be called the International Fisheries Commission, a group with limited regulatory powers but with plenty of authority to begin the needed research into the state of the fishery.

Before the treaty, halibut fishing had been a year around affair, an endeavor limited only by the weather and the courage of the men who had to face it. But the treaty did empower the IFC to impose a universal closure from every November 16 to February 15 to protect spawning halibut.

But research was the commission's main charge and research was being carried on to the full extent men and money would allow. On this research would be based new agreements between the two nations by which the halibut fishery hopefully would be revived.

The commission picked up considerable muscle in 1930 when a new convention was ratified. Essentially, the new document gave the IFC authority to change or abolish seasonal closures; to establish regulatory areas for closer attention to the needs of a particular stock; to license and regulate vessel departures and require the sealing of gear; to specify gear and to close grounds where juvenile nursery stock predominated.

Another convention, this one in 1937, gave the commission the power to ban the retention of incidentally-taken halibut such as that caught by trawlers. This convention did allow troll boats to keep such halibut up to a 2,000-pound maximum as long as the halibut was taken in a legal area during the open season.

The latest treaty, that of 1953, changed the name of the IFC to the International Pacific Halibut Commission to distinguish it from a multiplicity of commissions such as the Pacific Marine Fisheries Commission, the International Pacific Salmon Fisheries Commission and the International North Pacific Fisheries Commission.

Many men have worked hard and well with and for the IPHC. Over almost 50 years, the personnel has changed several times. But through it all, commission members and their staff have not lost sight of the goal they originally were assigned:

The "maximum sustainable yield" from the halibut fishery.

The Pacific halibut fishery reached its all-time low in 1931 with a catch of only 44 million pounds. But in 1931, the commission had hardly begun to practice what it had learned. By 1962, matters had changed. During that season, the catch hit its all-time high—75 million pounds.

By 1962, of course, the commission's years of work were beginning to pay off. But by 1962, the damage caused by the foreign trawl fleets was being felt too. The 1967 catch was down to 55.6 million pounds while in 1968, it was 48.3 million pounds. The 1969 catch came to 58.3 million pounds, a prosperous year for many fishermen because of high prices.

Without undue exaggeration, the American halibut fishery may be characterized as a fishery of aging men in aging boats. A federally-financed study of the economic status of the halibut fishery, published in 1963, reported the average age of American vessels then in the fishery to be 29.5 years.

Only in Canada had there been addition to the fleet or replacement by new boats. (The Canadians since then have launched in British Columbia a couple of dozen large combination vessels adaptable to two or more fisheries, among them the halibut fishery. The only significant big-boat addition to the American northern fleet has been for the Alaska king crab fishery.)

As for men—the study showed 25 percent of the active fishermen of the halibut fleet to be more than 60 years old at that time. There does not appear to be any appreciable change in the age ratio in the years since and the authors concluded that income from the halibut fishery and the related black cod fishery simply was not enough in an era of increasing "affluence" to attract young men to the industry and keep them there.

They added that "the family relationship in the predominantly Scandinavian halibut fleet will continue to attract young recruits to fishing but the effect of this relationship is bound to weaken in time."[1]

Their latter phrase is too close to the truth, that money is thicker than blood. Men, especially men with growing families, will not go where there is not enough money potential to make a hard and often dangerous life at least tenable. They will decline increasingly to enter a fishery where the same old problem has come up again...too much gear catching too little fish.

Suggestions, of course, are easy to come by; solutions, or at least generally palatable ones, are harder to find. But the report is correct in its assessment of the economic ills of the halibut fishery and in their proposed cures, just as William F. Royce was in his later and controversial analysis of the validity of gear limitation by some means in the Puget Sound salmon fishery.

The hook and line fishery for halibut would appear to be one of relative simplicity in a world where much or even most of the commercial catch of fish is taken by nets of varying and exasperating degrees of complexity.

Appearance here however is just as deceiving as it is in so many other settings. All manner of things can become disagreeable in the halibut fishery.

For one, the longline is as prone to mishap as any other kind of fishing gear and it even can find trouble peculiar to itself.

Weather, on occasion, may be tolerable but too often it ranges from bad to god-awful. This fishery now is confined almost entirely to the middle months of the year, the months from April to about the end of September. But it is carried on in ocean areas—the Gulf of Alaska, the Bering Sea and the top of the Pacific—where even the brightest summer day can turn in a couple of hours into a time of bitter storm. It is a rare year when there is no disaster.

The halibut fishery is a hard fishery. It demands the most precise teamwork from its men. The skills and their timing are not learned in a week or a month

---

1. Crutchfield, Dr. James and Arnold Zellner, "Economic Aspects of the Pacific Halibut Fishery," *Fishery Industrial Research*, April, 1962.

or in a season. Old-line halibut skippers like to say it takes three seasons to turn a green hand into a reasonably competent halibut fisherman. This may be a matter of debate among skippers and crewmen but no one who has seen a halibut crew at work can deny that the fishery is more exacting, perhaps, than any other of the major fisheries of the North Pacific.

# PACIFIC SALMON

In all the world where men fish, whether for fun or for money, they fix upon a favorite fish, a fish of substance such as the herring that has started wars and changed kingdoms, or a courageous fish like all of the trouts with their honored history of valiant resistance to the artifices of the fisherman.

Such a fish of the Northeastern Pacific can be none but the Pacific salmon, several sleek fishes that among themselves combine the merits of the herring and the trouts and all their kind. The salmons have created great wealth like the herring has created; they resist the rod and lure as stoutly as any trout ever has resisted them.

The great commercial fishery for the Pacific salmon dates only from 1864 but over all the years since then, the salmons, members of the genus *Oncorhynchus*, have been for fishermen and packer alike the single most valuable fish of the North Pacific.

Five of the six species of *Oncorhynchus*—Greek for hooked nose—are native to North America. The sixth, *O. masu*, occurs only in Japan and along the nearby mainland of Asia.

The salmon of North America are found from the Sacramento River north and west to the river systems flowing into the lower Chukchi Sea above the Arctic Circle. Major among the productive streams through that range are the Columbia, the Skagit, the Fraser and the Copper; the network of rivers that feed into Bristol Bay—the Nushagak, the Naknek, the Kvichak, the King Salmon, the Togiak, the Alagnak, the Dog Salmon, the Ugashik, the Egegik and the Meshik and the far-north Kuskokwim and the Yukon. There are others; from mid-California to Point Hope, some 2,600 air miles to the northwest, deep in the Arctic, literally thousands of lesser streams support the salmon of the western rim of the continent.

The Pacific salmon clearly is not as abundant as it was when it first came to the attention of the white man scarcely a century and a half ago. Man, reshaping and polluting his environment, has remade and poisoned that of the salmon too. It cannot be hoped with man around that the salmon will ever be restored to its old numbers. But the salmon is a tough and resilient fish with a strong drive for survival and it may well be thriving long after man has managed to remove himself bag and baggage from this earthly scene.

The five species of Pacific salmon are:

*Oncorhynchus nerka,* the sockeye, called red salmon in Alaska and blueback on the Columbia River; *O. tshawytscha,* the chinook salmon, known too as king, spring, tyee and blackmouth; *O. gorbuscha,* the pink or humpback salmon; *O. kisutch,* the coho or silver salmon, and *O. keta,* the chum salmon, named also the dog or fall salmon.

In dominant cycle years, these varying with race and stream of origin, the sockeye with its high price is the most valuable en masse to the fisherman. The sockeye run in weight from three or four to about seven pounds throughout their range. They mature chiefly at four years but three- to six-year-old mature sockeye are found. The sockeye normally spends a year in fresh water, this in lakes downstream from its spawning area, after emerging from the gravel in which it is hatched. Historically, British Columbia's Fraser River and the Bristol Bay area of the Southeastern Bering Sea have been the most important contributors to the fishery.

The chinook is the largest of all salmon and specimens graded "large red" bring the highest price per pound of all salmon. It averages from 15 to 30 pounds at maturity but individuals weighing more than 120 pounds have been taken in Alaska. Most young chinooks spend about a year in fresh water. Maturity usually comes at from three to five years in the southerly part of its mid-California to Arctic Alaska range. The Columbia and Yukon rivers are major suppliers of these salmon with some Yukon races known to have a life span of eight years or more.

The pink salmon, *O. gorbuscha,* is the most abundant of the five species. The young pinks go to sea immediately after emerging from the spawning gravel and mature without apparent exception at two years. The pink weighs from three to six pounds at maturity. In Washington's Puget Sound, the general southern boundary of its range, commercial runs occur only during odd-numbered years. Washington's catch many times has exceeded 5 million pinks during good years. In British Columbia and Southeastern Alaska, this small and prolific fish appears in commercially-important numbers every year.

The coho salmon, *O. kisutch,* is commercially-valuable throughout its entire range and is the salmon taken most often by sports fishermen. It is found from California to the sub-Arctic. The coho young spend about a year in fresh water and mature usually at from two to four years although older cohos are taken in its northern habitat. The adult coho weigh from four to 16 pounds.

The chum salmon, *O. keta,* ranges from Oregon to the Alaska Arctic. Young chums go to sea almost immediately after appearing from the gravel beds. From three to six years is the usual age of maturity. The adults weigh from eight to 12 pounds. The chum took many years to become appreciated as a canner's fish since its light pink flesh becomes almost white during the canning process. Early-day packers using the chum reportedly informed their consumers on can labels "This salmon guaranteed not to turn red." The chum's nutritive value, of course, equals that of any other salmon, except for a lower oil content, and it smokes especially well.

## The Wanderers

In the main, it has been only since World War II that ocean movement of the various species and races of the Pacific salmon has been well understood. It was rather generally assumed during the early decades of the fishery that most salmon stayed to maturity in salt water close to the mouths of the rivers they had descended.

Bit by bit, however, years of research experiments began to show that most salmon—not all—range widely from their home streams. This was startingly demonstrated during the 1950's by tagging conducted by the International North Pacific Fisheries Commission, a three-nation body representing the United States, Canada and Japan. This work, still underway, shows that North American salmon stocks, most specifically the reds of Bristol Bay, range far west of what was once believed to be their ocean feeding ground centered around the mid-Aleutians. There, in the far Aleutians and beyond, they mingle with Asian stocks and are subject to the indiscriminate Japanese high seas net salmon fishery.

But at the time the treaty governing the INPFC was written, it was believed that North American salmon stayed east or mostly so of 175 degrees west longitude. On this faulty assumption, the Japanese were allowed to pursue their fishing right up to 175 degrees west, putting them well into an area where they repeatedly have cut deeply into North American-bred salmon runs.

This lack of knowledge by the American and Canadian treaty negotiators set up friction that has grown more abrasive with each Bristol Bay salmon run.

Other long-term studies show lengthy ocean migrations by other salmon. Columbia River chinooks feed south to San Francisco Bay and northward past Vancouver Island into the Gulf of Alaska to at least the 59th parallel, about the latitude of Yakutat Bay. Sacramento River chinooks migrate north of Vancouver Island and Puget Sound chinooks are caught off Southeastern Alaska and Northern California. These examples can be repeated in greater or lesser degree for every salmon stream.

Conversely, there are resident populations of salmon that may move only a few score miles into salt water from their rivers. Puget Sound and the waters immediately north and west are home to large numbers of coho and chinook that feed and mature in those areas and contribute to the sport fishery the year around and the commercial fishery in season.

Salmon migration studies are nowhere near complete. These studies and the conclusions from them are one of the many factors on which optimum management of the imperiled Pacific salmon resource must be based.

All species of the Pacific salmon are anadromous. They are hatched in fresh water. In their own time, they move into salt water where they make most of their growth. At maturity, they move back into fresh water to spawn.

Salmon have a remarkably developed homing instinct. Almost all return to the stream of origin, even to the same creek and the same gravel bar where they were hatched. This ability of the salmon to find its way "home" has not been satisfactorily explained although a number of hypotheses have been offered. Most deal with the fish's reaction to the varying characteristics of fresh waters.

Hatchery-reared fingerling salmon, freed in a stream supplying the hatchery ponds, will return to the hatchery weirs or outfall gates. Others, hatchery-reared but released in other and distant streams will return to those streams to spawn. This had been shown conclusively by experiments with the eggs of Alaska salmon hatched and bred in Columbia River hatcheries, the young then released in tributaries of that river. At maturity, the marked fish make their way back to the stream of release.

There has been distinct success, especially in British Columbia and Western Washington, in restoring depleted or extinct runs or in creating brand-new runs by this method. Individual fish races must be most carefully matched to conditions of the new home stream, however.

The ability of the salmon to adapt itself under the right conditions to a new stream and thrive there can be no better illustrated than by the way a race of sockeye established itself through some 30 years in the Cedar River of Western Washington, a river that lies south of the normal range of the sockeye on Puget Sound, one that had been barren of sockeye until man put them there.

In 1935, sockeye fingerlings from Baker Lake in the Skagit River drainage were planted in the Cedar, a mountain stream that flows into Lake Washington with its waters eventually finding their way into Puget Sound through Lake Union and the Lake Washington Ship Canal. (These sockeye, incidentally, shun the fish ladder over the dam side by side with the Ship Canal Locks and seem to prefer to be lifted through the locks into fresh water in company with canal traffic.)

There was nothing scientific about this first planting and there seems to have been nothing scientific about subsequent plantings. Someone in the Washington Department of Fisheries, initially anyway, appeared only to believe that there was nothing to be lost by trying sockeye in the Cedar where small populations of kings and silvers had barely managed through the years to maintain themselves.

The sockeye project on the Cedar appears also to have been largely forgotten by the Fisheries Department under the pressures of World War II and the turnover in personnel inevitable in such an agency. Nevertheless, one or two biologists did manage to keep an eye on the river and it was noted yearly that spawning sockeye could be seen in the Cedar.

But no one got excited about the Cedar River sockeye until 1965 when it began to become apparent that a major sockeye population was building up in the stream. A race that had experienced difficulty in sustaining itself for years after 1935 found itself in the 1960's with an edge on nature for the first time.

An experimental commercial fishery was authorized in 1967 and 1968 with some 40,000 fish taken by gillnets during each of those seasons along a migration route extending some 50 miles along the Strait of Juan de Fuca and Puget Sound itself to Seattle. No fishing was allowed in 1970 and an escapement of about 350,000 fish was estimated.

In 1971, the first large scale commercial fishery was permitted with seiners allowed into an area along the east shore of Puget Sound north of Seattle where they had not fished (except on state test charters) since 1934 when a state law barred them from the area. Reaction from sports groups and operators of for-rent fishing establishments was unfavorable, to phrase it mildly. In total, the fleet, about 200 seiners and some 350 gill-netters at its peak, took no more than 265,000 of the run during the last two weeks of June and the first half of July. Sports fishermen in Lake Washington caught about 20,000 of the sockeye on various lures fished deeply and slowly. A special Indian fishery in the lake accounted for 9,000 fish while 80,000 others were caught in the usual commercial grounds along the Strait of Juan de Fuca. Escapement was estimated to meet or exceed the catch.

Commercial fishermen found it difficult to explain the low catch by the fleet through a four-week period. Many seine boats pulled out of the fishery by the end of the second week and returned to their usual grounds centered around the San Juan Islands or geared for Alaska. Some fishermen attributed the low catch (or a catch lower than hoped for) to the comparatively small area open to them and the apparent tendency of the fish to scatter across that portion of Central Puget Sound between the city of Edmonds on the east shore and the Kitsap Peninsula to the west. To the fishermen, at least, this was a violation of the long-held belief that salmon are supposed to follow well-defined spawning migration routes. In general, this is true but departures from this rule are numerous. An example is the occasional movement of Fraser River sockeye southward to that stream through the waters between Vancouver Island and the Canadian mainland rather than by their "usual" route through the Strait of Juan de Fuca.

It is hoped that the Cedar River run will average 1 million or more fish in cycle years. If this figure is reached, it may be due as much as anything to the nature of fresh waters which they traverse and in which they are spawned after leaving salt water. Lake Washington is a clean lake with neither sewage or industrial wastes going into it.

The Cedar itself is something of a rarity in an age of pollution. It is a relatively clean river. The upper river supplies water for the city of Seattle and a large suburban area to the north, the south and to the east of the city proper. Access to the watershed is restricted to official business. The lower reaches of the river lie mostly in timber and "stump ranch" lands. There is almost no industry along the lower river and the dumping of wastes into the river has been well controlled.

That closed cycle cultivation of silver (coho) and chinook salmon for the commercial market is physically practicable (with its long-term economic feasibility uncertain) has been demonstrated on Puget Sound. In 1970, a private firm, operating in cooperation with the National Marine Fisheries Service, the University of Washington and the Washington Department of Fisheries, initiated a commercial-scale demonstration in Yukon Harbor, a shallow indentation on the west shore of Puget Sound. The harbor has a high water exchange rate and the water is relatively clean, factors that led to its choice.

The commercial effort was predicated on an earlier NMFS demonstration that chinook and silvers could be reared in floating pens in salt water. It began with acquisition of 700,000 silver eggs from Washington State. These were hatched in incubators and the following fresh-water cycle of the salmon's life was pursued in the conventional hatchery manner except for the use of heat to speed growth. At the same time, 500,000 chinook fry were added to the experiment to arrive at a comparison between the two species.

During the summer of 1971, the fingerling salmon were moved to salt-water pens measuring 50 by 50 by 25 feet each. Natural tidal movement was the only method of circulating water through the enclosures and no artificial aeration was used. In salt water, growth rates picked up faster than those found normal for wild fish because of a dependable supply of artificial feed. Mortality rates were low while dogfish, finding their way through the wire, and fish-eating birds proved to be the only predators. The first fish were harvested and sent to market in January, 1972, in the eight- to 10-ounce size range.

The success of the pilot program, on which a large-scale salmon culture program subsequently was outlined, led inevitably to questions of importance to the entrepreneurs, the Washington Fisheries Department and commercial salmon fishermen. One matter considered by the private developers was the possibility of freeing a percentage of their young stock into Puget Sound to follow the accustomed life cycle of each species with the expectation that survivors would return to the pens as mature and sizeable fish, ready for market or for use as breeding stock.

The potential legal problems that such a procedure might generate appeared to be enough to baffle even a legal Solomon. The Washington fisheries authorities adroitly sidestepped any such complications, as far as salmon culture was involved, by setting up rules forbidding use of eggs other than those purchased from the state and directing that private culture of salmon be conducted "in a manner that avoids environmental competition with wild or (state) hatchery-produced fish."

At the time this private plan was in progress, the state itself was up to its knees in a program designed to develop and propagate salmon strains that would stay at home, i.e., salmon that would spend all or most of their life cycles within Puget Sound proper. This endeavor was wholly for the benefit of sports fishermen, who, with justification, complained that their pursuit of salmon within Puget Sound was usually a waste of time, money and effort. The sportsmen blamed this paucity of their quarry, silvers and kings, mostly and unjustly on commercial net fishermen. But most net fishing, with the exception of the so-called fall "inside" season for chums, not a sports fish per se, did and does take place outside of Puget Sound as that body of water is defined by federal law, essentially all of that water south of line due easterly from Point Wilson at the head of the Strait of Juan de Fuca.

The decline of the sports fishery (and the commercial fishery too or, at least, some phases of it) began to be pronounced by 1955 and it was, of course, due to many things, the major of them the despoliation of spawning grounds by human activities, abetted by stream pollution from domestic and industrial sources. (What can be done for salmon in a clean river is apparent in the preceding account of the genesis of the Cedar River sockeye.)

The serious and sometimes-biased discussion of the decline in sports catches during those years (during this period, the extensive and immensely productive ocean sports fishery for salmon reached its full development) evolved one hypothesis that appealed strongly to its adherents: to wit, that sportsmen themselves had unwittingly done themselves out of their sport. The hypothesis held that two or three generations of sports hook and line fishermen had succeeded in catching and decimating beyond return the salmon strains susceptible to the sports lure. This left then, the hypothesis held, only those salmon whose progenitors had been able to resist the sports hook in one way or another and that they passed unheeding and uncaring through a labyrinth of lures as they proceeded to their spawning grounds.

## The Way Home

The beginning of the spawning migration from often-great distances at sea, triggered by the first urges of approaching sexual maturity, varies with the species of salmon, the differing races among each species and the differing races of each stream. Generally, the shoreward movement begins in early spring and continues through autumn. The first of the Columbia River chinooks appear in fresh water in late spring, the last of them in early autumn. The runs are precisely timed with a give or take of only a few days. The Bristol Bay red salmon migrants bound for Iliamna Lake and its tributaries by way of the Kvichak River can be expected to peak at the river mouth about July 6 each year, this after a journey from their Westward Aleutian rearing grounds of almost 2,000 miles at a rate of about 30 miles a day.

With some important exceptions, spawning cannot be too long delayed after the fish come well into fresh water. The years at sea have supplied each salmon with just enough strength to find its home spawning ground that there is almost none to spare. A hangup of no more than a week can cut so sharply into a year of a cycle that it may take the races of that year many years to recover.

Some salmon races, especially the chinooks of the great river systems such as the Columbia and the Yukon spend many months, as many as six even, in the rivers before they spawn. Their journey upstream to their spawning areas is long and arduous and these fish have been found more than 3,000 miles from the river mouths in creeks and rivulets without name on any map.

The pink and the sockeye are most susceptible to undue delay on their spawning runs. The pink tends largely to spawn in small streams and many times they fall victim to the low water stages of the late summer when they begin their spawning runs from fresh water. The pinks of Puget Sound were hit hard in 1957 and 1959 by the high temperatures and low water of those cycle years.

Obstacles, man-made and natural, are too many and sometimes both man and nature combine against the salmon. Witness the Hell's Gate obstruction on the Fraser River where a naturally-constricted, fast-flowing stream, was further narrowed and made faster in 1913 by the rock slides set in motion by unknowing or careless railroad builders. In that 1913 season, perhaps 98 or 99 percent of 10 million sockeye spawners bound up the Fraser died fruitlessly downstream from Hell's Gate. Hundreds of thousands of the fish apparently tried to spawn in downstream tributaries of the Fraser but these streams were not sockeye country and the spawning effort was fragmented and useless. It took the 1937-created International Pacific Salmon Fisheries commission a lot of hard work and some 20 years to restore the Fraser River sockeye runs to a fraction of their natural abundance. It should be added that about 25 million sockeye had been taken from that 1913 Fraser run before the 10 million spawners got past the last net.

In fresh water, all Pacific salmon stop feeding and begin to display dramatic physical changes. Their bright silver turns to shades of brown and black and red, this last coloration noted most sharply among the sockeye. Body configuration changes too with the males developing hooked snouts and humped backs. From this latter characteristic, the pink salmon gets its familiar nickname "humpie."

The spawning procedure is almost as ritualistic as the mating dance of the whooping crane. Upon arrival on her native gravel bar or one close to it, the female salmon uses her tail and almost the last of her strength to scoop out in a well-percolated area a nest called a "redd" in which to drop her eggs. The nest is commensurate with the size and species of the fish herself and for a husky chinook it can be up to 18 inches deep.

When she is satisfied with her work and the male is hovering at her level in the current and a head's length or so ahead of her, she drops a handful of eggs while the male expels some of his milt with its life-giving spermatazoa. The eggs separate as they drift into the hard-won nest and the female again brings her tail into use to cover them with a layer of protective gravel. The mating may be repeated a dozen times with often a second nest scooped out before the female has exhausted her energy and her 2,000 to 8,000 eggs. Then she and her male companion or several companions give way to the tug of the current, drifting with it without fighting back as their lives come to an end.

All Pacific salmon die after spawning. But not uselessly. Their carcasses feed numerous animals and birds who come to seek them along the river banks before and after death. And in the water, the decaying salmon becomes another link in the food chain, one that nourishes the animalcula that will, by the coming spring, furnish the first food for the progeny now embryonic in the eggs safe beneath the gravel.

That is—under normal conditions they are safe in the gravel. The hazards are many. Spawning females arriving on the beds after the first have completed their task may ruin the labor of those who came before them as they witlessly scoop out their own nests. Flooding may scour away the sheltering gravel. Erosion or slides may cover the nests with silt and starve them of oyxgen. Low water and cold weather may freeze them out. Jagged, drifting ice may gouge across the beds like bulldozers.

But if matters go as nature intended, the eggs will hatch in three to four months; the young fish will live in their gravel shelters for several weeks, subsisting on the yolk of the maternal egg still attached to the belly. When the yolk is gone, the young come forth from the gravel to eat and be eaten. For those who survive, ahead lies the sea. And for those who survive the sea, there awaits their own "return to the river."

Roderick L. Haig-Brown once wrote:[1]

"The salmon are, perhaps, the most exciting of the world's fish. They are fish of grace and beauty...under natural conditions, they are prolific fish, multiplying to abundance; they grow to great size and their flesh is rich but delicate enough for epicures; and their predictable but always dramatic return to fresh water to spawn brings the wealth of the sea within reach..."

## History Of The Fishery

Although the Pacific salmon fishery is comparatively new among the great commercial fisheries of the world, it is the oldest of all commercial fisheries of the North-

eastern Pacific. But before it became a money fishery, it was as important to the native peoples of the region as was the buffalo to the tribes of the Great Plains. Indians, Aleuts and Eskimos from the Sacramento to the Yukon prized the salmon above all other fish.

Had it not been for the dried salmon of the Nez Perce, Lewis and Clark and their men might never have made it to the Pacific when they stumbled, starving, out of the Rockies into the Snake River country in 1805. Salmon was major among the foodstuffs of the Indians of the Columbia River basin and the average annual river catch by the Indians in the early 1800's is guessed to have run around 18 million pounds.

The Hudson's Bay Company began in the late 1820's to export salt salmon from the Columbia to Hawaii, Australia, China, England, California and the markets of the East Coast of the United States. So did their compatriots somewhat later on Puget Sound and at the mouth of the Fraser River.

These early-day British Columbians on the Fraser were practically sitting on top of the continent's second-richest red salmon fishery but neither they nor anyone else could see further than the chinook. It was well into the 80's before the sockeye began to get the attention he deserved from anyone anywhere, either on the Fraser or in Alaska. The king salmon was truly the king...but no one yet had thought to put him in a can. But his time was coming down on the Sacramento.

Young William Hume, who came west from Augusta, Maine, to mid-California in 1852, was seeking his fortune in the Golden State just like all the thousands who had ridden across the plains and mountains in the years since the strike at Sutter's Mill in 1848. But Hume seems to have had no intention of finding that fortune in the gold fields. He brought with him a gillnet. His home town lies along the Kennebec River and the use of gillnets was an art well understood on the Kennebec where the Atlantic salmon still ran strongly in those days.

It is unclear just why Hume lugged that gillnet all the way across the country. He may have known about the chinook of the Sacramento River, then a salmon stream as productive as any other along the coast except for the incomparable Columbia. Or he may only have been hoping.

Whatever his original design, Hume didn't wait long to put his gillnet to work among the salmon of the Sacramento. He and two one-time classmates from the East, James Booker and Percy Woodson, set up a commercial fishery on the river, peddling the fat chinooks of the stream where they could among the men of the gold fields and the villages being built along the valley. The home of the three young men was a jerry-built cabin beside the Sacramento in the community of Yolo in

1. Haig-Brown, Roderick, "British Columbia's Salmon," Canadian Department of Fisheries, 1959.

Washington County. Every man who ever has made a dime from packing salmon should remember this cabin with reverence. It wound up as the warehouse of the world's first salmon cannery.

In 1856, Hume went back East long enough to talk two of his brothers, John and G. W. Hume, into venturing out to California, presumably to help in the gillnet enterprise which was getting pretty big for its time and place. In 1863, G. W., in turn, went back home to see the folks and ran into an old classmate, one Andrew S. Hapgood, who then was superintendent of a lobster cannery on Fox Island. The two men swapped talk about their respective specialties and presently Hume convinced Hapgood that his future lay on the banks of the Sacramento in far-away California.

Hapgood agreed, on condition that William Hume, the leader among the brothers, join the venture too. William, with some money to spend and the entrepreneur's ability to sniff out a good thing, easily obliged and Hapgood came west in March, 1864, with a supply of gear and know-how to set up a salmon cannery on the Hume property on the river. The Humes, chiefly William, were to finance the operation and supply the raw material and the market. The brothers had bought an old scow and moored it on the river down the bank from the old cabin. On it, with the arrival of the first spring salmon, Hapgood presided over the packing that season of 400 cases of chinook talls in red, hand-soldered cans. Red was the only can color the market would accept. No one other than the Humes and Hapgood was allowed to witness the final mysteries of the canning rites, the sealing of the cans. Hapgood himself took care of this with the cans handed to him one by one through a small opening in a wall that closed off one end of the scow.

It was the smallest of small-time operations but, unknowingly, Hapgood, Hume and Co. had touched off its own industrial revolution, one that made some men rich and bankrupted others, one still going strong although long removed from the banks of the Sacramento. The only way was north.

By 1866, the Humes were beginning to feel crowded on the Sacramento and they wanted a greater supply of chinooks than the Sacramento could afford them. They looked to the Columbia and the Columbia looked good. So, during the summer runs of that year, William Hume started the first salmon cannery in all the Columbia country, another Hapgood, Hume and Co., plant at Eagle Cliff on the Washington side of the river 40 miles upstream from Astoria. The cannery prospered for some years along with several others that were built there. All are vanished now although the Eagle Cliff area still is a productive gillnet drift.

Again the Humes had started something big. John N. Cobb wrote that in 1881 there were 35 canneries strung along both banks of the lower Columbia with the brothers Hume founders of at least half of them. A 1966 Bureau of Commercial Fisheries report adds:

"The canning of salmon...grew rapidly at first...by 1883, the number reached a peak of 39. The combined pack of these 39 canneries in 1889 was 629,000 cases valued at over $3 million. (About $22 million at 1972 wholesale prices—Ed.)...by 1890, the number of canneries had declined to 21. ...The reduction was due to a number of factors, including an apparent decline in the abundance of chinook salmon."[1]

Along the Columbia, at least, no one yet was seriously interested in any other salmon but the chinook. But up on Washington's Puget Sound, someone had been thinking about the others.

Puget Sound had shipped its share of salt salmon around the world from the time the Hudson's Bay Co. had come into the country. But the Puget Sound salmon industry was beginning to hear of another and better way to process salmon. The saltery was to be largely displaced by the cannery.

No business register today lists the firm Jackson, Myers and Co. But the small company should be remembered for two firsts in the salmon industry on Puget Sound. Jackson, Myers built the Sound's first cannery at Mukilteo on the mainland on the east shore of the Sound north of Seattle in 1877.

More importantly, perhaps, their entire pack that first year was silver salmon, 5,000 cases of it, the first time any company had shown major interest in anything but the lordly chinook. A few years later, Jackson, Myers pioneered again. They packed a few hundred cases of pinks.

Jackson, Myers merely pointed the way to their fellows. Puget Sound then was home to some of the continent's biggest salmon runs. By 1915, there were 41 canneries on the Sound and some few of the company names still survive although most of them will be found now in the fine print of corporate mergers.

The enterprising Canadians up on the Fraser were a bit ahead of the Jackson, Myers people—seven years ahead. In June, 1870, Alexander Loggie and Co. built a small cannery at Annieville on the Fraser three miles downstream from New Westminster. The story there was the same as it had been on the Columbia in the 60's, the same as it was to be on Puget Sound a few years later—expansion as rapidly as possible to crop all the salmon that could be used from the runs that, for a couple of decades, seemed to be in inexhaustible supply. By 1901, there were 70 canneries between the Fraser and Portland Canal. Today there are no more than two dozen.

## North To Alaska

British Columbia and Alaska are the two areas where the drama of the Pacific salmon industry has been played out to its bitter end. British Columbia's coastline is long and Alaska's is longer. In that province and

---

1. Pruter, A.T., "Commercial Fisheries of the Columbia River and Adjacent Ocean Waters," *Fishery Industrial Research,* Vol. 3, No. 3, December, 1966.

in that territory, the giants of the industry had almost free rein for decades to act their parts as they wished. At times, matters came close to armed warfare, piracy was a fact of existence, the big got bigger mostly, and hundreds of salmon canning companies were born and died or were merged over the years. In Alaska, at least, there still survives a residue of resentment against the salmon packers from "Outside." The doings on the Columbia and Puget Sound were merely the prologue to the bigger performance.

Russians had dabbled in salt salmon in the areas they occupied so tenuously in Alaska, shipping some of it back to Siberia but using most of it to feed their native serfs. Americans and Canadians, careless of title to almost anyone's land when they could get away with it, had invaded the Russian territory for decades, interested in fur poaching mostly, and along the way they had learned about the region's salmon. But it took the Americans 11 years after the Yankee occupation to get a salmon cannery going.

In 1878, the North Pacific Trading and Packing Co. built and worked a small cannery at Klawock on the west coast of Prince of Wales Island deep in Southeastern Alaska. Later that same year, the Cutting Packing Co. started up a cannery at Sitka on the west coast of Baranof Island.

In a little more than a decade, the industry had spread through all of Alaska where they were enough salmon to gamble on, all the way from Portland Canal to the Bering Sea.

The other pioneers were the Alaska Packing Co., of San Francisco, at Kasilof on Cook Inlet in 1882; Smith and Hirsch at Karluk on Kodiak Island in 1882 also; the Central Alaska Co., on Little Kayak (Wingham) Island at the southeast approach to Prince William Sound in 1889; the Arctic Packing Co., first on Bristol Bay with a cannery at Nushagak in 1884; Point Roberts Packing Co. across the way at Koggiung on Kvichak Bay in 1895; the Alaska Packers Association at South Naknek in 1895 too. (Of all these, giant APA still is the only one in business at the same stand and its South Naknek cannery with five lines reputedly is the biggest salmon cannery in the world.)

On this sometimes insubstantial base was built the single most valuable fishery of the North Pacific Ocean. The total value of Alaska's salmon industry since 1868 (there were 70,000 pounds salted by Americans in that year) has been estimated by the United States Bureau of Commercial Fisheries at about $3.2 billion, a figure that increases every year and will continue to increase as long as the salmon runs can be properly maintained.

This return was not earned easily nor peacefully and, quietly, the struggle for position and power still goes on. This conflict was not always so sub-surface. It was noisy and sometimes scandalous. The battle of the titans of the railroads in the decades when the roads were the pawns in a cut-throat game have been accepted as the epic business battles of the United States. They may have been waged on a grander scale than those of the big men of the Alaska salmon industry (rail tycoons and salmon tycoons sometimes were the same) but they were no more bitter. Companies were formed and were wiped out; fortunes were made and lost; stockholders from all the United States and much of Western Europe found their investments worthless, their dollars, pounds, francs, marks and kronen gone forever.

Outstanding, perhaps, among these corporate brawls, was the "Salmon War of 1903," a messy affair that proved once again that the David and Goliath match was a fluke. In this encounter, the Alaska Packers Association, then a lusty 10 years old, met head-on with the upstart Pacific Packing and Navigation Co., going on three years of age. APA was big, well-backed, wise in the ways of the Alaska industry and the back-alley fighting surrounding it. And APA had plenty of ammunition for a battle in the marketplace—1,223,000 cases of the previous season's pack.

Pacific Packing had neither the ammunition, the experience nor the capital to survive much more than a skirmish or two with APA. The company had plenty of backing on paper (Guggenheim money, reports in the street said) but it turned out to be only paper after the first shots were fired.

APA really didn't have too much trouble putting Pacific Packing away when things got going. The company simply used its pack (APA's 1,223,000 cases almost equaled the total 1967 Alaska pack of 1,464,000 cases) and principally its 1,007,000 cases of reds to drive the market down and Pacific Packing to its knees along with the market. APA dumped at prices Pacific Packing could not conceivably meet.

At bottom, chums were down to 16 cents per dozen one-pound talls, not much more than it cost to put up a single can. In less than four months, Pacific Packing was in receivership. The holders of the company's common stock lost all they had invested although the owners of some bonds did get back a few cents on each dollar when Pacific Packing was liquidated.

The next year, an APA man, writing of the Salmon War, commented:

"...(it)...illustrates two principles which are necessary to be applied to maintain living prices;

"First, the supply of an article must not materially exceed the demand;

"Second, those interested in maintaining prices should not begin the cutting lest they lose control of the knife."

One may fault his syntax but certainly not his reasoning.

## A Way To Get Rich?

Any venture in salmon canning large or small, well-heeled or not, was chancy even in the best years in that tumultuous era, both in the cannery and in the market place. Salmon canning still is chancy, a business dependent upon the vagaries of hordes of mindless fish subject to myriad natural and man-made disasters.

The blasted hopes of thousands of persons—working fisherman, eager investor, decision-beset corporate officer—over the years since Hume went to work on the Sacramento in 1864 were summed best by the esteemed John N. Cobb who, in his time, knew more about salmon and the salmon industry than any other man. In 1911, he wrote:

"The many idle plants which dot the Alaska coast are silent monuments to the hazards of the salmon canning business."

Many of those "silent monuments" have crumbled away under the weight of the years and the elements. Fire has destroyed others. A few have been converted to other uses. On Bristol Bay, the worn old cannery complexes at Libbyville and Koggiung and Becharof testify to the trend toward corporate merger as stronger or luckier management prevailed and the weak or unlucky sold out. Abandoned cannery sites are sprinkled through the rest of Alaska's salmon country too. Not all canneries went dark, however, when the operating companies pulled out. In a few cases, entire towns backed formation of cooperative associations to keep the canneries at work during the season. Under green management, the co-op in a few instances only delayed the inevitable. Again, cannery operations themselves sometimes have been shut down permanently but related cold storage or processing facilities like smokeries or salteries have been kept going. Even these small endeavors are valuable to the often-tiny communities where they are located.

In 1929, there were 156 canneries in operation in all of Alaska. They packed 5,370,159 cases. That was tops in cannery numbers for the Alaska industry and even in that year of cheap salmon and cheap labor, most of those canneries probably did not make much money. Matters had not changed too much 38 years later in 1967, another year in which much of the Alaska salmon industry didn't especially prosper. The poor season had been predicted but it was even poorer than most forecasts and some cannery operators got badly hurt even though everyone hedged his bets. The big canneries, the companies with more than one cannery, consolidated cannery operations. The long-standing pool arrangements under which one cannery packs for one or more other companies were expanded so that any operating loss would be spread over several sets of books.

That year, 62 companies operated 71 canneries between the lower Yukon and Ketchikan. Four of the canneries were floaters working variously from the Yukon to Capa Yakataga. Six large companies operated 15 of the remaining canneries. Some other companies with efficient canneries of their own chose the pools. Many of the other companies were small firms with one canning line only and at least one was a hand operation packing highest quality red salmon in quarter-pound cans for European export.

Among them, the 71 canneries put up 1,464,006 standard cases of salmon, the smallest Alaska pack since the 1880's. Nobody paid big dividends off salmon operations that season and some of the weaker companies did not come back at all the next season. Their canneries were empty and silent.

John N. Cobb would have recognized the situation immediately.

## Salmon Gear

The Pacific salmon is taken in a variety of ways— by purse seines, by drift and set gillnets, by trolling gear and, on a most limited scale, by traps and fish wheels by some native fishermen of Southeastern and Western Alaska.

The salmon trap has been called the most efficient and economical means ever devised for catching salmon. But it has been a source of bitterness wherever it has been used.

The trap first appeared in the Pacific salmon fishery in May, 1879, when O. P. Graham, out of Green Bay, Wisconsin, built a stationary or pile trap on the Columbia. It was modeled after those then used on the Great Lakes and it worked efficiently. The first trap appeared on Puget Sound about 1880 or 1881 when one was built off Cannery Point on Point Roberts.

The trap went to Alaska in 1885 with construction of a stationary trap in Cook Inlet. It spread fast over the territory where it was to fish for almost 75 years. In 1907, J. R. Heckman, of Ketchikan, devised the first practical floating trap and it was placed in areas where the pile trap could not work. Traps were used in a limited way in British Columbia after 1904.

These fixed and floating traps all worked on the same principle, that of intercepting migrating salmon on their favored routes close to the beach and diverting them into a chamber from which they could not escape. The fish were kept alive and untouched until the moment they were brailed out to be moved by packer boat to the cannery.

In Alaska especially, trap piracy was sometimes a harsh fact of a harsh existence, a deadly game in which trap watchmen had to be dealt with, either by bribery or by violence, since disguising the pirate tender was hardly a practical matter in a land in which almost every man knew every vessel by sight.

Trap sites were highly valued and were the object of stiff competition wherever they were used. The sites were allotted by permits issued by the Army Corps of Engineers and an oft-heard complaint among Alaskans especially was that the Engineers discriminated against residents in favor of the big salmon packers from "Outside."

Boat fishermen everywhere liked the traps no more than might be expected and, for these reasons, the trap was a constant irritation among these opposing factions of the industry and became, in time, a factor behind Alaska's long drive toward statehood. More than almost anything, the Alaskans wanted to control their own fisheries, the territory's major economic support. Traps were used in Alaska long after they had been banned in other regions.

One of the first acts of the Alaska legislature after statehood became reality in 1959 was to outlaw all traps in the state. The new state promptly found itself embroiled with the federal government over the legislative order. The trap ban was disputed by the Indians of Annette Island who claimed treaty rights to fish as they wished in their "traditional and accustomed" fishing areas. The Indian claim was upheld by then-Interior Secretary Fred Seaton.

The case eventually went to the United States Supreme Court with the court upholding the right of the Indians to use a limited number of traps. Similarly, although traps have been outlawed for general use in Washington State for decades, the Swinomish Indians of that state are allowed to use them in their tribal grounds near the mouth of the Skagit River.

The Indian trap catches in Washington and Alaska are insignificant in relation to the overall salmon catch, however. In 1965, all salmon gear in Alaska took 274 million pounds of salmon with the Metlakatla traps at Annette accounting for only 173,910 pounds; in 1966, the comparative figures were 333 million pounds and 3,040,569 pounds, and in 1967, they were 138 million pounds and 57,363 pounds.

Since the general abolition of traps in Alaska, purse seines and gillnets have taken the major portion of the catch as they have in British Columbia and on Puget Sound.

## PACIFIC SARDINE

The Western Hemisphere's greatest fishery died lingeringly in the 1940's. The cause of death is still undetermined although postmortem examination goes on in the laboratories of the National Marine Fisheries Service and other agencies.

In the 1936-37 season, American fishermen landed about 1.5 billion pounds of Pacific sardine in California ports. In 1969, fishing vessels put ashore in those same ports an "incidental" catch of 125,000 pounds of sardine. What killed the sardine fishery and what is its chance, if any, to come to life again?

The Commercial Fisheries Branch, British Columbia Department of Recreation and Conservation, unhesitatingly blames the end of the fishery on over-fishing by the California fleet with an assist by that of the Pacific Northwest.

California sports fishermen, just as unhesitatingly, also blame the disappearance of the sardine on over-fishing. They like, too, to use it as a likely example of the fate of West Coast anchovy stocks if catch quotas are raised or eliminated.

The full answer to this oft-posed question probably is not as simple as over-fishing. Other factors, equally important, appear to have played a part too in the events that led to the disappearance of the sardine from its long-time haunts.

The Pacific sardine, *Sardinops caerulea* once abounded from Baja California to Southeastern Alaska. It must have been at least as abundant as the anchovy presently is through much of the same area.

The fishery was pursued throughout most of the waters where the sardine was found, although landings were larger by far in California than the total of all other catches. In seasons of abundance, great schools of adult sardines migrated north to supply big summer and fall fisheries off Washington, Oregon and British Columbia. The Canadians had begun taking substantial catches in the 1920's. Washington and Oregon fishermen followed the fishery too but made most of their landings in California because of laws in the two northern states that prohibited the use of sardines for reduction.

But these laws were repealed in 1935 and landings for the two states jumped to 65 million pounds that season from a previous yearly average of only 10 million pounds. In 1938-39, Oregon-Washington landings peaked at 80 million pounds but began to decline in the following years. The last commercial landings in Washington came to 2,000 pounds in 1951.

The story was almost the same to the north. Canadian landings in 1936 totaled 86,300 long tons but they declined thereafter just as rapidly as in Washington and Oregon.

A change in the makeup of the sardine stocks began to show up about 1940. Until that year, sardine catches off Oregon-Washington-British Columbia were almost entirely four-year-old fish, the dominant age group throughout its range and the one thus most vulnerable to the fishery effort. But during the later years of the expiring fishery, five-year-olds and older became the dominant age groups of the declining stocks. This shift in age groups reflected the intensive California fishery and its decimation of the strongest age group among the entire sardine population.

It is clear that over-exploitation of the sardine, an exploitation that no supervisory agency in its right mind ever would allow today, did contribute much to the death of the fishery.

But it is not entirely probable that a fish species of such numbers and such wide range can be fished past the point where it can no longer sustain and eventually replace itself. The ability of a fish stock to recover from overfishing when given a decent breather is demonstrated by the comeback of the Pacific halibut stocks under the shield of the International Pacific Halibut Commission. That the commission's policies were responsible for this comeback is evident; that the halibut is under threat again, not by Americans and Canadians but by foreign trawl fleets and their carelessly-used gear is not the fault of the commission.

It seems evident too that environmental changes may have been as hard on the Pacific sardine as were man and his purse seines. Among them may have been lower water temperatures. The post-1940 disappearance of the Pacific sardine coincided with a 15-year period of unusually cold water in the sardine's normal habitat. Other fisheries were affected adversely by it, including the coastal albacore fishery. Those small tuna stayed far to the south of their usual northward range during those years. Another element to be considered might be disappearance of or a change in the composition of the sardine's food supply caused by the lower temperatures. The long cold may have unfavorably affected sardine spawn on the grounds off California and Baja California. Conceivably, disease could have ravaged the spawn or mature fish themselves to such an extent they have not been able to recover.

If the sardine were going to make a comeback, it would seem that some evidence of that renaissance would be apparent. But there are no more sardines to be found on the spawning grounds, the one place in all the sea in which they could be counted upon to appear at the accustomed time, than there were in the late 1950's. If overfishing were the sole answer to the disappearance, the rebuilding stocks would be showing there. There must be many things together to be blamed for the vanishing sardine.

## Nature At Work?

Among the latter, perhaps, may be counted a monstrous quirk of nature itself. It has been postulated to the satisfaction of some researchers that, for reasons unfathomed and perhaps unfathomable, the Pacific sardine appears and disappears along the West Coast over great swings of the geologic pendulum, that it comes and goes in cycles ranging from about 500 to about 1,700 years.

According to this theory, the sardine's presence in great numbers is for a comparatively brief time only, then it vanishes for these much longer periods. This belief is based on examination and dating of fossil remains of the sardine from strata of known ages. During the gradual disappearance of the sardine, the ecological gap is filled by the anchovy, a state of affairs that seems to be coming into existence in this present era with the buildup of the anchovy stocks of the Pacific Coast.

This theory would seem to fit the facts, as presently known, of the most recent disappearance of the sardine, that is, the sudden growth of the stocks around the turn of the century, the gradual depletion of those stocks for one reason or another, and the burgeoning of the anchovy. If this theory should be correct, it would seem that the sports forces of California, particularly, militating against a commercial fishery for anchovy, might as well save their energy and let nature take its own course and own sweet time.

It would seem too that Monterey might be back in the sardine processing business in no more than 1,600 to 1,800 years if anyone cares to wait around for the revival.

The sardine fishery was the great money fishery of the California coast just before World War I to the 1950's. Two generations of fishermen lived by it and some grew wealthy on it. Monterey was its capital and John Steinbeck wrote fondly of Monterey and its people and its Cannery Row in the years when the fishery was still great. Monterey was a tough, restless and vital town in that time before it grew to depend more on tourists, tourism and the retired than on the sea and its harvest. Few of the men and women of Monterey who lived on the fishery, other than the fishermen themselves, saw a sardine actually taken because most of the fishing was distant from the town, an operation in the night in which long, heavily-corked purse seines of fine mesh sometimes took scores of tons of sardines in a single haul.

The Pacific sardine fishery was divided roughly into two parts. One, the major segment, was centered off California where, in big years, 1 billion pounds or more were landed. The other segment was based on Washington, Oregon and British Columbia but this fishery produced only a fraction of the California catch.

Some sardines were canned at Astoria in one-pound talls and a few were put up in British Columbia under the name "pilchard," an Anglicism sometimes applied by Americans to the Pacific sardine. The northern fishery ran from early summer well into the winter. The California fishery had two seasons—August 1 to February 15 in the San Francisco-Monterey region and from November 1 to March 30 in the San Pedro-San Diego district during the later years of the sardine era.

The sardines put a lot of people out of work when they took the deep six and they put a lot of boats out of work too. In its heyday, Monterey's Cannery Row had 18 plants and 78 seine boats fishing for them. Over the years, the fishery had developed high-capacity vessels similar to but larger than the familiar northern multipurpose vessel. When the sardines were plentiful, these boats had as much work as they could handle. But with the sardines gone, there was no place for most of these boats to go. They could fish salmon in Washington but that was scarcely a paying proposition with the fishery over-geared even then. They could not fish salmon in Alaska, where they might have worked profitably, because they were too big to meet the 58-foot overall "Alaska limit." A few were converted to tuna bait boats but they were too small to do much more than work the fringe of that fishery. Some few were sold south to Latin America and a handful found jobs as summer charter vessels for the Bureau of Commercial Fisheries, the University of Washington's Fisheries Research Institute and a couple of other agencies. Others lay idle until the Alaska king crab fishery began to grow in the early 1950's. This fishery still employs some of them although they gradually are being replaced by larger and tougher steel ships designed for this most exacting of fisheries.

Various nets were tried out in the early years of the Pacific sardine fishery. Net efficiency depended to some extent on vessel efficiency, that is, the ability of the vessel to put her net quickly around a school of sardines and begin the haul before the fish could escape. By 1905, dependable power had combined with the lampara net to produce the first really efficient catch method and the lampara ruled the fishery until about 1930. In that decade, the purse seine went into the fishery and by

World War II, the seine was taking three-quarters or more of the catch.

The sardine purse seine closely resembled the salmon seine from which it had been developed. These sardine seines ran around 250 fathoms long—the same as the present Alaska maximum—and from 20 to 30 fathoms deep as against the 18 or 20 fathoms for the average salmon seine. The sardine seine had no end taper and its mesh was much smaller, usually 1-1/2 inches stretched. The seine was fished much as in the salmon fishery although California fishing took place at night. Northern sardine men worked by day, however. The night fishermen made their sets on the phosphorescence stirred up by the dense schools of sardine as they coursed through the water.

Good fishing brought good pay days to the sardine fishermen. Loads of 140 to 150 tons were not uncommon although they were not the rule either. Not all sardines were acceptable for human consumption and here again, the State of California imposed standards as stringent as those in force then and now in the tuna industry. Neither Washington nor Oregon worried as much about sardine quality as did California since 99 percent plus of their catch was not intended for human use.

# POMFRET

The old-line fishery resources of the Northeastern Pacific—those like salmon and halibut and probably the crabs—are being used at or near their maximum yield. Only shrimp and some bottom fish of the species currently being used seem to be able to stand heavier fishing pressure. Any real expansion of northern fisheries must take place among those species not now being exploited, for one reason or another, for either food or industrial purposes. It seems apparent that growing population and a consequent need for protein will make acceptable the table use of many fish presently discarded as trash. Most of these species of fin fish now ignored by Americans and Canadians are valuable food fish elsewhere in the world and necessity probably will make them food fish in North America too, sooner perhaps than now is expected.

Among this gallery of "worthless" fishes is a bright image, one completely unknown to almost all fishermen of British Columbia and the West Coast of the United States because they do not fish with the proper gear where this fish is found. This potentially-valuable newcomer to the commercial fisheries of the Northeastern Pacific is the pomfret, *Brama japonica,* a sunfish-like creature of the high seas with remarkable quality as a food fish despite its rather unlikely appearance. The pomfret has a deep, compressed body and a narrow caudal peduncle with a deeply-forked tail and finlets in the apex of the fork. The long, serrated dorsal and anal fins form an almost symmetrical pattern. The forehead is arched, the eyes are set well forward and the lower jaw is underslung. Large specimens from the Northeastern Pacific have run to 20 inches or more in length with a weight of three or four pounds.

The pomfret of the Pacific closely resembles the African pompano or Cuban jack in structure and hopefully it will one day closely resemble the African pompano economically. On its range off the southeast coast of the United States the pompano brings the highest ex-vessel price American fishermen get for any fish with an average of more than $1 a pound in the round. It is regarded as the most succulent of all food fish taken in North America and this opinion and the ex-vessel price are reflected on the menus of those places where it can be found as a white, flaky fillet.

The pomfret is widely distributed through the warm and temperate waters of the world. There are fisheries for pomfret off the Atlantic coast of Spain and off the coast of Japan in the Pacific. American and Canadian contact with the pomfret of the Northeastern Pacific has been almost solely through incidental catches of the fish during high seas salmon investigation. Vessels of various research agencies have taken pomfret through all the area east of 175 degrees west longitude, roughly the longitude of Adak in the Aleutians, and north of 42 degrees north latitude. This covers a vast expanse of ocean that has barely been sampled as far as the pomfret itself is concerned.

No pomfret have been taken by the longline or trawl gear used by these research agencies but large numbers have been found in gillnet and purse seine hauls. Albacore trollers often take them too. This latter gear was fished in this exploratory work at or near the surface. One seine haul made south of the Aleutians by a vessel of the Fisheries Research Institute of the University of Washington College of Fisheries contained more than two tons of pomfret. (This research net fishing, incidentally, is the only legal pelagic net fishery conducted by Americans and Canadians in this part of the world and this explains why the pomfret is almost completely unknown to commercial fishermen of these nations.)

Since a scrap fish happens to be any fish people are not fishing for, all pomfret catches were routinely thrown over the side until personnel of the Bureau of Commercial Fisheries Research Center in Seattle, Washington, developed an interest in them and began a tentative study. Among the facts established by this research (and a subjective one) is that the pomfret is fully as tasty as the African pompano in any way it can be prepared. The fish used in this test cooking were specimens that had been frozen at sea and held that way for a considerable period before reaching the laboratory. The men who did this work are satisfied that the pomfret will be quickly acceptable as a food fish, fresh or fresh frozen, whenever and if a commercial fishery can be instituted.

Best catches of pomfret were made by gillnets in water with a temperature of 50 to 57 degrees F. No pomfret were taken in the Bering Sea during the seven years covered by the BCF study. No pomfret were caught before June and the best catches came during August and September. The pomfret is believed to spawn in waters south of the area covered by the survey and heavy concentrations of the fish might be found on those grounds if they can be located.

## SAURY

The handful of species of food fishes from the Pacific presently favored by Americans and Canadians is not unlimited to ultimate sustained yield despite today's relatively small use of most of them. The day must come, for many reasons, when that list of food fishes will be lengthened to take in other species presently unknown to the people who live along the eastern rim of the Pacific. Among such is the pomfret, one of those that are better on the table that many of the fish now used for food. Another of the candidates for table honors is the Pacific saury, *Cololabis saira*, a long-bodied little fish highly favored as food in much of the world but completely strange to most North Americans. It may not be more than a generation in the future that the saury will be as familiar to fish consumers of the western continent as it is in the homes of Japan where it is an important fish, one caught in substantial quantities by that nation's fishermen.

As far as looks are concerned, the saury is the exact opposite of the toothsome pomfret. The pomfret, as unknown to the public as the saury, looks first cousin to the sunfish with its compressed, deep body. The saury resembles a smelt or a flying fish or the needle-fishes, all smallish fishes of good food qualities. And, like the flying fish, the saury spends a minute fraction of its time out of the water. Although it cannot glide for distance like the flying fish, the saury distinguishes itself by its habit of repeatedly leaping two or three feet above the surface of the water.

The saury runs up to 14 or 16 inches long. A marked characteristic is the single dorsal fin set far back on the body, well behind the first rays of the ventral fin. Between the dorsal and the tail stand six small finlets much like those of the tunas or the mackerels. The lower jaw is long and protruding and quite flexible. It bears the typical marine fish coloration, greenish above and silver below. It is a pelagic mass spawner.

In the Eastern Pacific, the saury is found from Alaska to Baja California and everywhere it is regarded as a trash fish just as was the albacore three-quarters of a century back. It is a fish of massed schools and it is believed to be abundant over most of its range, although a lot of work must be done to establish abundance and assess a commercial potential when North American markets are ready to accept the saury as food or as an industrial fish. The saury has been reported in good concentrations off Oregon and Washington by American and Soviet scientists studying pelagic fishes, and a commercial fishery might well begin in these areas. Because of its schooling habit and its apparent preference for surface or near-surface existence, the saury seems to be a fish well adapted to purse seine or lampara net fishing.

The Japanese and a few Americans have experimented with fish pumps with some degree of success. In this process, the saury schools are attracted to the fishing vessel's side at night by the use of strong lights lowered to the surface. When the schools are sufficiently massed, the pump is lowered into the school and activated. The fish caught in its pull are piped into the hold while water sucked in by the pump is diverted back over the rail into the sea. Saury taken by Americans using this method have gone to reduction plants at a low price per ton.

## SCALLOPS

The long history of commercial fishing is a history of ceaseless search for better gear, more fish and new species of fish to exploit profitably. Once, beginning back in the 1820's with the Hudson's Bay men on the Columbia, the salmon (and the chinook only of all the Pacific salmon) was about the only fish of the whole coast taken for money. It was well into the 1850's, most particularly around San Francisco Bay, before any other species of fish began to appear in the commercial fishery. Now there are some 70 or 80 kinds of roundfish, flatfish and shellfish worked in varying degree from Southern California to the Arctic. That number can only continue to grow as the pressure of population and the maximum use of presently-popular species make it necessary and profitable for West Coast fishermen to seek and use many of those species now regarded—by Americans and Canadians, anyway—as trash fish. It must be remembered that the saucy and valuable albacore once was held in comparable esteem.

One of the Eastern Pacific's previously-unharvested resources is the sea scallop, now the object of the West Coast's youngest fishery, one still uncertain of its future. Large scale fishing for this creature, the weathervane scallop, *Patinopecten caurinus*, began in the Gulf of Alaska in mid-1968. (A very minor scallop fishery existed on Puget Sound from 1935 to 1952 but petered out when costs began to outrun returns.) By 1970, Alaska scallop fishermen began to appear to have fished themselves onto a thin edge. Some of the more accessible scallop beds were beginning to run lean. Others were closed to the scallopers because their dredges were damaging the king crab stocks. Boats were laid up, sold or converted to other fisheries. The fishery slowed as boat operators and processors took a breather and attempted to look into the years ahead.

The scallopers had something to look back at too. During the short bonanza period, the small fleet had set national scallop catch records. In three months in 1968, seven or eight scallop vessels delivered 1.1 million pounds of scallop meats. The entire fleet of no more than 18 boats landed, among them, 1.9 million pounds of meat. This means the fleet took a lot of scallops because the usable portion of the scallop runs to about seven percent of round weight. Predictably, the initial success of the fishery was followed by a rush to get in on a good thing. But for some of the late entries, it was too late, at least in 1969.

The presence of the scallop all along the coast in fairly easy-to-work waters (they are taken in depths up to 60 fathoms or so) had been known almost since the time the first white men settled in the country. But along most coastal strips, at least those workable by present methods, the scallop did not and still does not seem to exist in potentially productive amounts. The only known exceptions are the bottom areas along the curve of the Gulf of Alaska and westward into the Aleutian chain.

The scallop is found the world over; its name in English is used every day in every land where the language is spoken by persons who would be hard put to connect the wavy hem of a sewn cloth or an uncomplicated recipe for potatoes with the shape of the rim of the shell of an unobtrusive creature of the sea bottom.

To zoologists, the scallop is a bivalve mollusk of the genus *Pecten* and related genera. The word "valve" in this instance has nothing to do with hydraulics and is used in its zoological meaning of "one of the two or more separable pieces composing certain shells." This means simply that the scallop has two shells, the bottom one rather flattish, the upper rounded and distinctly marked with a series of ridges radiating from the shell hinge to the "scalloped" edge of the shell. Unlike most other mollusks, the scallop cannot live long out of water because it cannot close the shell tightly to hold in coolness and moisture. This means it must be shucked and cleaned as soon as possible after it comes aboard the dragger. Americans and Canadians use only a smallish meat portion trimmed from the larger of the two hinge muscles but all of the scallop is used in other lands just as North Americans consume all of the oyster.

The scallop, like most other mollusks, is an indiscriminate spawner. Scallops are single sexed and males and females expel rather considerable amounts of milt and ova into the sea around them from late summer to October. Fertilization takes place immediately and hatching follows in a week. Almost nothing is known of the larval stage of the scallop's life cycle except that the tiny juvenile settles to the bottom about the time the shell begins to develop.

The scallop prefers firm sand or rocky bottom where it will not be bothered by silting. The scallop can do something about bottom conditions too, an option not open to many others of its kind. It can swim in an uncertain kind of way by expelling water fore or aft from its shell. This gives it the ability to move around a bit, and scallops are known to shift positions when disturbed although the trip is not far. There appears to be sometimes-considerable movement within a bed but there is no discernible picking up and packing off to a distant clime by scallops in search of a better life.

The scallop fishery of the East Coast of North America is a hundred years old or more and its men have accumulated specialized skills and gear to work it. No such knowledge existed on the West Coast in early 1968 when the Alaska Department of Fish and Game and the United States Bureau of Commercial Fisheries decided on a joint survey of the state's scallop potential. A few king crabbers, seeking other work because of the 1966 decline in that fishery, had been scratching around at shrimp and scallop and had done enough exploration by late 1967 to generate considerable official interest about scallops. New Bedford, Massachusetts, has been an East Coast fisheries capital for generations, clear back to the days before government was transferred from London on the Thames to Washington on the Potomac. New Bedford men were especially deft at the art of scalloping and to New Bedford Alaska turned when it needed scallopers.

Around through the Canal along about the end of March, 1968, came the 90-foot *Viking Queen,* a wood-hulled veteran of the Atantic scallop grounds, complete with her gear and her men. The *Viking Queen* was a novelty to eyes accustomed mostly to West Coast combination hulls with their houses far forward and a long sweep of working deck aft. In the Pacific, only the old-type halibut "schooner" and the power scow carry their houses aft like the *Viking Queen* and the time-honored schooner design has begun to disappear as the boats got older and more efficient multi-purpose boats are built. But the *Viking Queen* had been built for a specific fishery and the boat and her men knew their business well, no matter how much vessel and gear differed from what was the usual on Puget Sound and in the Gulf of Alaska.

On a 40-day charter ending late in June, 1968, the *Viking Queen* confirmed that substantial beds of big scallops lay along the coast of the Gulf of Alaska from Cape Spencer north and west past the bottom of Prince William Sound to Kodiak Island. The ground was unknown any farther to westward then but the New Bedford men were easily willing to bet that other good scallop beds lay along the Aleutians and northward into the Bering Sea.

The *Viking Queen* stayed in the Gulf to get down to work on its own account. Early in the charter, the *Queen* sent word back home to come and get it while it lasted and three more boats came west as soon as they could fuel, provision and get underway toward the Panama Canal. The trio came into Puget Sound on June 12, 1968, traveling in single file along the Strait of Juan de Fuca, three blue-hulled little ships with over-sized gallows and heavy twin booms announcing to all who recognized them that they meant business in the Golden West. Eight more New Bedford boats came late in the year but it was the *Viking Queen,* the three that first followed her and three or four locals (with New Bedford men aboard) that caught enough scallops in those three months to shuck out that 1.1 million pounds of meat. The fleet skimmed off the best beds between Cape Spencer and Cape St. Elias, and in September most of the fleet and some late starters moved out to Kodiak where they fished extensive beds around the island until weather blew them out. Almost all early landings were at Seward where the East Coast people had set up their own receiving plant.

The 1.9 million pounds of meats the pioneer fleet shucked out that first season came to eight or 10 percent of the entire United States scallop catch that year. Normal U.S. consumption of scallops runs from 25 to 35 million pounds a year. Canadian imports account for about half that amount. The Alaska landings made a heavy impact on the total market and apparently were one of the causes of market stagnation that unfavorably affected the entire industry after 1969. Alaska scallops were high priced that year, too much perhaps for consumers who were feeling pinched by all parts of the national economy.

By late summer, 1969, only nine boats were working in the Alaska scallop fishery while a half-dozen or so of the New Bedford hopefuls were tied up in Seattle and other ports or were being converted for the shrimp fishery and other endeavors.

While the slimmed-down scallop fleet had been fishing through the year, the Alaska Department of Fish and Game was becoming coldly aware that some of the gilt had been rubbed off the lily and that king crab and other bottom dwellers were being harmed by the scallop dredges. When such damage was pretty well confirmed, some beds, particularly in the Kodiak area, were closed to fishing.

By that time too, the ADFG concluded that the scallop beds were not as extensive as first believed and that fishing quickly was reducing scallop concentrations below a commercial level while, at the same time, scallops being taken were showing smaller sizes.

The report added:

"...it is unlikely the known populations of scallops within Alaska can support an increased harvest. An annual harvest of 1 to 1.5 million pounds of shucked meats appears feasible but the take from individual beds will bear close watching as will conflicts with associated species..."

With this statement, the department made it clear that the crab fisheries come ahead of scallops in its favor, a logical assessment of the economics of the respective fisheries.

Thus another boom fishery went bust.

All hands had a great time during that 1968 season anyway, unhampered by such matters as licenses. Not a vessel of the fleet paid for a gear license that season because legislation requiring them did not exist.

But the 1969 Alaska legislature took care of that situation in a hurry.

## SHAD

Man has come to realize in recent decades the perils of transplanting a living creature from its native heath into another that superficially may be very much the same as the old one. Too often it has been learned the hard way that animals, birds, fishes and plants usually are best left where they are found naturally.

The sorry examples are easy to remember. The English sparrow and the starling of Europe have made themselves continent-wide nuisances in North America.

The rabbit is the scourge of the Australian grasslands.

The American gray squirrel is driving its cousin, the native red squirrel of England and Scotland, close to extinction.

In 1877, the carp of Asia and Europe was brought to the United States where it spread across the country ineradicably in that era's well-intentioned but misguided farm pond program.

Not for nothing is the mongoose of Southern Asia barred from the countries of North America. Good fortune more than any other factor has kept the piranha of South America out of the warm waters of Mexico and the American South.

The successful transplants (successful in that they have been beneficial in one way or another) are harder to come by.

In the United States, the ringneck pheasant of Asia happily has filled a niche that nature seems to have left vacant just for him. The chukar partridge, also an Asian, has been a welcome newcomer into the states of the Pacific Northwest from which it has begun to infiltrate British Columbia. The Pacific coho salmon seems to be doing nobly in his adopted home in the Great Lakes. Some other Pacific salmon races have been moved around successfully within their native sphere.

Nor can anyone fault the Atlantic shad which now might just as well be called the Pacific shad too. And the man who first brought the shad from the mid-coastal states of the East merits at least a gold star opposite his name in the chronicle of sport and commercial fisheries of the West Coast. This was in 1871 and along the Pacific Coast, the shad fits comfortably into an ecological spot much like that one the ringneck took over.

Fittingly enough, the tentative 1871 shad transplants went into the Sacramento River, not too distant in time and place from that first salmon cannery put together in 1864 by William Hume and his associates. Along about 1876, the first shad from this planting began to show quite a bit to the north among the nets of the salmon fishery of the Columbia River. The shad found a home in the Columbia, a main-stem river with dozens of tributaries suitable for spawning and the rearing of the fry. With the virtues of the Columbia system so apparent, the United States Commission of Fish and Fisheries scooped up 1 million shad fry from the Potomac and Susquehanna rivers in 1886 and popped them into the Columbia and its two major tributaries, the Snake and Willamette. The transplant thrived remarkably and the shad displaced no native species.

The American shad, *Alosa sapidissima*, is the heftiest member of the family of herrings. An adult female may weigh from 10 to 13 pounds although the average is less. The shad is anadromous. It spawns and is hatched in fresh water but most of its life is spent in salt water. The ocean segment of its life cycle is little known and its areas of oceanic concentration have not been determined. The shad, like the Pacific herring, is a relatively long-lived fish when it is lucky enough to escape the many mishaps that may come its way from the moment the egg is dropped until it reaches maturity. First-spawning females of the Pacific races usually are five or six years old while most males mature a year younger. Both sexes return to the ocean soon after spawning. The shad has been known to spawn a half-dozen times or more.

Columbia River shad spawn from early May to mid-July. Spawning occurs in many tributaries along the main river and upstream as far as the Snake. Preferred spawning areas appear to be shallows just above the mouths of the branch streams. Spawning takes place at night—from 25,000 to 150,000 eggs to a fish—with the eggs being quickly fertilized as they drift near the surface. After fertilization, the eggs sink to the bottom where they adhere to the first surface they touch. The fry spend most of their first summer in fresh water and descend to the sea late in the season.

The shad, where river pollution has not chased it away, is a principal sport fish of the mid-East Coast. It is well-regarded both for its somewhat bony flesh and for its roe in those areas where it still appears. The shad of the West is looked upon favorably, principally for its roe, in its chosen rivers from the Sacramento to the Fraser. But it does not inspire the same high-hearted devotion that it does in the East, mostly, perhaps, because the western sports fisherman so far is blessed with much more in the way of game fish than is his compatriot of the Atlantic seaboard. Still, it is a sporty enough fish on light tackle and the devoted shad fisherman of the Columbia forsakes most other endeavors during the brief river appearance of the shad.

The first commercial catch of Columbia River shad, 50,000 pounds, was landed in 1889. The largest catch from that river was 1,535,000 pounds, taken in 1963. Total shad production from the Pacific Coast averages about 3,000,000 pounds a year. Fishing pressures never have been excessive because of the low market value of the fish, usually less than a cent a pound. There is almost no commercial demand for the flesh of the shad and use is directed solely at the roe. Since females only can be utilized in the fishery and the net return is small, few fishermen bother to gear up for shad.

On the Columbia, the small fishery is further complicated by regulations designed to protect salmon and salmon-like fishes running concurrently with the shad.

The State of Washington limits the fishery to drift gillnets confined to small areas below Bonneville Dam. Legal gear is restricted to single-wall floater nets with no leads or trammel webbing with mesh size set at 5.5 to 6.5 inches.

## SHARKS

During World War II when meat and meat products often were rather hard to come by, hundreds of thousands of Americans ate millions of pounds of good-tasting fish sold to them under a variety of names.

Some of these names were as exotic as "swordfish," and "seabass" and some were as prosaic as "grayfish," "fillet of sole," "cod," and "halibut." The fillets and steaks were solid, flaky, of fine flavor and were in plentiful supply. They could be baked, broiled or fried.

Business was brisk in the seaside cities where these products of the sea were found. They were inexpensive enough for the consumer while the fisherman made money too because demand was steady or rising even when the seller labeled the flesh for what it really was— the flesh of sharks, sold widely in American markets for the first and only time in the nation's history.

Sales of the flesh, even under its right name, held up in the protein-hungry economy. It was good food then and it still is good food but try to convince the average American or Canadian of that fact, one well-known to most of the rest of the world for as long as people have been eating fish. In the United States and Canada, only a scattering of gourmet types and some few ethnic groups treat the shark as a fine fellow at the dinner table.

Raymond Cannon, in his excellent work, "How to Fish the Pacific Coast," discusses fondly the food qualities of the sharks and admonishes:

"Because of the commercial value and the growing enthusiasm of anglers for the shark as a game and food fish, we are forced to conclude that none of the common species found along our coast should be wasted. Most of these fishes are wholesome, of excellent flavor and are superior to many of the bony (common) fishes sold at markets…Anglo-Saxon prejudice against this group of fine fish is ridiculous. Through the rest of the world, they are prized as a delicacy and cooked in the same manner as any other fish."[1]

That sharks may possess a positive value seems to be increasingly recognized. In 1972, the National Marine Fisheries Service began a study of large sharks of the Atlantic, concentrating on migration, distribution, food and reproduction. Of the program, NMFS said:

"As more people turn to the sea for recreation, sport fishing grows in popularity every year. The number of U.S. marine anglers rose from 8.3 to 9.5 million between 1965 and 1970. Projected estimates for the year 2000

---

1. From the Sunset book, "How to Fish the Pacific Coast," by Ray Cannon, first edition, Copyright© 1953, 1967, by Lane Magazine and Book Co., Menlo Park, California.

run as high as 29 million. Meanwhile, it is thought that some stocks of the more traditional big game fish are beginning to show the effects of overfishing. The changed attitude towards sharks may be due, too, to our growing concern for everything in our environment and the realization that sharks can be an important food resource. In any case, it is generally agreed that a large shark can provide a fisherman with exciting sport, and few sport fishes can equal the spectacular leaps and swift runs of the mako."

Some 30 species of sharks and related fishes are found along the Pacific coast of the United States and Canada. About 15 may be regarded as good food fish. This number includes 10 or so of the sharks and several species of rays and skates. A lot of "scallops" have been carved and still are being carved from the wings of these skates and rays and sold at the same price as true scallops. Similarly, a lot of "crab" with a strong strain of halibut cheek in its ancestry appears in Crab Louis in eating places where profit is regarded more highly than honor.

Sharks and all their relations (among them some of the world's least attractive fish) are members of the class called *Elasmobranchii,* fish with cartilaginous skeletons rather than the bony skeletons of the common fishes. These elasmobranchs are found in all the world's salt water and all are most abundant in warmer waters. None of them is scaled and the sharks themselves have a rough hide covered with bony, toothlike projections like oversize sand paper. The sharks range from 18 inches or thereabouts in length at maturity to 50 or 60 feet for husky specimens of the whale sharks, the largest of living fishes.

The breeding of sharks is an imitation of the breeding of mammals in that fertilization takes place within the body of the female. The spermatic substance is transmitted from the male to the female through copulatory organs called claspers, protruding modifications of the pelvic fins. Some sharks lay horny eggs but most retain the eggs within the body until they are hatched and the young are delivered. This means of procreation is fairly common among fishes and the most familiar example is the pugnacious little guppy of the home aquarium. A sport of interest to becalmed mariners of the sailing ship days, men who had a deep and immediate concern with sharks and their habits, was to fish for sharks, bring the creatures aboard and gleefully slit the bellies of the females so they might kill the young carried therein. The only good shark was a dead shark.

Most sharks, not all, are ravenous feeders highly sensitive to blood and are attracted from considerable distance by the taste of it in the water. The world's literature of the sea is loaded with stories about man-killing sharks and it is apparent that some species will attack man or any other creature that comes their way. Among these dangerous sharks are the white shark, the tiger shark, the larger species of the hammerhead and the lemon shark. None of these come much farther north than the waters of the sub-tropics although

Southern California has recorded several shark attacks and at least one death in its waters since 1960. And in Australia, a shark lookout must be posted on some beaches because of shark intrusions into swimming areas. But not all sharks are so vicious. The whale shark and the basking shark are stolid and slow-moving and a threat only to the plankton on which they feed.

Despite their bad reputation, most sharks are not much more than a nuisance as far as the commercial fisherman is concerned. They mutilate his long-line and troll-caught fish, they play hell with fish in gillnet and seine and then they rip up the net too.

Not all of the shark species along the Pacific Coast go into northern waters. Most of them are found only south of Point Conception, a headland north and west of Santa Barbara, a natural line of demarcation between the coast's warm and cool waters. The species found most commonly off Oregon, Washington and British Columbia are the soupfin, the blue shark, two species of cowsharks, (the gillnetter's friend), the common thresher, the salmon shark, (one of the two species also called mackerel shark), and, of course, the small sharks called dogfish—millions of them. The big basking shark is fairly common in the summer months off the Oregon-Washington coast.

The shark, bad name and all, is one of the most versatile of fishes when it comes to putting him to good use. His flesh—or the flesh of many sharks—makes good eating. Industry has about two dozen uses for him. His skin can be turned into an exceedingly fine leather, an expensive one, much used for billfolds, for handbags for women and for other items where a tough and long-wearing material is needed. His cartilaginous skeleton can be converted to gelatine and glue. His liver can be processed into an oil with a multitude of uses. His flesh becomes fish meal.

A generation of babies was raised on the Vitamin A found in shark liver oil and other generations might still be if someone had not thoughtlessly compounded a synthetic Vitamin A, thereby sparing the lives of untold millions of sharks and shooting a booming fishery right between the eyes.

Until that synthesization, shark and cod liver were the world's major sources of Vitamin A, a johnny-come-lately to the ranks of the vitamins, discovered and classified in the late 1920's. The vitamin seems to exist in its most concentrated form in the liver oils of certain fishes and the liver oil of the sharks is exceptionally rich in it. The fish oils still are important in many countries where, in some cases, they are used as a food or food supplement for themselves alone, with no thought given to the vitamin value in them.

The soupfin shark was an important figure in the longline and gillnet liver fishery from its beginning off California in the mid-1930's to its collapse—bang!—in 1949, the year the synthetic appeared in quantity. Washington landings of soupfin shark peaked at more than 5 million pounds round weight in 1942 when shark abundance began to decline because of fishing pressure. No agency now bothers to record soupfin shark landings in Washington.

Similarly, the spiny dogfish, *Squalus acanthias*, small in himself, was big in the liver fishery and Oregon and Washington vessels landed 45 million pounds round weight of dogfish in 1944. That fishery stumbled on its face after 1949 too because of the synthetic A. It still continues to some extent in Washington's Puget Sound where draggers take about 1 million pounds each year for reduction.

## SHRIMP

The North American shrimp fishery is the only commercial fishery important clear around the compass from Nova Scotia in the northeast to Alaska's Aleutian Islands far to the northwest.

The shrimp, in its various species and sizes, supports vigorous fisheries along the Atlantic coast, through the Caribbean Sea, in the Gulf of Mexico (where it is the major fishery), and along the entire West Coast of the continent with Alaska becoming the dominant fishery area of the west.

About 300 million pounds heads-on weight of shrimp are landed by United States fishermen annually with these catches supplemented by some 200 million pounds of imports. There is a strong and growing demand for shrimp and consumption has kept even with supply.

The total United States catch effort embraces many species of shrimp through the East Coast-Gulf of Mexico range. The fishery of the Northeastern Pacific is devoted to five species of pandalid shrimp. These are:

The spot shrimp, *Pandalus platyceros*; the coon stripe, *P. hypsinotus*, also known as *P. dani*; the side-stripe, *Pandalopsis dispar*, and two species of small pinks, *P. jordani* and *P. borealis*.

The spot, side-stripe and coon stripe, with body sizes of from four to 10 inches, are also called prawns and are the types commonly used for batter frying or for boiling. The pinks have an average body size of from 3 to 5 inches and their meats are the so-called cocktail size.

Canned shrimp have been popular since early in the century but increasingly a larger proportion of the catch, because of new processing techniques, goes into the market either fresh or fresh frozen.

The two varieties of pink shrimp make up the bulk of the western commercial catch. *P. jordani* is the dominant species from California to Vancouver Island where its range overlaps with that of *P. borealis*, itself the dominant variety thence to the Aleutians.

The side-stripe is found over the entire range from California to the Aleutians with the spot and the coon stripe ranging north from the Washington coast to the Aleutians.

The pinks and the side-stripe are the species dominating the trawl fishery since they prefer relatively smooth sand and mud bottoms where trawls are most efficient. The other two species are also taken by trawls but in minor quantities. In Alaska, they make up the better part of the pot fishery used on rough grounds.

The former United States Bureau of Commercial Fisheries contributed greatly to the efficiency of the trawl fishery during the 1960's. The BCF's Seattle, Washington, Exploratory Fishing and Gear Base developed a selective trawl in the latter years of the decade which resulted in almost-clean catches of shrimp in experimental work. Its efficiency finally was refined to the point where catches averaged 98-99 percent shrimp. Standard trawl catches often run as high as 50 percent scrap fish. Catch depths range up to 250 fathoms with pinks most common between the 50- and 100-fathom curves.

The shrimp is a bottom dweller, one that except for the absence of claws, resembles a runty version of its relative, the lobster. Its paired antennae may stretch up to a foot from a shrimp six inches long. In life, the shrimp runs through the spectrum of various shades of green, blue, brown, gray, white and pink. Cooking converts these colors to various shades of red.

Spawning takes place in spring. The minute eggs cast by the female remain near the bottom after fertilization and hatch within 24 hours. The larvae rise to the surface where they are subject to the same wind and current drift that plays such a large part in the dispersion of other species of fish and shellfish. Like other crustaceans, the larvae go through several body changes before arriving at the adult stage and settling to the bottom.

The shrimp fishery of the lower West Coast of the United States is not a significant factor in the national fishery. That of British Columbia is a minor affair.

An Oregon fishery based chiefly on Coos Bay and Newport contributes up to 14 or 15 million pounds a year in good years. A fishery on grounds some 25 miles west of Washington's Grays Harbor came to life in 1957 and peaked with total landings of 8 million pounds in 1960. By 1963, it had collapsed for reasons that have never been satisfactorily explained. Overfishing does not seem to be the cause.

It is Alaska that has become the great western producer of shrimp with landings rising steadily each year since the 1950's. The Alaska catch by American fishermen approaches 100 million pounds a year with most of it landed at Kodiak. *Pandalis borealis*, one of the small pinks, makes up the bulk of the catch. Japanese and Russian shrimpers take large tonnages every year.

Alaska has large and still mostly-unexplored shrimp grounds of unknown potential. The Bureau of Commercial Fisheries has estimated that some 200 million pounds of shrimp can be harvested in Alaska each year on a sustained yield basis. This estimate is based on the so-far inadequate exploration of shrimping grounds as well as the fact that almost all American shrimp fishing has been done on proven grounds close inshore.

Despite this rosy prediction, the Alaska shrimp fishery has some internal problems that must be handled before the state's shrimp fishery can reach its maximum growth.

A University of Alaska study of the state's shellfish industry published in the December, 1968, issue of the school's periodical, *Review of Business and Economic Conditions,* said this of the shrimp fishery:

"Foremost is the need for better domestic shrimp processing to yield a product that will be more acceptable in the marketplace. Quality must be improved to compete with production from abroad and from other areas of the United States where labor costs are lower than in Alaska.

"Improved mechanization is needed in processing as is better handling of the raw product between the times when shrimp are taken from the sea and when they are marketed. There is almost universal agreement on these points in the industry and among government agencies concerned with fisheries.

"Except for the hand-peeled shrimp that are processed in limited volume in Alaska, most of the production is from shrimp that are peeled by machine. The machines are capable of removing meats from bodies and shell only after the shrimp have been conditioned by being held to the point where they are no longer strictly fresh. This conditioning is sometimes termed 'curing' or 'ripening.'

"The result is a change in the color of the product and a distinct alteration in flavor. Both results make for marketing resistance."

Improvement in processing procedures in some plants was reported later by the same author who wrote:[1]

"The peeler manufacturer has come out with a new model that generally does a better job. It works on blanched shrimp that are not 'cured' (as described in the preceeding quotation—Ed.). About one-third of Alaska production is done by the new machines but the emphasis is on frozen rather than canned shrimp."

Alaska's commercial shrimp fishery began in 1916 in Petersburg in the Southeastern Alaska Panhandle, utilizing the small pink shrimps so desirable for cocktail purposes. This was entirely a handpicking operation in the beginning and it still constitutes a fraction of the total output. Its product has excellent flavor, shape, color and texture and is a prized item wherever it can be obtained. With the introduction of machine peelers from the Gulf of Mexico industry in 1959, processing began to shift from Southeastern Alaska to westward and it now centers around Kodiak.

Fishing techniques changed somewhat with the shift. The beam trawl had been satisfactory for the Southeastern Alaska fishery because of its adaptability to the poor bottom found over much of that part of the state's waters. But to westward, much bottom was better suited to the otter trawl with its flexibility and greater size. In the first flush of expansion, the beam trawl seemed to have no place in the Kodiak fishery.

But exploration showed that much of the inshore area where shrimp abounded had steep slopes like those back in Southeastern, the kind of bottom where the beam trawl does its best work. These bays and narrow inlets are suited also to small vessels and the small vessel and the beam trawl have teamed up to produce an appreciable part of the Kodiak catch. The major part of the catch, nevertheless, comes from otter trawls because there are more of them and they cover much more ground.

In the Kodiak shrimp fishery and, presumably, in Alaska shrimp areas still to be explored, the "pollution" of shrimp catches by scrap fish—smelt, the herrings, etc.—has turned out to be an even greater problem than in other West Coast shrimp fisheries. If the experimental BCF shrimp separator trawl just mentioned does become economically functional, shrimp landings should increase even if no more gear entered the fishery simply because of the time saved by not having to hand-sort catches. Catches often are still being sorted when the trawl should be hauled and dumped and the scrap fish problem thus compounds itself. The BCF trawl works admirably at separating shrimp and scrap but, tow for tow, its catches usually run less than those of standard shrimp trawls.

Much or most of the Alaska shrimp resources appears to be safe from foreign exploitation since the establishment by the United States of the 12-mile contiguous fisheries zone. Since the coming of the CFZ, Asian catches of Alaska shrimp have dropped about to the point where the fishery has become unprofitable.

The world-wide demand for shrimp has run ahead of the supply since World War II. With proper development under realistic regulatory measures, the Alaska shrimp fishery might easily become a profitable export fishery, a fishery type not altogether common in the United States where increasing imports seem to be the order of the era.

## SMELT

Size in itself does not necessarily indicate lack of quality in anything man encounters. This is as true for some species of fishes as it is for the finest of diamonds. One of the most palatable of fishes caught commercially in the Northeastern Pacific is among the smallest. This is the smelt, taken in two species solely for human consumption. These are the silver or surf smelt, *Hypomesus pretiosus,* and the eulachon or Columbia River smelt, *Thaleichthys pacificus.*

Neither is much more than two good bites for a hungry man. If the little fish, from five or six inches up to 12 or 14 at most, have not been overcooked, those two bites are as tasty as any food fish and any healthy appetite should be able to account for at least a dozen at one sitting.

These smelt range from Northern California into Western Alaska but the commercial fishery is pretty

1. Wiese, John, "Alaska's Shrimp Industry," *Review of Business and Economic Conditions,* University of Alaska, Vol. 8, No. 2, July, 1971.

well concentrated in the Columbia River for the eulachon and in Washington's Gray's Harbor and Puget Sound for the surf smelt. There are large, virtually-untouched eulachon stocks in many rivers over its range available for commerical use any time they are needed. British Columbia's Fraser River probably has eulachon runs equal to those of the Columbia but only a few thousand pounds are taken each year. The surf smelt, too, is plentiful wherever it is found.

The surf smelt rather resembles a small herring although it does not develop the noticeably greenish-blue back of the herring. It is the smaller of the two Western species and spends its entire life cycle in salt water except for brief ventures by some races into the mouths of tidal streams to spawn. Sexual maturity comes usually at two years but spawners may be found at one year and at three. Most surf smelt spawn along the high tide line in relatively-sheltered coves and bights where the beaches are characterized by gravel of uniform size. The eggs of the smelt, like those of the herring, are adhesive and cling to the gravel until the larvae hatch in two to four weeks and are swept to sea by the ebbing tide.

The eulachon supports a rather substantial fishery, both sport and commercial, in the lower Columbia River system. The commercial fishery reached 6 million pounds in 1945 and every year since has exceeded 1 million pounds, a catch weight more than that of any other commercial species of the Columbia except the chinook salmon. The value of the eulachon does not approach that of the chinook, however.

The smelt is a fish of regional use chiefly and the fishery is limited more by market demands than the abundance of fish. The smelt is susceptible to mishandling more than most common food fishes (although all are vulnerable) because of its size and its tendency to bruise easily. Smelt, because of their lack of size, are sold in the round almost entirely. Careless handling often results in unattractive trays or packages of fish with blood and viscera, bulging eyes and twisted shapes all too common. Many would-be buyers are repelled by the offerings and only the knowledgeable consumer can see past the mess to the real fish.

The intrinsic fine flavor of smelt is not improved at all by careless cooking and too many home and restaurant cooks tended to be off-handed with the smelt and other delicately-flavored and tender-fleshed fishes. The Pacific hake, with a flavor as good as that of the smelt, has a similarly soft flesh and consequently faces almost impossible odds for acceptance by United States and Canadian consumers.

For these reasons, there is no great pressure on commercials to enter a fishery of small financial promise. Commercial smelt licenses issued by Washington and Oregon each season average fewer than 300 apiece for each state.

The sport or personal use fishery is something else. No accurate sport catch records are kept by either of the concerned states and it is entirely possible that the personal use catch in most seasons exceeds that of the commercials. The personal use limit is 20 pounds per person. It is common in good seasons for many thousands of persons to descend upon the smelt's favored tributaries equipped with all manner of catch gear from standard dip nets to wire waste baskets in pursuit of the smelt. With the 20-pound limit, it is quite possible for a family of four to take home 80 pounds of the fish. Wastage of these sport-caught fish, easily taken in bank-side shallows, must be significant. They do not freeze particularly well although they can be quickly and easily smoked for short-time preservation. It is known that many gardens along the lower Columbia River are fertilized in part each each year by the lowly eulachon.

The eulachon is an unassuming creature, larger on the average than his cousin, the surf smelt. Adult females are silvery while males display the greenish back of the herring and a skin texture like that of the finest sandpaper. (The name eulachon, incidentally, has been twisted into oolichon and hooligan by those more interested in the fish than in its proper nomenclature. Old river hands call them candlefish too because these quite oily fish once were dried, threaded with a wick and lighted up at dark in the absence of more efficient and fragrant means of dispelling the shadows.)

The eulachon is anadromous, with most of its life spent at sea where it feeds on plankton material strained from the water by gill filaments. In turn, everything bigger than the smelt feeds on it; the smelt, with the herring and the anchovy, is an important element in the sea's scheme of eat and be eaten.

Spawning runs begin in late winter and last through the early spring in the Columbia. The eulachon enters the river in vast schools like those of the spawning herring and turn from the main river to such tributaries as Grays River, the Cowlitz, the Kalama, the Lewis and the Sandy, chief smelt producers of the river system although the smelt do spawn in other streams. The spawners travel fast, often from 25 to 35 miles a day under the drive of the reproductive instinct, and a school worked over hard by fishermen one day may be many miles away over its spawning grounds by the next.

Spawning is conducted in the mass fashion of such school fishes with the eggs shed and fertilized over gravel beds where the sticky eggs settle and cling until hatching time, about 30 days. The river current immediately carries the almost-invisible larvae seaward, the newly-hatched smelt nourished by the stored egg yolk as they ride the river to the sea. Most spawners probably die although research opinion is not unanimous about this.

The commercial fishery for the surf and eulachon smelt is about as simple a fishery as any of commercial importance anywhere. It is about as complicated as hand seining a farm pond for carp or sunfish.

The Indian fishermen of the Quillayute Reservation of the Washington coast fish their smelt as comfortably as any people fish commercially anywhere. Their fishery is mostly a family affair with some of the catch diverted to household use and the bulk sold to buyers for a few cents a pound. Their equipment is as simple as the fishery dictates—a slim canoe hollowed from a cedar log because there still are men who know the art; a popping outboard engine that may not be used in the actual set, and a fine-mesh beach seine 20 fathoms or so in length and no more than a fathom deep.

Their river is the Quillayute, a full-running stream in the springtime, formed by the junction of the Dickey River a couple of miles inland with the already-joined Soleduck, Calawah and Bogachiel rivers, all rising in the glaciers of the Olympic Mountains.

The river upstream to the point the Dickey enters the main stream is much subject to tidal influence and the surf smelt like the brackish waters above its mouth for spawning. The north side of the river is sheltered for a half-mile above the mouth by a sand and gravel spit some 20 feet deep from low water mark to crown, just what the surf smelt is looking for.

The smelt enter the river on the low slack and the flood of the tide from the schools that mill around in the shallow bight curving along the mainland southward from James Island at the mouth of the river. Actually, most of the smelt that come into the bight spawn on the beach along the surf line and relatively few, stragglers only, come into the river and move along the north shore across from the village called La Push, under the eyes of the Indians whose credentials as smelt watchers extend a thousand or more years into the past.

The Indians must meet the conditions imposed on them by the erratic weather of their seaside home and the first task is to build a fire, a new one on a spot where hundreds of driftwood fires have burned in the years since the spit was formed by nature and shaped into its present form by man. In most smelt months on the Quillayute—May, June, July—the perennial coastal fog hangs damply along the shore and for miles off it, impartially enveloping all fishermen, white, Indian, commercial and sport, wherever it reaches. Or the wind backs southerly and brings with it the rain and the rain falls sometimes for days seemingly without end. The women and the girls tend the slow fires and the people huddle over them to keep a little bit warm. Most of them disdain the rain gear of the white fishermen and seem not the worse for it. Favorable tides come during the night as much as in the day; the fires glow in the dark across from the village and the white men's fishing boats lying there; the sweet wood smoke follows the wind and almost no one heeds the silent fishermen on the spit over the river. The Indians are accustomed to being ignored.

By tradition, the fishing areas are well defined, family by family, and the fisher people a quarter-mile

along the spit from the river entrance cannot rush downstream to work when the first smelt dimple the water as they come out of the sea. They must wait until the smelt come abeam of their own site before starting to set their seine.

When the fish do begin to show in a proper spot, the people of that location do not explode into action. This is not because of any disinclination to move. Rather, the smelt spook easily and undue fuss in the water might easily drive them out of reach of the short, shallow seine. The Indians move slowly and smoothly about their job.

The dugout canoe of the Indians of the Pacific Northwest was not designed originally with the outboard engine in mind. But the ancient configuration of the canoe allows an outboard to be hung from a bracket at the port quarter or, in a few larger ones, in a well, dory-style, just forward of the stern. (Indian youths, racing their canoes with good-size engines and no mufflers in the river across from the village, can be heard for a couple of miles downwind.) Either method of powering allows room for the seine to be piled close to the stern for the set. The beach end is anchored by any handy weight or simply is held at the water's edge by the shore party as the canoe is pushed out from the beach.

The set, obviously, always is made downstream to cut off the upriver-bound fish. Mostly the engine is not used because of the chance of fouling web in the propeller. With the short distance to be traversed, the paddle in the hands of a man whose people have used it for generations is just as effective.

The set may be held open but often it is not. The man in the canoe uses his own good judgment. He has fished in that same spot for years. To hold or not to hold is his responsibility but no one says anything if his guess proves out bad once in a while. There will be more smelt along pretty soon or by tomorrow, anyway.

There is no great strain to the haul with the little seine. The canoeman hands off his towline and all hands put their backs into it. They need only moments to drag the seine onto the beach and on a good haul, it takes a lot longer to remove from it 300 to 400 pounds of wriggling smelt. Whatever the catch, the Indians return unemotionally to their watching and waiting for as long as the tide is on the flood. On occasion, the fishermen find a bonus fish among their catch—a steelhead trout, an unwary salmon or some other food fish having occasion to be in the river when the tide is running full.

The white commercial fishery for the smelt of the Columbia and its tributaries is neither as colorful as nor much more efficient than the Indian fishery of the Quillayute. In the main stem of the Columbia, the commercials were restricted until 1970 to floater gillnets of small mesh fished along drifts where smelt are known to pass in their upstream journey. Fishing is done by night and by day because the river usually is muddy during the smelting season and the dense schools seem oblivious to the threat of the web. Strict closures apply to the smelt fishery as well as to all other commercial and sports fisheries of the river.

A new gear was introduced experimentally into the Columbia River smelt fishery in 1969 and proved itself so efficient on smelt and so harmless to the salmon and steelhead of the river that its general use was licensed for the 1970 season and thereafter. This was a small version of the shrimp separator trawl developed by Seattle BCF personnel for the ocean shrimp fishery. One trawl of one-inch mesh, 12 feet across the mouth and 14 feet long, designed specially for a one-man, one-boat operation, was used during the 1969 smelt season by permission of the Washington State Department of Fisheries. The smelt fishery had been moribund for several years and the department had visions of reviving it with more efficient gear, with the hope too that its product might find wider distribution than it had.

This wish has still to be realized. But the trawl takes smelt at high catch rates and turns out fish that are not bruised and bloody like those of the gillnets, a step in itself toward finding wider use. After study of the results, the state people approved continued use and some 40 of the smelt trawls were built for the Columbia the following winter from the BCF design. These commercial nets were allowed a maximum size of 25 feet along the head rope.

On the tributary rivers, commercial fishermen are confined to dip nets and here a bit more effort must be shown by the fisherman. Drifting a gillnet downriver for smelt is one thing; hours of working the long-handled dip net on a river such as the Cowlitz turns quickly into hard work, especially when the dips are many and the smelt are few. In these streams, the smelt often are more scattered than lower along the Columbia where the schools have not begun to turn off into the branches. Although the river reaches most often used by the smelt are well-known, the fast-moving little creatures pass along them quickly unless in the spawning act and dip netting frequently turns into hours of seeking blindly for the smelt, a job merciless on unaccustomed arms, shoulders and backs.

## SQUID

Several generations of readers have worked their way through Jules Verne's "Twenty Thousand Leagues under the Sea" and for most readers the great scene in that incomparable tale of adventure is the *Nautilus'* struggle to escape the tentacles of a giant squid.

Verne, who often worked on rather firm scientific ground, knew pretty well what he was writing about when he got around to his squid although there may be those without the artist's eye who quarrel with the size of the Verne squid.

That giant squid, squid of healthy sizes, do exist has been well-documented at least since the days of the HMS *Challenger* Expedition, the Royal Society's worldwide oceanographic survey, during the 1870's, three years long and as deep as the seas. The *Challenger's* squid may not be as big as the Verne squid but no matter.

The tales of the giant squid must date as far back as the time lost in history when man first went to the sea and began his never-ending adventure with its inhabitants. The whispered tale of the kraken, the giant squid, heard above the sea wind and the breakers on the outer reef, has sent seaside children to bed, round-eyed and fretful, for millennia. The squid, the were-animal and the ghosts of the dead are the elements of man's oldest horror stories.

The squid serves several useful purposes beyond his contributions to the world's folk literature. There seems, among these uses, to be a sufficient supply of extra-large squid, up to 50 or 60 feet, deep in the seas, to feed the sperm whale population of the world. The storied sperm whale seems to prefer squid above all other feed for his entree and only the ocean deeps know what mighty battles have raged across them all through time as squid and whale sought to determine who was going to have whom for lunch.

But most squid are small creatures of from one or two inches to 10 or 12 in length, fierce enough certainly to sea life in their respective weight classes but of little concern to man other than as a food fish and an attractive bait for many other fishes.

The squids are shellfish, members of the class *Cephalopoda*, meaning they have feet (in this case tentacles) around their necks. Since they need not cope with gravity forces on land and can move freely in a supporting medium, these feet around the neck pose no particular problem to them as they might if they had to shuffle around ashore. According to the genus, the squids have either eight or 10 sucker-bearing appendages by which they grasp food. It was freeing the *Nautilus* of these suckers that gave Cap Nemo such a hard time with that squid of Jules Verne.

The squids, all sizes of them, travel by jet propulsion, water expelled forcefully from the cavity of the squid's all-enveloping mantle. This mantle is a conical membrane covering the entire structure of the squid and is open only at the forward end. The squid contracts the mantle sharply to force water from the cavity through a rubbery siphon. It maneuvers either fore or aft by reversing the direction of the thrust of the siphon.

The squid is known too as the inkfish because of its ability to expel an inky cloud to protect itself from predators. Another name for some squids is cuttlefish, the name derived from cuttle bone, the name applied to the chalky skeleton of the squid and its beak. It is most commonly known to laymen as the ground lime supplement fed to canaries and other cage birds. Like many other sea creatures found near the bottom, the squid can change its color to match almost any shade or rock, sand, mud or coral.

Various species of squid are found throughout the Northeastern Pacific, all the way from tiny ones to the brutes that wrestle to one fall only with the sperm whale. A common species is *Loligo opalescens,* a 10-armed squid from six to 10 inches long. It occurs in great schools from California to Alaska. It is seen usually at night when it rises to or near the surface.

*Loligo opalescens* comes to sexual maturity at from one to three years with the males often precocious sexually as are the familiar jack salmon among true fishes. Spawning takes place through most of the year, usually in inshore waters and at night. Sperm is transferred from the male via his left ventral arm into the mantle cavity of the female during periods of pair contact. A male may go from one female to another during his mating frenzy. The sperm lies dormant in the mantle cavity and is used as eggs mature successively and the sperm is needed. The eggs are shed in gelatinous capsules and often are found in big clusters on the bottom. These clusters have been found in five-foot masses in Monterey Bay, California, a favorite spawning ground of the squid on the Pacific Coast. The eggs hatch in 20 to 35 days according to temperature, and the young immediately begin an active search for food. The elders die soon after spawning.

No commercial fishery is conducted specifically for the squid north of California. But they turn up in seines and otter trawls and may always be found in markets serving Oriental communities in the coastal cities of those regions.

California waters support a commercial fishery for food, for bait and for reduction. California landings of squid in the early 1970's came to around 20 million pounds a year.

The California commercial squid fishery was begun on Monterey Bay around 1863 by homesick Chinese seeking a favorite food for themselves and their land-bound brethren. In 1905, Italian fishermen introduced the lampara net into the fishery and they and their descendants have been predominant in it since.

The squid catch, like that in many other fisheries, is entirely dependent on market demand and vessel activity fluctuates accordingly. Smaller vessels use the lampara and the few larger ones tend to work with purse seines. Also found in the fishery, a development of the middle 1960's, is a fish pump called among fishermen, the "squid slurp." Most fishing is at night with sets made on visual sightings of squid shoals. The densely-packed squid, approaching the surface with the coming of the night, disturb the upper layers of zooplankton and these startled animalcula glow in the water under proper light conditions. Squid may be distinguished from schools of fish by the more uniformly-diffuse glow effect given off by the plankton above and by the squid tendency to move steadily on one course. Fish, in their pursuit of food, dart and wheel erratically.

A typical squid lampara net has a wide, deep bunt (as distinguished from the long, narrow codend of the otter trawl) about 30 to 40 fathoms wide and 25 to 30 fathoms deep, all of 1-1/2-inch mesh. The wings are from 40 to 65 fathoms long and run successively toward the tapered ends through four-inch, eight-inch and 16-inch mesh. The lead line is shorter than the cork line and when the net is hauled the lead line acts much like a purse line as it closes under tension.

## STURGEON

The largest fish routinely taken in the commercial fisheries of the Northeastern Pacific (with the exception of the halibut and certain sharks) is something of a leftover, a member of a family nature seems to have forgotten when the present order of marine life was evolving.

This fish is the sturgeon, an ancient fish widespread through geologic history but perhaps on its way now to extinction. Overfishing can be blamed for much of its decrease in its range through the fresh and sea waters of the Northern Hemisphere. But there exists a suspicion too that a slow, long-term change in natural conditions may be contributing to its disappearance. Like the dinosaur, the sturgeon may not be able to cope with this subtle ecological shift.

The sturgeon is among the oldest of existing fishes and one of the homeliest. Scientifically it is described as "a primitive fish…the head is covered with bony plates and the body armed with five rows of bony bucklers with the skin between these large plates more or less roughened by small irregular ossicles (bones). The conical snout bears four barbels on its under side and the mouth is protrusible, an adaptation to bottom feeding."

In other words, the sturgeon has some of the outward characteristics of the cod, the sucker, the shark and the turtle. It is not a pretty fish like the salmon or the tuna.

Most sturgeons are fresh-water dwellers but some, like the sturgeons of the Pacific Coast, are anadromous, spending much or most of their lives in salt water but coming into fresh water to spawn. Spawning takes place during the spring months. In both fresh and salt water, the sturgeon scours the bottom for shellfish and other tidbits tied down or too slow to escape its clumsy approach.

Two species of sturgeon are found on the Pacific Coast and both range from Northern California to Western Alaska. These are the great white sturgeon, *Acipenser transmontanus,* and the green sturgeon, *Acipenser medirostris.* The white sturgeon is one of the world's largest fresh-water fishes—by far the largest in North America—and one weighing 1,285 pounds was taken from the Columbia River near Vancouver, Washington, in 1912. One weighing 1,800 pounds or more reportedly was caught in the lower Fraser River near Vancouver, British Columbia, in the 1880's. The queen sturgeon, the less preferred of the two, reaches about seven feet at its biggest with a weight of about 350 pounds. Modern-day commercial catch weights average

*Text continued on page 93*

# PLATE 1

*FIGURE 1*

William L. High

*FIGURE 2*

International Pacific
Halibut Commission

*FIGURE 3*

*FIGURE 1— Dressing fresh-caught halibut.*

*FIGURE 2— The white underside of an adult halibut caught on sport-fishing gear.*

*FIGURE 3— An 18-year-old female Pacific halibut (Hippoglossus stenolepis), 60 inches long.*

ALASKA® magazine staff photo

# PLATE 2

National Marine Fisheries Service photo by William L. High

*FIGURE 1*

NMFS photo by William L. High

*FIGURE 2*

FIGURE 1— *The* Norsel, *a longline sablefish boat in Southeast Alaska.*

FIGURE 2— *Bringing aboard a sablefish trap.*

# PLATE 3

Washington State Dept. of Fisheries

*FIGURE 1*

MFS

*FIGURE 2*

*FIGURE 4*     NMFS

MFS

*FIGURE 3*

*FIGURE 5*     NMFS

*FIGURE 1*— *Albacore*, Thunnus alalunga

*FIGURE 2*— *Pacific herring*, Clupea pallasii

*FIGURE 3*— *Columbia River smelt*, Thaleichthys pacificus

*FIGURE 4*— *Sablefish (juvenile)*, Anoplopoma fimbria

*FIGURE 5*— *Pomfret*, Brama japonica

*FIGURE 6*— *Shad*, Alosa sapidissima

*FIGURE 6*     NMFS

# PLATE 4

*FIGURE 1*

*FIGURE*

*FIGURE 1— Reeling in a hake net*

*FIGURE 2— Spilling hake from the net*

*FIGURE 3— Starting the split*

*FIGURE*

# PLATE 5

*FIGURE 1— Pacific hake,*
Merluccius productus

*FIGURE 2— Pumping hake out of
the vessel hold*

NMFS

FIGURE 1

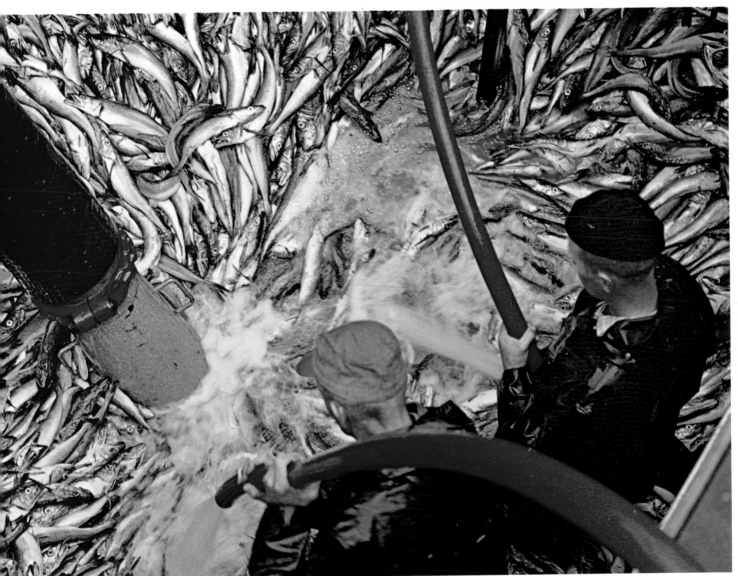

NMFS photo by William L. High

FIGURE 2

# PLATE 6

*FIGURE 1— Aerial view of drum seiners
pursing and setting the net*

*FIGURE 2— Another view of the set shown
in Figure 1*

FIGURE 1                                                    NMFS photo by William L. High

FIGURE 2                                                    NMFS photo by William L. High

# PLATE 7

MFS photo by William L. High                    *FIGURE 1*

NMFS                                        *FIGURE 2*

NMFS                                        *FIGURE 3*

NMFS                                        *FIGURE 4*

Oregon Fish Commission                          *FIGURE 5*

*FIGURE 1*— *Closing the purse*

*FIGURE 2*— *Surf perch*, Embiotocidae species

*FIGURE 3*— *Spiny dogfish*, Squalus acanthias

*FIGURE 4*— *Lingcod*, Ophiodon elongatus

*FIGURE 5*— *White sturgeon*, Acipenser transmontanus

# PLATE 8

*FIGURE 1*

WSDF

*FIGURE 4*

*FIGURE 2*

WSDF

*FIGURE 5*

*FIGURE 3*

WSDF

FIGURE 1— *Chinook salmon*, Oncorhynchus tshawytscha

FIGURE 2— *Pink salmon*, Oncorhynchus gorbuscha

FIGURE 3— *Coho salmon*, Oncorhynchus kisutch

FIGURE 4— *Sockeye salmon*, Oncorhynchus nerka

FIGURE 5— *Chum salmon*, Oncorhynchus keta

# PLATE 9

*FIGURE 3*

WSDF

*FIGURE 1*

WSDF

*FIGURE 4*

WSDF

*FIGURE 2*

WSDF

*FIGURE 5*

WSDF

*FIGURE 1*—  *Sockeye salmon in spawning colors*

*FIGURE 2*—  *Spawners seined from hatchery pond*

*FIGURE 3*—  *Female tested for egg ripeness*

*FIGURE 4*—  *Eggs stripped for fertilizing*

*FIGURE 5*—  *Milt from male fertilizes eggs in bucket*

# PLATE 10

*FIGURE 2*

NMFS photo by William L. High

*FIGURE 1* — Big skate (*juvenile*), Raja binoculata

*FIGURE 2* — Sablefish trap goes overboard in
Southeast Alaska

*FIGURE 3* — Jack mackerel, Trachurus symmetricus

*FIGURE 1*

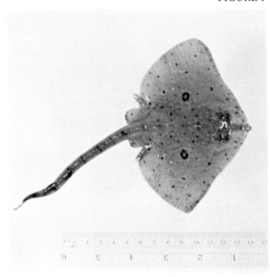

NMFS

*FIGURE 3*

NMFS

# PLATE 11

*FIGURE 1*

NMFS

*FIGURE 2*

NMFS

*FIGURE 3*

NMFS

*FIGURE 4*

NMFS

*FIGURE 5*

NMFS

*FIGURE 6*

NMFS

*FIGURE 7*

NMFS

*FIGURE 1*— *Pacific cod*, Gadus macrocephalus

*FIGURE 2*— *Rougheye rockfish*, Sebastes Aleutianus

*FIGURE 3*— *Orange rockfish*, Sebastes pinniger

*FIGURE 4*— *China rockfish*, Sebastes nebulosus

*FIGURE 5*— *Green striped rockfish*, Sebastes elongatus

*FIGURE 6*— *Quillback rockfish*, Sebastes maliger

*FIGURE 7*— *Red rockfish*, Sebastes ruberrimus

# PLATE 12

*Gulls descend on a surfacing hake net*

NMFS photo by William L. High

# PLATE 13

*FIGURE 1*

WSDF

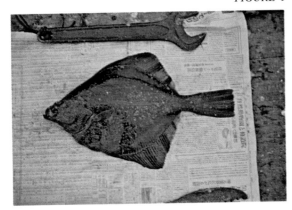

NMFS

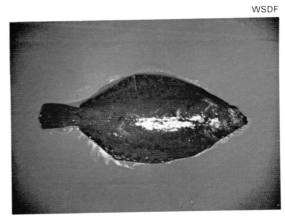

NMFS

WSDF

*FIGURE 4*

WSDF

*FIGURE 5*

WSDF

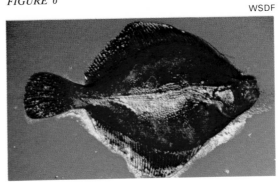

*FIGURE 6*

WSDF

*FIGURE 2*

*FIGURE 3*

*FIGURE 1— Starry flounder,* Platichthys stellatus
*FIGURE 2— Petrale sole,* Eopsetta jordani
*FIGURE 3— Flathead sole,* Hippoglossoides elassodon
*FIGURE 4— Dover sole,* Microstomus pacificus
*FIGURE 5— Arrowtooth flounder,* Atheresthes stomias
*FIGURE 6— Rock sole,* Lepidopsetta bilineata
*FIGURE 7— Curlfin sole,* Pleuronichthys decurrens

*FIGURE 7*

# PLATE 14

NMFS

FIGURE 1    FIGURE 2

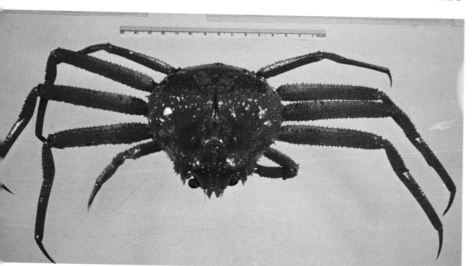

Alaska Dept. of Fish and Game

FIGURE 3

FIGURE 1— *King crab,* Paralithodes camtschatica

FIGURE 2— *Dungeness crab,* Cancer magister

FIGURE 3— *Tanner crab,* Chionocetes species

FIGURE 4— *Pink shrimp,* Pandalus jordani

FIGURE 5— *Sidestripe shrimp,* Pandalopsis dispar

FIGURE 6— *Spot shrimp,* Pandalus platyceros

FIGURE 7— *Squid,* Loligo opalescens

NMFS

FIGURE 4    FIGURE 5

NMFS

FIGURE 6    FIGURE 7

NMFS

# PLATE 15

*FIGURE 1*—  *Razor clams*, Siliqua patula

*FIGURE 2*—  *Butter clam*, Saxidomus giganteus

*FIGURE 3*—  *Eastern softshell clam*, Mya arenaria

*FIGURE 4*—  *Cockle*, Clinocardium nuttalli

*FIGURE 5*—  *Horse clam*, Schizothaerus nuttalli

*FIGURE 3*

WSDF

*FIGURE 1*

Nancy Simmerman, reprinted from *ALASKA*® magazine

*FIGURE 4*

Cockle

WSDF

*FIGURE 2*

Butter Clam

WSDF

*FIGURE 5*

Horse Clam (nuttalli)

WSDF

# PLATE 16

FIGURE 1— *Abalone,* Haliotis species

FIGURE 2— *Weathervane scallop,* Patinopecten caurinus

FIGURE 3— *Pacific oyster,* Crassostrea gigas

FIGURE 4— *Oyster spat on oyster shells*

FIGURE 5— *Oyster drill, predator from the Orient*

FIGURE 6— *Starfish, another predator*

W

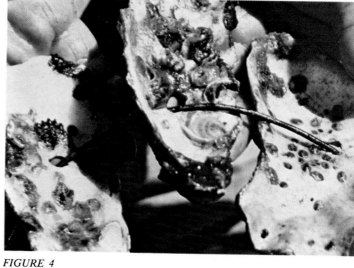

FIGURE 4

WSDF

FIGURE 1

W

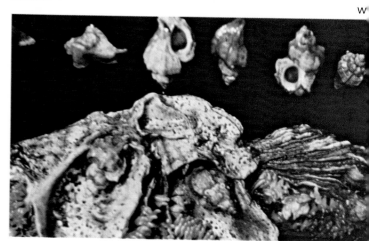

FIGURE 5

WSDF

FIGURE 2

WS

FIGURE 6

WSDF

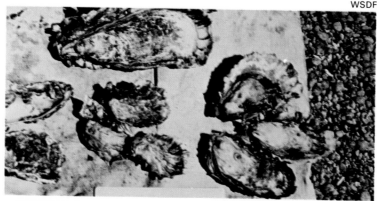

FIGURE 3

# PLATE 17

CDF&G

*FIGURE 3— Pacific saury,* Cololabis saira

ifornia Dept. of Fish and Game

*FIGURE 1— California bonito,* Sarda lineolata

CDF&G

*FIGURE 4— California barracuda,* Sphyraena argentea

F&G

*FIGURE 2— Blue shark,* Prionace glauca

CDF&G

*FIGURE 5— California yellowtail,* Seriola dorsalis

# PLATE 18

CDF&G

*FIGURE 1— Pacific sardine*, Sardinops caerula

CDF&G

*FIGURE 4— White seabass*, Cynoscion nobilis                CDF&

CDF&G

*FIGURE 2— Anchovy*, Engraulis mordax

*FIGURE 5— Swordfish*, Xiphias gladius                CDF&

CDF&G

*FIGURE 3— Pacific mackerel*, Pneumataphorus japonicus

# PLATE 19

*FIGURE 1— William Hume*

*FIGURE 2— G.W. Hume*

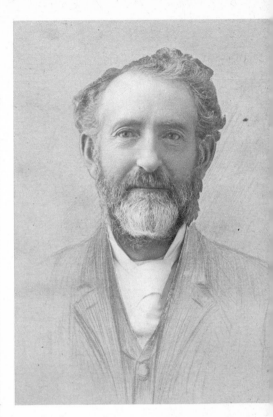

*FIGURE 3— R.D. Hume*

Pacific Fisherman

*FIGURE 4— The world's first salmon cannery, the shack and barge
on the Sacramento River where the Hume brothers, above,
with Percy Woodson, in 1864 began the most valuable
fisheries industry of the Pacific Coast.*

# PLATE 20

*FIGURE 1— Salmon-canning labels from the late 1800's. Top, William Hume; bottom, Badollet & Co.*

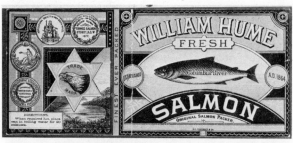

*FIGURE 2— The 1904 model of the Iron Chink, one of 12 built that year. It went to Bristol Bay.*

*FIGURE 3— The first Jensen can filler.*

# PLATE 21

*FIGURE 1— Pre-season view of the interior of the Aleutian Canning Co. cannery at Karluk on Kodiak Island in the middle 1880's.*

*FIGURE 2— Chinese workmen hand form cans at an unidentified cannery, probably in Alaska, soon after establishment of the first cannery in the territory in 1878. This job had to be done before the fish started arriving en masse.*

# PLATE 22

*This scene at a Pacific American Fisheries cannery on Puget Sound*
*shows young women industriously filling one-pound talls by hand.*
*Note that the supervisors, background, were men.*

# PLATE 23

*FIGURE 1— Girls place tops on one-pound talls to go to topper at right.*

*FIGURE 2— Labeling crew in warehouse at Columbia River Packers Association cannery at Astoria.*

# PLATE 24

*Pallet of one-pound talls goes into early-style retort in Alaska cannery.
Retorting principles have not changed although they have been refined
in the years since this picture was taken.*

# PLATE 25

*FIGURE 1— Picking up gillnet on Desdemona Sands
just inside the Columbia River entrance in 1894.*

*FIGURE 2—Native gillnet fishermen on Bristol Bay haul shackle of gear
into their cranky and dangerous sailing boat in the 1890's. This perilous rig
was to last more than a half-century longer at a heavy cost in human lives.*

# PLATE 26

FIGURE 1— *Men and horses haul beach seine on the Columbia River. This mode of taking salmon was banned in the 1930's.*

FIGURE 2— *Columbia River beach seiners and their lively catch.*

FIGURE 3— *Aleut beach seiners at Karluk on Kodiak Island had to make do without the help of horses in their beach seine fishery.*

# PLATE 27

FIGURE 1— *The three-mast bark* Flint, *with a load of canned salmon from Bristol Bay, moves up the Columbia River to Astoria with the help of two tugs working at her port side.*

FIGURE 2— *Unidentified bark lies at anchor in Bristol Bay, riding high and light, as she awaits her load of canned salmon. The picture was taken in 1929; the vessel is not identified.*

# PLATE 28

FIGURE 1— *Waiting for the tide at Egegik on Bristol Bay.*

FIGURE 2— *Canadian gillnetters work in the mouth of the Fraser River
sometime in the 1890's. Despite this gear concentration,
millions of sockeye managed to work their way past each season.*

# PLATE 29

FIGURE 1— *The Bristol Bay Packing Co. cannery at Dillingham on Bristol Bay, about 1947. This cannery layout was typical of the largely self-contained installations on the Bay.*

FIGURE 2— *The lonely graves of Bristol Bay. . . the dead are Puerto Rican, Italian and Mexican. All died in June, 1932, at ages 23, 34 and 49. The circumstances suggest they may have been shore-side workers since the bodies of lost fishermen often were not recovered. Homicide was not uncommon among the men of the Bay in those years.*

# PLATE 30

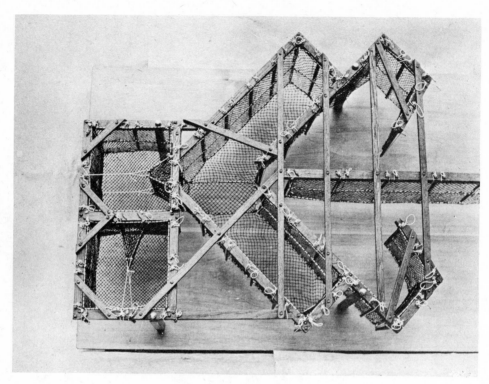

FIGURE 1— *The fish trap capitalized on the mulish dislike of ripening salmon to turn back in their drive toward their spawning beds.*

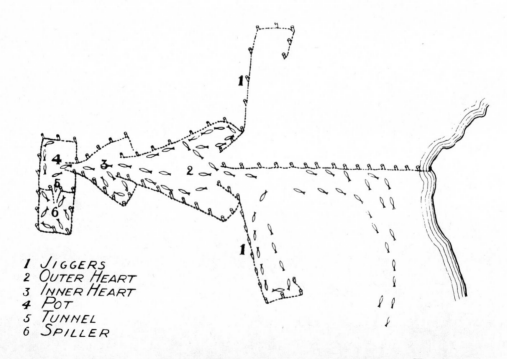

1 JIGGERS
2 OUTER HEART
3 INNER HEART
4 POT
5 TUNNEL
6 SPILLER

FIGURE 2— *J.R. Heckman's engineering of the first practicable floating trap in 1907 made possible the use of the trap in water where pile traps could not be placed. The diagram above explains major features of the floater.*

# PLATE 31

FIGURE 1— *Workmen empty the spiller of a fixed trap near Sooke in Southeastern Alaska. The suited and white-shirted men lounging on the catwalk seem to be above such labor. They probably are visiting company executives from Seattle, headquarters of the salmon industry.*

FIGURE 2— *Two barges and two tenders make their slow way to the cannery with 30,000 salmon on the barges. More than 15,000 were left in the trap, set near Point Higgins in Southeastern to await the next trip.*

# PLATE 32

FIGURE 1— *The* Edith, *in 1912, represents one of the earliest powered trollers. Trolling on the West Coast began with small sailing boats and came into its own with the gasoline engine.* Edith *is fishing lines from bow poles and poles at each side of the man at the wheel. Puget Sound appears to be the area.*

FIGURE 2— *The steam tug* Queen, *a vessel of Alaska Packers Association, seen on Bristol Bay in the era before World War I.*

FIGURE 3— *The first diesel-powered fisheries vessel, the Pacific American Fisheries tender* Warrior, *makes a triumphal run in Seattle's harbor after delivery to the company in 1914.*

# PLATE 33

FIGURES 1 & 2— *Two eras in the California sardine fishery. At right, lampara boats, loaded above their rails with deck loads of sardines, wait to load off at Terminal Island, during World War I years. Below, the purse seiner* Wanderer *lumbers toward her cannery with a full load in 1950.*

# PLATE 34

FIGURES 1 & 2— *The first true tuna "clipper," Patricia, left, is seen soon after her launching in 1924. She had refrigeration and could keep her bait in live tanks. Below, crewmen of a small tuna seiner tidy up their gear after a haul. The round corks are made of true cork. The time is in the 1940's.*

from 40 to 50 pounds for both species although specimens up to 500 pounds or so appear on occasion. Regulations wherever the sturgeon is sought commercially limit the catch to fish longer than four feet and less than six feet. Those under four feet are immature; those over six feet are potential spawners.

The white sturgeon abundance of the Pacific Coast is centered on the Columbia River with the Fraser next in line and smaller populations found elsewhere over its range. The green sturgeon appears to be scattered rather more generally through the area it is known to inhabit. Nowhere are the numbers great for either species.

The sturgeon was fairly common when the white man first came to the Pacific Coast. But when the commercial fishery for salmon began in earnest after 1864, salmon fishermen went out of their way to destroy the sturgeon wherever they found it because of its habit of blundering into gillnets. On the Umpqua River of Oregon and on the Columbia too, intensive efforts were made to eradicate all sturgeon, hunts something like the rabbit drives of the West in the 1920's and 30's.

Neither the Indians nor the first settlers thought much of the sturgeon as food although a minor commercial fishery began to develop about 1880 between Vancouver and Astoria on the Columbia. In 1888 a small rail shipment of frozen sturgeon got to a more sophisticated market in the East where it soon was realized that the roe of the sturgeon of the Columbia could be processed into caviar every bit as tasty as that coming from the sturgeon of the Volga and the Caspian Sea. Besides, the flesh smoked well.

Thus the ever-suffering sturgeon suffered some more. It took such a beating that the peak catch of 1892 on the Columbia, 5.5 million pounds, rapidly fell off to a level of around 300,000 pounds a year. This hell-for-leather exploitation of the sturgeon caused a Washington State Fisheries Department officer to write in 1968:

"There was neither proper management nor regulation of the fishery...this disastrous overfishing resulted in the almost-complete elimination of the breeding stocks."

It was this destruction of potential breeders that hit the Columbia River sturgeon so hard. Sturgeon are not like the prolific and fast-maturing pink salmon, for example. The pink grows rapidly and invariably spawns in its second year. But in its second year, the young sturgeon hardly has begun to grow. In keeping with its anachronistic family history, the sturgeon matures more slowly than any other fish of the North Pacific with the possible exception of the halibut. Estimates vary but it would seem that sexual maturity comes to the female sturgeon between the 15th and 20th year. The male matures earlier and does not grow as big as the female, again just like the halibut. And, again like the halibut, it is this slow coming of age that makes the sturgeon so terribly vulnerable to over-fishing.

When the commercial fishery had skimmed off the cream of the sturgeon crop of the 1880's and 90's, there was still heavy fishing pressure because of the demand for the roe and the flaky smoked meat. Most young sturgeon had no chance at all to reach breeding age. By the time the indiscriminate slaughter was ended and control put on the fishery, it was almost too late.

The sturgeon stocks not only of the Columbia but of the other rivers between it and Northern California, had been reduced, like the halibut stocks of the North Pacific by the time of World War I, almost to the point where they could not replace themselves, to say nothing of increasing their number. Almost but not quite. Sensible regulations now govern the taking of sturgeon for any purpose and if nature does not undercut the sturgeon as it may be doing around the world, the ungainly fish may again establish itself in something like its aboriginal abundance.

## SWORDFISH

Most of the fish sought by the commercial fisherman of the Northeastern Pacific are school fish of many sizes and uses. Among them are the salmon, the smelt, the herring, the anchovy, the mackerels, fish that cling together through their life span and are taken by nets in numbers beyond count. The fisherman pursues few fish that swim alone. As with all rules, there are exceptions to this one—the swordfish, a proud and solitary creature sought with equal fervor by sport fishermen as a trophy and by some few commercial fishermen who get more for his flesh per pound than any other fish of the Pacific Ocean.

There can be no confusion about the identity of the swordfish. He is almost unique among the fishes. There is only one species of swordfish no matter where in all the world he roams, the broadbill swordfish, *Xiphias gladius*. His name is unique even, one that is something of a redundancy emphasizing the sword, his distinguishing characteristic. The word *Xiphias* is a Greek term meaning sword-shaped while *gladius* is the Latin name for the short sword carried by the infantrymen of the Roman legions.

The swordfish is found around the world wherever the waters are warm. It ranges from coastal areas to the open sea. Nowhere is it plentiful and most fish sighted are alone although at times several may be seen sunning together on the surface.

The average weight of the adult swordfish runs from 200 to 400 pounds or thereabouts, although some weighing well above 1,000 pounds have been taken by the harpooners who work in the commercial fishery.

Superficially, the broadbill resembles the marlins and the sailfish, close enough relatives and often next-door neighbors in the waters they inhabit. But the swordfish has no pelvic fins and no scales like the marlins. (Juveniles under one-half pound have plates and scales that soon vanish.) It has a tall, sickle-like dorsal fin set well forward on the back while the marlins have dorsals reaching most of the way to the tail. The swordfish has, also, a wide lateral keel on both sides of the body just forward of the tail. The back color may range from gun-metal to bronze while the sides and belly are silver and white.

As with so many solitary marine fishes, not all is known about the life cycle of the swordfish. It appears to spawn in late spring and early summer when the loners pair up and head inshore. Here, minute eggs are shed and quickly fertilized. The eggs drift on the surface and hatch in from three to four days.

The sword of *Xiphias gladius* is a broad, flat, heavy extension of the upper jaw and may run to a third of the total length of older fish. The sword is an efficient weapon for the fish's purpose, that of slashing its way through a school of fish to kill and cripple so it may turn and feed at leisure without having to chase down a morsel at a time. The swordfish does not shrink from combat with objects larger than small school fish and it has been known, when hooked or harpooned, to torment its tormenters by charging hell-bent into any craft they may be using.

The swordfish has no natural enemies other than the killer and sperm whales. Since it is not accustomed to running scared, the swordfish becomes vulnerable to the commercial fisherman when it lies finning or moving slowly on the surface of the sea. The harpooner, abetted by the tendency of the swordfish not to move from something it does not identify as a menace, usually comes close to his target before he need launch his weapon from the bow pulpit where he works.

The small American West Coast swordfish fleet bases mostly in San Pedro and San Diego. It fishes waters off Southern California and upper Baja California. Vessels range up to about 45 feet. A trip may run to two weeks. The boats are distinguished by the long, narrow bow platform, ending in the harpooner's pulpit, that resembles the long bowsprit of mid-size sailing vessels. The catcher gear is a 12- or 13-foot harpoon of three-quarter-inch pipe with a 24-inch lily iron of soft, hot-roll iron. This head flexes when the fish rolls after the harpoon strike and does not easily tear from the flesh. The harpoon may be thrown or pushed as deep as eight to 10 feet into the water to reach the fish.

The federal government's rigid restrictions on mercury levels in fish makes it necessary that much or most of the United States catch goes into foreign markets. It has been learned that mercury concentrations in fish increase with age and size. This has been found true with large swordfish as well as with older halibut and sturgeon. The United States standards ended a Japanese swordfish market in the United States that had flourished since World War I. Imports from that country had ranged to 15 million pounds a year. Most of the American catch has been diverted to Italy.

# WHITE SEABASS

The mackerels and the mackerel-like fishes seem to hold the record for world-wide confusion, both popular and scientific, over their proper names.

Another candidate for at least a share of this small honor along the Pacific Coast of North America, anyway, would appear to be the white seabass, a favorite fish for hundreds of thousands of sport fishermen, and a market fish of merit. California, center of the commercial fishery, such as it is, averages landings of about 1 million pounds a year although the fish is not as abundant as it was around the turn of the century. It is used solely as a food fish.

The white seabass, *Cynoscion nobilis,* is known to fishermen as the white croaker, the king croaker and the weakfish, names not too far off the mark because the white seabass is a croaker and some ichthyologists do classify the weakfish as a croaker. But the young of the white seabass are sometimes called sea trout and regional names, none of them scientifically correct, are attached to the white seabass wherever people fish for it. Matters are complicated a bit further by two other species of the croaker family masquerading in the market as the white seabass. These are the Gulf corbina, *C. othonopterus,* and the totuava, *C. macdonaldi,* both natives of Baja California and its teeming waters. This impersonation of the white seabass does not harm anyone since both the corbina and the totuava are fish of good quality on the table.

The range of the white seabass overlaps that of many of its relations, among them the shortfin seabass, *C. parvipinnis,* which bears small dark spots all over its body while the white seabass does not.

The white seabass runs up to four feet in length and has posted a weight of 80 pounds. The common weight is from five to 25 pounds with most market fish at the lower weight. They have the familiar double dorsal fin of their kind with spines in the short forward dorsal. The back of the fish is typically steely blue to gray, the pattern of hundreds of marine fishes, with the sides and belly fading to silver and a silvery white.

It is a school fish and usually is found around kelp beds which furnish both food and shelter. It prefers inshore water most of the time but seems to avoid estuaries with their infiltration of fresh water. It occurs along the Pacific Coast from lower Vancouver Island to the Gulf of California. It is most abundant from Point Conception southward. It is another of the mass pelagic spawners.

The California commercial fishery runs the year around with summer and fall months showing best landings. The commercial fishery uses gillnet and troll gear and this fishery is another of those utilizing several species simultaneously rather than being directed at one species as the salmon or albacore fisheries are. Seabass, barracuda, bonito, mackerels and others support this fishery.

# GROUNDFISH OF THE NORTHEASTERN PACIFIC

The word "trawler," used carelessly or unknowingly, has become a word handy to many persons, including some seafarers who should know better, to describe any and all fishing vessels, no matter the true nature of any one.

The trawler is an individual, a distinct type of vessel performing a specific job with a specific kind of gear. Essentially, it tows a net through the water, on the bottom or close to it usually, to trap within it groundfishes too sluggish or not alert enough to escape its jaws.

In the Northeastern Pacific, the trawler unfortunately becomes confused with the "troller," a hook and line vessel also working at its own job with its own kind of gear as it seeks out such pelagic species as salmon and albacore.

This confusion of terms is due partly to the similarity in sound of the names. It is due also to the abysmal lack of knowledge about commercial fisheries and the men who man them among the greater part of the population of the continent.

This misuse of the terms trawler and troller probably amounts to little more than academic annoyance anyway (except in press and radio-television news stories) since both trawler and troller continue to plod ahead on their respective jobs, those of catching as many as possible of desirable commercial species as quickly as possible.

The trawl net in various forms is used the world over in fisheries concerned with species that live mostly on or near the bottom. But there are some exceptions to this general rule such as the longline halibut fishery of the Northeastern Pacific where trawls are barred to United States and Canadian fishermen.

Despite this world-wide use of the trawl for many kinds of fish, there are only a score or so fish species of commercial importance in the Northeastern Pacific trawl fishery. These include a dozen or a few more round fishes and perhaps a half-dozen flatfishes, not counting the Pacific halibut. These few species of fish, culled from more than 100 trawlable varieties, are all the market wants either for human consumption or industrial use. The United States and Canadian market, that is... other nations are hungrier and not so fussy.

Just the same, this still-limited demand keeps alive and healthy a United States and Canadian fishery that accounts for about 130 million pounds in landings in ports from San Diego to Kodiak every year. (This figure does not include shrimp landings.)

Americans and Canadians are not the only nations fishing these groundfish stocks. The Soviet Union and Japan have worked them massively for almost 20 years. South Korea entered the Bering Sea-Northeastern Pacific area in the middle 1960's. Other Asian nations, notably North Korea, may be expected to join this high seas endeavor.

No one knows even the approximate abundance of these fish stocks. They cannot be unlimited. Alverson (1968) estimated the maximum sustainable yield of the 10 most important trawl species in the area between the Oregon-California line and the Bering Sea to be from 1.2 million to 2.4 million tons a year.

These 10 species are the yellowfin, the walleye pollack, the Pacific Ocean perch, the rock sole, the Pacific hake, the arrowtooth flounder, the flathead sole, the Dover sole, the Pacific cod and the spiny dogfish. Only a half-dozen of these are presently used as a food fish in the United States, and this list of those in most abundance does not accurately reflect the importance of others with a high value as food fishes.

Among the important species of round fish in the West Coast trawl fishery are the true cod, the sablefish and the lingcod as food fishes and the Pacific hake as a future potential source of fish protein concentrate (a whitish powder for use as a food additive), and as frozen fillet blocks for the several products prepared from them. These four fishes have no special relationship other than the accident of frequenting the same waters, in general the area from Southern California to Western Alaska.

Each of these species is taken by otter trawl over most of their range although there is a late summer and early fall long-line fishery for sablefish from Oregon into the Gulf of Alaska.

The unprepossessing lingcod, one of the homeliest of fishes, redeems his ugliness by bringing to the fisherman a fair price, comparable to that of all species presently being taken by United States and Canadian West Coast draggers.

The true cod, *Gadus callarias,* must be numbered among the most fabled species of the North Atlantic all the way from the Lofoten Islands to the banks off North America. The earliest Europeans (after Leif Ericson) to venture into the westerly waters of the Atlantic found the grounds from Newfoundland to Labrador rich with cod. Probably even before the Columbian discovery of the New World, a flourishing trans-Atlantic traffic in salt and dried cod was underway. It still exists and even today the Portuguese dorymen still come under sail to the Grand Banks to spend six months filling their holds with their highly-aromatic cargo.

The Pacific cod, *G. macrocephalus,* something of a Johnny-come-lately to the world's attention, has never achieved the high place held by its Atlantic cousin although it has contributed to a substantial fishery through the years.

There are three other members of the true cod family, *Gadidae,* found along the Pacific Coast. These are the pollack, tomcod and very rarely seen, the long-finned cod. Of these three, only the pollack promises a substantial fishery for Americans and Canadians although, to date, it has not been exploited by either of them. But the Asians, especially the Japanese, take tremendous tonnages of the pollack every year from the waters both north and south of the Aleutians. It is distinctly possible that these foreign fleets will cut the pollack resource to the point where a healthy fishery will be unattainable unless international agreement on permissible catch preserves the stocks. The Pacific hake fishery appeared, as late as the mid-1960's, to offer both a food and industrial fishery for Americans and Canadians when they felt a need for it. But this promise vanished when the Asians moved south of the 55th parallel in the early 1960's. The Russians concentrated on hake to the point where abundance dropped so sharply that even they acknowledged, however obliquely, that the species was being overfished.

The other species commonly called "cods" are not cods. This group includes the sablefish, misnamed the black cod; the rockfishes mistakenly called rock cods; the greenlings similarly mislabled kelp cods, and the lingcod which is neither a ling nor a cod.

The true cod varies in color from brown to gray with numerous brownish splotches along the back and sides. The belly is white or grayish. Its outstanding distinguishing characteristic is a whisker-like barbel on the lower jaw. Although other fishes possess the barbel too, it and the three distinct dorsals make identification easy. Spawning takes place in the later winter and early months of spring. The eggs float free until hatching.

The largest known true cod weighed 211 pounds. Weights of from 50 to 60 pounds are common enough but fish from heavily-worked grounds average mostly from eight to 12 pounds. The cod is caught in depths down to 250 fathoms.

The sablefish, *Anoplopoma fimbria,* is another deepwater species, a winter and spring spawner like the cod with several hundred thousand free-floating eggs shed by the female. Large fish average from 30 to 40 inches in length with a weight of from eight to 10 pounds. Its back runs in color from dark green to black with light gray on the throat and belly. The lining of the gill covers is black. The sablefish is taken also in a growing pot fishery.

The sablefish smokes well and frequently is marketed as "black cod" or "finnan haddie," both misnomers. Genuine finnan haddie comes only from the flesh of the haddock. From 6 to 8 million pounds of sablefish are taken each year in Pacific Coast fisheries in depths paralleling those of the cod. East Asians take more than 60 million pounds a year from the same waters.

The lingcod, *Ophiodon elongatus,* belongs to the family *Hexagrammidae* along with a number of other fishes called greenlings. They possess slender bodies, double nostrils and large mouths with heavy, sharp teeth. Back coloration varies from kelp brown through black, blue or green, depending upon habitat. The bellies are whitish or cream-colored. They are voracious feeders, devouring almost anything small enough to swallow. They are rated among the most palatable of marine fishes. They have been taken up to five feet long in the Northeastern Pacific.

The lingcod is a shallow water fish comparatively. It prefers areas of strong currents in the inter-tidal zone near reefs, kelp beds, jetties and breakwaters. During the December-March spawning period, the female deposits clusters of adhesive eggs in nests normally found below the low tide line. A single fish may cast from 200,000 to 500,000 eggs which stick to the protective rocks around the nests. The males stand watch over the nests after fertilization, both to guard against intruders and to aid in water circulation through the nests by use of their fins. Frequently, a male will watch more than one nest if the male or males that should be on station there are killed or forced away.

The Pacific hake, *Merluccius productus,* swarms in great schools off the Pacific Coast from California to the Gulf of Alaska from spring to early autumn.

Despite its numbers, it is among the least-used of all major species as a foodfish in the United States and Canada. Elsewhere in the world, the hake in its various but quite-similar species, is well-regarded as food although its flesh softens quickly if not quickly handled and frozen.

Among users of the hake, both for reduction and for human consumption, are the Russians. The Soviet fleets operating off the coast from British Columbia south to California take it in large quantities. The Russians have reported an annual catch of 168,000 tons and more from American and Canadian coastal waters.

Experiments sponsored by the United States Bureau of Commercial Fisheries show the hake to be a prime source of high-quality protein concentrate resembling wheat flour in color and texture. This concentrate has a potential value as a dietary supplement for the people of the so-called "under-developed nations" where protein lack is serious.

The hake has a large mouth and projecting jaw. It has two dorsals, the first short, the rear one long and notched. In color it runs from dark gray to black with heavy splotches on the back. It has silver sides and belly and a black mouth lining. It runs up to three feet in length.

Hake from the Pacific Coast states and British Columbia apparently spawn off the coast of Central California in late winter and early spring. This spawning concentration coincides with its virtual disappearance from its summer range.

Although the Pacific cod, the lingcod, the soles and the sablefish are the drag fisherman's "money" species, the total landings of these varieties generally is topped by the catch of rockfish. Through the entire area of the West Coast trawl fishery, rockfish landings usually are two to three times greater than those of the other roundfishes.

# THE ROCKFISHES

The Northeastern Pacific is home to at least 56 species of the genus *Sebastes**, the rockfishes. Of this number, four are of real commercial importance as food fish although other rockfish species are caught in the same trawls and reach retail outlets under various names.

The prime quartet includes *S. alutus*, the Pacific Ocean perch (not a true perch by any means), the most valuable of the rockfishes; the orange rockfish, *S. pinniger;* the red rockfish, *S. ruberrimus*, also called red snapper, a misnomer, and the black rockfish, *S. melanops*, popularly called black seabass.

All are taken by trawl gear although there is a long-line fishery in some areas for the red rockfish. Most rockfishes range from California to Central and Western Alaska although the most important commercial fishery takes place in the waters from Northern California to Dixon Entrance.

The rockfish often are mistakenly called rock cod or sea bass although they have no close relationship to either of these fishes. Attempts have been made by private and federal agencies during the past 30 years to standardize the popular names of some of the more common rockfishes but there does not seem to have been any appreciable effect among fishermen, either commercial or sport.

The rockfishes are stout, thick-bodied fishes with large heads, large eyes and heavy scales. The forward dorsal fin has strong, sharp spines that can inflict painful and even serious wounds if carelessly handled. All species closely resemble one another in body shape but differ sharply in coloration. As a rule, those found in lesser depths are more drably colored than those from deep water. These latter, down to 200 fathoms and even below, often display bright colors highlighted by various shades of red. The red rockfish is an example of the deepwater coloration phenomenon.

Rockfish are among the few marine fishes that bear living young. Females of some species may carry more than 1 million developing eggs. The larvae usually are discharged in the spring and in that stage and the juvenile stages, they are one of the chief food supplies of other and bigger fish. There is little definite knowledge of the early life cycle of most of the rockfishes although it is known that the young are found scattered from inshore tidal pools to areas far at sea.

None of the rockfishes is exceptionally big. The Pacific Ocean perch and the rose rockfish seldom weigh more than three pounds. The red rockfish weighs up to 20 pounds and most others of the rockfishes fall somewhere between that weight and the three pounds of the Pacific Ocean perch.

* Formerly called genus *Sebastodes.*

# THE FLATFISHES

The third group of fishes important to the trawl fisherman are the flatfishes with some 300 species scattered throughout the world's seas. They vary in size from the tiniest of sand dabs to the 800-pound halibut of the North Pacific and the North Atlantic.

Of this great number of species, fewer than two dozen, perhaps, are of commercial importance through the entire Northern Hemisphere. In the Northeastern Pacific, those of interest to the commercial fisherman are scarcely more than a half-dozen.

The halibut is the favored flatfish of the North Pacific and had been for ages before the first commercial halibut fishery began off Washington's Cape Flattery in 1888.

The other species, as abundant as they are, were of little or no value for other than personal use or limited local sale until advancing technology made it possible to harvest them in large numbers and to preserve them properly for distribution across a continent-wide market area.

The emergence of the otter trawl as an efficient and flexible method of fishing in the years between the great wars coincided with refinement of refrigeration and freezing techniques that, together, revolutionized a number of fisheries and the handling and marketing of the products of these fisheries. One of the beneficiaries was the flatfish fishery.

The members of the order of flatfishes appear to be much alike but scientists recognize at least five families among them. Some have no value in any type of fishery. Those of some importance, however little, besides the halibut, are a few of the soles, flounders and turbots.

No true soles, the family *Soleidae*, are found in the Northeastern Pacific fisheries. But common usage in this area has fastened the name "sole" onto flounders and other flatfishes, a distinction that is of interest chiefly to the ichthyologist, not to the fisherman who faces up to a rough sea to catch them.

The distinguishing characteristic of the flatfishes is exactly what their name implies—the extreme flatness of the body in comparison to length. The body of the adult is tightly compressed from side to side although the larvae appear very much like the larvae of many other kinds of fishes, including the round fishes.

Another distinction among the flatfishes is the presence of both eyes on one side of the head. The larvae carry eyes on both sides but as growth progresses, the eye of one side migrates around the head to join its mate on the opposite side, the one that will be the uppermost as the fish lies on the bottom. At maturity, the eyed side is variously colored, usually in shades of olive or brown, while the blind side is white or grayish.

All flatfishes are characterized too by the long dorsal and anal fins running most of the length of the body from head to tail along both edges. In the halibut, these fins come to an apex in the midline that give the fish its distinctive diamond shape.

The flatfishes of the Northeastern Pacific usually spawn during the late winter months and into early spring. Each female produces up to several hundred thousand free-floating eggs subject to strong wind and current influences, as are the larvae during the period between hatching and assumption of the adult form and descent to the bottom to live.

About 50 million pounds of the various species of flatfish—not including halibut—are taken by Americans and Canadians each year in the area from California to Dixon Entrance in depths from five to 250 fathoms. The principle species, in order of their value to the fisherman, are the petrale, English and Dover "sole"

starry flounder, rock sole, turbot and sand sole. Most other flatfishes in the catches are regarded as scrap fish.

The joint US-Canadian fishery is a small-time affair in relation to the massive effort of the trawl fleets of the Soviet Union in the Gulf of Alaska and southward along the Pacific Coast as far as Baja California.

In 1967, for example, the Soviet fleet took more than 120,000 tons—240 million pounds—of these common flatfish species. Earlier, in 1961, the Russians and Japanese together are known to have harvested more than 1 billion pounds of flatfishes from the eastern Bering Sea, mostly yellowfin sole.

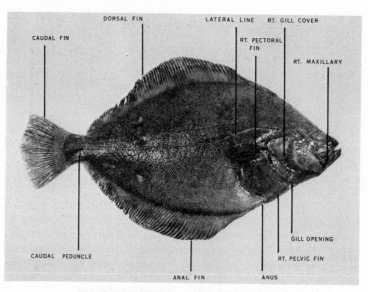

**FLOUNDER ANATOMY**

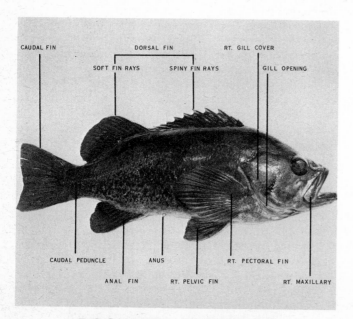

**ROCKFISH ANATOMY**

*The anatomy of fishes—schematic renderings of physical features of the three major types of fishes.*

Courtesy, Stanley N. Jones, publisher,
Saltwater Fishing in Washington.

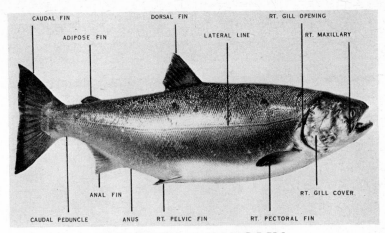

**SALMON ANATOMY**

Robert Browning

*Purse seiner crewmen near end of haul with waist deck filled with chum salmon. Rain-geared crewman facing camera has been tending lead line and purse rings. Ring bar stands on rail at port quarter. Seine skiff lies off, awaiting orders.*

St. Dominick, *all-steel western combination vessel launched in 1970 for multi-fishery work from Puget Sound to Bering Sea.*

Ray Krantz, National Fisherman

# 3/ Fishing Vessels of the Northeastern Pacific

When man first began to fish, he went about this business from lake shore, river bank or ocean beach, forced by technological deficiencies to make do only as far as he could wade or reach out with spear, arrow, cast net or hook and line. This gave the fish something of an advantage and it forced the fisherman to give some thought to this problem of getting closer to the fish. The obvious answer was some sort of device that would give him mobility on the water, something to allow him to float to the fish or allow him to pursue them if need be. Even the most unobservant of fishermen should not have had too much trouble in arriving at the answer. Again just as the origin of fishing gear is hidden in pre-history, so is the first use of water craft hidden. But it is easy enough to guess that the drifting log led to the raft of two or more logs lashed together with vines, flax, papyrus or strips of rawhide. There followed, at different times over the world where fishing was conducted, the coming of the skin boat or the dugout canoe and, finally, the planked hull which was to evolve in time from boat to ocean-going ship. There were variations on this latter development such as the dead-end boat of papyrus bundles of the Egyptians or the balsa wood craft of the West Coast of South America but largely the sequence followed this pattern although at varying periods of human development.

The Northeastern Pacific was one of the last of the world's seas to feel the full impact of total creativity on the water. It was well toward the end of the 18th Century before the white man thought to establish himself for keeps in this part of the world although it had been occupied by Asian immigrants for something like 25,000 years or perhaps longer. These people, the Indians, Aleuts and Eskimos, were fine fishermen with complex gear that had served them pretty well for millennia. They were skilful on the water but their development of water craft had reached its own dead end where it stopped with the dugout canoe, the kayak and the oomiak. Their ability as sailors with what they did have in the way of boats was exemplified by the sea-going Indians like the Tlingits and Haidas of Southeastern Alaska and British Columbia and the Makahs of Cape Flattery in Washington State at the northwest corner of the nation. All these Indians were offshore fishermen and whalers, navigators of some precision, and raiders and slaveholders whose appearance on the horizon in their canoes shaped from cedar logs was quite enough to send more peaceful folk fleeing pell-mell into the forest, carrying with them their treasures, followed by their male children and then their daughters and women and lastly their slaves, all in that order of descending value. As late as the 1850's, news that these Indians were coming their way sent the whites of the country in search of a hole to hide in too.

## From Canoe To Steam

Not all of the old arts of the coastal Indians have been lost or forgotten. Some artisans among them still hew canoes out of cedar with power tools fresh from the mail order house and some of the canoes may still be found at work in the odds and ends of commercial fisheries of the Indians such as the smelt fishery of the Quillayute and Quinault and other coastal rivers of Washington. And in Washington too, the slim reef net boats of Puget Sound, mostly Indian operated, are patterned after the graceful canoes of the old days although these reef net vessels are flat-bottomed where the old canoes had something of a vee bottom. The canoes of today are not as big as those of a century or more ago but they are fully as seaworthy and they take an outboard engine as well as any boat ever designed specifically for one. A memory worth keeping must be that of a young Indian boy trolling for salmon from a battered canoe with a battered five-horsepower outboard five or six miles off the mouth of the Quillayute in June, 1968, in a Force 4 breeze, enough to send the white sports fishermen heading for shelter inside the river mouth. But not the little Indian...just before dark he charged up the river with three cohos and a feeder chinook.

It is a matter of some surprise that the Indians of North America never got around to learning the secret of the sail, a tool that astounded them when the first whites came to their coasts. The answer probably is that they left Asia long before the sail came into use there and, once in North America, only a comparative few of them lived on salt water where the sail could be of some use. Here they had no compulsion to go across the horizon and they got as far as the dugout and no farther. The sail drove the first ships that came to the Pacific Coast and it actually is only one fairly-long lifetime ago that power began to appear in almost all fishing vessels. This was after 1900 at the time when the gasoline engine came into practical being. It is only a short lifetime ago that the last of the sailing ships was retired, this the codfish schooner *C. A. Thayer*, the last vessel in the once numerous dory fleet of the Gulf of Alaska. She was put to pasture in 1950.

Because of the late opening of the Northeastern Pacific, followed fairly quickly by the discovery of steam power, sail played a much lesser role in the fisheries than it did in the ancient fisheries of the North Atlantic, the Mediterranean or the Western Pacific. On the West Coast of North America there were no real commercial fisheries until the 1880's because population was small and demand for fish was equally small. Too, it was impossible until that time to keep fish in good condition on any large scale for more than two or three days after they came from the water and smoked and salt fish supplied most market demand. So by the time that interest in fish began to build, steam power was handy and was put into quite small vessels, boats much smaller than most people today realize. Steam tugs chased sail out of the California drag fishery soon after

1880. By 1910, gasoline power was good enough to have replaced steam on the small steam vessels and by the 1930's, diesel engines were nudging gasoline aside. By World War I, all the fisheries of the Northeastern Pacific were under power except the Gulf of Alaska/Bering Sea dory schooner fleet and the Bristol Bay gillnetters.

A score or so of good-sized steam vessels dominated the North Pacific halibut fishery after the industry learned in 1888 how to ice halibut for the trans-continental rail journey, the magazine *Pacific Fisherman* reported. This moved the halibut fishery from its position as a limited regional fishery and made it second only to salmon as a money fishery. Among this fleet were the *New England, Onward Ho* (lost with 37 men in the winter of 1916), *San Juan, Zapora, Chicago, Independent* and *Weiding Brothers, Columbia, Flamingo* and *Grant*. These vessels had a considerable capacity, up to 400,000 pounds, but 200,000 pounds was regarded as a good fare. The *New England* was 130 feet long and carried 36 men who fished from 12 dories. On one trip, she came home to Vancouver, B. C., with 173,000 pounds of halibut, with a man share of $315 for a round trip of about 17 or 18 days with nine or 10 days of fishing time. She last fished in about 1929 and was sold for scrap during World War II. Dory fishing for halibut was outlawed in 1935.

## The Halibut Schooner

The look of the halibut fleet changed rapidly in the first years of this century. Commercial fishing on the West Coast had come of age by then as the fisheries of today had taken shape or were beginning to do so. There was no place in them for the "smoke boats," becoming too expensive to run. The sailing vessel, too, was coming toward the end of its centuries of domination of the sea although it still had some good years ahead of it.

Fishermen (as always) were seeking a way to fish that required less vessel and fewer men. The boat came first and was followed by the method. The method still is evolving fruitfully although the boat has passed her prime. Longlining was the name of the new game and the new boat was the Pacific halibut "schooner," a unique vessel that sprang logically from the true sailing craft when gasoline power made its appearance in the West.

(The dories of the sailing and steam vessels used longline gear but that use was cumbersome and inefficient because the dories could work only relatively short lengths of line at a relatively slow pace. The power schooners started out with the dories but, quickly, the halibut men learned that the gear best could be worked from the boat itself. This knowledge cut crews to fewer than they had been on the steamers and the sailing vessels and the result was a safer and more efficient fishery. Longlining now has been extended to strings of pots and a flourishing fishery may arise from this application.)

Gasoline power came first to the halibut sailing fleet in 1903 and by 1905 nine or 10 vessels had installed auxiliary engines. At the same time, power was going into vessels of the salmon fishery. In 1905, *Pacific Fisherman* related, the first halibut boat to rely primarily on power was built. This was the *Northland*, a 60-footer, whose owners installed two 25-horsepower engines of two cylinders each. The *Northland* carried sails but sail alone would not have been sufficient to allow her the 25 or 30 trips of 600 miles or so that she could accomplish each year with her little engines.

From the *Northland*, over the next 20 years, sprang 200 or more of the new vessels. The *Northland's* hull had been essentially that of a sailing vessel, a descendant of the traditional Gloucester schooner of the banks fisheries of the Northwestern Atlantic. But under power, these sailing hulls were slower while, at the same time, their net capacity was less than that of the developing hybrid.

The new vessel, the halibut schooner, became a pure power boat with appropriate lines—a long horizontal keel; a plumb or vertical stem and bluff bow; house aft, two stubby masts; stern rounded, lines low and beam narrow in relation to length. These craft were (and are) heavily planked and addicted to rolling. Many used a steadying sail. In most the gas power eventually became slow-speed diesel power of the ka-chunk, ka-chunk, ka-chunk type, and later still this was replaced by high-speed diesel of double the power at about half the weight and space needs of the old-line rigs. Those vessels were solid and husky and as seaworthy as any boat of that size anywhere but they were uncomfortable to live with. The largest of the type reportedly was the *Dorothy* of Seattle, 103 by 21 by 10 feet. More commonly, length ran between 65 and 85 feet. The last of the line was built in 1927.

Indubitably, these schooners were stout ships. Good timber was easy to get in the ports, Seattle and Vancouver, B. C., where most were built. They were designed for the rigors of a year-around fishery of the North Pacific and with cheap timber to be had, it was plentifully used:

"Fairly typical was the *Progress*, built in 1914... reported to have had frames four inches square, doubled and placed on 12-inch centers, although this is narrowed considerably forward! Her planking was two and a half inches, dressed inside and out, 'and from 50 feet aft from her stem she is sheathed in iron bark to protect her against floating ice on the Southwestern Alaska halibut grounds.'

"The wood available to the boatbuilders during this period came from the virgin fir forests of the Northwest and it was possible to get the long, straight-grain timbers which have long since disappeared from the markets. It was not surprising, then, when a year or two ago a schooner rammed a new steel crab boat, put an 18-inch dent in the crabber's side just aft of the bow and buckled the deck like a tin can. The schooner backed off with no more damage than some chipped paint.

"But a fishing vessel must be more than durable; it must be compatible with the fishery and the gear. The evolution of halibut schooners, the gear they use(d) and the grounds they fish followed parallel paths so that the boats, as we know them today, while apparently almost antiques, are in reality matched to the fishing method used and the regulations governing the fishery."[1]

Of a total American and Canadian halibut fleet in 1972 of about 350 deep-water vessels, only 19 or 20 were Pacific schooners. At that time, it was estimated that another 40, somewhat modified topside for other fisheries, were still at work. The schooner makes a good side trawler. She is not a good seiner because of her house-aft configuration. Most of the survivors of the old fleet found their new home in the albacore troll fishery because this fishery requires precisely the qualities that distinguished the schooner originally—seaworthiness and durability.

On November 1, 1972, 16 of this scant score of schooners still in the longline fishery lay, empty and silent, within 200 yards of each other along the east piers of Fishermen's Terminal in Seattle, Washington. The halibut season was barely ended; it was too soon after those arduous months to turn yet to the black cod fishery for those vessels that would work during the off-season. The wind from the south urged them uneasily against their mooring lines; all wore scarred white paint and the work marks of the long season. The youngest of them could have been no less than 45 years old; the eldest first touched water in 1913. Whence came their valiant names and who were the men that named them—*Yakutat, Masonic, Northern, Chelsea, Seymour, North, Majestic, Republic, Resolute, Eclipse, Atlas, Lindy, Constitution, Atlantic, Grant, Vansee?*

## The Western Combination

Only a few years younger than the halibut schooner is a versatile and easy-to-work vessel that has become the special fishing vessel of much of the Eastern Pacific Ocean, not merely its northeastern shoulder. This craft is the Pacific or western combination vessel, sometimes called the "northern" seiner for reasons unclear because it may be found seining wherever there is a seine fishery, with geography no determinant.

In one gear configuration or another, it is the dominant boat of the west coast of the American continents. It takes anchoveta and related species off Chile and Peru, and tuna from the entire eastern tropical Pacific; it fishes mackerel and anchovy off California and shrimp clear to Kodiak; it purse seines salmon from Puget Sound to the Unimak country; it trolls albacore from Guadalupe Island to above Vancouver Island (in some years), and from California to the Pribilofs it

---

1. Philips, Richard H., *The National Fisherman*, "Pacific Coast Halibut Schooners," September, 1972.

fishes the several species of Pacific Ocean crabs. It has been described as the world's most efficient fishing vessel although this must be taken partly as hyperbole, even though the western combination, properly geared and handled, must be regarded as one of the world's better boats.

Until the coming of the stern trawler, most fishing vessels from Scandinavia to Japan were much of a type, vessels of all sizes and fisheries with super-structure aft and deck working space forward. It was this traditional design that was midwife at the birth of the Pacific halibut schooner when western fishermen were looking around for a boat better suited to the longline fishery than the big steamers or the awkward vessels of sail. Similarly, when fishermen a decade or so later needed a vessel suitable for more than one fishery, ingenuity, abetted by economics, rather quickly showed them what they were looking for. The old salmon and herring seine fisheries had developed a small boat of 40 feet or so with a cramped wheel house set just forward of midships with a cramped working area aft. As seines grew bigger and the net fishery spread, the boats got bigger too and increasingly the wheelhouse was moved forward. The first of the true western combinations was launched late in World War I and her design was quickly adopted by builders along the Pacific coast. Except for refinements dictated by more sophisticated design, the shape has not been changed essentially. A handful of these early vessels, dating back more than a half-century, still are fishing.

The standing design of the western combination vessel puts the house tightly forward with the engine room directly below. The aft bulkhead of the engine room is flush with the aft bulkhead of the house a deck above. The galley is on deck. Crew quarters are on deck or in the focsle, depending on size of the vessel. All bigger boats of this design are marked by heavy mast and boom and flying bridge. The upper bridge is extensively used by salmon seiner skippers for spotting fish and for controlling the set. Older boats intended for the salmon seine fishery show a turntable on the stern above cutaway bulwarks for seine handling. New vessels do not need the turntable because the Puretic Block and/or the seine drum have done away with any need for it. Deck gear varies according to the fishery being worked. This easy switch of gear is one of the virtues of the western design. Among others are the handy deck working area and the ability to work nets from a natural position over the stern or at the quarter.

The western combination boat comes in several sizes without, however, any real variation in the basic design. The distant water seiners of the American and Canadian tuna fleet, new ones running up to 258 feet in length, are merely elegant versions of the 50-year-old design with more shelter space above the weather deck than there is total space in such vessels as the salmon seiner. The other extreme may be found in the range from 40 to 50 feet in length, vessels big enough to accommodate two or three men with some degree of comfort. These are the boats that work the salmon and albacore troll fishery and appear occasionally in the salmon gillnet fishery. They work the beach seine salmon fisheries of Prince William Sound and Kodiak and, with live tanks, they potfish for Dungeness crab wherever this creature is taken.

Somewhere below the middle of the size scale come those disappearing vessels of 70 to 85 feet or so that were built specifically for the Pacific sardine fishery in the glory days of that enterprise. When the sardine had vanished by the early 1950's, many of these boats were out of work. They were too big for the Alaska seine fishery although some did find a home in the emerging Alaska king crab industry. Others went to South America during those days when American fishermen were teaching the Latins how to fish modern boats and gear in the growing anchoveta fishery. A few existed and still do on dragging and research charters with a look sometimes into the halibut fishery. Others just mouldered away at their berths. Like others of the western combination, these boats have their own distinctive profile, one bulking a little higher above main deck because of the wheelhouse and skipper's stateroom on a bridge deck over the house. These vessels as they stand are as obsolescent as the halibut schooner and no more are being built to that exact design because a new look has come to western type vessels.

The king crab fishery of the Gulf of Alaska and points westward and northerly demands tough vessels and tough men and of the two, the tough vessel is the more important. The rush of the king crab fishery toward maturity after 1950 caught the industry with its pants around its ankles as far as proper boats for this rigorous work were concerned. Alaska-limit boats, over-limit sardine boats and power scows were the first boats into the fishery and until 1965, they were the only boats there with the exception of some odds and ends of craft like three or four converted military vessels and the *Deep Sea*, the Wakefield concern's pioneer catcher/processor vessel. Losses of boats and men were far heavier than they should have been even with allowance for the extremes of weather and one observer remarked bitterly:[1]

"...the king crab industry...(brought)...with it an urgent demand for larger and larger vessels that could withstand the rigors of winter in the Gulf of Alaska. As many fishermen's widows and the insurance companies well know, many of the vessels that were taken into the king crab industry were fit to be condemned and would have been condemned in any fishing nation other than the United States."

1. Fulton, George, notes taken by author from presentation of paper titled "The American Trawl Fishery in the North Pacific Ocean," at The Future of the Fishing Industry in the United States conference, March 24-27, 1968, University of Washington College of Fisheries.

In 1966, the insurance people imposed rigid stability standards on the king crab fleet and some vessels were weeded from it, this action following a period of almost back-to-back accidents amounting to six or even boats and more than a score of men lost. Naval architects had not been dragging their feet the while, however, and on the drawing boards, in the mold lofts, or on the ways were the makings of new vessels intended for this most arduous of fisheries.

The first boat to be built specifically to fish with the large, rectangular pots that had become the standard gear of the fishery (the first gear in the post-World War I fishery were trawls and tangle nets) was the *Peggy Jo*, launched in 1966. She was 96 feet, six inches long, with 28 feet of beam, all steel, designed by Ben F. Jensen, of Seattle, built by the Martinolich Shipbuilding Corporation, of Tacoma.

The first class of vessels (as distinguished from a custom-built craft like the *Peggy Jo*) to go into the fishery resembled the North Sea trawler or a refined power scow, with house aft and deck working area forward. First in this group was the *Sea Ern*, a 1967 product. Later boats of the class went to 91 feet. All were designed by Ben F. Jensen and built by Pacific Fishermen Inc., of Seattle.

Next out of the yards were the 1968 models of the two-deck western combination vessel, boats as handsome in their sizes as the big tuna ships are in theirs. These boats were built initially in 81-, 86- and 94- and 104-foot lengths. Lead ship in the 94-foot class was the *Olympic*, built by Marine Construction and Design Co., of Seattle, Washington, and delivered in September, 1968. The *Olympic's* principal dimensions were:

Overall length, 93 feet, 11 inches; waterline length 85 feet, 9 inches; molded beam, 25 feet; molded depth, 12 feet; maximum draft, 13 feet, 7 inches; fuel capacity, 25,000 gallons; main engine, 725-hp. diesel; speed, loaded, 11 knots; speed, light, 12 knots.

The *Olympic*-class hull is an all-welded, double chine design, one that considerably exceeds present international standards for fishing vessels as these standards were expounded by the United Nations' 1960 Safety of Life at Sea Conference. Safety is added by outboard buoyancy tanks running the length of the fish hold. Icing in the bitter weather of the Gulf of Alaska and the Bering Sea is a constant danger to king crab boats and the *Olympic* and her sisters display a "whaleback" raised deck. This, with a crown on exposed foredeck surfaces, is intended to cut down on ice accumulation in this area. (This intent is reinforced by a shield forward of the anchor windlass which is designed to deflect surplus water overboard.) Standing rigging has been cut to a minimum by use of a tripod mast rather than the conventional mast with cable stays. The tripod mast, incidentally, appears to be standard design on all new construction of the western combination vessel.

Safety has been built into the design of these vessels. But safety has not been the only factor built into them.

Versatility is as much a part of them as safety and they are quickly capable of converting from crabbing to drum or block seining, trawling or longlining. But their primary mission is to catch king crab and keep them alive and to this end, the *Olympic*-class boats have two tanked holds with a capacity of 5,500 cubic feet, enough for 110,000 pounds of crab. A third dry hold gives them a total of 7,400 cubic feet of hold capacity. The refrigeration system supplies either chilled or circulating sea water.

Two of almost everything needed for navigation and communication is standard practice among king crab boats and these new vessels are no exception. The vessels of this fishery work most of the time in stinking weather far from other vessels or a base of supplies and when they need something, they need it right away. Routine equipping calls for two radars, two recording depth sounders, a depth flasher, three radio-phones, magnetic and gyrocompass, direction finder, loran A and C, autopilot, hydraulic steering, intercom and loud hailer and single-lever engine controls.

At its best, life is uncomfortable most of the time in the high latitudes of the king crab fishery but designers have added amenities to these new vessels that dull somewhat the sharp edge of discomfort. Crew quarters are heavily insulated against the winds that rage through this part of the world, along with temperatures well below zero. These vessels need only three or four men and each man has his own stateroom with the skipper's, true to tradition, situated, complete with head, on the bridge deck. The galleys have walk-in freezers for food storage and refrigerator-freezers for daily use; the cook works over an electric range and all hands have a washer-dryer combination to use. Both heads have showers and under the shelter deck, there is a compartment where the men working the deck can duck between pots for a smoke and coffee. This is high living; the codfish schooners that worked these same waters only a generation ago had not changed essentially from the days of Columbus.

By 1972, about 40 of these new vessels had gone into the king crab fishery and all, despite the sag in production and the general assumption that the maximum sustained yield had been reached, still were in it with no evidence that owners had any intention of diverting the new craft into any other fishery.

The Canadians of British Columbia, with no king crab of their own and no particular interest in king crabbing anyway, have developed a fine dexterity of movement of their big new vessels from one fishery to another as the season dictates. The Canadians were a decade or so ahead of the West Coast Americans in their construction of combination boats in the large sizes. In the early 1960's, yards in Victoria and Vancouver began to turn out a series of multi-purpose vessels ranging from 89 to 103 feet. These craft look much like the later American vessels of the *Olympic* design although they lack something of the streamlining that is an American characteristic. These vessels were intended generally for the halibut and herring fisheries with a sideline as salmon tenders. And, like other new vessels of the era, they can easily be shifted over to dragging whenever that fishery appears to warrant more capacity.

An early member of this Canadian fleet is the *Milbanke Sound*, delivered in July, 1963. She measures 89 feet by 26 feet by 12 feet with a 550-hp main engine and a loaded speed of 10 knots. She is tanked for work as a salmon packer and has a hold capacity of 210 tons of fish. It was the coming of the *Milbanke Sound* and others like her that lowered the average age of the Canadian halibut fleet, as compared to that of the American fleet, and raised its efficiency and safety factor considerably. By 1969 with the collapse of the British Columbia herring fishery, some of these new vessels had been sent to the Canadian East Coast to fish herring. Two or three tried purse seining for albacore off the West Coast of the United States in the mid-1960's but soon abandoned the venture because albacore do not normally school densely enough or stand still long enough to make seining practical.

## The Alaska Limit

In the mid-1920's, the United States Bureau of Fisheries, predecessor agency to the former Bureau of Commercial Fisheries and at that time overlord of Alaska fisheries, imposed something called the "Alaska limit" on vessels working in the Alaska purse seine fishery. As finally honed down, the regulation limits the size of these vessels to 58 feet overall length. The merits of the rule are debatable but it stands and must be observed by all who wish to seine for salmon in Alaska. Actually, there probably no longer exists any great sentiment in favor of revocation of the Alaska limit because of the great investment in vessels built to conform to it. Almost 50 years of experience with construction under the rule has resulted in a compact boat as efficient as it can be under regulations meant to favor inefficiency. Development of the western combination boat may have reached its finest state in this small boat.

The first western boats were designed especially for the seine fishery and all other western boats stem from the seine boats. (British Columbia salmon seiners, unhampered by any length limit, have tended to settle for two sizes, one looking much like the Alaska-limit boat but reaching into the mid-60-foot length, the other resembling the old American sardine boat with a two-deck house and a length above 75 feet. Vessels like these are found also on Puget Sound where no limit on length exists.) Over the years, designers have sweated over their boards to come up with the optimum combination of favorable characteristics. But the arbitrary maximum length of 58 feet imposes a certain limit on the talents of even the most ingenious designer and the only way to go is sideways. Thus, these latter-day limit boats tend to revert to a design popular about the time of Sir Francis Drake—short, wide and bluff-bowed. These boats of the 1960's and later do have somewhat better power than that enjoyed by Drake and his freebooters and consequently, they make a bit better time through the water than those people did.

This trend toward wider and handsomer Alaska-limit boats can be illustrated by four vessels built between World War II and 1969. The first, *Ocean Mist*, launched as *Midway*, all wood, delivered in 1949, is 58 feet long (57 feet, 11 inches actually, to meet the technicalities of the regulations) with a beam of 15.5 feet and draft, light, of six feet. The second, *Patty J*, delivered in 1957, wood also, is 58 by 16 by seven feet. The third, *Josie J*, welded aluminum, perhaps one of the first applications of this metal in American fishing vessels, measures 58 by 18.5 by 7.5 feet. The fourth is *Jamie C*, welded steel, launched in 1969, with dimensions of 58 by 20.2 by 10 feet.

The concurrent evolution of good looks in fishing vessel design is quickly apparent among these four boats. The *Ocean Mist*, first of the four, with her nearly-vertical stem and relative lack of curved lines through her bow, shows the influence of a design period when utility, not beauty, was the rule. But along those 20 years between *Ocean Mist* and *Jamie C*, a softening of angles crept through the drawings, a corollary, perhaps, of the increasing American concern with beauty and purity of environment. This interest in good looks was not universal when *Jamie C* was built. Another limit boat, laid down and launched almost simultaneously with the *Jamie*, resembles the *Ocean Mist* far more than she does the *Jamie*. Her hull is welded steel too but she measures only 56.5 by 16.5 by six feet. Her bow lines are but slightly refined from those of *Ocean Mist* although her gunwale does run aft in a smooth, sweeping line. The *Jamie's* hull profile is easier to look at than that of the other boat, however, with its sharp break at the focsle deck and the long sweep aft. The vessel has scarcely a straight line in her except for the rectangular windows of the wheelhouse.

The *Jamie C*, after outfitting, represented the peak of limit boat design. She was built to the specifications of an experienced fisherman interested in more than summertime salmon seining. The vessel is quickly convertible to trawling, crabbing and albacore trolling. She has been equipped more elaborately than many owners are willing to spend or able to spend. Her complement of electronics gear is as nearly complete as that of the new king crabbers. She reportedly was the first fishing vessel in the United States to be built with a hold of stainless steel where more than five tons of that semi-precious metal were used. Her stanchions and pen boards are aluminum. Her refrigeration system allows for brine spray for salmon and circulating sea water for crab. Hold temperature and outside water temperature are recorded simultaneously on a seven-day graph on a monitor mounted on a galley bulkhead, a position where it is almost impossible to avoid an up-to-the-minute appraisal of the vital temperature of the hold. Stainless steel was not confined to the hold; the net shield and bulwarks to a point amidships are faced with it to reduce web chafing and the time spent patching it. She carries 5,200 gallons of oil and 1,500 gallons of fresh water. Two 20-kilowatt auxiliaries supply electric power. Her wheel is 58 by 44 inches, stainless steel.

The owners of vessels built as bulkily as the *Jamie C* have necessarily to pay something of a penalty when it comes to main engine power. It takes more horsepower to move these big boats at the accepted cruising speed of 10 knots or thereabouts. Bigger engines cost more going in and bigger engines use more oil per mile of travel. The *Patty J*, almost 13 years old when the *Jamie* went to work, cruises at a comfortable 10 with a 220-horsepower engine, one of the "Jimmies" so popular with West Coast fishermen. The *Jamie*, with her beam needs a 472-hp engine of the same make to achieve a similar speed. Men experienced in the handling of vessels of similar beam and semi-cruiser stern report they steer awkwardly in a following sea of any size.

The *Jamie* and limit boats contemporary with her or later make older vessels of their kind look something very like slave ships. Just as some designers gave little thought to looks, good or awkward, neither did most owners pay much attention to crew comfort. Focsles were (and a lot still are, of course) cramped and crowded, uncomfortable the year around, warm and stuffy and smelly in summer, cold and damp and smelly in winter. Gear stowage was mostly non-existent and the focsle deck seemed paved with boots. Living conditions were especially poor below decks in the boats that fished eight to 10 men in the days before the Puretic Block or the drum slimmed crews down to four to six men. Sanitary facilities were exceedingly primitive on older boats and on some consisted of no more than a tin wash basin and a pail that doubled as deck bucket. Most boats built after World War II have sink and running water in the focsle and some have showers. People playing around with design of these limit vessels might well show a bit more ingenuity in allotting position and reckoning size of the head but here is a place where most seem to have developed a blind spot. In any event, though, even the skimpiest of heads is superior to the drafty bucket. A handful of newer vessels such as the *Josie J* have two staterooms on deck, enough bunk space for a four-man crew if fishing with the drum.

## No Limit

To the south of that Pacific Northwest area so much oriented to the salmon seine fishery, Oregonians and Californians build to the lengths they wish, free from restrictions imposed by the Alaska limit. Vessels designed for the fisheries of those states run well over the Alaska limit because all fishing along that coast is offshore. All salmon seining from Puget Sound to Central Alaska takes place in usually sheltered waters. The only open water seining is done around Kodiak Island, Unimak Island and the south side of the Alaska Peninsula although even there it still is a close-to-the-beach operation because of the salmon's preference for routes along the beaches after coming inshore on the spawning migration. But no matter the size of these southern vessels, they show the characteristics bequeathed them by their western combination ancestry. In mid-California, however, there appear smaller vessels with lines reflecting the Mediterranean inheritance of the men who fish them. These are the Monterey hulls with clipper bow, canoe stern and rakish lines that would look pretty much at home with a lateen sail rigged above them. This Monterey influence appeared in the north for the first time about 1966 when a big troller built at Moss Landing on Monterey Bay showed up in the Southeastern Alaska salmon fishery. Similarly, a stranger appeared on the Kodiak shrimp grounds in 1969 with the coming of a trawler built on Gulf of Mexico lines. This boat was built in Mobile, Alabama, and its proving out resulted in more construction orders for the same firm, mostly because the vessels could be built in Alabama for less money than they cost on the West Coast.

As for new-boat construction on the West Coast, Puget Sound yards built more boats through the 1960's than California, Oregon and Alaska together. This did not mean necessarily that Washington's fisheries were that much healthier than those of the other states. It indicated merely that the Puget Sound yards were building most of Alaska's boats as well as its own. Included in this construction was that for the distant water tuna fleet, based in Tacoma, where the first of the super-seiners, *Royal Pacific*, was launched in 1961. This trend toward distinctively bigger seine vessels was typified by the Hornet class of 167-foot seiners built from 1962 in Tacoma. The *Hornet*, when she was launched, was the largest tuna vessel in the world. She and her sisters had a beam of 35 feet and drew 21 feet. They cruised at from 12.5 to 13 knots loaded. They carried from 750 to 800 tons of fish in 14 wells. In practical use, they were world-ranging. Bigger vessels have since been built. The largest was the 258-foot *Apollo*, designed to carry 2,000 tons of tuna. She went on the ways in Tacoma in 1969 and hit the water in 1970. Before her launching, the converted navy vessel *Day Island* held the world's record for capacity with a capability of 1,000 tons, a load she filled many times to the confusion of those who had said wisely it couldn't be done.

There was little demand for new construction in California during that period because the demise of the sardine fishery left a surplus of big, capable vessels for any work that offered except for distant-water tuna fishing. Mackerel occupied some of these but that fishery fell off too toward the end of the decade because of decrease in Pacific and jack mackerel stocks attributed solely to over-fishing. By 1970, a doubling of the anchovy quota offered employment to a few boats although the future of that fishery appeared doubtful because of management unwillingness to invest in necessary reduction and cannery facilities, due to extreme pressure by sports groups against any commercial anchovy fishery whatsoever.

## The Pack Mule

One other big vessel, big at least as West Coast fishing and support vessels are counted, has played an important but undramatic part in development and supply of fisheries of the Northeastern Pacific, especially in the fisheries of British Columbia and Alaska. This vessel is the homely and humble power scow, the self-propelled pack mule of western fisheries, destined to spend its working days lifting bundles and toting bales from Puget Sound to the Bering Sea. The power scow has not even the good looks of the least attractive fishing vessel; its profile is as unglamorous as any marine profile anywhere with its two-deck house squatting on the stern, its deck running almost flatly forward to its snub nose. There is minute rise only along the bulwarks from house to bow, its bottom is almost as flat as its deck and it waddles across the water as ungracefully as a duck on land. But it is a mule for work, comparatively inexpensive to build, blessed with a long working life.

The power scow comes in several sizes and a typical one measures 81 by 26 by 7.7 feet, that draft being about that of the average Alaska-limit boat. This particular scow had twin 230-horsepower diesels, enough to move her way to about 7 or 8 knots with a following breeze. The scow is found wherever salmon are fished from Puget Sound north as well as in a few other capacities where an inexpensive freight-hauler not needing licensed officers can be used. On Bristol Bay, a fleet of smaller scows works freight and packs salmon from catcher boats to cannery sites. These small ones are particularly welcome on the Bay because they can be hauled out easily enough in that inhospitable land to spend the winter above the ice.

In the usual course of events, most power scows get around a good bit during the season, especially those scows belonging to canneries. The scows going from Puget Sound to Bristol Bay must leave their southern base no later than late April or early May for the plodding trip through the Inside Passage, along the arc of the Gulf of Alaska to the Aleutians and through False Pass into Bristol Bay. Northward, they carry odd lots of freight for their canneries and on the Bay they work as lighters because the sea-going barges that carry the rest of the supplies must lie several miles off the land, cautious about the shallow water and shifting shoals of the Bay. At season's end, the scows carry the pack out to the barges for the trip south or some of the scows themselves may lug a load of canned salmon back to Puget Sound. But not all scows are condemned to the pack mule existence. They are steady platforms in rough water and some few have been used successfully in the Alaska king and Dungeness crab fisheries. And a highly-refined version of the scow was among the first vessels built specifically for the king crab fishery.

## The Mosquito Fleet

From San Diego to Bering Strait, more than 20,000 vessels work some part of the year in fisheries other than those involving longlining, seining, trawling and crabbing. This figure is six or seven or even eight times the number of bigger boats that spend their time in the so-called major fisheries. This more numerous fleet is that which depends on the troll and gillnet fisheries, big fisheries in their sum, and others like the smelt, shad and abalone fisheries that are not to be denigrated because they ordinarily do not fit the public image (or the fisherman's own mental image) of the old Gloucester fisherman holding hard to the spokes of his ship's wheel. This number of boats runs the whole gantlet of types from the Indian canoe to the offshore albacore troller, some of these latter being bigger and more seaworthy than many of the "big" vessels of the big fisheries. We have seen a certain uniformity of design among the western combination fleet, the halibut schooners and the power scows. But in this mosquito fleet of trollers, gillnetters and crabbers, uniformity falls apart because almost any kind of boat can do for some or even most of this work. The salmon gillnet fishery can be regarded as an outstanding example of a ragtag and bobtail fleet of boats because the gillnet boat can be and often enough is any boat that can stay afloat long enough to make a set and haul it.

The old Bureau of Commercial Fisheries issued a publication, Circular 48, called "Commercial Fishing Vessels and Gear," which tries to describe in standard terms the major vessel types of the United States and, by inference, those of Canada because those of the two nations came from the same background. For most kinds of fishing vessels, Circular 48 does quite well. But when it comes to a description of "Salmon Gillnetter, North Pacific Coast," Circular 48 is not up to the challenge. The circular prosaically lists a set of dimensions—the maximum being from 22 to 32 feet in length by from seven to 16 feet in beam with draft from one and a half to three feet—for gillnet vessels. But that flat description cannot begin to encompass the variety of boats of all descriptions that make do as gillnetters.

Those measurements may fit a good number of the boats of the gillnet fleets of the Columbia River, Puget Sound, British Columbia and Alaska but they come far short of "describing" those gillnet boats. Those measurements at their extreme may show the length of the Columbia River bowpicker or the Bristol Bay boat and a potful of the rest of them but there are a lot smaller and quite a few bigger. (Incidentally, if one correlates maximum length with maximum beam from the BCF specifications, one comes up with quite a boat—some-

...hing like a pocket-size power scow.) One need only walk the piers and floats of Fishermen's Terminal in Seattle, Thomas Basin in Ketchikan or the Aurora moorage in Juneau to be reminded that the gillnet boat is of many shapes. There are, to be sure, some rather standard hulls like the Columbia River boat or the Bristol Bay boat but these two came to be that way because of efficiency in the first place and federal fiat in the second. Elsewhere, economics, comfort, convenience and geography dictated the size and profile of the gillnet boat and some of the dictates are strange indeed.

The Columbia River boat is the archetype of gillnet boat design. It was among the first of the West Coast gillnetters and it evolved to meet the peculiarities of river fishing under sail in the days when river gillnetting was legal everywhere. It started with the Humes on the Sacramento in the early 1860's and moved north, river by river, with the canneries as far as the Yukon. The Bristol Bay sailing boat was simply a Columbia River boat made a bit bigger and a bit huskier and it worked under sail on the Bay until 1951, most of two generations after all other gillnetters had been fitted with gasoline power. On the Columbia, however, with the boat not frozen into sail by governmental edict as on Bristol Bay, it evolved logically into a power boat. With power, it became the "bowpicker" still found on the Columbia and on Washington's coastal Willapa Bay and Grays Harbor but almost nowhere else. There is an easy answer to the stern location of the engine in these boats. When engines first were fitted to them, common sense pointed out that installation would be least complicated at a point as far aft as possible, away from the sailing gear that stayed on most of the boats for some years after engines appeared. When sail, mast and boom disappeared, common sense again told the fishermen that setting and hauling of their nets still was best done from the bow where the chance of fouling the screw was minimized. The Columbia River boat still shows its sailboat ancestry in the lines of its hull, especially in the sharp upward sweep of the bow sheer. But the double-end of the old boats has given away to a square stern while a shelter deck and crackerbox wheelhouse have replaced the open stern half of the original. Forward of the house, the boat remains undecked so nets may be stacked there. A net hauler sits on the starboard bow rail.

The Columbia River bowpicker is admirably suited to the peculiar conditions under which it evolved because the demands of a river gillnet fishery are almost entirely different from those of a salt-water fishery. Thus, with the few exceptions just noted, it is not well suited to salt water. Salt water, in general, requires bigger and more stable boats with, in many cases, at least a semblance of living space and comfort.

A resurgence of interest in the bowpicker became apparent in the early 1970's when the first few boats appeared on Puget Sound and the type began to spread northward through British Columbia as far as Central Alaska. Since licensing agencies do not break down their licenses by type of boat within a fishery, it is not possible to estimate what proportion of the West Coast gillnet fleet is composed of the renascent bowpicker. (Licensing agencies, of course, do distinguish among such vessel types as seiners, gillnetters and trollers in the salmon fisheries, for example.)

These latter-day bowpickers are not the narrow-beamed, relatively fragile boat of the river fleet. Instead, they are stable, beamy, fast craft, mostly of fiberglass construction, with unloaded speeds up to 30 knots. In appearance, most of them, with their snub noses, superficially resemble the small infantry landing craft of the armed forces.

Their ancestry varies; one of the first stemmed from a deep-vee pleasure boat hull with appropriate conversion; the forerunner of another (the dominant version) was a salvage and log patrol boat developed for use on Puget Sound after World War II. Length averages 23 to 27 feet or so with a beam of about eight feet. Power covers the whole range of possibilities—inboard engines, both gasoline and diesel; inboard-outboard, pure outboard and even jet drives.

Most of these little bowpickers have accommodations almost as skimpy as those of the Columbia River boat. But they have what many fishermen (particularly those who might be called "locals") want—speed enough to move fast from one fishing ground to another. This is their virtue. Their disadvantage is their short range. They are much favored by part-time fishermen, those who work a shore job by day and fish a gillnetter by night. The counterpart of this bowpicker type is the kelper or day boat of the salmon troll fleet.

The deck gear of these bowpickers no more resembles that of the Columbia River boat than does the hull and power unit. Here, the light, rail-mounted roller for a net pulled by hand has been replaced by the hydraulic-powered reel of the standard gillnetter while heavy fairleads have been built in forward to control the net on the haul.

The present-day Bristol Bay boat was developed under much the same stress that has accompanied the evolution of the Alaska-limit vessel. When the ban on power in the gillnet fishery was lifted in 1951, the Bureau of Commercial Fisheries fixed a maximum length of 32 feet for the new generation of gillnetters. Not all boats, of course, were built to this maximum. Lengths of 28 and 30 feet are popular on the Bay while there is an entire fleet of skiffs in the 18- to 22-foot range abounding on the Bay. These most often are powered by outboard engines and are used commonly to tend set nets although some do use drift gear.

The first power boats built under the Bristol Bay limit were heavy wood craft with displacement hulls. Most of these boats are still useable and almost all vessels of the company fleets are of this type. Fishermen using their own boats have tended toward aluminum and fiberglass as hull materials, partly because of ease of maintenance in a region far distant from sources of supply and partly because of the resistance of the hulls to outside damage. Gasoline still is the major power source although diesel engines are to be found in an increasing number of boats.

The Bristol Bay limit has resulted in the development of a compact and efficient day boat of clean profile with minimum room for two men, a good boat for this touch and go fishery of two or three weeks peak duration but one of scant comfort for her crewmen. Bristol Bay fishermen do not use the power net reel found in most other gillnetters. Instead, a power roller on the transom helps manhandle the net aboard for picking out the fish. Most fishermen regard the reel as too slow for this particular fishery.

An older gillnet design and one fairly common still, especially in British Columbia and Southeastern Alaska, harks back to the days just before the western combination came to life. This is a slow, narrow-beamed displacement hull, with square stern, a short cockpit and pill box wheelhouse just forward of the midships point with a low cabin reaching forward almost to the stem. This is the design the BCF chose to illustrate Circular 48's Pacific Coast gillnetter, something of a poor choice because the Bristol Bay boat or one of two or three others might better show the modern gillnetter. These hulls, again, are of wood with gasoline mostly as the power source. Like the halibut schooner or the sailboat, these boats no longer are built because of other and more efficient designs.

Washington State's Puget Sound fishing district is home to a growing fleet of gillnetters of outstanding design, at least as far as good looks and living comfort are concerned. Gillnetting, like trolling, often is a husband and wife (or friend) enterprise, one in which women seem to insist on somewhat more civilized accommodations than most men do. Early in the 1960's there began to appear in Puget Sound ports pleasure cruiser hulls adapted to the needs of the gillnet fishery. These run to 36 or 38 feet in length for the bigger and, except for the reel mounted in the cockpit, they are hard to distinguish from craft of identical lines used solely for cruising and boozing. The hulls are beamy, planing types with gasoline engines rated upward from 300 horsepower. They are fast and maneuverable although not, with clumsy handling, as good sea boats as the steadier displacement hull in the same sizes. Nevertheless, many of these fancy boats make the round trip from Puget Sound to Southeastern Alaska every salmon season. Most hulls are of wood although fiberglass turns up increasingly. One boatyard in Vancouver, B. C., turned out a series of this type after 1966 with aluminum hulls. The only fault found with them was hull noise when underway in any kind of sea.

Fishermen, gamblers at heart to a degree, like to hedge their bets just as do other people who wager, and a result of this laying off the bet is to be found in the troller-gillnetter combination boat seen fairly often in the salmon fisheries. Like many hybrids, the cross is not a 100-percent success but it does allow the move from one gear to another to be done within minutes, if

need be. There is no standard boat for the combination but the usual procedure is to equip a troller with the optional gear. This combination boat usually is bigger than the boat intended only for gillnetting. The gear fishes just as efficiently whether it be troll or gillnet at work but the fisherman finds it harder to handle his gear, especially the troll gear. The gillnet reel can be placed in one position only, squarely on the center line within a yard or so at the most from the transom. This mandatory location then displaces the gurdy assemblies from their usual spot a few inches to port and starboard of the centerline where the fisherman is accustomed to working them at short range from the trolling pit. If the gurdies are to stay on deck, they must be moved right and left almost to the rail to keep feet out of them when the net is being worked, or they must be placed overhead on the pipe davit where the troll line blocks are secured. Usually, this puts them out of reach of anyone in the pit and the pit must be covered to give the fisherman a place to stand. This, in turn, makes it awkward to run lines and gaff fish and it introduces an element of danger since the fisherman is not protected from going overboard, as he is in the pit. Nevertheless, the quick-change ability of the combination is enough to make it attractive to many men.

Not all gillnetters are as big, as fast, as pretty or as versatile as some of these so described. There are gillnetters that fit no recognizable category as well as some that come as a surprise in the gillnet fishery. One small landing craft, big enough for a dozen infantrymen or so, fished on Puget Sound in the 1950's. Another vessel, fitting the Confederate description of the gunboat *Monitor,* "Cheesebox on a raft," fished part of one summer season. A Boston Whaler, working one shackle of gear, has been seen fishing in mid-Puget Sound off and on while Indian fishermen from two or three of the reservations on Puget Sound use their dugout canoes quite handily as gillnetters.

## The Troll Boat

The West Coast troll fleet is just as heterogeneous as the gillnet fleet with the same circumstances accounting for the variety of vessels. Circular 48's specifications for the West Coast troller come somewhat closer to the mark than for the gillnetter with its general dimensions of 25 to 60 feet in length, beams of eight to 18 feet, draft of two and a half to seven feet, hulls of wood or steel and gas or diesel power. And again, like those specifications for the gillnetter, this description can be made to embrace a wondrous variety of boats, some of them fit more for farm ponds than salt water.

Larger vessels in this troll fleet are often intended for more than one fishery because only in exceptional seasons can a boat earn its keep and deliver a decent profit in one only. The commonest duality covers the oft-mentioned salmon and albacore troll fisheries. The salmon troll business is usually a coastal affair with shelter close and easy to reach. In British Columbia and Southeastern Alaska, much trolling goes on in the channels and straits and bays of the Inside Passage and adjacent waters and is as comfortable a fishery as any in the world. But albacore fishing, bait or troll, is something else because these erratic and undependable creatures of the high seas stay well offshore most of the time when they are in range of West Coast fishermen and the fishermen must go to them. This may mean any distance from 20 to 300 miles to sea and this means equally that vessels going after albacore must or should meet rather high standards of seaworthiness. (During the 1969 albacore season, at least six boats with some 20 men vanished off Washington and Oregon without leaving any evidence of their passing.) Seaworthiness generally can be equated with size and stability and as a result most (but not all) albacore boats run over 45 feet in length. Western combination craft of the Alaska - limit size are frequently to be found in this fishery and a number of the bait boats are considerably larger.

It is a truism that almost any kind of boat can be used to some extent for a little while in almost any fishery. This particularly is true of the gillnet and troll fisheries. The open outboard with one or two engines on the transom fits well in the troll fishery because it can work closely inshore among the reefs where the chinook salmon likes to lie up between feeding periods. Similarly, the comfortable cabin cruiser is at home in this special fishery where neither ice nor distance is a concern of the fishermen. Often enough, a couple of sport rods do for gear on these boats. (Both British Columbia and Washington State moved in 1968 and 1969 to eliminate these "kelpers" from the salmon fishery and leave it to professional fishermen. British Columbia's intent was to cut its salmon fleet of all types in half in the attempt to create a financially-healthy harvesting segment of the industry. In Washington, the kelpers fished with "sports-commercial" licenses and the state moved against them by prohibiting the use of "angling" or "personal use" gear in the commercial troll fishery. The action was challenged in the courts but in time was upheld.)

## Senior Citizen

An older troller, common in Washington, British Columbia and Alaska, dates back to a time before World War I when the western combination was under incubation. This boat, in several variants of profile, developed when power first began to be put into fishing vessels and design called for less beam than is common now. The hull runs along rather severe lines with vertical stem. A good number of older boats of this type are double-ended, a reminder of the days of sail. The boat is among the more primitive of fishing vessels still in use, with a wheel house ordinarily just big enough to hold two men if one of them stands up. The galley is below and forward in the berthing area. Most of these boats do not bother with bulkheads between the galley

space and the engine compartment and on some vessels the galley stove has only several thicknesses of asbestos between it and one or the other of the fuel tanks. This provides for a certain amount of breathlessness when that tank happens to be filled with gasoline. An enclosed head is rare on these boats.

These vessels, as mentioned, are narrow for their length. A representative of this type is *Amalie L,* 36 by 10 by three and a half feet, hull of wood, power a six-cylinder gasoline engine. The *Amalie* was built in 1939 by an Alaska fisherman for work specifically on the Fairweather Ground of the eastern Gulf of Alaska offshore between Cape Spencer and Lituya Bay to the north. Boats like the *Amalie,* with two or three tons of ice in their pens, are good sea boats but uncomfortable in a beam sea even with stabilizers out.

The troll fleet appears, in the main, to be made up of boats older than those of any other fishery of the Northeastern Pacific. There is economic reason for this just as there is economic reason for the generally advanced age of all United States fishing fleets. There just is not enough money in fishing for most men to make it worthwhile to build new boats. There are stories, and some of them are true, of big earnings in the troll fleet. But for the most part, the trollerman fishing from mid-April through early autumn nets less money than he probably would make in a shore job for the same period. One inquiry into the economics of the troll fisheries put the average gross income of salmon and albacore trollers at about $3,000 a season. This figure has to mean that a lot of trollers gross considerably less than $3,000 every season. Nevertheless, troll fishing possesses a certain glamor for many men, a charisma that takes them from their winter jobs every spring and puts them to work anywhere from California to Alaska for almost half the year. Even the laziest or the most inefficient troller is reasonably sure of catching enough fish to pay for fuel, food and ice. If matters get bad enough, he can take his food from the sea. The choice is wide...filet of rockfish for breakfast, filet of lingcod for lunch, halibut steak for dinner, ad infinitum.

Despite the lack of financial incentive in the troll fishery, new boats do go into it every season. New construction appears to be centered around the 36-46-foot size range, boats that can be converted from one fishery to another with little effort. A lot of this new construction is being done by owner-operators to save boatyard costs. There are stock designs for new boats of all sizes and uses. Two or three builders in the Pacific Northwest, for example, can supply stock hulls in fiberglass with optional power, deck gear and interior arrangement. These boats lend themselves easily to trolling, gillnetting, Dungeness crabbing, shrimp potfishing, halibut longlining and any other fishery where one, two or three men at most are required.

On the West Coast in the late 1960's, wood still was the most popular material for small fishing vessels although steel was the only material being used for boats from the Alaska-limit class upward. Wood's popularity was due mostly to its modest cost in comparison to steel, fiberglass and aluminum. The ferro-cement process appeared in commercial fishing vessels for the first time in the west in 1969 when a 56-footer appeared in the Alaska troll fishery. The skepticism aroused by the initial reports of the ferro-cement process in 1965 appeared to be well-dissipated by the end of the decade when a number of starts in ferro-cement construction were reported along the Pacific coast. The center of this interest originated in Vancouver, B. C., where most of the first experimenting on the West Coast took place.

## The Seine Skiff

No summary of the fishing vessels of the Northeastern Pacific would be complete without a mention of the once-humble seine skiff, in the beginning a wooden boat powered by weary men pulling at sweeps as they towed one end of a seine uptide along the beach.

The modern seine skiff is several craft now and the ones of interest here are those that work with the seiners of the West Coast salmon, herring and mackerel fisheries. The seine fishery for salmon employs the majority of the skiffs to be found in the West.

In some of the world's seine fisheries, various means have been found to do away with the skiff. But in the Pacific salmon fishery, the seine skiff is still alive and well, big and husky and, in some cases, more heavily powered than its mother craft.

Much effort has been expended over the years on reducing the size of fish boat crews to the number needed only for "steaming" the vessel. This, on the average West Coast boat, can be done by a mere handful of men. But under the apparent policy of "legislated inefficiency" more men than needed for running must be aboard such vessels as seiners to handle the gear on the haul. The salmon fishery truly enough has seen some reduction in crew size through adoption of the Puretic Power Block and the hydraulic drum. The eight or nine men of the old table seiners have been cut to six or even five on vessels using the Power Block, while the drum can be fished handily by four men and, in emergency, by three, with the skiffman still doing his traditional job and the deck gear being run by his two companions.

Nevertheless, the skiffman has been the object of the long cold stare for some years and his position appears to be the next to come under attack. But no one yet has devised a practical means of replacing the skiffman and his skiff. Most salmon seine fishermen still begin their careers in skiffs that resemble those of the old days only in the respect that each floats.

In the salmon fishery it is the skiff's job to secure the bunt end of the seine while the set is being made, to tow that end while the set is held open, to hold the seiner out of her seine while the haul is underway and, finally, to buoy the cork line during brailing.

The need for the seine skiff in these jobs has been eliminated in the Iceland and Norwegian herring fisheries by adoption of several expedients, including the installation on the Atlantic seiners of bow thrusters for positioning themselves in relation to their seines. These boats drop their nets quickly and the off-end, heavily corked, holds its position well enough as the seiner roundhauls quickly on the school.

These measures work in ocean fisheries where currents are broad and sweeping, obstacles are far away and the water is deep. And they might work in such salmon areas as the Salmon Bank shoal off San Juan Island at the head of the Strait of Juan de Fuca where the reefs of the island itself are the major threat to the well-being of the seine.

But they will not work in the Inian Islands of Alaska and in most other salmon waters of the whole coast because of the narrow passages the salmon follow, the many obstacles all around, the tidal currents that race to eight or 10 knots and a seiner, her seine and her skiff may be swept for miles during the set and haul.

The thrusters would be happy additions to the equipment of any salmon seine vessel big enough and well off enough to take them. But the Alaska-limit regulation (with every prospect that the limit may be cut to 50 feet over-all) rules out this expensive gear for this specialized fishery. The seine skiff still must be called on to do the job it has done for so long.

The skiff has progressed from the wood hull through steel and into aluminum and back to steel. Many new skiffs of the salmon fishery still are built of aluminum construction that was pioneered on the Pacific Coast by Alfab, Inc., of Edmonds, Washington, and Marine Construction and Design Co., of Seattle, Washington. The first aluminum purse seiner in the Pacific, the biggest fish boat of that metal when built in 1960, was constructed by Alfab.

The first aluminum skiffs measured out at 17 feet, six inches, with diesel engines of 60 horsepower and up. New skiffs approach 19 feet by nine feet with engines of up to 225 horsepower.

Smaller and lighter skiffs are still built for seine fisheries such as that around Kodiak where vessels smaller than the Alaska limit are the rule. Wooden skiffs are still seen at work in such fisheries.

The virtue of the big skiffs is their ability to handle their end of the the net in fast-running tides such as those of the Inian Island group as handily as the mother vessel handles hers. But the bigger skiff, by virtue of its size and great weight, as much as 5,000 pounds, compounds the dangers the seine skiff has always brought to the salmon fishery with its relatively small vessels. A skiff of almost any size is awkward to launch and retrieve and it presents a constant peril in bad sea conditions because of the upward shift in the vessel's center of gravity when it rides aboard.

The situation can be bad enough on the conventional seiner and it is worse on the drum seiner. Many of these vessels were not designed as drummers when they were built and the addition of the heavy drum and its auxiliary gear often has a highly adverse effect on stability.

With a big skiff riding against the drum, even a small sea can cause an uncomfortable and often dangerous roll.

A skiff not properly secured can be the greatest threat of all to the seine vessel, even in fairly smooth water. In rough water, a shifting skiff usually can be blamed for the disappearance without trace of seine vessel and an entire crew.

## Life Saver

Some mention should be made also of a protective device found quite often among fishing vessels of the smaller classes, especially those that work offshore where a run for protection may be too long for any good or impossible because of sea conditions. Call this thing sea anchor, drogue or parachute...under any name, it works.

Many pieces of gear deservedly have earned the name "fisherman's friend" over the years. Anything that adds to ease of work or safety or comfort of life aboard a fishing vessel deserves that distinction. In these respective categories recent or fairly recent additions to the list might include the Puretic block for net hauling; the self-inflating life raft in sizes to fit all fishing vessels, and the addition aboard of refrigerators, food freezers, showers and other amenities of life for the men who man the vessels.

Not to be ignored among these aids to work, safety and comfort is the sea anchor, a device that must date back in increasingly crude form to the first vessels to venture out of the river estuaries into the uncertainties of the unsheltered seas. It could not have taken long for these early sailors to learn the hard way that the chances of riding out a blow and its high seas with a whole skin were increased immeasurably if a ship's head could be held into the weather until things calmed down enough to make maneuvering a relatively safe matter.

Commercial development of sea anchors did not begin until late in the last century and sailors who needed them before they became a market practicality thus were forced to make do with almost anything with sufficient flotation and a place to secure a line or a bridle. Over all the centuries that ships have put to sea, their men have became remarkably adept at rigging sea anchors from spars and sails. Emergency sea anchors could be jury-rigged for the largest of sailing ships (although their efficiency decreased in direct relation to the size of the ship) and today sea anchors are available commonly for vessels up to 100 or 150 feet in length. Most are sold for vessels under 100 feet.

The sea anchor employs the principles of the flier's parachute to keep a vessel's head into wind and sea and to slow the rate of drift. This latter can be a factor to be reckoned with seriously, especially on a lee shore where sea conditions make it impossible for a vessel to move ahead into the weather. (Drift can be important also to vessels wanting to wait out the night hours on or close to a favored fishing area. Ocean currents often reach relatively high rates of movement in specific areas and these speeds may be abetted by winds prevailing in the general direction of current movement. In the area west of the Umatilla Lightship on the upper Washington coast, the southerly drift of a vessel riding without a sea anchor has been estimated repeatedly at from six to eight miles for the period between 10 p.m. and 4 a.m., normal non-working hours for salmon troll "trip" boats.) Properly rigged, the sea anchor does these two jobs quite effectively and adds considerably to the safety and stability of a boat riding out bad weather. Albacore and salmon trollers like to use them in conjunction with their stabilizers during their few idle night hours to steady the craft so that crewmen may get at least a little sleep without the usual roll of a vessel using stabilizers only in waves of any size at all.

Sea anchors used in these times by American and Canadian fishermen range from simple, cone-shaped affairs to devices of considerable complexity, complete with buoys, tow lines, bridles and other paraphernalia. The elementary form, the one most used by small coastal fishing vessels and carried, if not often used, by many pleasure boats, is the cone-like anchor about four feet long and about two and one-half feet across the mouth. The mouth is held open by a steel ring. At the tapered end, some models—not all—have a small opening to permit passage of water in order to soften the impact of the vessel pull against chute and tow line under the force of big seas. At the tapered end there is secured a simple trip line by which the anchor can be collapsed and pulled aboard with little effort. For drifting, the tow line is bent to a shackle at the apex of four to six shrouds sewn into the heavily-reinforced leading edge. The tow line normally is secured singly to the bow deck cleat and run through a chock or port at either side. Some users prefer bridles run from the cleat to each side. Care should be taken when setting or hauling the parachute that its lines do not become fouled in a turning wheel or in stabilizers and their cables.

Other sea anchors increase in complexity of design and handling in proportion to the ingenuity of the designers and the willingness of users to spend money on them. Superficially, this complexity may seem needless but the requirements of a 100-foot vessel are somewhat more demanding than those of a 36- or 40-foot craft. (See sketch on following page.)

Several of these designs were tested by the then-Bureau of Commercial Fisheries in 1968 with its research vessel *John N. Cobb*, 93 feet by 25 feet by 10.6 feet. This veteran research craft, named after the first of the Pacific Northwest's eminent fisheries scientists, has ridden out a lot of heavy weather during her workhorse career in which she has crisscrossed the Eastern Ocean from Baja California to the Line Islands northwest of the Hawaiian group; from the Pacific Coast proper to Southeastern Alaska, out along the Aleutians and far up into the Bering Sea...her crews became experts on bad weather. Their tests of the sea anchors were spurred by genuine interest in their sea-keeping qualities.

One chute was a type launched in its container, opening automatically after buoy and tow lines were paid out. Within 10 minutes after setting, the *Cobb* swung bow on to wind and sea. In 26-knot winds off Cape Flattery, vessel drift was cut from 2.6 knots to 0.3 knots.

In tests with another and bigger chute in the same area with winds at 38 knots with gusts to 63 knots, the *Cobb* rode steadily although her bow displayed a tendency to swing away from the wind with two to three minutes required to swing back. After tow line length was increased to 400 feet, the swing almost disappeared and the experimenters reported they were inclined to believe that had the tow line been bridled both to port and starboard there would have been no swing at all.

Both of these chutes can be collapsed and retrieved manually with slight effort under normal conditions by two men. An independently-turning gypsy on the anchor winch, such as those used for cork line hauling by tuna seiners, probably could be utilized also with the gypsy storing the lines during periods when the sea anchor would be used frequently.

Despite their usefulness, these big chutes used on craft as large or larger than the *Cobb* pose a certain amount of danger to men working with them under heavy sea conditions. Tremendous tensions are set up on the tow lines as line length and wind velocity increase. Tow line tensions up to 4,000 pounds were reported on the *Cobb's* tests with the larger chute in the 63-knot winds. A line parting and recoiling under lesser strain than this can injure critically or kill quickly...

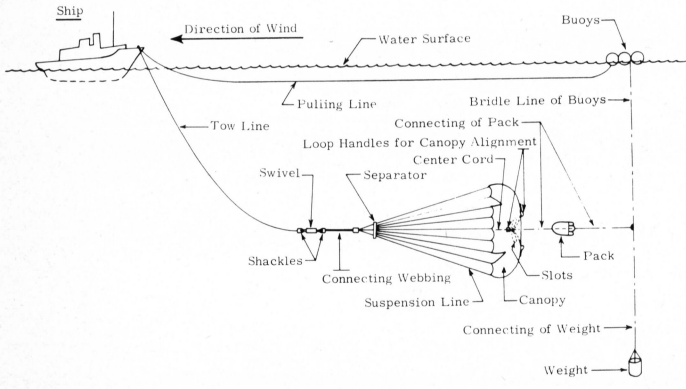

Fred W. Hipkins

*Cradle rocker—sea anchor cushions shock of stormy sea when properly rigged and matched to vessel. This Japanese-designed anchor is rather more complicated than most sea anchors but does its job well for craft running above 100 feet in length.*

Robert Browning

*Puget Sound purse seiners, using Power Blocks, at work on the West Beach grounds off Whidbey Island. Skipper of the Seagull waits at his flying bridge wheel to see what his competitor, in haul at left, comes up with. These vessels are designs of the pre-World War II era.*

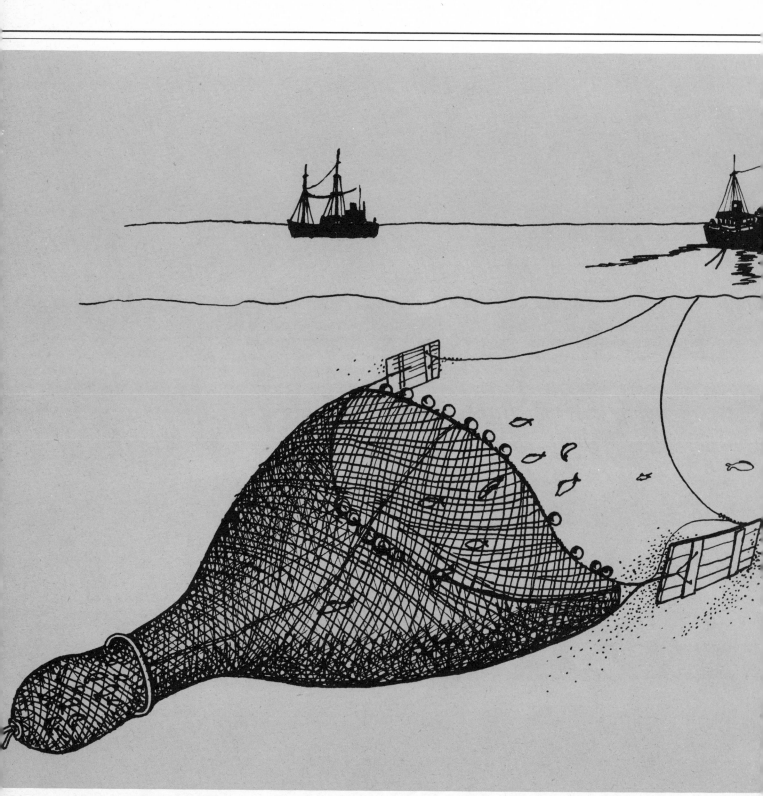

Linda Rogers

# 4/ Fishing Gear of the Northeastern Pacific

People have been fishing for fun and for money for quite a few thousand years now with a growing degree of efficiency and over that time they have managed to put together quite an assortment of things more or less useful in the pursuit of fish and shellfish. This gear runs the whole spectrum of equipment from hooks of bone to illegal explosives. Some of it, like that bone hook, has not changed at all since it first was fashioned. Some is as modern as the exploration of space and, to fishermen, almost as expensive.

Despite the hundred thousand or so years that men have been fishing, the total body of fishing gear is surprisingly small and quite similar from continent to continent. There really are only four ways to catch fish—with spears, with hooks, with nets and with traps or pots. In all the world wherever fish are caught and by whatever means, one of these four basic methods is used to take them. The actual tool is merely a variation on one of these major themes. This is true also for the harvesting of immobile shellfish. The rake or the tong is merely a variation on the spear, and the dredge, in principle, is a form of the bottom trawl.

The manner in which the first fish was taken by the first man to go fishing can never be known. But it might be guessed that the first fish was a fish trapped in a tidal pool or a river shallow by ebbing tide or falling water level and the fisherman used his hands to snatch it out or a sharpened stick to impale it. (It might be guessed too that this first fisherman went back to his cave to describe the big one that got away.) Fish are still being taken by essentially these same methods although all the gear but the hands has been improved since then. The spear was a logical step from the pointed stick for fishing and for fighting and men used it for both with equal fervor until they found other methods more effective for each task.

Actually the trap may have been used even before the sharpened stick and beyond doubt the trap preceded the spear. The traps first used by design must have been natural traps, rock or reef formations or beaver dams against which fish could be corralled by a body of people advancing through the water toward the trap site. Indian bands of the Amazon Basin still fish like that although men who have lived among them report that women look with high disfavor on water any deeper than mid-thigh. Some of the streams harbor a small fish that unerringly finds its way into the sexual organs of women, attaches itself to tender flesh and must be removed surgically. And surgical methods of the Amazon Indians are primitive indeed.

The hook came late in the unwritten history of fishing. It was an invention of considerable sophistication because for most purposes it had to be combined with a lure of one kind or another to dazzle the quarry into striking at it and the man who first whetted one out of bone or shell was a genius of his time. (He might even have been genius enough to confine himself to making hooks for an expanding market and let other people fish them in the cold and the rain and the spray. The people who make fish hooks today may count on a more stable income year in and year out than the people who use them.)

## The Net—The Big Gear

When men first began to fish in earnest, they learned quickly that spears and hooks have inherent limitations when fish are needed in quantity and that traps, man-made or natural, usually may be counted on to produce only certain species of fish and only at certain times of the year. Nets first were weaved as soon as the technology of their time made available the materials needed for them. There were many materials to be used and all were used at one time or another in many places —flax, papyrus, cedar fibers, vines, strips of animal hide...

The fish net, of course, was not the discovery of a single man nor of a single people at a special time in pre-history any more than was the fish hook. Rather, the net was devised in many forms by peoples scattered across many lands over a great span of years as they turned more and more to the lakes and rivers and seas around them in their eternal search for food.

Today, fish nets in their various modifications take most of the world's commercial and subsistence fisheries catch. No one yet, in an age when men walk on the Moon with confidence, has arrived at a better method of catching fish in large number.

Fish nets—seines, gillnets and trawls—do not make up the greater part of the world's fishing gear despite their pre-eminence in the world's fisheries. There are scores of different kinds (not types) of fishing gear used around the world under hundreds of names although the differences may be of small degree. There are far more names for fishing gear than there are kinds of gear. Names for gear kinds vary sharply among nations and sharply within a nation. Washington State's "ring net" is Michigan's and Wisconsin's "dip net" while in other states it is a hoop or drop net. The National Marine Fisheries Service calls this same net a "lift net"; the true ring net or the half-ring net is something else again.

There is no federal law defining fishing gear in the United States or Canada although the states and provinces have their own systems of gear terminology for license purposes. These systems, as noted, are mostly confusing and illogical. The old Bureau of Commercial Fisheries had its own system of nomenclature, the one its successor agency still uses in its statistical reporting. It has no standing in law but it is a good system and it will be generally followed here. Generally, but not in all instances. This BCF system reduces all gear to understandable denominators no matter where or how it is used in the United States. This system lists some 50 kinds of gear, covering everything from snag lines to fish wheels, under 170 or more local names. Much of this gear has no use in the Northeastern Pacific and will not be considered here. At least two kinds, traps (pound nets) and paranzella nets, are of historic interest only for most fishermen. Another, the fish wheel, is legal only in Alaska and is not even listed as a li-censed gear item by that state. Another, the reef net, is used only in Washington State.

## Gear Nomenclature

Any system of naming fishing gear necessarily must lump together all those that do the same or essentially the same kind of job. The NMFS gear system breaks down the four major types of gear previously defined into eight general groupings. These groups include encircling or encompassing gear (with the purse seine as the archetype); entrapment gear (pound nets, crab pots); entanglement gear (gillnet or trammel net); lines (troll or longline); scooping gear (reef net and fish wheel); impaling gear (harpoon, spear); shellfish gear (a class that begins with dredges and runs clear through to crowfoot bars), and a miscellaneous category that lists everything that can't be fitted into other groups and that includes the still-unimproved human hand.

Gear customarily found in the Northeastern Pacific includes haul (beach) and purse seines, lampara nets, beam and otter trawls, trolling lines, longlines, drift and set gillnets, crab pots, oyster and scallop dredges and a variety of hand instruments concerned with the taking of clams, most of it on the order of the clam "gun" or shovel. There is, also, a small amount of other gear, a part of the above "miscellaneous" category, such as the abalone gear of California. But this specialty gear amounts to little in the total of commercial fishing gear used by Americans and Canadians of the West Coast. Seines, gillnets, troll lines, longlines, trawls and crab pots make up 95 percent or more of the gear used over this vast area.

The definitions of these various gears are simple enough. The construction of some of them is complex and their use sometimes more so. The clam gun is about the only piece of all of it that is simple in design. (Digging razor clams successfully with it is another matter.) The purse seine is the most massive of all these as well as the most complicated and the most maddening. The purse seine costs money too, more than most of this gear although all of it is expensive. A deep Puget Sound seine 300 fathoms long with all new webbing and all new and most modern lines and fittings cuts deeply into $25,000.

Of all this gear, the ubiquitous gillnet with its tendency to be found in clusters is the gear most disliked by non-gillnet users of the waterways and this goes for other fishermen too. Troll gear is even more mobile than the easily-moved purse seine but is unselective when it comes to fish and can be as annoying as any type of commercial gear when it acts up. Longlines, trawls and scallop dredges are chiefly gear of the deep sea and they are little seen at work except by the men who use them. They fish lonely waters.

## Gear That Vanished

The gear just described so briefly is used in the Northeastern Pacific and no major changes appear in sight in any of it. But successful fishing gear can be replaced anywhere at anytime for several reasons just as two once-important gears were driven out of business on the West Coast by pressures they could not withstand. Obsolescence doomed one, politics did away with the other. The first, the paranzella net, opened up the West Coast trawl fishery in 1876 but by World War II, it had been replaced by the otter trawl, no more efficient but a lot cheaper because the paranzella needed two boats and the trawl only one. The second of these vanished gears is the salmon trap, a device that made possible the economical, large-scale harvest of salmon from the Columbia to the Alaska Peninsula. The traps had almost 75 years of productive life before they were banned for reasons based on the political muscle of sports fishermen, pseudo-conservationists, states-righters and people who had no economic stake in the traps but who owned boats, seines and gillnets. There are several Indian traps in Alaska and Washington fishing for salmon and the latter state has at least one herring trap working in Holmes Harbor on Puget Sound.

The paranzella vanished from United States fisheries most of a generation ago but a modification of it still is being used just off the East Coast by Spanish trawlers. The paranzella is a bag-shaped net towed by two vessels that run at various distances apart to keep the mouth open and at various speeds according to the depth desired. A small paranzella can be worked by two quite small and rather slow boats as was done in the old California bottom fishery. The Spanish trawlermen, however, have refined paranzella fishing into a peculiar art. Their vessels at work on Georges Bank and other Western Atlantic areas are in the 130-foot range with crews of 20 men or so. United States aerial patrols over coastal fishing areas have counted up to 30 of these Spanish "pareja" vessels on Georges Bank at one time. Their nets run about 300 feet between the wings and they can be fished from the bottom to the surface. The practicability of the pareja fishery on the Western Atlantic grounds appears questionable, at least in United States terms of economic return. Many American draggers with small crews find it hard to make money there under present conditions. Apparently the willingness of the Spanish fishermen to accept miniscule wages and poor working conditions makes the fishery at least marginally profitable to the Spanish owners.

## Salmon Traps and Other Gear

The pound net came to the West Coast from the Great Lakes and promptly was made over into the device known as the salmon trap although it was used for herring too wherever that numerous fish was taken during the heyday of the fishery. The pound net got its name from the "pound," the chamber in the trap complex where fish were held until needed after they had blundered their way into it. In the West, however, the pound got to be called the "pot" and it still is called that in the few places where the traps still stand. The first traps used in the Pacific salmon fishery were stationary affairs built around piling. But in 1907, the first practical floating trap appeared off Ketchikan in Alaska's Southeastern Panhandle. This device let traps be placed in water too deep or over bottom too rocky to allow piling to be driven. On the floater, logs substituted for floats and head ropes and the whole affair was held in place by anchors.

There was no standard salmon trap and design was based on the requirements of each site and was more or less elaborate as the situation dictated. The essential features of the trap were the lead, the hearts, the pot and the spiller or spillers. The lead was a wall of webbing or wire and webbing built at a right angle to the beach along which the migrating fish customarily traveled. The lead diverted the fish from their intended path into the "heart" or hearts of the trap. The heart was not the heart of the trap, however, but was merely the first chamber the fish entered and a trap might have two or three or four hearts. These were vee-shaped enclosures (from which the name stemmed) with the lead passing through the apex toward the pot. The trap took advantage of the unwillingness of the salmon to turn back over their own wakes when they ran into an obstacle. Because of this disinclination, the salmon groped forward along the web of the lead and the hearts toward the fatal pot from which there was no escape for any creature not able to fly or climb. In simple traps, the fish could be brailed right from the pot into a tender. Usually, however, the fish were driven into the adjoining spiller. The spiller had a vertically-movable floor of wire or webbing and this could be raised by manpower or steam to spill the fish into the receiving vessel.

Gillnets and seines were the first nets to be used on the West Coast and they were being used effectively a long time before the first white men came poking up from Mexico or along the Aleutians from Siberia. Indians of the Columbia River and Puget Sound and the coastal lands to the north used beach seines (the BCF's haul seines) of cedar and other fibers for salmon, smelt, herring and whatever other fish they could round up with them. The use of gillnets is not specifically documented by the first whites to write about the Indian fisheries but the gillnet is an eminently obvious development among the peoples of the Northwest who depended on fish for much or even most of their food supply and were quite aware of the habits of their quarry. Gillnets were a practical tool for almost all the waters the Indians fished and there is no reason to believe the fishing Indians were not as familiar with the gillnet as was William Hume when he came West in 1852. Hume was a third-generation river gillnetter and the only real difference between his gillnet and the Indian gillnet was Hume's more sophisticated materials.

Gillnets were being used by whites on the Columbia at least as early as 1825 when Dr. John McLoughlin built the Hudson's Bay post at Fort Vancouver, Washington. The gillnets were followed by beach seines, fish wheels and, in the 1870's, the trap. (For a few years after the Humes built the first cannery on the Columbia in 1866, purse seines were used at the river mouth.) The old Indian beach seines were weaved of wild hemp or cedar with cedar floats for the cork line and stones with holes bored in them for the lead line. The first seines used by the whites were not much of an improvement over those of the Indians because in the 1820's, the nearest centers of supply were months away and cedar floats and stones on the lead line worked just as well for the white men as they did for the Indian. By the time the Humes began working the river, though, the seines had grown with the times. They ran from 100 to 400 fathoms long and were set by boat off the sandbars of the river and hauled by horses working belly deep in the current to get the seines out of the water before the catch got away. These nets were legislated off the river in the early 1930's but in their final years they still were taking 15 percent or more of the Columbia salmon catch. The drift gillnet is the only commercial gear now used on the river for salmon.

The purse seine was never of any great consequence on the Columbia but it found a home on Puget Sound and northward through the salmon country of British Columbia and Alaska. In most years, it is the primary method of taking salmon in all these waters. The purse seine, even in the old days, was mobile and relatively easy to fish everywhere salmon were taken. It appeared on Puget Sound in the late 1870's with the building of the first canneries and it moved north with the canneries although it played second fiddle to the trap through the first half of this century. The great virtue of the purse seine in the Pacific salmon fishery always has been the ease with which it can be moved from an unproductive area to a ground more promising. This virtue got a boost in 1903 when the first gasoline powered seine vessel appeared off Puget Sound's Point No Point in the early summer of that year. This fisherman was the aptly-named *Pioneer,* about 30 feet and sporting a five-horsepower engine. Power installation spread fast after the coming of the *Pioneer* and by 1912 all of the salmon and halibut fleets were getting from here to there by gasoline power.

The lampara net is a relatively uncomplicated forerunner of the purse seine, a gear with an honorable history of work among fisheries of the Pacific Coast with most of that effort put in over the years off California. The lampara, so-called from the Italian word "lampo," meaning lightning, was introduced by Italian fishermen in California late in the last century. It has been used chiefly for those fish like California sardines, anchovies and the mackerels where speed in setting and closing is of utmost importance.

The lampara is shorter and shallower than the purse seine and can be set and hauled in less time and with less power. The net has a large central bunt and short wings with wing mesh larger than that of the bunt. It has a cork line and lead line. The net is set with one tow line secured to a buoy or to a skiff, the other to the fishing vessel itself. The set is made rapidly around a school of fish with the haul quickly begun to keep the catch in the net. The closing of the lead lines as the haul begins forms a floor through which the fish cannot escape. The lampara's major use in the Pacific Northwest has been for taking bait and it is sometimes called the California bait net.

The lampara found its greatest use in the glory days of the California sardine fishery. In the beginning of the fishery, the lampara was used almost exclusively and it took billions of pounds of these fish before the encroaching purse seine began to crowd it out. But even at the end of the fishery, the lampara still was doing a lot of work. There are more lamparas licensed in California than in all other areas of the Pacific Coast.

The Indians of Puget Sound were fishermen as astute as any fishermen anywhere ever were and they were the inventors of one gear used nowhere else in North America. This was the reef net, first seen by white men on Point Roberts Reef and at Village Point on Lummi Island. These nets then used the same primitive materials of the other Indian nets although the white man's cordage soon drove the native webbing out of the fishery. The net is fished among the reefs that give it its name, set out horizontally in the narrow passages the salmon must traverse to get into fresh water. To be worked properly, the water must be clear enough that men stationed on a low watch tower built on boat or raft can watch the salmon come over the net. The fish are guided by leads over the webbing into the bunt of the net. When the lookout believes the time is right, the lead line of the net is raised and the fish are trapped in the bunt and can be brailed from it. The whites didn't think much of the reef net and its dependence on the instincts of migrating salmon and gradually the reef net came into disuse as the Indians turned to other methods of fishing. John N. Cobb reported in 1930 that he knew of no reef nets then operating on Puget Sound. But the nets did meet a specific need and gradually both whites and Indians began to use them again.

## Good Fishermen They

The fishing Indians of the West Coast were innovators, not second-guessers, and the first coastal trollers of record were the Indians living around the mouth of the Columbia. They had invented the fishhook at some time remote in their dimly-lighted past and were skillful in its use for salmon and for halibut. The Columbia River Indians used their canoes as troll vessels and fished profitably for salmon over much of the estuary of the big river. There was a small amount of trolling by the

Indians at other places along the coast although most had learned ages ago that nets were more fruitful than the hook when it came to laying by a winter's supply of salmon. The first West Coast trolling by whites was on California's Monterey Bay where it began to be practiced in the 1880's, long after the haul seine and the gillnet had been put into use by the Humes and their contemporaries. These early trollers used sailboats and a couple of hand rods, each with a single line and hook, to catch king salmon for the fresh market where the going rate in that day was three cents a pound, or less if there was a glut. The market for fresh salmon had a definite limit and trolling languished all along the coast until discovery in 1898 of the mild-cure process of salting salmon. This allowed salmon to be held for uses beyond those of the fresh market and hard-salting and by the early 1900's, trolling began to assume the proportions of an industry. By the middle of World War I, say 1915 or 1916, salmon trolling from gas-powered boats was carried on from Monterey to the Aleutians. The southern trollers quickly learned after the first albacore were canned in 1906 to go after these will-o'-the-wisp fish whenever they appeared far enough inshore to make their pursuit practical. Albacore trolling spread to Washington and Oregon and off British Columbia after 1936 and many men with boats big enough for this off-shore fishery still start their seasons with salmon and switch to albacore when they show off Oregon in July.

Troll gear can be about as simple or as complex as a man wants to make it. Some albacore rigs, with no limit on lines, are marvels of ingenuity. As many as a dozen lines or more can be fished for albacore if a place can be found to rig them and men are willing to work them. There is no standard system for naming these lines and most fishermen, if they bother to refer to them with anything more than an epithet, have differing names for them. Some trollermen merely number the lines to the right from the port outboard line. Modern salmon troll lines usually are of stainless steel.

The troll line and the longline are almost contemporaneous in the Northeastern Pacific with both dating substantially from the 1880's. The commercial halibut fishery began in 1889. The longline is not the same thing to all people. In New England where it once was heavily fished, the longline is called a "trawl line." Elsewhere, it is called a "set line" and in the South and on some inland waters, it is known as a "trotline." Whatever its name, though, it is a main or "ground line" carrying a series of baited hooks on short attached lines called "gangions." The gangions vary in length but they are set at intervals of 13 to 18 feet or more apart on the ground line. A complete unit of longline gear is called a skate and it is either 250 or 300 fathoms long. This length is that which could be worked without too much strain by two men fishing from a dory in the old days.

Two or more skates of gear can be joined to form a string or set. Halibut vessels fishing the Gulf of Alaska and the Bering Sea customarily carry 50 skates of gear

or more. Each set is anchored to the bottom at each end with the anchor marked by a plastic buoy and a bamboo pole with a bright plastic flag and flashing light. The hooks are baited with herring or other fish or with octopus, a bait that found great favor in the early 1960's. A baiting machine has appeared on the market. Most ground line and gangions now are made of medium-lay nylon, replacing the hemp and Manila that was standard material until after World War II. The lines are hauled by a powered gurdy.

This gear is extremely vulnerable to other gear being used around it and during every halibut season, much American and Canadian halibut gear is lost to careless or uncaring trawlermen of the Soviet, Japanese and South Korean fleets. Longline gear is mandatory for the halibut fishery controlled by the International Pacific Halibut Commission. Longline gear catches fish other than halibut and much of the sablefish taken in the Northeastern Pacific is caught on longline gear set for halibut. There is, in addition, a longline fishery specifically for sablefish.

## Crab Pots

The Indians of the long Pacific Coast didn't miss much when it came to getting food from the sea but they seem largely to have overlooked the crabs that abounded in all the waters from Baja California to the Bering Sea. They were familiar with abalone, oysters, clams and mussels because they left mounds of mollusk shells wherever they camped long enough for kitchen middens to get started. But these ancient garbage heaps are singularly free of remains of crab shell, the Indians unwittingly leaving it up to the white man to make a good thing out of the crabs. The use of crabs was traditional, of course, with the whites whose ancestors had been eating crabs since before the days they painted themselves blue and dressed in animal skins. Crab pots came to the Pacific just about as soon as the first white men did although development of a commercial fishery had to wait a century or so.

There are several species of Pacific Ocean crabs big enough to be worth the trouble of catching but the easiest one to get at is the Dungeness crab, *Cancer magister,* abundant almost everywhere along the continental shelf. The commercials started fishing Dungeness in the late 1880's but the fishery didn't amount to much until the 1920's when canned crab began to be a highly acceptable seafood item.

These crabs are taken in pots in a fishery complicated by weather more than by any inherent difficulty in the pursuit of crab themselves. Dungeness pots are round with a diameter of 42, 48 or 60 inches. King crab pots are seven to nine feet square and 30 to 36 inches deep. All are built of steel rod with web of nylon or wire.

King crab pots are heavy, 300 to 400 pounds empty, 2,000 or more full, and dangerous in a rough sea although the most modern of deck gear can reduce the chance of accident. Some few vessels in the fishery, unfortunately, are not equipped with this handling gear. Pots are baited with razor clams, squid or herring. The pots are set in strings and worked from downwind. The Alaska Dungeness fishery runs from May through September; the king crab fishery begins in early August and ends in mid-February in most areas. Dungeness fishermen in British Columbia and the American states to the south begin their season in December and usually fish out all the eligible males in eight or 10 weeks.

## Statistically

In records somewhere, every fisherman, every boat and every piece of gear on the West Coast is properly logged as a part of the functions of federal and state agencies concerned with commercial fisheries and their regulation. The federal government is especially zealous at this collection of statistics, many of them relating to matters of great obscurity although others do have a genuine validity.

Among these sets of statistics is the BCF's annual summary of "operating units," meaning the above men, boats and gear reduced to numbers of a rather respectable magnitude. These summaries show that fishing along the Pacific Coast, despite its vagaries, does employ a lot of men and a lot of gear. This overlooks the fact, of course, that fishing should be using even more men and more gear over that vast section of the Pacific that lies between Baja California and Bering Strait. (The following figures are United States compilations for the four Pacific Coast states and do not include those for British Columbia fisheries. The province lists about 6,500 fishing vessels of all classes every year, these manned by some 13,000 fishermen. Both figures are expected to decrease as the Canadian government pursues its policy of limiting entry into the salmon fishery. Average BC landings come to about 180 million pounds of fish and shellfish a year with a value of something like $60 million. British Columbia landing figures have become unbalanced because of the failure of the herring fishery in the late 1960's. This fishery has provided up to 250,000 tons in some years.)

The federal summaries show the four Pacific Coast states to have some 45,000 full-and part-time fishermen working aboard about 20,000 "vessels and boats." This figure includes the youngest boy in the smallest skiff as well as those tuna men fishing off West Africa, a fair distance from the Northeastern Pacific but an area, nevertheless, that seems bound to loom larger for West Coast fishermen as tuna stocks of the Eastern Pacific are taken at their sustainable maximum yield.

By states, average figures for the men read like this: Alaska, 18,700 fishermen; Washington, 10,000 fishermen; Oregon, 5,000 fishermen, and California, 11,300 fishermen. The figure for Alaska would seem to show that about half the eligible males of that state concern themselves with fishing in one form or another for at least a few days or weeks each year although Alaska Department of Fish and Game figures do not reflect that assumption. The NMFS figures claim to be without duplication, meaning that the number of out-of-state fishermen in Alaska has been winnowed from the Alaska total. Alaska's own average summaries show 14,872 licensed resident fishermen and 6,487 non-residents for a total of 21,359.

The kinds and quantities of gear used in the various fisheries of the four states may come as something of a surprise to many fishermen. Most fishermen, not all by any means, spend their working years in a single fishery or a related fishery in the same general area. The reasons are obvious enough—North Pacific salmon, as an example, are not usually found off Baja California any more than swordfish frequent the waters of Shelikof Strait. To some extent too, fisheries tend to be something of closed corporations in the sense that men of certain ethnic groups tend to be dominant in them. . . Norwegians in the halibut fishery, Slavs in the salmon seine fishery, Portuguese in the tuna fishery. Men prefer to stay with the known and familar and thus many may have little knowledge of the gear and practices of a fishery only a few miles distant or even right next door. Most salmon seiners know almost nothing about trawling or trolling and they know even less about the use of the lampara net for squid or the fine points of drift netting for barracuda or sea bass. (This provincialism is not universal. Fishermen display a certain restlessness more than do most men and every man who has fished knows the man who has fished, it seems, everywhere a man can fish...tuna off Africa, anchoveta off Peru, mackerel off Florida, lobsters off Maine, cod on the Grand Banks, salmon on Bristol Bay.)

## The Net Fishery

The operating principles of the seine and the trawl are exactly the same as they were when their web was weaved of cedar or papyrus or hemp or whatever was used for net building along the coasts where they were first fished. Only the materials and the motive power have been much changed.

The trawl and the seine as worked in modern fisheries have separate functions although that distinction may tend to blur somewhat as gear researchers work toward nets that will do two or three jobs.

Presently, the purse seine, the seine seen most often in major fisheries, takes fish at or near the surface. Its specialty is the capture of such species as Pacific salmon, mackerel, the tunas, anchovy, sardine, herring and menhaden, all fish found normally within a comparatively few fathoms of the surface.

The trawl fishes deeply. In its commonest form, it fishes on the bottom or within a few feet of bottom for flatfish, shrimp, cod, haddock, rockfish and the like, towed by any one of a variety of large and small fishing vessels. The trawl can be deadly for immature fish and for such species as the Pacific halibut when it is worked over grounds they inhabit. There is no legal trawl fishery for halibut in the Northeastern Pacific although foreign trawl fleets apparently take halibut in sometimes-considerable amounts.

Much effort world-wide has been expended since World War II seeking an efficient mid-water or pelagic trawl to harvest those fish that customarily are found off the bottom and below the surface. Northern Europe, the United States, the Soviet Union and Japan have been most active in this endeavor. Ideally the perfected net would fish at any selected depth from the bottom to the surface. A high degree of success has been attained for some mid-water trawls working under specific conditions. The ideal trawl, one able to fish effectively at mid-depths, has not yet appeared commercially on the Pacific Coast although such trawls are found in limited use on the Atlantic Coast.

The former Bureau of Commercial Fisheries Exploratory Fishing and Gear Research Base in Seattle, Washington, has been one of the major American centers in this search. Base personnel have designed and assembled a series of experimental trawls since 1960. Design of the first of these, the Cobb pelagic trawl (named after John N. Cobb, the great American authority on Pacific fisheries) was based on the belief that a large trawl towed at a relatively slow speed might catch more fish at midwater than a small net moving somewhat faster. The Cobb trawl proved unwieldly and after a period of testing, illuminated by many hours of direct observation by scuba divers (with all the nets), the best features of the Cobb trawl were melded with the 400-mesh Eastern trawl to produce one designated the BCF Universal Trawl Mark I. This net too proved unhandy for small trawlers in severe tidal conditions and it gave way to the Mark II, a design still being tested and modified. Initial results showed that this net appeared to be a distinct improvement over its predecessors.

While the Seattle BCF group was testing its trawls off the Washington-Oregon coast during the latter 1960's the Soviet Union was testing mid-water trawls in the same area too. Unlike the Americans, however, the Soviets characteristically claimed complete success for their designs and published catch figures aimed at proving their assertions. No one or no Americans, anyway, can disprove the Soviet boast.

The Soviet test vessel, the factory stern trawler *Novaia Eva*, worked from September through November, 1968, outside the 12-mile limit from 43 degrees north latitude to 49 degrees north, about the latitudes of Cape Blanco, Oregon, and Amphitrite Point on the West Coast of Vancouver Island. The ship used a universal trawl 50.8 meters long with a vertical opening of 17.5 meters and a horizontal opening of 18 meters. These measurements, give or take a few inches, are 166 feet by 55 feet by 58 feet. The trawl was fished for hake at depths ranging from 135 fathoms down to 235 for a total of 95 tows. These efforts yielded 1,000 metric tons (a metric ton is 2,204.62 avoirdupois or standard pounds, 35,38 pounds short of the Anglo-Saxon long ton) for a 10.5-ton average. The next month, the *Novaia Era* moved northward into the Bering Sea where the same trawl was used for bottom and pelagic trawling for herring. Here, the reported catch was 4,300 metric tons with a per-haul average of 10.3 tons (against a 5.1-ton average with conventional trawls) with peak hauls of 20 to 40 tons.

## The Old Days

The history of the early days of the trawl fishery of the West Coast is sketchy although it is known that Italian fishermen introduced the paranzella net on San Francisco Bay about 1875 with the net towed by two sail boats. In the years right after 1880, steam pair trawlers, small tugs by type, were working out of San Francisco with the paranzella net. Pair trawling did not vanish from the West Coast until after World War II when the paranzella net was replaced by simpler and less expensive otter trawl gear.

To the north, the schooner *Carrie B. Lake* began fishing off Oregon with a small beam trawl in 1884 but was lost at sea the next year. The June, 1903, issue of *Pacific Fisherman* reported an unnamed vessel was experimenting with an otter trawl in the halibut fishery around the Queen Charlotte Islands of northern British Columbia. These pioneer attempts at trawling were limited by markets and by the absence of practical means of preserving fish for more than a few days after the catch. Like so many of the world's fisheries, the West Coast drag fishery had to wait until technology caught up with it.

But population increase along the Pacific Coast began gradually to result in more demand for fishery products other than salmon and halibut and by 1925, a stable trawl fishery had begun to shape up in the Pacific Northwest and British Columbia. By 1941, the fishery was well-established although it was largely a winter fishery carried on by halibut vessels, salmon seiners and trollers seeking an off-season income.

World War II gave West Coast otter trawling—the beam trawl was never really in the running except for certain specialized jobs—its biggest push, the one that has made it into a big fishery. There was an immediate and continuing demand for protein and for fish oils from shark livers for obtaining Vitamin A. The small draggers of the fishery—still the same seiners, halibut schooners and trollers—did the work neatly. Most were too small for military impressment and they needed only a few crewmen each. This was no real drain on manpower because most of the drag fishermen of the war years were old hands of middle age and upwards.

While the West Coast drag fishery was getting its leg up under wartime impetus, filleting, freezing and packaging processes were being perfected, an advance that kept the stimulus of the war years going after the war and its demands had ended.

Although many drag vessels still spend part-time in the winter fishery, there is an increasing number of year-around vessels at work. The Canadians of British Columbia, with large, easily-accessible trawl banks lying just off their coast, have led in the design and construction of good-sized combination vessels suited, among other employment, for the drag fishery.

The trawl fishery of the Northeastern Pacific is an important contributor to the total value of the commercial fisheries of that area. But in this part of the Pacific, the continental shelf is only from 15 to 50 miles wide at most and nowhere do there exist the great offshore banks such as those of the North Atlantic. Thus, the fishery economically does not lend itself to the development of large, long-distance trawlers with bigger crews such as those fished by Americans and Canadians and all the fishing nations of Northern Europe in the Atlantic.

The Soviet Union has deployed notably massive fleets of big trawlers off the West Coast of North America since the 1950's. These ships fish into Central American waters, taking great tonnages of hake, rockfish and other trawler species. Such large-scale operations obviously are not practical for Americans and Canadians who expect at least a modicum of return for time, work and money expended. But the Soviets, concerned with tonnage quotas rather than money profits, apparently find their great effort at least tolerable. There were rumblings in the official Soviet press in the late 1960's of major changes in top-level fisheries because of a less-than-optimum performance by the USSR's world-wide fleets, but changes, if any, were not announced to the western nations.

(The first appearance of the post-World War II fleets from East Asia off the coasts of Alaska were of immediate concern only to fisheries professionals, including men of both various government agencies and the industry. Comment and speculation on the future activities of the growing fleets and their continued long-term presence in what always had been regarded as "American" waters was confined solely to the professional publications of the United States and Canada. The public was unaware of the fleets and highly unconcerned about them in any event. Fish were fish and the sea was the sea. . .

(It took a Seattle newspaperman to come up with a factual and well-documented account of the intruder fleets for the layman and a prediction that the day was not far distant that the Asians would show up off the coasts of the Lower 48 states.

(The reporter was Don Page, marine writer for *The Seattle Post-Intelligencer*, the first newsman from "Outside" to view the Russian and Japanese operations in the Gulf of Alaska and understand exactly their portent. In 1962, after a week aboard the Coast Guard cutter *Winona*, as it surveyed the foreign ships and their integrated fishing, processing and supply methods,

Page wrote for his paper—without qualification—that the Russians would show up off Washington in no more than "a couple of years."

(That was August, 1962; the first Russian ships appeared off Cape Flattery in the following summer.)

Because of limiting geographic and market factors—American and Canadian consumers are interested in a bare dozen of drag fishery food species—the American West Coast dragger typically is an aging boat of 50 to 75 feet, owned and skippered by one man. It works with a crew of three to five men, preferably with three or four. It is this small crew factor that makes West Coast dragging economically possible. There are, to be sure, a handful of larger, modern draggers fishing the West Coast but they are a minority despite rather workable federal subsidy programs aimed at encouraging modernization of the generally over-age American fishing fleets.

This typical boat may fish anywhere from Southern California to Southeastern Alaska on proven grounds where a financially-acceptable minimum catch, at the least, is a probability. (The United States Bureau of Commercial Fisheries and its predecessor agencies must be credited with the greater part of the exploratory work in the bottom fishery. Whereas the halibut fishermen of the early years of the century did most of their own discovery of new grounds, federal agency vessels for two generations have undertaken most bottom fish exploration because of the reluctance of commercials to commit time and money to a potential will-o'-the-wisp venture.)

This boat may operate out of any one of a number of coastal fishing ports with relatively short runs to fishing grounds and, happily, in view of the small size of most vessels, a short return run to shelter in foul weather. The only vessels making fairly long trips are those from the ports of Puget Sound, Washington. These boats are forced, because of the scarcity of productive grounds just off their own coast, to work off the northwest coast of Vancouver Island and as far north as Hecate Strait. This makes for round trips of from 500 to 600 miles, seven to 12 days port to port. These vessels, many of them approaching obsolescence, are slow with nine or 10 knots considered good speed for them.

## The Trawlers

American and Canadian trawlers of the West Coast use two types of gear, the beam trawl and the otter trawl.

The beam trawl is a net of the same relative construction as the otter trawl. But the wings of the trawl are held open by a rigid horizontal beam or spar rather than otter boards. The unwieldy beam limits the size of the net because of the obvious difficulty of dealing with a spar up to 40 feet in length on the rolling, pitching deck of a small dragger. Hence, the beam trawl is of little significance in the ground fisheries of the area.

Its chief use is in the shrimp fishery of Southeastern Alaska and Kodiak.

The otter trawl, first used in Great Britain between 1860 and 1870, is by far the more flexible of the two and is found in sizes up to 100 feet across the mouth with a depth of 20 feet and a length of 150 feet. These are the average measurements of the so-called Iceland trawl of the North Atlantic which, with regional modifications and different names, is the standard groundfish trawl of the North Pacific.

The otter trawl is named after the otter gear of the world's navies, the gear used to tow paravanes to support the cables by which mine-sweepers cut mines loose from their anchor cables.

In place of the paravane, the otter trawl uses two foils called otter boards (or doors) attached by bridles to the wings of the trawl. These otter boards hold the trawl mouth open horizontally because water pressure on them under tow tends to move them diagonally away from the drag vessel's heading. The boards may be of wood, aluminum or steel.

The mouth of the trawl is kept open vertically by a varying number of floats of aluminum, glass, cork, or synthetics seized to the headrope and by a lead line at the foot rope. The net ends in a narrow, tubular bag called the cod end into which the catch is gathered as the net is towed through the water.

The net itself is a complex assembly of panels—as few as six or eight or as many as 18—that must be hung exactly right to be fished properly. Until the development of scuba diving gear, exact knowledge of a trawl's behavior out of sight deep below the surface was mostly a matter of "by guess and by God." The Bureau of Commercial Fisheries Exploratory Fishing and Gear Research Base in Seattle again proved itself a front-runner in trawl research with its use of divers to study trawl action during development of a mid-water trawl. The facts of trawl action learned there have been passed along to commercial operators who have made profitable changes in their net makeup because of this visual and film study.

This particular BCF base pioneered too in the experimental development of a shrimp trawl of a peculiar design that cuts by 97 percent, on the average, the catch of scrap fish during most tows. In many cases, test tows on BCF vessels and on charters showed shrimp catches clean enough to be iced down with no sorting of scrap fish needed. Other trawls normally retain two or three times more scrap fish than shrimp.

West Coast otter trawls are fished in two ways. One method, the older, is side trawling in which the net is set and hauled over the side of the vessel, usually to starboard. Side trawling has mostly disappeared in the West Coast drag fishery as operators turn to reels or drums to handle the trawl over the stern. With the drum, only the cod end need be strapped over the side. It is estimated that 90 percent of American and Canadian West Coast draggers use the drum.

In this West Coast area, vessels still using the side trawl method are mostly the halibut schooner type with the house aft and deck working space in the waist. The starboard side mounts a pair of heavy A-frame gallows, one well forward, the other on the quarter. Blocks secured to the gallows handle trawl warps, otter boards and the trawl itself. The trawl winch is mounted just forward of the house.

With the end of the tow, the haul begins with the warps brought aboard by the winch. When the trawl comes alongside to starboard it is strapped aboard in sections through a block on the boom until enough is aboard so that the cod end can be hoisted over the rail by a haul line running through another block.

The operation is awkward enough in good weather. In rough weather and much trawling is done in bad weather—the strapping procedure is time-consuming and dangerous with the net and floats swaying overhead subject to every influence of wind and sea.

## The New Way

Drum trawling first was used on the West Coast in 1954. The drum turns a cumbersome and hazardous method of fishing into one safer, faster and more economical. Many drum trawlers use only three men instead of the four or five needed by the side trawler. The vessel most used is the western combination type, often a salmon seiner picking up a winter income in the off-season drag fishery.

The trawl drum, hydraulically powered, can be easily installed at modest cost with no major modification to boat or gear on any vessel with enough deck space aft to work it. When the drum is used, the gallows are mounted aft to port and starboard. Most drum draggers use a stern roller to reduce chafing of lines and net.

A welcome advance over either method of trawling is the stern ramp trawler, a type almost unknown among Americans and Canadians of the West Coast but one common across all the Atlantic and one found increasingly every year among the Soviet fleets fishing off the Pacific Coast of North America.

The stern ramp trawler, no matter its size, can be picked out by its square stern, the inclined ramp or slipway leading forward to a protected working area and by one or more gantries at the stern to which blocks and other standing and running gear are secured.

With certain exceptions such as a few Netherlands stern trawlers, the house is set well forward. Some vessels feature stacks mounted side by side at starboard and port to permit a long, unbroken working deck.

The stern trawler shoots its net, tows it and hauls it entirely over the stern. During the haul, the entire net is winched up the ramp until stern gantry gear can lift the cod end into the working area. There is no overhead or side handling of gear. Working areas are protected by high bulwarks usually higher than the tallest man.

Fishing, sorting and net mending can be done with a minimum of the discomfort and inconvenience endured by crewmen of other trawler types. Most of the danger element is missing and the vessel can fish in water as rough as her size will permit.

General plans for two vessels designed specifically for the somewhat-limited trawl fishery of the West Coast have been projected by the Bureau of Commercial Fisheries' Seattle gear base.

One vessel would be an 84-foot seiner-stern trawler, the other a "middle-distant" trawler-seiner, combination boats able to work the year around in at least two fisheries.

The smaller boat would be aimed at the Vancouver Island-Hecate Strait grounds while the other, with a range of about 1,400 miles, would be capable of working all areas now being fished as well as the Gulf of Alaska banks now used almost exclusively by the Soviets and

the nature of the trawl and the manner in which it is intended to work.

By comparison, the hanging of a 300-fathom salmon seine, as complicated as that job may be or appear to be, is easier to learn and easier to accomplish because, unless there is a taper hung into one or both ends of the seine, the webman need deal only with linear strips of webbing of no more than three widths, long, narrow rectangles 25 or 50 or 100 meshes deep.

The otter trawl, whether it be the 400-mesh trawl or any other trawl, must be cut from webbing strips along a series of angles with all the problems attendant on this procedure. This can be done by rule of thumb or by the use of one or another of several formulae designed to allow this cutting on the bias with a maximum of accuracy and a minimum of wasted webbing. (For an effective application of one such method, see "A Method for Tapering Purse Seines," Appendix II, Part 3.)

*Nylon line backspliced to take steel thimble. This configuration in various sizes finds many uses aboard fishing vessels.*

Japanese, grounds where bad weather is something to be considered with concern the year around.

Although no new vessels embodying the exact plans of the above vessels have been built by Americans, there do exist craft more than capable of fulfilling the objectives of the projected BCF boats. These are the vessels constructed for the Alaska king crab fishery.

## The Construction of Trawls

The assembly of an otter trawl is, perhaps, the most complex net fabrication the fisherman must face. The standard 400-mesh trawl that has come to be much used in the western drag fishery consists of from five or six to 12 or so pieces of varying shapes and sizes as well as cod end chafing gear, a variety of floats and bobbins and the other necessary rope and metal rigging components. To put all this material into an efficiently functioning net requires a complete understanding of

Of the varying number of webbing components in the 400-mesh trawl, all but the intermediate and cod end sections must be cut along tapers for rather considerable lengths. To cut these tapers precisely appears to require a smooth working surface, accurate measurement and a wealth of professional skill. As an actuality, most trawls are cut and hung originally in the lofts of net manufacturers according to the desires and needs of the man for the net. In some instances adequate working space does not exist and measurements and professional skill may sometimes be less than optimum standards.

The number of webbing pieces in a trawl depends upon several variables, most important among them being the intent of the net designer and the availability of webbing supplies. Mostly, however, the number of trawl segments is contingent on the depth (in meshes) of webbing strips commonly available from suppliers. The depth most commonly stocked is 100 meshes. This forces the net designer to accommodate his design to that depth and the result is, in most cases, a trawl of more segments than the designer would prefer. For example, a trawl design to be considered shortly requires an overhang and top 123-1/2 meshes deep. But, since the available webbing was the standard 100-mesh depth, the added 23-1/2 meshes had to be cut separately from the strip and sewn to the aft end of the top segment. Ideally, webbing would be available in any desired depth but this is a matter of practical ordering and purchasing only for net manufacturers who build many trawls of one design. It is apparent that the fewer pieces there are in any trawl, the less expensive becomes its production since sewing webbing segments takes much time and expensive hand labor.

Every net used by fishermen must be balanced to fish at its optimum capability. This is true of the gillnet and the purse seine and it is even more pertinent of the otter trawl because the otter trawl must work within limits more sharply defined than those of the surface nets. The trawl must move slowly along on the bottom, a matter treacherous in itself; the leading edge of the upper portion must have a measure of buoyancy to offer an opening for the mouth while the lower portion must be weighted enough to keep it in contact with the bottom. The webbing sections must be hung properly so that no seam or piece of webbing experiences undue strain; the whole may be fitted with bobbins of hard rubber to allow it to be dragged as easily as possible across an unseen rough surface, often by a vessel of insufficient power.

Until the post-World War II development of self-contained diving gear, along with the improvement of underwater photography, the workings of the trawl and other nets were largely a matter of guesswork, abetted by intuition which, in turn, is based mostly on experience. Matters have improved...divers and photography have combined to give researchers and fishermen for the first time an understanding of the stresses affecting the trawl as it works and for the first time in the long history of trawling, designers and fishermen have arrived at a point where they can correct deficiencies and improve fruitfully on long-standing designs.

Seasleds towed beneath the surface alongside the net were first used by divers to observe trawl action. The sleds were cumbersome and limited the movements and observations of the divers. When it became apparent that most of the webbing of properly-hung trawls was stable and taut in the water, divers began to work directly on the net surfaces. This considerably expanded their own ability to observe the nets at work and served equally to extend the range of their cameras. Passage to the net is accomplished by following the towing warps downward.

Obviously, the procedure is a tricky one, heavy with danger, and the possibility of trouble is high, even for experienced divers. But because skilled commercial divers might not understand the full import of what they might see of the net, at least one BCF gear research base, that at Seattle, determined to train its own biologists to become qualified divers, a decision that since has coupled the skills of capable divers with those of men wise in the ways of gear.

Diver examination of midwater trawls is somewhat simpler than the study of bottom trawls because of the silt stirred up by the running gear of the latter. Unsuspected net action has been observed with all types of trawls but especially in connection with bottom trawls. One report notes:[1]

"Observations were made of a 400-mesh eastern bottom trawl rigged by a commercial fisherman to be most efficient for flatfishes. Fish were herded by the sweep lines and forward net wings which were in contact with the bottom. However, the center of the footrope was 10 inches off the bottom, permitting a large percentage of the flounders to pass beneath the net.

"Neither roundfish nor flatfish resting on or near the bottom were seen to react to a trawl headrope overhang if it passed at a height of 10 feet or more above the fish. When flounders passed well aft of the footrope (after entering the net), they usually drifted quickly to the cod end, facing the direction of the tow. Roundfish more often oriented to a section of the web and swam with the net. Some then turned and swam across to the other side panel but were eventually carried back toward the cod end. With small catches, individuals of all species swam for long periods in the cod end because of the low water velocity."

One phenomenon investigated was the relatively slack water in the cod ends of some trawls, at least, as observed in the preceding paragraph. Forward web of the trawl was seen to be taut from water pressure while the terminal of the cod end was collapsed with only slight indication of current flow. Trash picked up from the bottom as the trawl passed over it was seen to drift in eddies and the experimenters quickly learned through instrumentation of nets that water flow near the webbing inside the trawl had greater velocity than that outside the webbing. The difference reached a quarter-knot in some readings.

It was decided further that slack water in the cod end gave fish a chance to rest and renew sometimes-successful escape efforts. That the slack water is present in some, if not all trawls, was shown by one cod end that retained its 15,000-pound catch after the divers pulled the puckering string while the net was being towed about 12 fathoms beneath the surface. It took about 10 minutes for the catch to drift free of the net.

The trawl is many things to many persons in every fishing nation where the trawl works. This discussion will be confined to the 400-mesh eastern trawl of the West Coast drag fishery.

1. High, W.L., "Scuba Diving, A Valuable Tool for Investigating the Behavior of Fish Within the Influence of Fishing Gear," experience paper, Food and Agricultural Organization of the United Nations, 1967.

The 400 eastern was popular on the West Coast from about the time drag fishing reached maturity there, in the years just prior to World War II, to the mid-1960's. Then a Norwegian trawl, still with some degree of use, took its place for two or three years. But by 1968, the 400 eastern, with some modifications in design, regained its place in the affection of fishermen, particularly in Washington and Oregon. The appearance of more heavily-powered vessels in the fishery has allowed the 400 to be enlarged by about 20 percent while at the same time five-inch mesh has mostly replaced the old four- or four and one-half-inch mesh.

But no trawl design (or any net design, for that matter) can be assured of any permanent place in the affections of the fishermen who use it. Any net that appears to promise more fish and more money may promptly replace a net that has served a man faithfully and well (and sometimes mulishly) for years. Similarly, continuing design changes in the 400-mesh trawl have been underway since the first version of that design was introduced in the western fisheries. (An example of such modification will be used to illustrate this continuing trend.)

Fish and Wildlife Circular 109, written in 1961 by the then-Bureau of Commercial Fisheries, describes the otter trawl thus:

"The otter trawl is a device for catching bottom fish. It is constructed of twine webbing so that when fully assembled and rigged it will take the shape of a huge funnel while towed along the bottom of the ocean. Floats and weights are utilized to keep the mouth of the net open. To spread the mouth so that it will cover the largest possible area, each wing is fastened to an otter board or trawl door. Each door is fitted with chains for attaching to a towing cable from the trawling vessel. The resistance of the water to the forward motion of the boards, as they are towed at different angles, forces them to pull in opposite directions and thus keep the mouth of the net opened. When the Vigneron-Dahl (V-D) version of this gear is used, the otter boards are attached at some distance from the tips of the wings."

This is a simple description of a complicated piece of catcher gear, one with a long and honorable history dating back at least to the 14th Century. The actual history of the trawl is as obscure as the history of any fishing gear but the first written mention of it sounds familiar.[1]

"Some arrangement of beam and bag net, essentially a trawl, was used in England, causing a complaint to Parliament in 1376 that certain fishermen had subtly devised a contrivance to which was attached a net of so small a mesh that no manner of fish could escape and that such practice was to the great damage and destruction of the fisheries of the Kingdom. No wonder, for the meshes were 'of the length and breadth of two men's thumbs.' After the lapse of (almost) 600 years, we still hear the same complaint but in different language, often neither so restrained nor so quaint."

That ancient and disliked piece of gear has evolved into many versions of the bottom fishing net but the principle on which all these nets are built is exactly the same as it was in 1376. Not all trawls sweep the floor of the sea as they used to but a trawl is still a net towed through the water by a fishing vessel, whether it be a 40-footer in the Gulf of Mexico shrimp fishery or a lumbering stern trawler-factory ship of the Soviets.

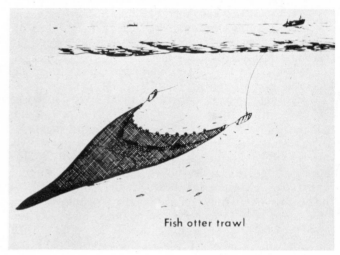

Fish otter trawl

National Marine Fisheries Service

## The Trawl Itself

The body of the trawl may be reduced essentially to three major units. These are, from forward, the wings, the body and the cod end. In turn the wings may be divided into upper and lower (or top and bottom) elements, the body into upper and lower (or top and bottom) sections while the cod end may be one piece, woven as a single unit, or it may be split into one or more intermediate pieces, hung between the trailing edge of the body and the cod end proper.

The wings again are composed of right and left top wings and right and left bottom wings for a total of four pieces. Each pair is identical in length and width, that is, the top pair are twins and so are the bottom pair.

Immediately behind the top wings lies the net overhang or "square," the widest piece of webbing in the trawl. It is not a "square" but is, instead, a rectangle. The overhang is designed to prevent the escape of fish flushed from the bottom by the ground lines or by the foot rope or tickler chain if the latter is used. In effect, it is an umbrella designed to prevent things from going up instead of things from coming down.

Next aft along the trawl are the belly and top sections. The belly section and the top section, exclusive of the overhang, are of the same dimensions. These sections may be assembled from as many pieces as the

---

1. Scofield, W.L., "Trawl Gear in California," California Department of Fish and Game, 1947.

net maker desires or as forced on him by the depth of his available webbing, or each may be composed of a single piece of webbing if his trawl is small or if he can order webbing as deep as he wishes it to be. If the top can be cut from a single piece of webbing, the overhang becomes an integral part of the top half, the case in the modified 400 design to be discussed in detail.

Although the use of a single piece of webbing in either of the two major sections eliminates much webwork in the initial hanging, such use does pose the possibility of more work in repairing a tear if one should occur in either of these sections, simply because of the greater distance such a tear could run. If this should happen, a rip may run the length or the width of the piece. If two or more pieces are used to assemble either of these segments, there may be some slight chance that a tear would end at the junction of the next piece, although this is a factor that cannot be predicted. The only near-certainty is that a tear probably would stop at the laceage and rib line selvage joining the halves of the trawl.

The terminal section of the trawl may consist of one, two or three pieces. If it is one piece only, that piece is the cod end, hung to the aft end of the body. If one or two intermediate pieces are used, they lie between the cod end and the body. In any event, intermediates and cod end are of one piece only, each folded and sewn along the single seam thus formed. It is in this section of the trawl that the heaviest webbing is used.

The catch comes aboard in the cod end and to free it, it is necessary to open the end of the bag. To this effect, a series of rings is lashed into the selvage meshes of the cod end terminal and a line is threaded through them, much as a purse line is passed through the rings of a seine. The line, called puckering string by most

trawlermen, may be secured by a knot or by one of several patented metal closing devices called clips. The usual puckering string knot is the hangman's knot with fewer turns than the fatal 13 of the gallows. As for the clips, one tested by the former BCF worked flawlessly through 170 experimental tows off the Oregon-Washington coast. Time saving was estimated at about 30 percent for each split.

## The 400-Mesh Trawl

The 400-mesh trawl is so named because of the mesh measurement around the throat of the net, the edge just aft of the overhang. The webbing is not cut to that exact number because of the need to reinforce the side seams by bunching and lacing meshes. One standard design is cut 12 meshes oversize on each side with the extra meshes laced into the side seam or gore with the true 400-mesh circumference achieved when the halves are laced together.

All webbing is subject to cutting error at some point or other, sometimes even by the most skilled of webmen. Purists insist that the maximum allowable error in cutting for trawls is plus or minus one mesh. Most net houses show errors greater than that one mesh while the average fisherman is not skilled enough in cutting tapers, for example, to approach the net house standard. Errors, unless gross, can be buried (like the medical profession's mistakes) in the pickup of unwanted meshes if there is an excess or the hand weaving of additional meshes to make up a deficiency. The most precise cutting and fitting of nets probably may be found among the webmen of such research units as the gear bases of the National Marine Fisheries Service.

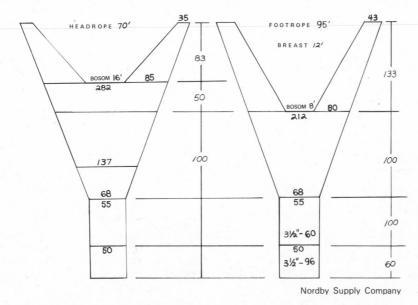

Nordby Supply Company

*The 400-mesh Eastern trawl has been a standby of West Coast drag fishermen since the period just before World War II. This is a representation of its earlier and heavier configuration.*

The essence of the art of cutting web consists of this —all segments must fit precisely when it comes time to put them together.

The webbing specifications of the standard design just referred to are as valid as any for the 400-mesh trawl. This net is to be hung on a head rope 70 feet long from wing tip to wing tip and a foot rope 95 feet long over the lower span. Both lines are to consist of three-eights or one-half-inch diameter (the size is optional to the owner) galvanized preformed wire of 6/19 construction, that is, six strands of 19 wires each. Both are to be served with one-quarter-inch or one-half-inch polypropylene line snugged as tightly as possible. The bosoms, the stretch on head and foot rope between each pair of wings, are 16 feet and eight feet in length respectively.

The breast lines border the wing tips, running from the inboard corner of each top wing (viewed on a schematic layout) across its length to meet the lower wing. At first glance, this would appear to be at an angle or a semicircle. The actual configuration of the wing tip in a trawl towing normally shows the breast lines to be nearly vertical with a slight concavity facing inward. These particular breast lines are 12 feet long and are made of one-half-inch or five-eighths-inch 6/19 wrapped wire with an eye splice in each end.

The webbing segments of this design may be counted as few as seven or as many as 11 or 12. The seven-count would include the four wings, a one-piece overhang and top and a two-piece belly. This figure excludes the entire cod end section and its usual three or even four segments, a common practice. The maximum count would encompass the four wings, the overhang, the top, a two-piece belly, two intermediate panels and either one or two pieces in the cod end.

For present purposes, it would be assumed that this trawl consists of eight pieces of webbing—the four wings, the overhang, the top and a two-piece belly with the cod end assembly not counted. This is a valid compromise between too few and too many pieces, one that takes into account the expensive hand labor of the net loft that must go into the initial fabrication of the trawl and the exigencies imposed on a dragger crew trying to patch a net made of too-large segments on a cluttered and weather-racked deck somewhere at sea. Some types of otter trawls used in the North Atlantic run up to 18 pieces. This would seem to be running a good thing right into the water if an old figure of speech may be paraphrased a trifle.

All webbing in this trawl, with the exception of the cod end segments, is four-inch stretch measure, woven either of twisted or braided nylon of No. 36 or No. 42 strength. Varying with the manufacturer, No. 36, new, is rated at a breaking strength of from 275 to 290 pounds while No. 42 should run from 330 up to 340 pounds. Sewing twine may be as heavy as No. 60, about 600-pound test, or as light as No. 36. Here, the net maker must strike a balance between strength along his lace-

ages and damage to his webbing. Something must give —preferably the lacing twine.

The top wings in this design are 83 meshes deep, tapering from 35 meshes wide at the wing tip to 85 meshes at the junction with the overhang. The overhang runs 282 meshes wide across the wing junction. It tapers down to 212 meshes at the junction with the top through a depth of 50 meshes.

This two-piece top has a total depth of 100 meshes. The forward piece of this segment obviously must be 212 meshes wide at the overhang junction where it, in turn, tapers to 137 meshes at the junction with the second piece. From 137 meshes, this piece tapers to 68 meshes at the junction with the intermediate piece.

In final form, the cod end segment is a cylinder although it narrows somewhat at the terminal. The intermediate, 110 meshes in circumference, and the bag end, 100 meshes around, are woven of heavier twine, No. 60 in the intermediate and No. 96 in the cod end itself. Both have three and one-half-inch mesh. The intermediate is 100 meshes deep, the bag is 60 meshes. The entire trawl, from wing tip to puckering string, would be about 125 feet long if all meshes were fully stretched fore and aft.

To fit the 100-mesh intermediate to the 136-mesh circumference of the last section of the top, 26 meshes must be picked up in the sewing, a baiting (from abating) rate of one mesh from the 138-mesh side to every fourth mesh on the 110-mesh side. Similarly, 10 meshes, or one in every 11, must be picked up from the 110-mesh side of the intermediate to fit into the 100 meshes of the cod end.

The bottom half of the trawl differs from the top in the shape and size of the wings, the absence of any component resembling the overhang and, in this particular net, the use of the two-piece belly.

The bottom wings are longer and narrower in relation to their length than the top wings. The top wings are 83 meshes deep while the bottom wings are 133 meshes deep, running the entire distance from wing tip to belly junction, taking up the 50 meshes used by the overhang in the top section of the net. These bottom wings taper from 43 meshes at the wing tip to 80 meshes at the belly junction. The belly comes to the same size overall as its topside mate—100 meshes deep and 212 meshes wide at the throat with a taper to 68 meshes at the intermediate junction. The only extra component is the addition of chafing gear to be secured beneath the cod end. The chafing gear is made up of seven-inch, hog-ringed mesh of five-sixteenths polypropylene, a tough substance that does much to protect the more fragile webbing of the cod end from the rigors of the sea bottom. Not all trawls carry chafing gear. It is most commonly found on those nets intended chiefly for the taking of the bottom-hugging flatfishes.

The webbing segments of the otter trawl are only a part of the entire net assembly. The trawl cannot fish without proper rigging any more than can the purse seine or the gillnet. Essential to its performance are

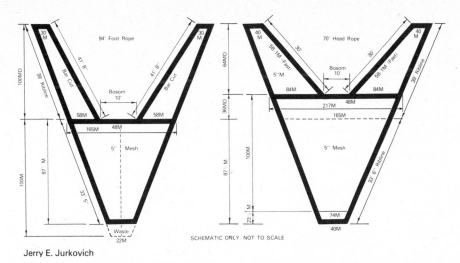

Jerry E. Jurkovich

*Here is a 1972 version of the 400-mesh Eastern trawl, designed at the NMFS Seattle gear base. This design uses five-inch mesh webbing rather than the four-inch of the standard design, and is built with wider wings to achieve deeper vertical opening.*

various support lines secured along the body; sweep or ground lines to connect the trawl to the all-important otter boards; bobbins (if they are used) of two designs to support the foot rope in its quest along the bottom; floats to support the head rope so the trawl will have a proper mouth opening. In addition, there is a variety of small hardware to connect this line and board gear and allow it to function as it must to fish the trawl in the desired manner. This gear and all other gear from towing winch aft will be described. The fishing vessel itself can be of any design as long as it carries enough power to handle the awkward trawl on the shooting of the net, the tow and the haul.

The endless modification of the many designs of the otter trawl (as well as all other marine fisheries catcher gear) has been noted. A pertinent and successful example is a trawl designed and built by personnel of the Exploratory Fishing and Gear Research Base of the National Marine Fisheries Service Northwest Fisheries Center in Seattle, Washington. This trawl uses five-inch No. 42 mesh rather than the four-inch of the net just described and it is built with wider wings to attain a deeper vertical opening. Excluding the two intermediates and the cod end, it is an eight-piece net, a number of segments forced on the designer by the necessity of using webbing only 100 meshes deep. The pieces are the four wings, the one-piece overhang and top, a two-piece overhang and top, a two-piece belly and a bobtail piece, 74 by 40 by 23-1/2 meshes cut from waste webbing, to be fitted to the aft end of the top to fill out its design size and shape. If webbing 125 meshes deep had been available, this small separate section would not have been necessary and would have cut down a bit on the needed webwork.

The sewing of this section to the trailing edge of the top was the logical choice for positioning because the use of this narrowest junction saved the much greater amount of handwork that would have been necessary if the desired length of the top had been obtained by sewing in the needed increment at a wider part of the top. This trawl, disregarding its two intermediates and the cod end, measures out at about 80 feet in length with all meshes pulled tight.

The head rope is 70 feet, the head rope bosom is 10 feet; the foot rope is 94 feet, the bosom is 10 feet also. Both wires have the same specifications as those of the previously-described trawl, as do the breast lines, 14 feet in length in this case. The top segment of the net is 123 meshes deep from the 217-mesh width at the throat, with a five-bar, one-point taper to 40 meshes at the junction with the first intermediate. The rib lines are three-quarter-inch Dacron.

The bottom wings are 30 meshes wide at the tip, 58 at the belly junction. The belly, at that junction, is 165 meshes across with the taper to 40 meshes. The entire belly segment is 87 meshes deep.

This trawl, as with all trawls, can be used with or without bobbins, according to the discretion of the skipper and his knowledge of the bottom to be fished. From 12 to 24 eight-inch aluminum floats may be hung to the head rope, the number to be used depending mostly on the action of individual nets as seen by divers or assumed by deduction.

Three ground or sweep lines, rather than the two commonly used, are intended for this trawl. (Some confusion in terminology may occur because of the use of the same name for more than one piece of gear or the use of several names for a single piece as in this instance.) The bottom ground line is one-half-inch twisted wire while the upper lines are three-eighths-inch. The net is designed to tow with a 36-foot opening across the mouth and a vertical opening of from nine to 18 feet.

## Cutting the Trawl

As noted earlier, otter trawls on the West Coast customarily are cut from webbing 100 meshes deep. It is apparent, of course, that a trawl may be assembled from odd pieces of webbing as long as all pieces of the net, except for the cod end segment, are of the same strength and mesh size. This latter does not apply to certain specialty trawls such as the shrimp separator where, of necessity, web of differing measure must be used to achieve the desired effect.

Preferably, the cutting should be done in an area of smooth surface with room for the webbing to be lain out linearly, so the webmen have something of a picture of the cuts they must make. In practice, however, cutting and assembly often take place in close quarters where the webmen see only the material directly at hand. This situation is true of some net lofts; it is true of many areas where fishermen cut and assemble their own nets. Most certainly, it is true of the conditions in which fishermen at sea must make do as they try to patch badly-ripped nets on a deck that makes a cramped net loft assembly area appear, in comparison, to have the approximate dimensions of a tennis court. To boot, the net loft is not rolling and pitching and taking water through the scuppers or over the rail, nor is someone trying to sort fish while the net is being worked.

Only the most expert of webmen can be expected to achieve complete or near-complete accuracy in the cutting of trawl pieces under the conditions just described. This lack of adequate working space must be blamed for sometimes-considerable errors in web cutting, errors that may not be discovered until net sections must be matched for sewing.

The cutting of trawl pieces may be done with more hope than skill by putting down a piece of webbing and aiming the scissors toward the point where the webman wants the cut to end. This may or may not (mostly not) result in a taper of the desired angle and accuracy. This approach more often is that of the fisherman himself trying to cut and hang his own trawl, although some professional webmen are not immune to a degree of this sloppiness.

Therefore, unless a net designer and/or webman possess superlative skill, an instinctive feel ( a rare attribute but one that does exist) for the lie of the webbing and the angle of approach, a preconceived plan should be followed by all hands, regardless of the amount of available working space. There are two basic requirements.

The first is that the webman must understand the intended function of the trawl and the principle underlying the accurate cutting of web tapers, an art fully explained, as noted earlier, in Appendix II, Part 3. A study of this exposition should enable anyone with an elementary knowledge of the weaving of webbing and a grasp of simple arithmetic to learn the method easily. It does no harm, either, to have some instruction from a man versed in the use of these formulae for the calculation of the taper, both in working the mathematics and in practice cutting of tapers on waste webbing. Without such training there may eventually be more waste webbing than the webman intended.

The second requirement unless, again, the webman has the highest degree of expertise in cutting tapers, is that a cutting diagram be lain out. This diagram can be something as offhand as a quickly-penciled sketch on scrap paper or a carefully plotted schematic design on engineer's graph paper. In any event, the diagram (see sketch) should clearly indicate, not neccessarily to scale, the pieces to be cut and the tapers to be used in the cutting. The careful use of this cutting diagram eliminates all but an unavoidable minimum of waste web and most of this waste, if that be the proper name,

CUTTING DIAGRAM FOR 400-MESH EASTERN TRAWL
WITH WIDER WINGS TO ATTAIN GREATER
VERTICAL HEIGHT—5-INCH MESH

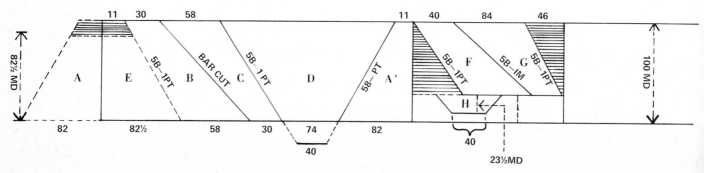

Linda Rogers

can be used when it is needed, as inevitably it will be, in patching tears.

The rule of thumb or shot in the dark method of cutting trawl pieces can only result in extra and needless work, the loss of expensive webbing, an ill-balanced trawl and abraded tempers.

The accompanying cutting diagram is the pattern from which the pilot model of the just-described NMFS trawl was cut. It is not drawn to scale but it does show what pieces must be cut and how they must be cut from the strip. In this instance, the entire cut along the strip took 504 meshes, a figure that includes one waste mesh or bar for each cut, in this case seven. The shaded area shows waste webbing, waste in the sense that it did not go into the net but was set aside for later patching.

This net was cut by its designer, a complete expert with scissors and needle, the author of the paper contained in Part 3 of Appendix II. The work was done on a concrete surface roughly 12 feet wide and about 18 feet long in an NMFS storage room containing fisheries research gear ranging from scuba diver air tanks to crab pots. Nevertheless, despite the designer's experience and intimate knowledge of the new design, one minor mistake in cutting was caught and corrected in a body taper.

From left, the letters A through H on this cutting diagram show the net segments. The tapers are indicated along each cut. Barred sections are waste webbing. Dotted lines show schematically the cutting and movement of pieces to their proper positions as in the case of A becoming A', one-half of the belly. Figures along edges of the strip show width of cuts.

A'—One half of the belly, cut at A and sewn to E to form the full belly.

E—One-half of the belly.

B—Bottom wing.

C—Bottom wing.

D—The one-piece overhang and top.

A—The section that became A' to fill out the belly.

F—Top wing.

G—Top wing.

H—The bobtail section cut from waste and sewn to the bottom of D to fill out the top configuration.

It will be noted that each angled edge bears a legend denoting the taper to be cut to achieve the desired shape of the trawl segment. If the reader understands the method of cutting tapers as predicated in Part 3 of Appendix II, the cryptic notations "5B-1PT," or "5B-1M (FAST)" or "BAR CUT" tell him all he needs to know to cut the tapers with the accuracy they must have to be joined together properly, and with a minimum of covering up of cutting errors.

If taper cutting is not understood, the legends mean nothing, not even if they are fully spelled out. The note 5B-1PT means a cut of five bars to one point while 5B-1M means a cut of five bars to one mesh. The note "BAR CUT" is not abbreviated; it means a cut directly along the bars of webbing without zig-zagging through points or meshes. Their meaning is illustrated in the accompanying diagram.

It is said:

"A netting fundamental is that two bars constitute one mesh."

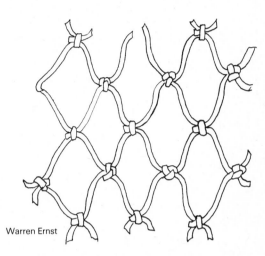

Warren Ernst

*Web sketch illustrates netting fundamentals—cleaned knot at upper left shows full mesh while cleaned knot above, right, shows two strands or points.*

This is demonstrable from the above sketches but it is a technicality mostly of interest to the webbing manufacturer and the designer of nets; to the webman who cuts tapers by rote, the statement has no real meaning. He cuts where someone tells him to cut. To the non-net-designer, the statement that two bars constitute one mesh is a matter of some wonder since it seems obvious that any single mesh must have four sides or legs or bars, call them what you will, to assume its closed shape. But only two of these bars can be cut and still preserve that shape as in the mesh cut or a part of it as in the point cut.

Every webman knows or is assumed to know that the bars must be cut in two ways to achieve the two cuts, the mesh or the point. If the mesh is cut across the "run of the knots," that is, across two bars parallel to the selvage, the cleaned knot will form a loop or one full mesh. If the cut is made with the run of the knots, at a right angle to the selvage, the cleaned knot will show two loose strands, the point. Each of these cuts has a vital place in the proper cutting of tapers and it is only by following the cutting precepts illustrated in the diagram, and in the much more complete exposition in the appendix, that tapers may be cut as they should be cut.

The third term, the bar cut, is the simplest of the three to understand and accomplish. If one looks from one selvage to the opposite with the selvages roughly parallel, it will be seen that the bars in all lines of meshes resemble the steps of a ladder or the ties of a railroad track running into the distance. A bar cut is simply a cut along this line of bars. The bar cut can be only a

45-degree angle cut whether or not the selvages actually are parallel—which they usually are not. The cut can be done in any configuration, i.e. in any manner the webbing happens to lie or in any manner the webman happens to grasp it, but the bar cut always is a 45-degree cut.

The trawl cutting diagram shows that the first cut, one half of the bottom belly, is made by means of a five-bar, one-point taper running from 82 meshes wide at the bottom selvage to 11 meshes at the top selvage, right and left respectively if, by chance, the webbing strip were to be lain out flat and squarely in the cutting area. To begin this cut, count 82-1/2 meshes along the right selvage and at the 83rd mesh cut through the double selvage, counting this as one bar. Cut through the next four bars. This brings the scissors to the first point cut. Move over one mesh to the left and cut through the two bars to make the point as shown in the webbing diagram. Cut through the next five bars, move over one mesh and cut the point. Repeat this procedure until the opposite selvage is reached.

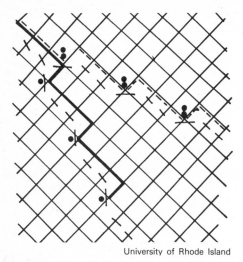

University of Rhode Island

*Taper cuts—note that cutting two side bars of one mesh leaves a point marked with one dot or a mesh, two dots, as illustrated in preceding sketch.*

The next cut will form a bottom wing. It is the bar cut which can be done from either selvage but with a taper from 30 meshes to 58 meshes, it is simpler to count 30 meshes than 58. On the 31st mesh, follow the line of bars from the left selvage to the right.

The third cut is a return to the five-bar, one-point cut, to produce the other bottom wing with the reverse taper, 58 meshes down to 30 meshes. With completion of this cut, the second bottom wing now is clear of the webbing strip and one side of the top belly has been shaped. The next step is to count 217 meshes along the strip to get the full width of the top and overhang. On the 218th mesh, cut to the right, five bars and one point, to form a taper 74 meshes wide at the right or bottom selvage. The second side of the top belly has been formed.

The cutting diagram calls for the movement of the second half of the bottom belly as indicated by the dotted lines at the left of the cutting diagram. Since one side of the section has been formed by the cutting of the top belly taper, all that is necessary is to count over 82 meshes to match the first cut of the bottom or belly. On the 82nd mesh, cut directly across the webbing at a right angle to the selvages to form the edge that must match the right angle edge of the first half of the bottom belly.

The adjacent barred section has no part in the initial construction of this trawl but, as noted, can be used later for patching. The next three cuts involve the top wings, shorter and wider than the bottom wings, and the cutting of Piece II, to be used to form the final configuration of the top belly.

The first cut involves counting 40 meshes to the left and on the 41st, cutting a five-bar, one-point taper through to the opposite selvage. This piece will be the top left wing. The next cut is a five-bar, one-mesh fast taper, the jib cut of the East Coast fisheries, illustrated in the webbing diagram (not the trawl cutting diagram). This is the only taper that may be cut from either selvage, although again it is quicker to count 40 meshes along the left selvage and cut from there than it is to count from the opposite selvage.

With this cut accomplished, a five-bar, one-point taper frees the right tip wing from the remainder of the webbing strip, shown as a barred segment. From Piece H, the waste from the cutting of the top wings, comes the tail of the top of the trawl, the small 74- by 40-mesh section needed to meet design requirements. This is the final cut of the trawl sections.

The cutting out of the trawl can be done by an experienced webman in as little time as two hours or even less for men who are true experts. It is the assembly of these segments into a functioning net that consumes man hours because of the meticulous matching of edges and the required mesh for mesh sewing that must be done if the net is to work as it is intended to work.

## The Assembly of the Trawl

The designer of the trawl under study, a man long experienced in both theory and practicality of trawl design and trawl fishing, once said:

"If your hands don't hurt when you get your net into one piece, you've done a sloppy job."

He was referring specifically to hanging the body of the trawl to the rib lines after all segments had been sewn together. But his casual remark underlines an element important to all webwork, that of tying down securely everything that must be tied down, that of making all knots secure, as tight as a man can pull them, no matter where they may lie in the net. A single slipped or lost knot may be of no importance whatsoever.

But when the stresses faced by the trawl, the greatest among all nets under usual conditions, are considered, it must be realized that a series of slipped or parted knots can be inimical to proper net action by allowing specific webbing areas to bulge or sag or even open a seam. The good webman must face the assembly of the trawl and all other nets, too, with strong hands and fingers to use properly the tools of his art.

A trawl accurately cut and as accurately assembled should lie, when stretched length and breadth on a net loft deck, almost mesh for mesh and bar for bar. Most trawls do not. That some trawls may display this symmetry can be construed as workmanship of the highest order, an art as demanding and as pleasing as any of the visual forms to which we are exposed and which we call "art" because someone says it is. As with many forms of art, this symmetry of the trawl has a functional importance; it means that the net, fishing under normal water conditions, will tow evenly without undue strain on any line, seam or webbing segment.

Assembly of the trawl falls into three phases: the joining of the just-cut segments; the joining of the assembled top and bottom segments and, finally, bending the net itself to the breast lines and the head and foot ropes. Of these three, the sewing together of the net segments is most demanding in skill, perhaps, while the hanging of the trawl halves to the rib lines may be regarded as the hardest work, the task of most monotony.

The putting together of this trawl is a matter of simplicity in relation to the assembly of certain other trawl designs, but its apparent simplicity does not in any way reduce the degree of skill necessary to do the job right. There is no single and invariable pattern of sewing trawl segments together. One noted East Coast trawl designer advocates beginning assembly of the net by working first on the belly. The net he describes has a two-section belly with each section composed of three segments; next he joins the trawl square or overhang (top), two sections only; then the top wings, each of two sections, followed by the lower wings, each of five segments. The wings, in turn, after their assembly, are secured to their respective segments of the net and all is followed by sewing and attaching the cod end, a mere two pieces.

The trawl under continuing discussion, however, with its fewer segments, may be assembled and hung to its lines by one skilled man, working conscientiously, in about the time the average West Coast trawlerman might get around to figuring out the procedure needed for the just-described North Atlantic trawl.

Assembly of this trawl was begun by sewing the top wings to the overhang. For the utmost in strength, this sewing must be done mesh for mesh. The segments are connected by a row of hand-woven half meshes formed with a No. 36 or No. 42 sewing twine. The sewing twine used in this trawl was a 36 soft lay twisted nylon of American manufacture, chosen by the designer because of his long familiarity with its stability when knotted. The sewing knot was the familiar becket bend.

(The hanging or selvage lacing knot is a hybrid clove hitch. The knot begins with a single turn formed into a half hitch. This is followed by a lock hitch lain on each side of the first turn. This knot is essentially the same as the jam or stop hitch although it has one turn less than used commonly in the latter knot. The discussion of knots that may be used in web work is a discussion that might run forever, since many kinds of knots under many names may be used with good effect. The becket bend, a favorite knot for securing the tail of the twine from one needle to the twine of a fresh needle, as well as sewing webbing edges, bears at least eight names. It is the knot fishermen and sailors know as the sheet bend but it is also the knot used in weaving webbing and the one used in patching web as the weaver's hitch or the weaver's knot. By any name, it works.)

Either of the top wings may be sewn to the overhang to begin the trawl assembly. The important step is that the sewing begin from the outer edge of the wing and proceed inward toward the bosom. Although some webmen may not take the trouble, it is wise to match the first mesh of the wing and the first mesh of the overhang and secure them temporarily, then count the necessary 84 meshes (in this trawl) inward toward the bosom and secure that last mesh with the corresponding mesh on the leading edge of the overhang. To do it even more exactly, especially for the novice webman, it may be desirable to count off and mark corresponding meshes at one or more additional points on this leg. After all, if these 84 meshes were to be stretched to their full length, this one junction would run 35 feet long, enough distance for even a passably-good webman to miscount his meshes. Under most trawl assembly conditions, as has been noted, working space is cramped and the webman sees only a portion of the leg he may be working, a factor that makes errors even easier to come by.

The connecting row of half-meshes is begun by knotting the twine to the first mesh at the outside edge of the overhang (using the becket bend), then passing the needle under and around the first outside mesh of the wing to begin the knot there. With completion of that knot, the needle and twine come down to the second mesh of the overhang, where the knot is secured, and the procedure repeated across the entire junction. The bars of these half meshes are slightly longer than the bars of the trawl webbing since there is a tendency toward shrinkage of the sewing twine, a favorable factor in that it tightens the knots even more snugly than the hand can do it.

When the first top wing (assumed for working purposes to be the left wing) has been hung, the right wing is sewn to the overhang in exactly the same manner as the left wing was worked. With the wings hung, there remains one more section to go with this top section of the trawl, that bobtail 74-mesh by 40-mesh section 23-1/2 meshes deep, cut from the top wing waste, that must be sewn to the trailing edge of the top and overhang section. This is sewn from left to right in the described manner with beginning and end meshes tied off and at least one reference point established along the seam. The attachment of this small segment completes the configuration of the webbing of the top half of the trawl.

It will be remembered that the belly of this net is composed of two sections, one coming from the first taper cut along the webbing strip, the other from the first cut after the top was cut. These halves may be joined by starting the sewing at the leading edge and proceeding aft with the customary half-meshes.

With the completion of this seam, the bottom wings may be sewn to the belly in the same manner as prescribed for the top wings by sewing inward a distance of 58 meshes from the outer edge of the trawl toward the bosom.

## Formula for Hurtin' Hands

The taper cutting of the trawl segments has resulted in the loss of selvaging from several pieces of the trawl. The bottom wings have retained their selvage on both ends since they are cut completely through the strip. The top wings retain selvage on one end only. There is no selvage along the sides of the top and overhang, nor is there any selvage along the left side of one part of the two-piece belly. The result, therefore, is a lack of strength along the side seams of the trawl, its Achilles heel, an area of critical importance and one that may be the weakest part of the trawl if it is not carefully tended during the next step in assembly of the webbing. The trawl body now exists as two halves which must be carefully joined by the most competent webmen available. If this not properly done, the net may just as well be left on the nearest pier...

(Many times in trawl assembly, the double selvages are removed when sewing is begun because they form knots too bulky to suit discriminating webmen. One leg of the twine is cut away between knots. In cutting the top wings of this particular net, the bottom selvages are lost automatically because they do not run the full depth of the 100-mesh webbing.)

The entire length of the body seams must be selvaged with twine customarily of lighter test than the body webbing because the light twine bites into or clings more snugly than twine of the same test as the webbing or heavier. The twine used in this procedure for this net was No. 15 nylon doubled. The designer chose the lighter twine because the rib line to be added would furnish the needed security to the seam. The choice, again, is a matter of discretion on the part of the designer or the webman. The aim is make the seam as strong as human hands can pull the knots.

The selvaging procedure is not the sewing of the familiar double twine selvage found along the edges of the strips. If such selvaging were wanted, it could be done, but it is not needed and a laborious and time-consuming job thus is avoided. The objective is reached by an entirely different procedure.

This selvaging is done by carefully matching meshes along the entire length of the seams, then bunching them and sewing them. This operation results in high integrity along the seam, which is further reinforced by the rib line. In effect the bunching results in a rope of one-half or five-eighths inch and is in itself about as strong as the rib line.

Matching of meshes in the top and belly halves of the trawl is most important. This can be done by finding the first mesh in upper and lower wing and securing them with a temporary seizing. Then, counting carefully, the webman can proceed from that point to the aft end of the body of the trawl, tying off matching meshes at any desired interval. To accomplish this ideally, the trawl should be spread flat with medium stretch on a net loft deck with the top placed carefully over the belly half, mesh for mesh, bar for bar. Unfortunately, most such work must be done in decidedly more cramped quarters.

Some writers on trawls advocate tying of first and last meshes on each side of the trawl, then locating the exact midpoint and marking it with a seizing. This is done if the mesh counts differ and the mesh of the long side must be picked up to match the short side. The matching can be done further by quarters and eighths, much as webbing being hung into a purse seine cork or lead line is pointed—if the job is being done by the book. But mesh for mesh exactness is not as important to the balance of the seine as it is to the trawl. (The bunching also allows intact meshes to be used to begin repairs if a tear occurs.)

With meshes located and marked, the seam may be selvaged in either of two ways; by bunching and lacing the meshes, then hanging the seam to the rib line, or by bunching meshes and sewing and hanging them to the rib line with the same knotting. The second method saves time and labor. The first results in a stronger side seam and the hurting hands of the conscientious webman, and was the method used in the trawl being described.

The bunching of meshes is exactly that but the actual sewing technique is a "knot on knot" procedure, that is, the sewing twine is reeved through the desired number of meshes from each edge but the twine is snugged against the side knots of the meshes, strengthening them to a degree and eliminating chafing of the twine against the bars of the bunched meshes. Any number of meshes may be thus bunched and three were taken from each side of this trawl. Up to 10, five from each side, is advised for trawls to be subjected to unusual stress.

To achieve the laceage, the webman (assuming he is right handed) takes in his left hand the necessary number of meshes and lines up the three knots (in this case) from each edge of the trawl. The needle is run through the six meshes and the twine is tightly knotted by use of the hybrid clove hitch or two half hitches, with the twine carefully positioned against the mesh knots. The needle is then passed at least once around the roll and the sewing is repeated with the next bunching of meshes, each three from the comparable bar of each half of the net. The selvaging may be started along either side, working aft from the wing tips. This is the most tedious part of the work of assembling the trawl but it is a part in which sloppiness cannot be tolerated.

Somewhere in the process of assembling the trawl, the intermediates and the cod end itself must be sewn and attached to the body of the net so it may be hung to the rib line when that work is begun. The intermediate pieces, two of them, and the cod end, are 120 meshes in circumference and 60 meshes deep. The three pieces are cut across the run of the knots in the strip to get the 120-mesh depth which becomes a 120-mesh circumference when the pieces are sewn. When the pieces are folded over, the knots then run with the stress of the trawl. Since these three segments are each of one piece, they are simply laid out, with one half folded to the left, then bunched and sewn or laced.

The side seams of the intermediates and the cod end are laced together since the edges are points rather than meshes. The junction seams are sewn together with a row of half meshes since they retain their double selvage. The lacing of the side seams is termed "lacing on the cross" and is done on the knots because otherwise they would untie.

Since the three pieces have exactly the same circumference and equal number of meshes, they are joined with no mesh pickup to concern the webman. When the assembly is sewn to the body of the trawl, there is a slight difference in measure since the 80-mesh circumference at the trailing end of the trawl body amounts to 400 inches, while the 120-mesh circumference of the first intermediate with three and one-half inch mesh comes to 420 inches. This leaves some six meshes of the intermediate which, in theory, should be picked up when it is sewn to the body, a pickup of one mesh in every 66. Actually, the slight pickup is immaterial and can be done in the first three or four knottings, or the extra meshes can be lost in the selvaging of the seam of the first intermediate.

The rib line, obviously, must be bent to the trawl in two sections, each running aft from the beginning of the laceage at the midpoint of the wings along the entire length of the net to the splitting rings. The rib line may be likened to the keel of a ship, the spine of a man, a bearing wall in a house. It is the member that gives essential strength to the entity behind the breast lines and the head and foot ropes. The seam laceage must be sewn to it as tightly as possible.

Three-quarter-inch Dacron line was chosen for this net because, unlike nylon, it stretches only minimally. A 24-hour soak before hanging will eliminate almost all shrinkage. Undue shrinkage or stretch would negate much of the function of the trawl by allowing bulging in one place or slack in another. The rib line must be fitted to the laceage so that it lies along it exactly, i.e., with no slack and with proper tautness so that the knots will be sewn along it at proper distances. Slack in the line would result in knots too close together

or knots too far apart. The structure of the trawl itself allows a necessary amount of flexibility through the entire net, enough to ease off somewhat on the tensions set up by a tow resulting in a heavy load of fish in the cod end and the intermediates.

The hanging of the trawl to the rib line begins at the mid-point of the breast lines, the beginning of the body laceage. The leading end of the rib line is rigged with a bronze or stainless steel thimble, a version of the eye splice with a concave groove on its outer surface to contain a line. The thimble is left hanging at the beginning of the laceage. In this trawl, it will be shackled to the middle one of the three ground lines.

Sewing the webbing laceage to the rib line again is a mesh for mesh procedure, that is, each group of knots in the laceage is hung to the line, sewn to it as tightly as possible with a knot of the webman's choice. To further strengthen the hanging, run the twine around laceage and line as many times as desired between knots, usually on every other bar or on each mesh.

The next bunching of meshes is picked up and similarly secured to the line and this process is repeated until the splitting ring assembly is reached. Here the tip of the line is double- or triple-knotted to the laceage and the excess is cut away and the end taped or tightly whipped to prevent fraying.

The hanging of the lacing to the opposite rib line is done in exactly the same manner, a matter of monotony and hard work and the hurting hands of the good webman. When one considers that the knots come out at about three inches apart all along the entire distance of the rib line on each body seam and along the single seam of the intermediates and the cod end, a matter of 197 feet (call it 200), the tired hands of the webman are entirely understandable.

The hanging of the trawl webbing to the rib line completes the assembly of the net itself. But there is other rigging to be done, with some of it such as that on the cod end being fairly complicated. Here the puckering string rings, the device that allows opening and closing of the cod end terminal, must be hung into the webbing and the even more complicated splitting strap rings must be put into the cod end between the puckering gear and the junction of the second intermediate.

For the puckering string arrangement of this trawl, 31 steel rings, two and one-half inches by three-eights of an inch, were strung around the cod end terminal on a one-quarter-inch nylon line. The edge of the cod end terminal retains the double twine selvage woven into it during manufacture. These selvage meshes are bunched three or four apiece, the line is passed through them, and a ring is bent onto the line with a girth hitch. This secures the ring firmly and the next is put onto the line and knotted seven inches away from the first. The seven-inch distance between knots is held for the entire circumference of the cod end. With use of the girth hitch, it is possible to remove any individual ring from the line by taking up enough slack with one hand to slip the ring through the bight with the other. Rings may also be replaced in this manner.

Formerly (and some trawlermen still follow the practice), it was usual to secure the running ends of the puckering string to close off the cod end with a knot of some variety, mostly a four- or five-turn hangman's knot. Under the stress of a tow, particularly a fruitful one, these knots displayed a great ability to jam and sometimes two or three men were needed to free them. In addition, the knot had to be retied after every split and the haul was slowed both by the jamming and the need to retie the knot after every split. Several workable metal clips are marketed and most trawlermen now use one or another of them to do the job easily and well.

Drag boats hauling over the side, usually from starboard, cannot lift a heavy catch aboard in a single effort. To do so would rip the net beyond reasonable chance of repair at sea, so the net is hauled and emptied in splits or sections, with splits averaging from 1,500 to 3,000 pounds or even more for some heavy-duty nets. (The stern ramp trawler, almost unknown on the West Coast, winches all its net aboard, including intermediates and cod end, since the catch need only be eased up the ramp rather than swung into the air on a boom end.)

(This NMFS trawl actually is remarkably uncomplicated in comparison to most otter trawls. The net works as well or better than older models while, at the same time, it is free of most of the clutter of lines found on many trawls no larger and some smaller than this advanced design. The British are especially adept at complicating their trawls with a variety of lines that has made its collective way across the North Atlantic to New England and the Canadian Maritime Provinces but are virtually unknown on the Pacific side of the continent. In addition, this trawl has fewer segments than most British trawls and most trawls of the Western Atlantic. However, the British historically have been responsible for the basic design of the otter trawl and many of its major advances. Nevertheless, the NMFS center in Seattle appears to have surpassed the British in some respects, especially in the development of relatively-simple trawl designs such as this one, in specialized trawls such as the shrimp separator, in the development of giant mid-water trawls and their opposite number, the tiny smelt trawl of the Columbia River. International competition aside, the Seattle base is the United States leader in trawl development.)

The splitting strap is designed to choke off a portion of the cod end to allow its load to be dumped on deck and it is the splitting strap gear that is hooked up to block and tackle to lift the net from the water when the split is begun. In this trawl the splitting strap is a one and one-quarter-inch nylon line reeved through seven rings distributed around the circumference of the lower third of the cod end. The bottom two are hung to the webbing at 19 meshes from the terminal, the next two at 20 meshes, the last pair at 21 meshes, in

an elliptical effect that puts less stress on the net during the lift than if the strap ran at a right angle to the circumference of the cod end.

The rings are set at equal distance around the circumference, however, about 20 meshes apart. The rings, six inches by one-half-inch, are not hung directly to the webbing but to five-strand rope patterns called "spiders." They resemble, in effect, the stylized Japanese design of the sun with rays streaming from it. The spider legs are sewn mesh for mesh at equal angles from each other on the bars and the rings are secured in the center by a clove hitch.

Each spider is 42 inches long (in this trawl) from end to end before the ring is secured.

The splitting strap itself has a tiny eye splice in one end and a shackle or figure-eight clip in the other so it will not slip through the splice. With this metal fixture, the splitting strap is connected to a second line, either heavy manila or nylon. This connecting line runs forward along the axis of the top of the trawl to a ring set in a spider some eight feet above the splitting strap installation. One or two aluminum floats of six- or eight-inch diameter are secured to this line. The ring is called the "pickup" ring and it is the ring used to hoist the cod end aboard to begin splitting. From this ring there runs forward to the leading edge of the overhang a lighter line called, variously, "tag line," "pull rope," "haul-up line" or "lazy line." It is secured to the head rope bosom by a double half hitch or clove hitch or any other knot desired. It also may have one or two floats attached to it. The only function of this line is to control the movement of the cod end so it may be guided into position for hoisting. The line is removed from the head rope as soon as that part of the net comes up to the reel, and is secured lightly until it is needed.

With puckering strings and splitting strap assembly complete, the body of the trawl must then be bent to breast lines and head and foot ropes. In practice, these jobs and any other work may be done simultaneously. Work procedure depends usually on the manpower available.

The breast lines are hung to the wing ends in somewhat the same manner that the net segments were joined by half-meshes. However, these hangings are longer and of heavier twine, ordinarily, No. 48 or heavier since the leading edges behind all of the forward lines are subject to great strains. This is most true of the webbing behind the foot rope since it is or should be in contact with the bottom. Each mesh is knotted individually to the breast line with the meshes lain evenly against the line so there will be no distortion of the wing tip configuration.

Since it is of importance, at least to the conscientious webman, that webbing and breast line come out even with no webbing left wild, the midpoint of the wing tip and the midpoint of the breast line are determined and matched before the hanging begins and hanging proceeds, in most cases, outward from the midpoint.

This meticulous matching must be carried even farther when the head rope is bent to the top wings and the bosom because of the much greater span to be covered and the consequent greater chance for error. The eye splices of the head rope are matched and its exact center is located and marked with a length of twine. The midpoint of the bosom is similarly located and marked as are the meshes that will become the junction points of the wing tapers and the bosom. The tapers are measured accurately and marked at desired intervals and are matched against the head rope between the junctions and the eye splices. If the tapers have been cut to the correct measure, the webbing will end at the eyes. If there is an excess, it must be picked up in the hanging. If it falls somewhat short, it may be left at that. Before the hanging begins, the marked meshes along the tapers are secured to the head rope so that the webbing can be divided equally among these key tiedowns. The hanging itself may be done from either the wing tips or from the junctions although most webman work inward from the eye splices.

The belly of this NMFS trawl was hung directly to the foot rope by the hangings already described but this is not an especially common way of doing the job. Most often used is the "bolch" line, an auxiliary line of three-eighths or one-quarter-inch wire to which the body webbing is hung by a series of mesh by mesh knots. The bolch line, in turn, is hung to the foot rope by hangings similar to those used to secure the lead line of the purse seine.

Some trawls use either of two chain devices by which to stir their quarry from the bottom. The shrimp trawlers of the Gulf of Mexico use a chain called a "tickler" which is cut from five to six feet shorter than the foot rope. It is secured to the foot rope the appropriate distance inside the eye splices and at the center point of the rope. So hung, it moves ahead of the foot rope in a triangular configuration at a distance calculated to flush shrimp from the bottom and into the mouth of the trawl. The tickler is a necessity for the southern shrimp fishery since fishing is done at night because the species taken burrow into the bottom by day. The West Coast shrimp fishery, including that of Alaska, is a daytime affair. But the tickler and variations have been adopted by western shrimpers, however, because the chain effectively stirs up the pandalid species of shrimp sought in that fishery.

The tickler chain seldom, if ever, is seen on western fish trawls. When such a device is desired, a "dropper" chain is used. This chain is seized directly to the foot rope between the wing tips at intervals of about 20 inches and drops from it in loops up to about 18 inches deep. It is used fruitfully on trawls fishing specifically for flatfish because many of these species tend to lie right on the bottom when inactive and many individuals may partly work their bodies into sand or mud when in this state.

Chain also may be used by trawlermen, in the absence of telemetry or photography, to determine the height of the foot rope above the bottom. This is done by use of single dropper chains of varying lengths hung from the foot rope. Observation of the "scrubbing" effect on the links after a tow gives the fisherman an accurate indication how his foot rope is riding.

Among trawl rigging long used on the East Coast is roller or "bobbin" gear which has come increasingly into favor with western draggers since the mid-1960's. These assemblies are intended to allow the trawl to move over rough or steep ground without excessive use of power and with a minimum threat of damage to the trawl itself from bottom obstacles. Bobbin complexes are cumbersome to handle on deck and they are not especially compatible with the drum trawling systems used by most West Coast draggers. But they do allow nets to work hazardous grounds more easily and the extra work they entail on deck during setting and hauling is more than offset by less net damage.

The bobbin has two forms, one to be used along the foot rope bosom, the other to be strung under the wings. The bosom bobbins look much like the thread bobbins of the sewing-machine but they have a flat outer surface five inches across or more rather than the flanges and sheave of the home-maker's bobbin. In actuality, trawl bobbins are heavily-structured hard rubber wheels with metal spokes and an axle passage, the opening for the cable on which they are strung. The rubber of the outer structure is much like that of heavy duty truck tires while the spoke and axle structure is built of high tensile bushing stock. These bosom bobbins run from nine to 24 inches in diameter. The 18-inch bobbin is the one most used. It weighs, in its standard size, 17 pounds on deck but rather less in the water.

The torpedo-nosed wing bobbins resemble exactly one half of an egg with their one flat surface and their egg-like nose, the forward or leading element of the bobbin. They have essentially the interior structure of the bosom bobbin, modified somewhat to meet the peculiar requirements for this bobbin. In standard sizes, they are 18 inches long and 18 inches in diameter and weigh 25 pounds on deck.

The wing bobbins are shaped and hung to move on the bias, that is, they must be able to move in the direction of the tow while, at the same time, they are working under a wing that may lie as much as 60 degrees off the axis of the bosom of the foot rope. Because of this angular movement, the wing bobbins do not roll as do the bobbins strung on the bosom. Instead, they are dragged along the bottom and their shape stems from this requirement.

The bobbins are hung independently of the foot rope or the bolch line. They are strung on a five-eighths or three-quarter-inch twisted steel cable which is shackled into the eye splice of the foot rope at one wing tip and runs the length of that line to the opposite wing tip. At intervals throughout its length and specifically at the junctions of the wings and the foot rope bosom, the cable is hung to the foot rope by 18-inch dropper chains.

The number of bobbins varies according to the width of the bosom and the length of the wings. The trawl under discussion, if it were to be fitted with bobbins, probably would take four along the bosom and five along each wing. The bosom bobbins would then be spaced roughly two feet apart. The wing bobbins would be spaced three feet apart.

The bobbins are separated by any one of several types of spacers, ranging from wood to lead-rubber compounds. Some are cylindrical while others resemble an over-sized and elongated purse seine cork. The commonest type used on the West Coast is a three-pound spacer of the latter design strung by twos and threes to attain the desired separation. These may be complemented by cable clamps, one on each side of the bobbin, which lock tightly onto the bobbin line and prevent the roller from slipping to right or left. The use of these clamps reduces the number of other spacers.

All trawls must use flotation devices of some type to keep the head rope high to achieve a desirable vertical opening. These may be hollow floats of glass, iron or aluminum and, of the three, aluminum is the one in standard use. This NMFS trawl uses 15 six-inch floats of that metal with one centered on the head rope bosom while the others are spaced equally at five feet in each direction from the center float. (The hoped-for vertical opening when the pilot model of this trawl was designed was put at nine feet. Diver observation and measurement on its test tows showed the vertical opening to average 10.5 feet with a horizontal opening of 38 feet. The designed horizontal opening was 36 feet. This relatively simple trawl was quickly adopted by fishermen of Washington and Oregon and went into commercial use on a number of boats in the spring of 1972 although the pilot trawl did not go into the water from the *R/V John N. Cobb* for its first workout until December, 1971.)

There are several methods of securing the floats to the head rope, among them the use of a short line from the float seized to the head rope. The most usual method and the least complicated is the one used on this trawl, a series of turns of twine around the head rope and through the float grip, secured by another series of turns at a right angle to the first at the float grip.

With the hanging of the floats, with the trawl bent to its framing, its assembly is complete. A bale of webbing, reels of line and wire, and bronze and steel fittings have been turned into a net of great potential if it is fished to its full capability by a man who knows exactly how to go about the job.

## From Trawl To Boat

There is only one relatively uncomplicated piece of gear (in form, at least) between the tip of the trawl wings and the vessel's drum and gallows on which the net and its auxiliary gear rest between tows. There are, actually, two pieces of this gear, a pair of otter boards or doors, terms used interchangeably in the same breath by drag fishermen. The boards are of many sizes and designs the world over and, as with all fishing gear, continual experimentation goes on wherever the boards are used to arrive at a more efficient form. Primitive versions of the otter board go far back in history as indicated in the Scofield anecdote earlier in this section. But the door did not begin to assume its modern forms (and the plural is used purposely) until the coming of steam power to drag vessels around 1860 in Great Britain.

It would not be stretching the truth unduly to assert that every long-time West Coast dragger uses a different otter board. Many are tailored in some degree to the fisherman's own design; others are production models that have, for a time at least, approached something like standardization. But these, too, may and usually do show something of the trawl skipper's own notions about what makes the door work best. In the early 1970's, six major types of doors with a myriad of modifications were under study in the world's major fishing nations. No national chauvinism was apparent in this endeavor; otter boards patented in Hong Kong, a design differing markedly from those most in use at that time, rated high on the list of the boards that worked most effectively.

Modifications of the Hong Kong design originating in South Korea appeared in the United States and, since these doors resembled outwardly a door then being built on the West Coast, patent infringement suits were filed in several instances and the builders of these boards became remarkably reluctant to discuss the ancestry of their doors in writing. But it appeared, after several years of legal jousting, that the design of the doors lay in the public domain and that was the end of the matter.

The first trawls worked with the boards secured directly to the tips of the trawl wings. Later, mainly to provide a means of herding fish toward the net, the boards were set at varying distances ahead of the wing tips and connected to them by ground or sweep lines. This was the Vigneron-Dahl method of trawling and it still is the one most widely used except for such specialized nets as the smelt trawl and most shrimp trawls. Most western fish trawls use about 30 fathoms of ground line between the wings and the boards. The scope of the warp depends, of course, on the depth being fished and it may run from two to one to seven to one with the least scope used in greatest depths.

The first otter boards used with powered draggers were simple rectangles of wood with an iron shoe on the underside of the door to add weight to that edge and to protect it from the bottom. By the 1970's, the door most in use was one of the modifications of the vee-design which, in turn, seemed to have a strong strain of South Korea in its background.

This door is of all-steel construction, in shape a cambered vee steel sheet with a torpedo nose leading edge, framed top, bottom and at the leading edge by a solid steel round which gives it great strength with a weight

Marine Construction and Design Company

*Modern steel trawl doors, with a weight of about 700 pounds, about six feet long and four deep. The boards lie in their fishing position, starboard board at left, port board at right.*

of about 1,200 pounds. Those used for offshore dragging with heavy nets are about five feet wide and about nine feet long. The vee is shallow with a rise from the center line to the framing of about eight inches. Under tow, the vee faces inward. These boards are not interchangeable; one is intended for starboard, the other for port. The Hong Kong design can be switched to either side but running gear and weights on the bottom edge make it difficult and dangerous to handle on deck.

Rigged to the vee-side of the door is a three-leg bridle. The aft pair is of equal lengths of chain secured by a ring and swivel to a U-bolt or padeye welded to a steel plate which, in turn, is bolted to the body of the board. The third leg, the shortest and leading leg, consists of two steel members, a male and female piece. The female or sleeve segment ends in a ring welded closed after it is passed through its padeye. The male member ends in a flat-shank opening shackled to the chain legs. It is secured in the sleeve by a heavy bolt passed through the entire diameter of both segments. Tension of the tow thus falls well forward.

On the offside of the board are welded two U-bolts or padeyes on steel plates set closely to the trailing edge. From these run a two-leg bridle or tail chain assembly secured to the ground line by a G-hook and flatlink when under tow. This off-set positioning of the bridles on the two sides of the boards produces, under tension, the angle necessary for the boards to pull away from each other and open wide the mouth of the trawl.

Some board designs call for a change in the towing rigging. They use a heavy steel bail in place of the tow-

ing bridle. The bail is shaped almost like that of a water bucket, running through a 180-degree arc. It is positioned along the apex of the vee and is fixed to the board at both ends and can swing on those points. The ends lie equidistant from the board ends. At about 45 degrees from each end of the bail, the steel is studded with from four to six holes through which warp shackles can be secured. This number of choices allows changes to be made in the towing angle of the board until experimentation finds the one best suited to the vessel and the net. It is a design liked by research workers because of the flexibility offered by the changeable locations for the warp fastenings.

## Use Of Trawls

The "shooting" of a net is one of those hoary phrases of the English-speaking world's fisheries that does not mean what it seems to mean—the explosive discharge of gear into the water to put it to work. Rather, the shooting or setting of any net—gillnet, purse seine or trawl (the term usually is used only for trawls)—is a relatively slow and carefully pursued endeavor with every care taken to insure that the net goes into the water cleanly and easily. Any net launched suddenly into the water is a net quite likely to be in trouble. The only gear set with any great speed is the longline and because of the inherent tendency of the hooks to foul the main line, whole skates of gear may go into the water in a mare's nest and remain there in that state until the string is hauled.

The appearance of the trawl drum on the West Coast in 1954 and its subsequent adoption by most draggers quickly transformed the traditional fishery into one of less drudgery and less danger. Setting and hauling became faster and easier. (This period of the early 1950's was a fruitful period for West Coast fishermen, and eventually many of the world's fishermen, because in the four years after 1950, Mario Puretic's Power Block, the seine drum, the trawl drum and nylon netting became proven and useful gear.) The drum, with its quick haul, made more tows possible during any given period of time while, at the same time, it reduced most crews from four or five men to three. Some smaller vessels cut down to two men while one-man operations on inside day boats are not unknown although they are not common.

The trawl drum resembles the salmon seine drum but it ordinarily is smaller. The reel itself runs from six to eight feet between flanges on most draggers while the flanges are about four or five feet in diameter. Some larger reels on big vessels have been divided by a center flange so they can carry two trawls, one rigged for bottom dwellers, the other aimed to those species that ordinarily haunt waters somewhat higher. Some reels have been fitted with two-piece flanges bolted inside each outer flange on which the ground lines may be level-wound by feeding them through a slot on the edge of the flange as they begin to come aboard. This arrangement allows less monitoring of the ground lines and manual level-winding as well as a more even distribution of the net on the reel.

Most drum trawlers use an unpowered stern roller built into the stern bulwark with a rise just above the level of the rail. It runs the length of the reel or somewhat longer and markedly reduces gear chafing as it is set and hauled. This stern roller system is used by drum seiners and the same roller conceivably could be used in both fisheries if necessary as a vessel moves from one to the other with the changing seasons. Some seine roller assemblies extend beyond the after beam of the vessel by as much as six inches or so to allow maximum latitude for smooth handling of the seine as it comes across the roller through the fairleads. This protrusion of the roller and its framing sometimes causes minor complications when similar vessels lie rafted together or when they must lie along the walls in lock chambers to await movement from one level to another. The roller does not, however, extend beyond the greatest width of the vessel.

Almost all drag boats use hydraulics for powering winches and drums. On the western combination vessel rigged for trawling (and this includes almost all West Coast trawlers), the trawl winch is located in the traditional center line spot just abaft the deckhouse. The winch may be of two or three types but most modern trawl winches can be used for seining as well. These winches coil each warp on its own drum. Two men can work these winches handily while on advanced models one man can do the whole job from a remote control console.

From their drums, the warps run athwartships to blocks abeam of the winch, thence aft along the bulwarks through one or more sets of blocks or fairleads, rising to a trawl block hung in the gallows or davits at the quarters where they are led to the otter board bridle or bail. The warps and the blocks are a major source of accidents aboard draggers. The trawl drum has its own power and control system but, on the rig described above, it too can be operated from the same remote station.

Setting of the trawl begins by freeing the cod end and pulling enough of it from the slowly turning drum to pass it across the roller and into the water while the vessel maintains way so following seas will not wash the floating webbing into the rudder assembly or the wheel. With the cod end in the water, it exerts enough tension on the freely rolling drum to pull the body of the trawl after it. Floats and bobbins must be watched as they come off the reel to avoid fouling. Engine speed is slow ahead.

When the net is fully in the water, the reel is checked until it is certain all components are riding properly and in balance. The brake again is released and the ground lines are run out until the flatlinks and swivels that secure them to the boards are on the stern roller. There comes into play then an auxiliary cable or chain that has only one function—to safeguard the transfer of the ground lines to the boards on the set and their removal on the haul. This inconspicuous but vital piece of gear is called an "idler" and in the succession of gear wound on the reel, it precedes the ground line, lying between it and the standing cable on the reel. The idler is G-hooked into the ground line flatlink at all times during the trawl cycle—when the net and its lines are on the reel, during the set, during the tow and through the haul. But in the course of that cycle, the G-hook on the other end of the idler undergoes a change of position.

At this point in the set, with ground line terminal gear steady on the roller, the G-hook on the tail chain of each board is hooked into the ground line flatlink, with the idler G-hook left in position. Now the ground lines are safe—they cannot be lost. The brake of the drum again is released and the idler runs off until the weight of the gear in the water rests on the tail chain assembly. This allows enough slack in the idler itself so that it can be unhooked from the reel and hooked into a padeye or U-bolt on the inner surface of the door. Most doors have this padeye near the upper edge. The board described earlier has it positioned just above the peak of the vee. During the tow, then, the idler dangles—idles—between the ground line and tail chain junction and the fitting on the door.

With ground lines clear and the idler in place, the winch drums up the warps enough to lift the boards slightly from their position on the gallows where they have been secured by a short length of chain. The free end of the chain carries a hook which bears the weight of the board. The winch action allows this chain to slack off so the hook can be freed manually. When the boards are clear, the weight of the gear shifts to them.

The doors are lowered away, the engine still at slow ahead, to a depth of from five to 10 fathoms where they ride while they are checked for conformity as they begin to spread under the pull of the warps. Engine rpms are then run up to two-thirds or three-quarters throttle, not a considerable speed when the drag of the gear and the probable maximum nine- or 10-knot unloaded speed of the vessel is considered.

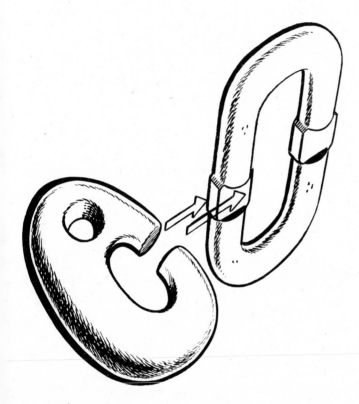

*G-hook, much used, heavy-metal line connector, most often found among trawl gear.*

The scope of the warp must depend, of course, on the depth to be fished and when the last 75 fathoms or so of this necessary length are still on the winch, speed is backed down easily to slow ahead to avoid any sudden loss of warp tension and the resulting probable fouling of gear or loss of balance. This speed is maintained until the last of the warp footage is in the water and the winch can be braked and dogged down. Engine speed then may be slowed even further for three to five minutes to allow the trawl complex to settle gently to the bottom. The virtue of this grace period has been learned through telemetry. Trawl fishermen no longer need work almost literally in the dark after their net drops from sight if the vessel owner wishes to invest in the needed elements. Telemetry, properly used and understood, allows a full reading of the attitudes of all major components of the gear. Almost all West Coast draggers operate by sense of touch, however, because of the cost of equipment.

When the doors reach the ground, warp tension and tremor or telemetry tell the skipper his net is down and fishing…if it has touched down in its normal configuration with the doors upright and functioning at their full capacity. In this connection, a knowledge as complete as possible of bottom currents contributes to a full fare. In comparatively deep water, current direction may not necessarily be the same as that at the surface on many grounds. These are areas of daily tidal extremes, or where rivers such as the Columbia influence current patterns over thousands of square miles or where inland salt-water bodies such as Puget Sound, its contiguous American waters and the Canadian waters north ebb and flood through the Strait of Juan de Fuca.

All manner of unpleasant things can happen to a bottom fishing net and sometimes they do. A loss of tension on the warps and ground lines can cause a door to fall onto its side to be dragged uselessly along the ground with its proper angle of approach to the direction of the tow lost and beyond recovery. In most cases with many types of boards, the gear must be hauled to reset the board properly. The possibilities of damage to the net itself from bad bottom need not be described.

And, paradoxically, a trawl with too many fish in it can be as upsetting under some circumstances as a trawl with not enough payload in it to balance off the cost of the oil used during the tow. Trawls (and purse seines too) suffer seam or webbing rupture because of overloads during the haul when considerable strain touches every element of the gear. This has been particularly true at times during the off-again, on-again, gone-again hake fishery of the Washington coast and Puget Sound.

## Hauling The Net

The length of the tow is the prerogative of each skipper, a decision to be based on his knowledge of the ground he is working, by the fish he "sees" on his recording echo sounder and by any problems that may be introduced by wind and sea. Tows may run from 30 minutes upward. (Accurately-timed tows of 30 minutes or equivalent periods are used in research fishing when comparative readings are wanted for different nets or other gear working as nearly as possible under identical conditions.)

Retrieving the gear begins by drumming the warps aboard while the vessel proceeds at slow ahead, enough speed to keep constant and even tension on the warps. But there is a minority of skippers, a very small minority on the West Coast anyway, who prefer to lose way and let the vessel be pulled sternways over the boards and trawl until the boards are clear of the bottom. But, except for this minority, this maneuver is used only in an attempt to free the net when it is snagged on some bottom obstacle. Sometimes this retrograde movement succeeds in clearing the net with little damage or none at all. At other times, the skipper may have no choice other than to use all the power necessary to rip the net loose with full knowledge that it may be ruined beyond the point of practicable repair or that hours of patching may be needed to make it usable again.

When the boards are off the bottom, they are hauled fairly rapidly until they are about 10 fathoms below the surface. The haul rate is slowed then until the warps bring them up tightly against the gallows as far they can move—"two-blocked"—so they can be hooked to their place on the gallows.

Since it would be difficult and dangerous to attempt to work the cod end with the warps taut and high from winch to gallows, the warp on the side of the vessel to be used (either port or starboard according to rigging and the skipper's preference) is slacked off until it can be tucked safely out of the way below the rail. The position of the warp when taut may vary from vessel to vessel, that is, in its run from winch to gallows it may ride higher above deck on one vessel than on another. On some vessels, the warp may run athwartships to a block, thence aft through one or more blocks almost at deck level or at least below the level of the rail before rising steeply to the gallows block.

When boards and warps are out of the way, the G-hook at the after end of the idler cable is released from the otter board fitting and hooked into its flatlink on the standing cable on the reel. The reel turns again until the idler goes around it to the point where the board's tail chain assembly can be reached and the ground line can be freed from it to follow the idler onto the ree. Reeling continues until the ground lines are aboard, to be followed by the wings and the body of the trawl until the junction with the intermediate is on the stern roller.

Two somewhat different systems may then come into use, the method dependent upon the type of trawl winch used. In the most modern of these installations, boom winches are not used and all handling of the intermediate and the cod end is controlled by the trawl winch. But most vessels in the West Coast drag fleet rely on the older of the two systems wherein vanging and topping winches control boom movement and auxiliary winches on the boom or gypsies on the trawl winch handle the boom tackle.

At this point a cable is led from a block on the boom end to a three-foot chain strap which has been passed around the intermediate at a point determined by its position in relation to the reel. This, in turn, is predicated by the apparent catch, that is, an empty intermediate or one nearly so will be wound further onto the reel than one heavy with fish. The winch controlling this cable is located, on most vessels, on the undersurface of the boom about one-third of its length from the mast. The winch also controls the tackle used for splitting.

When the cable is fast to the strap, any slack is taken up and the vessel is swung almost on its own vertical axis to starboard and downwind of the gear in the water. At this point, if necessary, the drum is backed off so the cod end and the intermediate can be walked around the gallows as the vessel swings to put it into position for working. The engine is shifted to neutral.

The intermediate is then hoisted gallows high or higher to determine the exact volume of the catch and to allow the skipper to decide if the whole cod end can be swung aboard (just as the lightly-loaded fish bag of the purse seine bunt is brought aboard if brailing is not worthwhile) or if it must be split one or more times to accommodate the catch.

If splitting must done, the lazy line is used to maneuver the cod end to a position where the pickup ring can be hooked into another cable running from the winch on the boom. At the same time, the line from the boom-end block is moved to a desired position on the upper or lower intermediate to hold it while the slack of the intermediates and the cod end is returned to the water with the surplus of the catch.

The splitting line hoists the loaded cod end terminal across the rail, the puckering string is pulled and the load is dumped on the deck. The string is pulled tight, the clip secured and the boom-end tackle hoists or bounces the cod end sharply to re-charge it. The procedure is repeated until the entire catch is aboard. Each split should require about two minutes, a time dependent upon crew skill, the smooth functioning of all gear and the state of the weather. When the cod end is clean and the puckering string and the lazy line are secured, the boom tackle is cleared and the remainder of the net is wound onto the reel to await the next set.

But meanwhile, the true hard work of drum trawling proceeds; the catch is sorted, "trash" fish, sometimes most of the catch, go over the side dead or injured beyond recovery and market fish are iced down. Sablefish usually are dressed and the wings of edible skates are trimmed from the carcasses for a ready market, sometimes to appear as "scallops" or as "halibut cheeks" or as the prime ingredient in a "crab" louis in second-rate fish shops and eating places.

The waste of fish from the "run of the sea" dragger catch is a shameful affair in a hungry world but the trawl fisherman of the United States and Canada has no choice. There is almost no industrial market for his reject fish while the American housewife, because of her refusal to buy and use more than a few choice species of fish attractively displayed, has been called the American fisherman's "greatest enemy."

## PURSE SEINES

The purse seine probably has made more money for its fishermen than any other gear used in the Northeastern Pacific. It appeared on the East Coast in 1826 and it is old in the salmon fishery, where it has earned many millions of dollars over the years for the men who

Robert Browning

*Skiffman, skiff and seiner starting haul; young man and heavy skiff pull vessel clear of her entangling net. The vessel,* Ocean Mist *of Seattle, shows end-of-season-scars.*

work it. The purse seine gave new life to the moribund American tuna fishery and it is being used profitably in the adolescent California anchovy fishery. The seine is a versatile and cranky piece of gear, one that can be moved right along with the fish with no more cost than some few gallons of oil. It can be fished from a boat as small as 40 feet or from one as large as the larger United States tuna vessels. The seine working smoothly can trap a school of fish quickly and certainly. Or it can foul itself up somewhere and haul nothing but water and junk. The seine is expensive to build but it can repay its creation many times in a single season. No matter the catch or its value, the seine has a fixed cost and the cost goes on, fish or no fish. The seine is big, awkward, clumsy and wonderfully efficient in the hands of a crew who know it and anticipate its moods and do not trust it too far.

Over an average of seasons, purse seines and drift and set gillnets account for about equal catches of salmon through the areas where this gear is fished. For example—in Alaska in years of heavy pink runs and average or less than average red runs, the seines tend toward bigger catches because their usual catch is the pink. In years when reds are dominant and pink runs are down, the gillnets may come out ahead. In the three seasons, 1965 through 1967, the Alaska seine catch came to 361 million pounds of salmon. The gillnet catch was 352 million pounds for the same three seasons. Both 1965 and 1966 were good red seasons on Bristol Bay while 1966 was one of the best pink seasons since World War II.

The troll catch always is small compared to the catch by the other gear. During the 1965-67 seasons, the Alaska troll catch totalled only 29.3 million pounds. The troll fishery, however, has a certain high value of its own because of its concentration on chinook, pound for pound the most valuable to the fisherman of all Pacific salmon.

Use of all this gear is severely regulated and some observers have described such regulation as "legislated inefficiency." Of this trend, one critic has remarked:[1]

"Through the years, laws have been written to perpetuate the salmon resource, provide a maxiumum sustained yield and, at the same time, keep the employment rate at the highest possible level. Purse seines have been limited in length and depth and mesh size as well as by area in which they may fish.

"These limitations are not necessarily governed by biological or economic factors except in the apparent attempt to decrease maximum efficiency and lower yield per unit of effort. Gillnetting is faced with the same limitations...a mandatory minimum mesh size of five and three-eights inches has been established for Bristol Bay. Not only is this impractical but the measurement is controversial in itself in terms of how it is done and under what conditions and what means. The effect of the law allows the harvest of the larger-size fish and theoretically provides for the escapement of the smaller members of the species. A prolonged course of action such as this certainly opens to question the effects of heredity.

"Law eliminated certainly the most efficient and non-selective method of catching salmon when, for all practical purposes, fish traps were outlawed. Ignoring the political implications, both fixed and floating traps in many areas undoubtedly provided not only the most efficient method of catching fish but also the most satisfactory method of regulating escapement. This method provided the lowest cost per unit of effort at the time when numbers of units of mobile gear had not been developed to the point where they are today.

"If efficiency is to be increased, legislation must be developed whose main consideration is perpetuation of the resource and let the chips fall where they may as far as economic factors are concerned. Historically, without the passage of restrictive legislation, the American economy has weeded out the inefficient and marginal operators in almost every industry. If our goals are to produce salmon at the lowest possible cost, it would appear that our laws are in need of drastic revision."

## From Muscle To Power

Mid-20th Century technology has given the present day seine fisherman advantages his father or his grandfather might not have been able to envision a generation or so ago. Perhaps the single most important advance has been the development of hydraulic systems that allow him to work his seine with fewer men and less sweat.

Hydraulic power has had two major applications in seining in the Northeastern Pacific. One is its use as

the heart of the Power Block invented by Mario Puretic in the early 1950's, a smoothly-running block that removed salmon seiner crewmen from their status as the successors to the galley slave. The first Puretic blocks were used in 1955 and their quick acceptance by salmon seiners marked the beginning of a revolution in the world's purse seine fisheries.

The other important application of hydraulics to seining was its use as the power system of the seine drum. The drum (its principles have been applied to trawling too) controls the seine, setting it and hauling it with a minimum of labor by crewmen, and on rainless days men need not bother with rain gear. They have no seine swaying overhead on the haul to shower them with water, jellyfish and all the debris the seine picks up.

The Puretic block and the power drum arrived almost simultaneously in the American fisheries of the Pacific Coast although the drum concept was rather old hat in British Columbia then.

National Marine Fisheries Service
*The salmon drum seine configuration, counterclockwise from upper left—the vessel, the set, pursing, hauling, setting.*

One Puget Sound seiner mounted a drum in 1950 despite the scoffing of fishermen who could not see that the still-awkward drum had an exciting potential for the taking of salmon. The scoffers were outnumbered by the believers after a couple of seasons and in the winter of 1952-53, 11 more Puget Sound boats converted and fished the drum successfully in the 1953 season. All of these early drums used mechanical power takeoffs but hydraulics engineers were already at work on new systems, and later conversions and all new construction intended for drum installation took hydraulic power.

Enthusiasm for the drum cooled somewhat when it was outlawed in Alaska because its most active proponents were seiners from Washington State who did and still do much of their fishing in Alaska. The original crackdown on the drum, first by the Bureau of Commercial Fisheries before statehood and by the Alaska Department of Fish and Game after statehood,

---

1. Turnbull, Ward, notes taken by author from presentation of paper titled "Where Management Requirements Impose Inefficiency," at The Future of the Fishing Industry in the United States conference, March 24-27, 1968, University of Washington College of Fisheries.

did not demand that the drum assembly itself be removed from the seiner while fishing in Alaska. Drum seiners merely worked in the conventional manner with the Power Block and stacked their seines between the flanges of the drum when they fished in Alaska. But in the early 1960's, removal of the entire assembly was made mandatory.

The drum was too efficient for Alaskan officialdom. It took too many salmon too fast. But seiner operators were not too much bothered by the Alaska ban because they could still fish the drum in Washington State. And they had another tool that was almost as efficient as the drum. With it, they could set and haul at least twice as fast as they had been able to do in the days of power rollers and hand-strapping of the seine to get it aboard. This tool was the Puretic Power Block.

Mario Puretic was a Yugoslav who left his homeland as a kid to make his stake in the United States. He worked at many things back and forth across the States before he applied his talent as inventor to the Power Block, a device that turned the world's seine fisheries upside down.

*Mario Puretic*

Before Puretic and his Power Block, it took eight to 10 men to work a salmon seine. It was work of the hardest kind. It took too much time too and time is ill-spared in the fleeting salmon fishery. And the average share on a boat with 10 men is a lot less than the share on a boat with five or six men, all it took when Puretic got done.

Puretic, who had been fishing sardines, mackerel and anchovy out of Southern California when he devised the original Power Block, was no stranger to hard work. It had been his lot for most of the time he had been in the country. But as a working fisherman, a hard-working fisherman, he had been becoming increasingly concerned about the difficulty of hauling nets. Where other men were content to complain about it, Puretic characteristically did something about it. He began designing a work saver.

It took him only several months of 1954 to come up with what eventually evolved into the Power Block.

This prototype was an oversize rope-driven block to be secured to a boom end or a gallows in a net vessel's working area. The seine was to be started through the block by a lead, up and over and down, cork line, lead line and web and rings, handy to the men on the stern waiting to stack it. No more hand-hauling, no more spine-twisting labor, work that made men old before they should have been...

The rope drive was not what Puretic had in mind for the prototype but economics interfered here, just as economics so often plays a part in the success or failure of any novel undertaking. Even Robert Watt, Robert Fulton and Alexander Graham Bell had money problems. In his only first-hand account of the development of the Power Block, Puretic wrote:

"The rope drive was a temporary way of motivating the prototype to see whether or not the net would behave as I visualized because even the most expert of fishermen insisted the net would go crooked. In my original patent, there was no rope drive mentioned and it was developed...for initial economy. Hydraulic components were very expensive for a new venture and an uncertain business. No one would put a boat at my disposal to experiment with hydraulics...and engineers whom I met at the time suggested I use a power plant landed (loaded) aboard ship so as not to involve the ship's installation and that would have added further costs."

The prototype got a trial on the *Anthony M* out of San Pedro and the system worked. It worked beautifully. But after the trial no one on Puretic's home grounds was remotely interested in it. No one would give it a look. Puretic truly was the traditional "prophet without honor in his own land." All this as he carried around with him the rough design for a device that would be as important to the world's fisheries as was Eli Whitney's cotton gin to the slave owner, as was Cyrus McCormick's reaper to the scythe-weary farmer. It should be remembered that most of the world's fish are taken by the purse seine.

Although Southern California's fishermen may have looked down their noses at the Power Block, men of the northern salmon seine fleet were further-seeing than their fellows to the south. Puretic wrote:[1]

"Early in 1955 after the first trial on the *Anthony M*...the news traveled fast and small groups of fishermen from Puget Sound and British Columbia flew down to San Pedro to see what they couldn't believe. When they returned home, they gave MARCO my address and immediately MARCO wrote me, inviting me to Seattle...I came to Seattle with the prototype Power Block in the trunk of my car."

MARCO was a young and enterprising shipyard and marine gear company called Marine Construction and Design Co. The company used the acronym

---

1. Puretic, Mario, letter to the author, October 25, 1969.

MARCO in its advertising and it was enterprising because it wanted to keep from starving to death.

MARCO had been founded two years earlier by Peter G. Schmidt Jr., of Olympia, Washington, who went into his enterprise fully aware that the old and accepted approaches were quite likely to result in failure. Seattle had been a shipyard town for almost a century when MARCO went into business, a town where competition was as fierce as in any shipyard town on the Pacific Coast or any other coast. Schmidt and his engineers were not waiting for business to come looking for them. They went looking for business and on their drawing boards there already were system and gear plans no one in Seattle or elsewhere had thought of before. But MARCO was not wrapped solely in its own thoughts; it was ready to listen to the thoughts of others too.

*Peter G. Schmidt Jr.*

In Seattle, the Power Block got the recognition denied it earlier. While Peter Schmidt and Puretic put together a financial agreement, MARCO engineers began to plan for production of the Power Block. The prototype "was suspended on top of the boom on the salmon boat *Bullmoose* (lost in July, 1969, in Alaska's Icy Strait district with two of her crew...Ed.) and, as the fishermen observed the performance, MARCO was taking hundreds of orders even before the Power Block was put on the drafting board..."

So many seiner men wanted the Power Block "right now" that MARCO had to assign numbers to impatient skippers on a first-come, first-served basis to keep things on an even keel. That was 1955. By 1960, a skipper in the northern seine fleets who could not boast a Power Block was regarded as something of a poor relation.

The final design, and still essentially the present-day design with varying modifications and sizes for various fisheries, was an aluminum vee-sheave with non-skid rubber facing driven by a medium-pressure hydraulic unit. The sheave was contained in an aluminum housing about four feet long, two and a half feet wide and 18 inches through. It weighed less than 200 pounds.

The story of the Power Block did not end with the salmon fishery. From the MARCO yard on Salmon Bay in Seattle, the Power Block has swept around the world. Modifications of it are used in every major seine fishery. The Power Block and the concurrent development of light and long-lasting synthetic materials for giant seines generated the conversion of the American tuna fleet from bait boat to purse seine in no more than three years after 1959. (The *Anthony M* was one of the first of the California boats to hang up a Puretic block.)

By 1972, well more than 11,000 Power Blocks were in use in 40 or more countries. This included Soviet Russia whose engineers "invented" the Power Block immediately after they first saw one outside the Iron Curtain.

The United Nations Food and Agricultural Organization in Rome estimates that more than 40 percent of all the commercially-caught fish in the world—both food and industrial fish—are taken by the Puretic block.

The collaboration among Mario Puretic, Peter Schmidt and MARCO might easily be called one of the most fruitful since Adam and Eve joined forces.

Marine Construction and Design Company

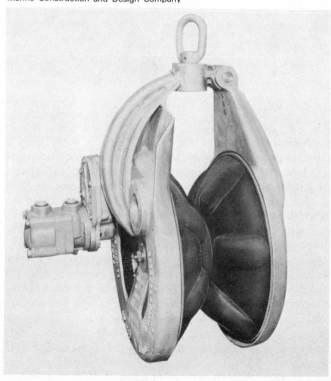

*Power Block, of the size used for maximum-legal salmon seines.*

## Salmon Seines

Purse seines in the Pacific salmon fishery tend to fall into three general types because of the somewhat different circumstances under which they are fished. The major design might be called the Puget Sound seine, a long, deep seine of a usually-standard build although it, like all nets, often is subject to a difference in detail according to the fisherman who will use it or to the net loft superintendent who may actually have the chief hand in its layout.

A second major type is the "jitney" seine of Kodiak and other Alaska areas, short, shallow seines tailored to regulations of the Alaska Department of Fish and Game. These regulations are realistic because these seines are fished usually in shallow, foul-bottom areas where seines of the Puget Sound variety would be sadly out of place.

The third type, the one least seen, is the drum seine of Washington State and British Columbia, a short, rather shallow seine built specifically for the drums of the vessels that have been built or converted for their use. This seine is built as a rectangle with cork and lead lines of almost equal length rather than with the shorter lead line of the conventional seine.

The legal requirements for the Washington seine are put forth in schematic form in that state's commercial salmon regulations:

"Lawful purse seine salmon nets in Puget Sound shall not exceed 1,800 feet along the cork line while wet, and purse seine and lead combined shall not exceed 2,200 feet. Neither shall contain meshes of a size less than four inches nor shall the meshes of the seine and lead be lashed together to form one continuous piece of webbed gear. It shall be unlawful to take or fish for salmon with purse seine gear which contains mesh webbing constructed of a twine size smaller than 210d/30 nylon, 12-thread cotton or the equivalent diameter in any other material. It shall be lawful as part of the purse seine to have a bunt 10 fathoms long and 200 meshes deep which may contain mesh of a size not less than 3.5 inches..."

In addition, Washington requires that salmon seines used in the fall fishery in certain Puget Sound fishery areas (designated by number under the state's system) be hung with five-inch mesh in the first 100 meshes below the cork line. This strip of larger webbing must run the entire length of the seine except for the bunt. The intent is to allow escapement of immature king salmon at a time when they may be found in large numbers in the fishery areas where the harvest is aimed chiefly at chum salmon. This fall season runs usually from the first Monday in October to Thanksgiving. The passage of sub-legal salmon through Power Blocks or onto power reels is prohibited also.

Alaska limits purse seines to a maximum of 250 fathoms and specifies shorter maximum lengths for some areas. British Columbia allows a length of only 220 fathoms with a maximum depth of 250 meshes, figured on a minimum mesh size of 3.5 inches stretch measure. Washington has no limit on purse seine depths but 350 meshes is the common maximum for seines worked by Power Blocks. Both states and British Columbia allow up to seven-inch mesh in the 25-mesh lead line strip to ease tension on the lead line.

Common Alaska specifications include one for Southeastern with a maximum permissible length of 250 fathoms and a permissible minimum of 150 fathoms. In that area, no seine may be less than eight and one half fathoms in depth and no more than 19.5. Prince William Sound rules call for a minimum length of 125 fathoms and a maximum of 150 with a minimum depth of 9.5 fathoms and a maximum of 17. In the Kodiak area, purse seine lengths from 100 to 200 fathoms may be used at the option of the operator. At least 50 fathoms must be 150 meshes deep with a minimum depth of 100 meshes.

The 250-fathom Southeastern seine noted in the preceding paragraph probably is as representative as any seine in the salmon fishery. It is five strips deep through the body for a total of 300 meshes of four- and six-inch measure. It tapers to two 100-mesh strips of the requisite 3.5 inch web in the bunt. Bunt webbing is of No. 36 or No. 42 nylon seine twine. The cork line selvage strip is 25 meshes deep, four-inch measure, No. 21 twine. The next lower strip is 100 meshes, four-inch measure, No. 15. The next strip, the mid-strip, the strip from which the taper is cut to the bunt, is 50 meshes, four inches, No. 15. The fourth strip is 100 meshes, four-inch, No. 15. The lead line is 25 meshes deep, six-inch web, No. 36 or 42 twine. The depth of this seine when fishing should be about 18 fathoms along the body.

## California Seines

The California seine fishery is confined chiefly to mackerel, both the scarce Pacific species, a true mackerel, and the prolific and abundant jack mackerel, one of the jacks and not a true mackerel but a fine food fish; the swarming squid and the masses of anchovy, the fish that took over the ecological spot left empty by the virtual disappearance of the Pacific sardine.

California has fisheries regulations more restrictive, perhaps, than any other state or province of the entire Pacific coast and banned to the purse seine are scores of fishes, including the salmon which cannot be taken by nets of any type anywhere in California.

California seines are cut from the same bale of webbing as their companion pieces of more northerly waters. Nylon and other synthetics have replaced cotton; nylon has replaced manila and plastic has replaced Spanish cork. Length and depth are regulated more by the exigencies of the fishery than by official fiat, in contrast to the strict rules cited above for salmon seines of the northern tier of fisheries.

Mackerel and anchovy seines are much like the herring seines of Washington, British Columbia and Alaska with mesh size much smaller than that allowable in any waters for salmon. California mackerel seines commonly use one and three-eighths-inch stretch measure mesh while the tiny anchovy calls for a stretch measure of 11/16 of an inch. An average mackerel seine measures out at about 275 fathoms on the cork line and about 36 fathoms stretch measure in depth.

The purse seine now is the dominant net of California waters although that was not always the case. The seine's clear-cut lead over all other roundhaul nets dates back to the beginning of World War II.

California's first purse seines were introduced into the state's waters in 1894 by the fishing vessel *Alpha* (a fitting name for a pioneer in a burgeoning industry) whose skipper was working sardines and mackerel for a San Pedro cannery. The *Alpha* fished one seine for sardines, the other for mackerel. The sardine seine was 120 fathoms long and eight fathoms deep with one-inch mesh while the mackerel seine was 135 fathoms by 17 fathoms with two-inch mesh.

The purse seine's pre-eminence in California fisheries did not come early or easily as it did in the northern fisheries for salmon and herring and, somewhat later, the sardine. Just as the clumsy paranzella net, towed by two boats, eventually was driven into oblivion by the more efficient otter trawl, it appeared for some 30 years that the purse seine might be driven out of California's fisheries by an interloper from the Mediterranean.

This was the lampara net (described earlier), introduced into California from Tangier in North Africa in 1905 by Italian fishermen seeking a net simpler and easier to handle than the cumbersome purse seine with its complement of purse line, purse ring bridles and rings, all elements of the net subject to fouling on almost anything they happened to brush.

The lampara was lighter, quicker to set and haul and it needed fewer men to work it, a virtue always uppermost in the minds of skippers and/or fishing vessel owners. The lampara was a supreme success in the sardine fishery of California, specifically at Monterey and San Pedro. In 1915, the lampara drove the last seine out of the Monterey seine fishery. In San Pedro, the lampara fished sardines for the next decade while some 125 purse seines there were devoted to tuna and the so-called white fish—barracuda, yellowtail and white seabass, all of them now "verboten" to the seine.

But the lampara was not destined to fish smooth waters forever as its devotees began to learn along about the tail end of World War I. Over the years, fishermen using the lampara experimented endlessly with variations on it in the interest of speed and economy of operation. The lampara spawned two offspring, hybrids that were fathered by the purse seine. Like many hybrids, one was of no particular value, while the second drove its lampara parent out of bigtime fishing.

In 1917 and 1918, sardine crews at Monterey and San Pedro experimented with little success with the placing of pursing gear on the bunt of the lampara. This was the "half-ring" net and in that World War I

form it ran into something of a dead end. But over a span of years, the use of the half-ring net began to spread as hanging techniques improved.

But this form of the hybrid did not meet the exacting demands of fishermen and by no later than 1930 the rings had been extended along the full length of the lampara's lead line. This was the true ring net, the one that drove its parent over the hill to the poorhouse by 1940. In 1929 at Monterey, there were 56 true lamparas, two purse seines and two ring nets. By 1940, the ring nets were in the ascendancy although their day in the sun was to be a short one too.

In the middle age of the sardine fishery when the grounds were close to the canneries, the light-weight lampara had things its own way because of its simplicity and its need for fewer men and smaller boats. But eventually the sardine had to be sought even farther from its old haunts and here the seine vessels had an edge because of their greater size and power. With the concurrent development of the power roller (not the previously-mentioned drum) for hauling the purse seine, its advantage over the hand-hauled lampara became insurmountable. By 1950, the purse seine was in the majority everywhere big nets were legal in California for fishing. By 1960, the lampara and its hybrid descendants had been driven into the bait fishery for sport and commercial use. No longer do they play any role in major fisheries.

What waits in the wings for the purse seine?

U.S. Department of Interior

*The purse seine: fully set; with rings coming up, (lower left); hauling by hand, the old way; hauling through Power Block, (upper right).*

## The Construction Of Purse Seines

The theory of purse seine fishing is simple enough; the practicalities are something else. The purse seine can only be described as one of the most complex and contrary pieces of gear used by fishermen anywhere. It exemplifies Murphy's Law reading "If anything can possibly go wrong, it will."

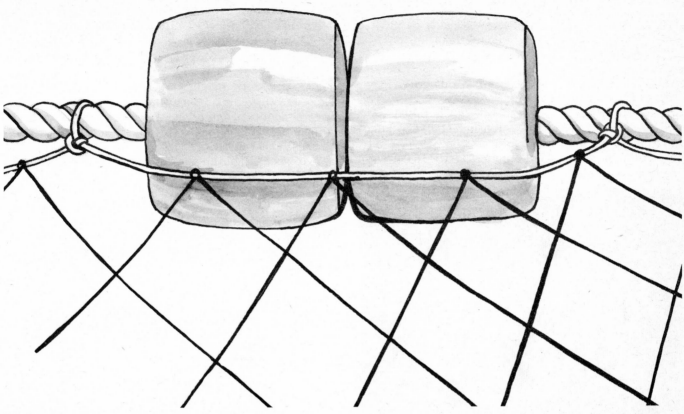

Linda Rogers

All purse seiners necessarily must work on the same principle, that of surrounding and trapping fish by closing the bottom of the net. But almost every seiner skipper has somewhat differing ideas on such matters as the details of construction of the seine, its balance and its handling when being fished.

Most men have experimented with various methods of working the seine. Some of these variations on a theme work out, some do not and, to an unfortunate extent, many fishermen stick to the old and tried but not necessarily the most productive methods.

Some of these varying practices are compelled by the size of the seiner or by the arrangement of her standing and running rigging. The conventional or block seiner faces one set of problems; the drum seiner faces an almost entirely different set even though both boats are using similar seines with the same principle to catch the same species of fish.

The seine is fished at the surface with the upper edge of the webbing held there by the buoyant cork line. Almost no cork is found in cork lines any more; it has been replaced by tougher and lighter synthetics.

The bottom edge of the seine is weighted and held down by the lead line, heavy lead sleeves set along a manila or nylon line, or line built with a core of lead. Attached by bridles, usually of one fathom, is a series of rings through which a rope called a purse line is threaded. The purse line closes the bottom of the seine when a haul begins, trapping within it the fish it has surrounded.

The depth of a salmon seine depends on bottom conditions and water depth to some extent. Foul bottom and shallow water would seem to call for correspondingly limited depths of webbing. But adding to or subtracting from the depth of a purse seine is a time-consuming affair. It can be done the hard way aboard the seine vessel but this takes even longer than it takes ashore. To offset this problem of bottom contact, many salmon seines are built with a taper in one or both ends. This tapering narrows that part of the net and allows it to be fished in shallow water close to the beach with a minimum of fouling.

Ripped webbing is an every-day affair in the seine fishery and even under good conditions, a rarity, much crew time often goes to patching web. A major hangup can result in hours or even days of lost fishing time, something ill-afforded in the often-brief salmon fishery.

Purse seine design and construction is a matter of simplicity to men who understand an apparently-complex procedure. But it is not a matter of simplicity to the novice and often enough, many young fishermen learn by rote the hanging and mending jobs they must do without having any real awareness of the principles they are employing in their work.

To the uninitiated, a purse seine reduced to its component parts, scattered seemingly at random in a net loft or in a net yard, appears as a bewildering array of lines, corks, leads, webbing strips and rings, an assortment of gear that looks as if it could never be put together into an efficiently-functioning taker of fish.

But the purse seine (or any other big net for that matter) is not as complex as it might appear to the untrained eye. It is a rather primitive device in comparison to other instruments or gear used at sea. The principles on which purse seine design is based were well worked out more than a century ago. The purse seine of that day has been improved mostly by the substitution of new and better materials for those used

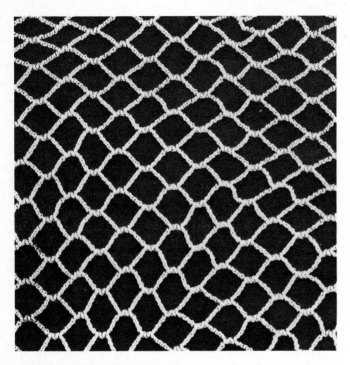

*Knotless webbing of Japanese manufacture. Small-mesh webbing like this can be as decorative as it is useful.*

in all nets through so many ages. Underwater viewing by divers during the years since World War II has resulted in some minor changes in design, but by and large, the purse seine of today closely resembles the purse seine of 1826, that year it first showed up in the United States.

Machinery does today what generations of fishermen did so laboriously by hand. Machinery weaves the webbing of the net and lays or braids the lines that work it. The ancient art of net weaving is still practiced in some parts of the world where machine-woven netting is not available or the fishermen cannot afford it. The machine does the job quickly with a degree of skill the hand weaver cannot approach. It would take a score of fishermen hundreds of hours to weave the webbing for a seine 300 fathoms long and 20 fathoms deep, par-

ticularly if that net were to be one like the anchovy net with its tiny mesh. Much of the netting material used in the United States comes from Japan. The Japanese machines produce netting materials faster and cheaper than most of that produced in the United States.

The Japanese, with their centuries of tradition as a fishing nation, have turned net making into a true art and American fishermen, whatever they may believe about Japanese fishing practices on the high seas, appear to have no qualms whatsoever about using Japanese-made net materials. Large advertisements for Japanese netting appear in American fisheries magazines and newspapers side by side with stories and editorials castigating the Japanese for such things as net fishing for American red salmon native to Bristol Bay. The incongruity bothers no one.

The Japanese produce both knotted and knotless webbing and this latter netting, weaved in almost all strengths, has found much favor with fishermen, especially for such heavy nets as seines and trawls where great strains must be endured. Knotless webbing appears, however, to have more slippage than the knotted. Slippage in heavy knotted webbing can be offset considerably by tarring, a process that firms up knots and lays the tiny fibers common to synthetic webbing. Light netting like that of the gillnet commonly no longer is tarred.

All webbing used in present-day fishing nets, at least in the nets of American and Canadian fishermen of the Pacific, is of nylon or related synthetic materials. These are light, strong and less susceptible to damage of all kinds. This virtue is apparent in the tropical tuna fishery where the useful life of a cotton seine was put at one year. Nylon's life is estimated at at least four years and usually more. The substitution of nylon for cotton has cut shark damage to nets in this warm water fishery because, size for size, nylon webbing is stronger and more resilient than cotton.

Manufacturers of netting supply web strips in lengths as customers order them. The commonest lengths on the Pacific Coast are 200 fathoms, 360 and 400 fathoms with the strips for salmon seines running 25, 50 and 100 meshes deep.

All seine and gillnet regulations make much of measure by mesh size, a subject that has always been something of a matter of disagreement among fishermen themselves and between fishermen and enforcement men. Mesh size "customarily" is measured from the inside of one knot of a mesh to the outside of the knot diagonally opposite when that mesh is stretched. This is the oft-mentioned "stretch measure" and the ease of obtaining the exact length of a mesh appears obvious.

But all is not as obvious as it sometimes seems and any man who would measure a mesh must consider the material from which the mesh is woven, whether the mesh be wet or dry as well as the poundage of the pull exerted on that mesh when it is being stretched. Any one or all of these elements, singly or in combination, can produce a difference in mesh measure that can be of considerable importance to the fisherman. This is especially true in gillnet measure where the difference, a very possible difference, of an eighth or a quarter of an inch in measure can contribute much to fish gilled or fish escaped.

The inherent complications in mesh measurement have been described like this in reference to the California method of measuring between two diagonal knots tightly stretched: [1]

"...a two-inch mesh (stretched measure) opens up into a square one inch on each of the four sides. This open mesh in a net usually is hung diagonally so that it appears as a diamond rather than a square. In this case, the distance between the diagonally opposite knots of a fully-opened two-inch mesh is approximately 1.4 inches. This has a bearing upon net length and depth because over-all measurement usually depends upon the degree to which the meshes are opened...a net of two-inch mesh that is 100 meshes deep would measure roughly 200 inches deep if pulled vertically until the meshes closed, but if pulled horizontally,

probably has not changed greatly for many hundreds of years although like most gear connected with fishing, it is now made of materials somewhat more advanced than wood or bone. Plastic became the preferred material after World War II because it is light, durable and inexpensive. It does the same job as that performed by the bobbin on a sewing machine—it holds an easily-renewable supply of twine, enough to let a webman work for five minutes or so without the need to refill his needle. And, if he wishes, he can carry a dozen or so needles in his pockets so that he may work without pause for a considerable time.

The oldest needles were flat and most needles still are and this is the type preferred by most West Coast fishermen. Needles in the shape of a flattened cylinder are seen too, these used most often in the lofts of net makers because they can be filled by machine in the time it takes an ordinary fisherman to reach for a roll of twine. The people who make a business of nets prefer these so-called "round" needles because they get more work per man per hour from them by virtue of the fast loading.

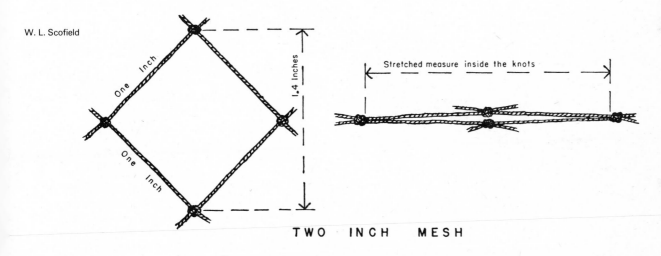

W. L. Scofield

One Inch · One Inch · 1.4 Inches

Stretched measure inside the knots

TWO  INCH  MESH

*A two-inch mesh, open (left) and stretched. This points up variables inherent in web measure and consequent difficulties. Common yardstick is "stretch measure."*

the depth would be about 12 inches, 100 times the depth of one knot. If the same net hung with each mesh squarely open, the depth would be 140 inches, 100 by 1.4. This means that net measurements of length and depth are generally only rough approximations..."

This is why net depths usually are specified by meshes rather than inches and lengths by fathoms because mesh size can be many things to many men.

## The Needle

One of the oldest tools of the fisherman is the needle or shuttle which he must use to weave or mend his nets. Its antiquity must approach that of the fishhook and it is an instrument of great ingenuity. Its essential form

Flat needles run in width from about one-half inch to about one and one-fourth inches with a length of from four to 10 inches. The round needles (and fishermen call them round even though they are flattened cylinders) tend to run somewhat longer although their widths across the flat face are comparable to those of the flats.

Both have dull points to facilitate the needle's passage through the web. The round needle has two converging tongues at the point and the twine feeds from

1. Scofield, W.L., "Purse Seines and Other Roundhaul Nets, in California," California Department of Fish and Game, 1951.

between them. The flat needle has a single tongue recessed in an opening proportional to width and length of the needle just behind the point proper. The twine feeds from this recessed tongue. Both types have recessed "heels," a detail that makes the off-end of the needles resemble a squared U. This shaping is necessary to keep the twine from slipping from that end of the needle. A filled needle, accidentally dropped, will lose no twine by unwinding. The needle itself is fool-proof; not so the fingers of the novice trying to use it.

All round needles and some flat needles have a small hole just above the heel. This is to engage the running end of the twine with a pass-through and a lay-over of the first round of the twine to secure the start of the loading. When flat needles do not have the hole, the loading is started by running a turn of the twine around the tongue and securing it with a layover of the first round of the twine.

To continue the loading of the round needle, simply run the twine from point to heel and back until the

*"Round" net needle.*

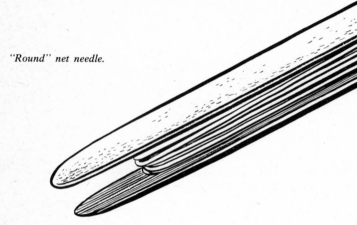

needle is filled. To continue the loading of the flat needle after the initial securing of the twine, run the twine down and around the heel and up to and around the tongue, movement which becomes, in effect, a series of U-turns. The twine must be pushed down on the tongue with the fingers of the hand holding it so that the tongue may take its full complement of twine. Reasonable tension must be maintained to fill the needle properly.

## The Biggest Saucer

The properly-designed purse seine, with the exception of the drum seine, is assumed to have a saucer shape in the water after the set has been closed and pursing has begun. This assumption may not necessarily be the real situation because tidal currents can twist a seine out of any resemblance to what its designer planned for it. This shape of the seine is dependent upon a lead line shorter than the cork line and in the ordinary seine this difference is about 10 percent, although this figure may vary somewhat according to the judgment of the de-

signer or the wishes of the owner of the net. This shaping of the net makes for quicker and cleaner pursing as well as for easier lifting of the rings after the pursing is completed. In addition, the shorter lead line helps to keep the cork line from sinking during pursing because of the lesser ratio of lead to cork. It is the tendency of the cork line to sink during pursing of the drum seine that leads some vessel operators to heavily cork the bunt end of the seine. Some drum seines have double cork lines along much of their length although this extra corking contributes to crowding of the drum by the time this stretch of the seine comes well aboard. This double corking may extend for as much as 200 fathoms of a 250-260-fathom drum seine.

## To Build A Seine

There were described earlier in this seine section the dimensions of the Icy Strait seine of Southeastern Alaska. This seine is used generally throughout the Alaska Panhandle and seiners from out of the state, from Puget Sound, for example, must substitute this seine for their longer and deeper Washington seines. There are more seines of this size, probably, than any other single seine of the Pacific salmon fishery. It has the 10-fathom bunt, the longest allowed in the salmon fisheries of the United States and Canada, while its other dimensions fall midway between the big Washington seines and the jitney seines of Alaska.

This seine, to be used here as a model, has, as noted, a taper toward the bunt end of the seine, not in the bunt itself but beginning between two strips of equal mesh size and strength at the junction of the bunt webbing and the webbing of the body of the seine. Tapers began to appear in salmon seines about 1955 when a growing number of fishermen began to realize that such a seine can be fished closely off the beach with less chance of fouling. A deep bunt end touching bottom almost always manages to foul itself in one way or another as it rolls across the bottom. Purse lines snag, webbing itself hangs up and too often a great mass of the net rolls up in the lead line, a mess that literally must be manhandled to clear up.

Among the materials needed to build this seine are cork line and lead line material of five-eighths-inch polypropylene seine line. The maximum length of this seine is 250 fathoms on the cork line proper but allowance must be made for an adequate amount of tow line at each end of the seine, a minimum of 15 fathoms on the seiner end and somewhat less on the bunt end, the exact amount of surplus at each end to be left to the discretion of the designer. The lead line, 10 percent shorter than the cork line, requires at least 225 fathoms of line. But if the lead line is extended to become the breast line at each end of the seine, a method sometimes used, some 20 fathoms more must be allowed for each breast. More commonly, the lead line terminates with the webbing, and nylon or manila line of five-eighths or one-inch diameter is used for the breast lines.

American and Canadian purse seines of the Northeastern Pacifc are constructed of horizontal strips of webbing. Salmon seines are laced together in strips running from 25 to 100 meshes deep with 100 meshes the maximum depth usually supplied by the web makers, although these latter, like the purveyors of almost anything found in the market place, will supply, within their ability to deliver, almost anything a customer may desire.

This net, with the webbing to be hung 12 fathoms to 10 fathoms of cork line, requires, then, 300 fathoms of webbing. But this webbing will not be delivered in 300-fathom strips because three mesh sizes are involved. The bunt is 3.5 inch measure; the lead line strip is six-inch measure and the balance is four-inch.

Jerry E. Jurkovich

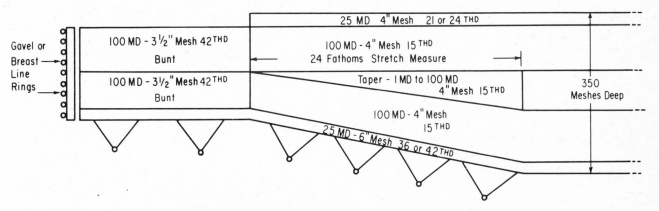

*Diagram of bunt end of 250-fathom "Icy Strait" salmon seine, the type most used in Southeastern Alaska. Note the taper ending at the beginning of the bunt.*

The nature of the purse seine requires that it be buoyed at the surface and this requires a flotation material of some kind. This seine was designed to be floated by plastic "corks" shaped in a shallow oval five inches long and four inches in diameter. This seine uses about 1,450 of them.

Conversely, the bottom of the seine must be held in the water by an off-setting weight to produce the perpendicular wall of webbing that hopefully will surround a substantial tonnage of fish. The lead line may be weighted with sleeve leads of five-eighths-inch diameter, weighing four ounces each. The Nordby design is preferable. If sleeve leads are used, this lead line will require from 2,700 to 3,600 of them to allow for a spacing of from 12 to 16 to each fathom of the lead line. Lead line weight may be supplied too by the newer, cored lead line, nylon line with a lead core weighing from three and one half pounds per fathom to four and a half.

The design of this seine calls for 12 fathoms of web to each 10 fathoms of the cork line, a standard ratio for seines. (Standard Pacific Coast gillnet designs use from 200 to 300 fathoms of web to each 100 fathoms of cork line, although 220 by 100 is the usual ratio for salmon nets.)

The bunt will be cut to 12-fathom length; the lead line strip may or may not come in its full 250 fathoms while the four-inch body webbing may be cut to length from longer strips or laced together from shorter.

This webbing and other materials break down into the following elements:

Two strips, each 12 fathoms long and 100 meshes deep, of 3.5-inch mesh woven of No. 36-42 nylon seine twine.

One double selvage cork line strip 25 meshes deep and 300 fathoms long of four-inch mesh of No. 21 nylon twine.

Two strips, totalling 290 fathoms each, of four-inch mesh 100 meshes deep, woven of No. 15 twine.

A double selvage lead line strip 300 fathoms long, six-inch mesh, 25 meshes deep, of No. 36 or No. 42 twine.

A taper strip 20 fathoms long running from one mesh to 50 meshes in depth, four-inch mesh, No. 15 twine. (The paper dealing with the formula for computing the taper will be found in Appendix II, Part 3.)

About 70 snap or round purse rings of bronze, aluminum or stainless steel.

About 420 fathoms of three-eighths-inch polypropylene line from which to fashion one-fathom bridles for the rings.

About 280 fathoms of purse line material with 250 fathoms of three-quarter-inch nylon and 30 fathoms of seven-eighths or one-inch diameter.

About 60 fathoms of three-eighth to one-inch nylon, polypropylene or manila line for breast lines if the lead line is not to be extended to form the breast lines.

A variety of small hardware, such as spring-loaded figure-eight clips and other items, to be placed in the seine at their various required positions.

Enough hanging twine, in sizes from No. 21 to No. 60, to properly sew and lace these seine components into a functional net.

To hang this seine, there is needed a work area long enough to allow work on a 10-fathom stretch at a time. This 10-fathom measure is a standard unit used by all purse seine fishermen and seine hangers and all elements of the lines and webbing are keyed to this length. The width of the work area is not particularly important

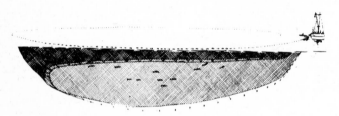

*Optimum shape of purse seine after set has been completed and before pursing has begun.*

other than to allow some freedom of movement of the webmen. One major net loft, for example, working with salmon seines, utilizes an area about 10 feet wide and 80 feet long, quite sufficient for the 10-fathom stretch and stowage of materials at each end. Men working in the open usually take up more space simply because it is available. On the other hand, much gear work can be done on the crowded deck of a seine vessel as it travels, with strips being added or removed or needed patching accomplished.

Auxiliary equipment useful but not necessarily required includes such fixtures as X-frames to support cork and lead lines for convenience while they are being worked; fixed steel posts with rings welded into the tops for the same purpose; low frames with horizontal cross-pieces at each end of the 10-fathom length for ease of measurement, and fathom length units of measure—wood doweling, lath or metal rods. The professional net loft may use hooks suspended from the overhead for line support.

All material for this seine should be placed at the end of the 10-fathom stretch at which work will begin. Webbing is supplied in cases; corks are cased too; lines come on reels; leads, if they are to be used, are shipped

in casks containing 50 pounds each; rings may be supplied in crates, barrels or sacks.

It is usual but not invariable for all web strips of a new seine to be laced together before the body is hung to the cork and lead lines. This practice is more common in the professional loft than among vessel operators doing their own work. Here, webbing ordinarily would be laced in 10-fathom lengths and hung to the lines on each stretch. This necessarily would be the case in any web work situation on a vessel at sea.

Webbing may be sewn or laced as the need is indicated. When sewing is done, webbing is picked up on the needle and knotted on each mesh or on a series of meshes picked up. When webbing is laced the needle is reeved through a series of meshes along the edges of two strips to connect them loosely with knots thrown in at the option of the webman at distances running from a few inches to two or more feet apart. Similarly, webbing may be hung to the cork and lead lines by lacing or by knotting at each hanging.

Many knots may be used in webwork but those most often found in lacing and sewing are the clove hitch and its variations and combinations, or variations of the half hitch. A double half hitch is the easiest of all lacing and sewing knots to use but it has a tendency to slip under strain. A most solid knot has been described by an experienced webman (but not an expert in knots per se) as "one half hitch with two half hitches thrown in front of the first to lock it." The clove hitch and the reverse clove hitch, or knots that appear at least to resemble them are seen too, but the truth of the matter is that many webmen, old hands though they may be, do not know the proper name for a knot they may have been using for a generation. This observation applies mostly to the knots used for sewing and lacing because the knots used in web patching, for example, are few and well-known. The commonest knot in web patching is the weaver's knot, a complicated affair at first view but one that may be learned easily by any man who knows he must learn it as quickly as possible.

## The Pickup

The webman must deal in some manner with the difference in lengths of webbing and lines as with the 12 fathoms of webbing to 10 fathoms of cork line designed for the Icy Strait seine under discussion here. An excess of webbing is accounted for by "picking up" meshes as the webbing is hung to the line. Meshes are picked up by overlapping two or more meshes on the same hanging so that webbing and line come out evenly on the stretch and on the entire length of cork line and lead line.

The rate of pickup of meshes to hang-in excess webbing often enough is a matter of eye expertise and this may work well enough for a few fathoms. But the exact pickup and the space between hangings or tie spaces

*Text continued on page 173*

# PLATE 35

*FIGURE 1— The graceful gaff-headed schooner, Jennie Decker, built in 1901, was prettier than many of the halibut dory schooners. Most were "baldheaded," carrying no topsails as the Jennie is flying. At 53 feet she was smaller also. She shipped an engine in 1907 and worked into the mid-1950's as a trawler.*

Pacific Fisherman

Pacific Fisherman

*FIGURE 2— Life was dangerous and sometimes short in the dories of the schooners and steamers. This sailing dory was 14 feet long and many men were lost from such as this one in the days of the dory fishery.*

# PLATE 36

Pacific Fisherman

Pacific Fisherman

*FIGURE 1—   The dory steamer,* Independent, *built in Tacoma, Washington, in 1901. She worked 14 dories and about 35 men. She was wrecked on Middleton Island in the Gulf of Alaska in 1916.*

Pacific Fisherman

*FIGURES 2 & 3—   Halibut fishing is still hard work as rollerman, above, struggling with reluctant halibut, shows. He is working at port side of a powered halibut "schooner." At left, the western combination boat* Presho, *seasonally geared for halibut, hauls her gear on a quiet day.*

# PLATE 37

Pacific Fisherman

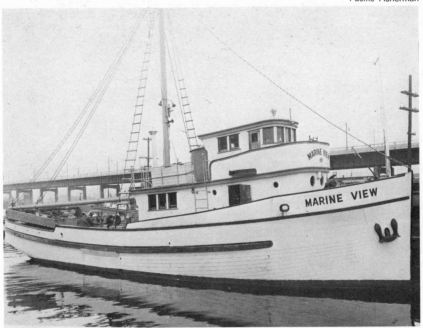

*FIGURE 1— The* M/V Marine View, *built as a sardine seiner, as she is here, found herself out of work when that fishery collapsed after World War II. She eventually was converted into a king crabber.*

Pacific Fisherman

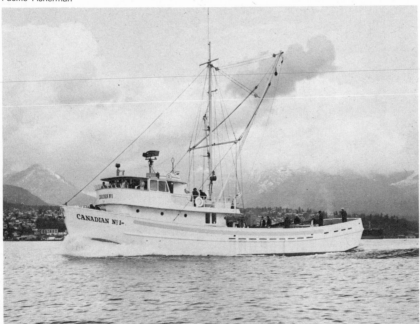

*FIGURE 2—* Canadian No. I *is representative of the big, versatile vessels the Canadians of the West Coast began to build after the mid-1950's. They were some 10 years ahead in that step toward fleet modernization.*

# PLATE 38

Marine Construction and Design Co.

Marine Construction and Design Company

*FIGURES 1 & 2— Assembly line techniques are seen in use in Seattle yard with eight steel Alaska-limit western combination vessels built at once. Above, hull is formed upside down on jig, then lifted, left, and set upright.*

# PLATE 39

Marine Construction and Design Co.

*FIGURE 1— Here, the combination hull on opposite page sits firmly on way, almost ready for launching as workmen apply final touches.*

*FIGURE 2— Salmon seine winch with gypsy and auxiliary gypsy.*

*FIGURE 3— Beach or "pocket" seiner, 42 feet, of type found mostly in Alaska. Power Block hangs from boom end.*

KAREN MARIE
KODIAK

Marine Construction and Design Co.

Ron Rawson, Inc.

# PLATE 40

Marine Construction and Design Co.

*The rigors of the king crab fishery are seen here as crabber crewman
works crab block. Vessel rolls to starboard as pot bridle appears.
Block automatically compensates for stresses.*

# PLATE 41

Robert Browning

FIGURE 1— *King crab pot, seven feet square and three feet deep, with marker buoys and lines coiled within, ready for loading aboard fishing vessel.*

Marine Construction and Design Co.

FIGURE 2— *Pot launcher at starboard rail of* M/V Royal Viking, *launched in 1972. This heavy framework receives and cradles pot between hauling and setting. Note crab block on davit with hydraulic lines sprouting from below decks. Lines service booms also.*

# PLATE 42

Tacoma Boatbuilding Co.

Tacoma Boatbuilding Co.

*FIGURES 1 & 2— Two generations of tuna super-seiners, Caribbean, 167 feet by 21 feet, with a capacity of 750 tons of tuna, launched in Tacoma, Washington, in 1962, and Apollo, 258 feet, 2,000 tons capacity, launched there in 1970. The Caribbean and her sisters of the Hornet class gave Americans supremacy in the world tuna fishery after 1962.*

# PLATE 43

Robert Browning

Robert Browning

*FIGURES 1 & 2—The shrimp trawler,* Dawn, *with the unlikely home port, Minneapolis, Minnesota, lies at pier in Seattle, Washington.* Dawn *was the first stern ramp trawler to enter the Alaska shrimp fishery (or any other Alaska drag fishery) when she appeared in the north in 1971 after a journey from the Midwest.*

*FIGURE 3—* M/V Olympic, *first of one class of vessels built specifically for the Alaska king crab fishery.* Olympic, *launched in 1968, was 94 feet long. Later crab vessels from the same yard went to 110 feet.*

Marine Construction and Design Co.

# PLATE 44

*FIGURE 1— Seiner crewmen hang their seine on pier at Fishermen's Terminal, Seattle, Washington. After breakdown for winter storage, the entire net must be re-hung. Cork line lie in foreground, web strips at left.*

*FIGURE 2— Seiner skipper knots hangings for web to new nylon lead line with lead core.*

*FIGURE 3— Fisherman hangs selvage strip to purse seine cork line. Hangings lie between each cork.*

# PLATE 45

Robert Browning

*FIGURE 1— Young fisherman displays cork line detail.*

Robert Browning

*FIGURE 2— Junction of purse seine cork line, heavy line at left, manila breast line, with davit loop held up.*

*FIGURE 3— Seine cork line hangs taut on 10-fathom stretch with selvage strip webbing "pointed" to proper position for hanging, right.*

Robert Browning

# PLATE 46

*FIGURE 1— Fisherman laces strips of webbing together.*

*FIGURE 2— Needle work in web strip lacing.*

*FIGURE 3— Round purse ring seized to one-fathom bridle with double half-hitch.*

# PLATE 47

Robert Browning

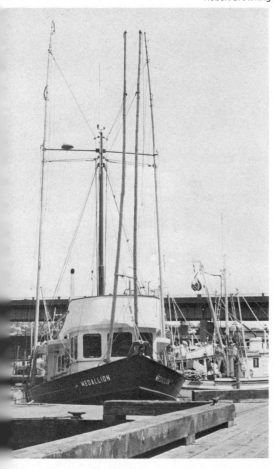

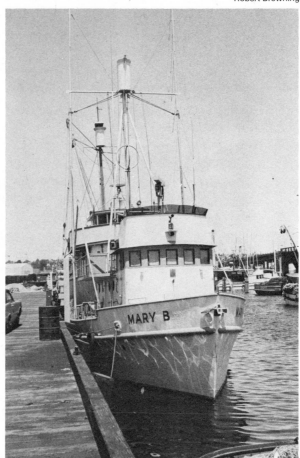

*FIGURE 1— Troller* Medallion, *52 feet, all steel, with aluminum poles.*

Robert Browning

*FIGURE 2— Albacore troller* Mary B, *a conversion from a Navy hull.*

Robert Browning

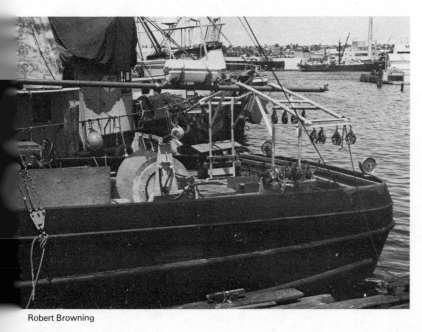

*FIGURE 3— The trolling pit area of the* Medallion, *showing port side gurdy assembly, troll blocks and albacore line haulers on transom.*

# PLATE 48

Robert Browning

*FIGURE 1— Port side gurdies aboard troller* Medallion. *They are hydraulically powered.*

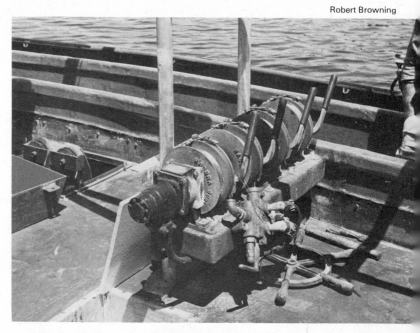

*FIGURE 2—* Medallion's *starboard troll pit area seen from forward.*

Robert Browning

*FIGURE 3—* Medallion's *checkers are all-aluminum and self-draining.*

Robert Browning

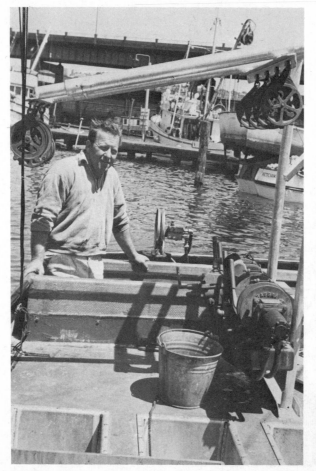

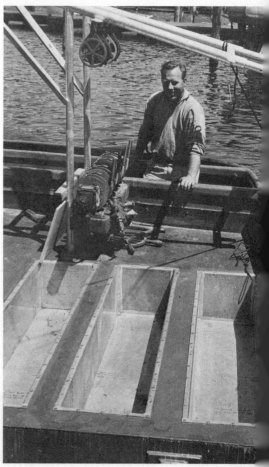

# PLATE 49

shermen's News

*FIGURE 1— The bowpicker gillnetter grows up. Fast and efficient craft like this began to appear in the fisheries after 1970.*

*FIGURE 2— Conventional gillnetters* Otter *and* Moose, *built for Alaska, are all aluminum with gasoline power. They are fast and durable.*

Pacific Fisherman

# PLATE 50

Robert Browning                                    Marine Construction and Design Co.

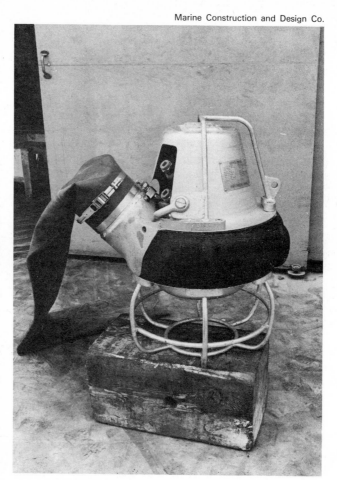

*FIGURE 1— Gillnet fishermen hang corkline.*

*FIGURE 2— Fish pump designed for any fishery where the quarry is small and caught en masse—squid, herring, saury, menhaden and the like.*

Robert Browning

*FIGURE 3— Tarp covers bull rail at Seattle's Fishermen's Terminal as turning gillnet reel winds net aboard after rehanging for start of the new salmon season.*

on cork and lead line can (and should be) calculated to the mesh if one wishes to pursue the matter that far. In the net loft, it is done. Outside the net loft, this exactness is not always achieved.

To be hung exactly, the webbing should be "pointed" or measured on each 10-fathom stretch by halves, quarters and eighths. This can be done for the cork line by stretching the line taut along the 10-fathom length, then tying on the first and last meshes of the 12 fathoms of webbing at beginning and end of the stretch.

Then the midpoint of the webbing is located by mesh count or by fathom measure and this mesh is tied to the line. Quarter points are next located and tied, and then the eighths. The webmen then hang the remaining web, picking up the necessary meshes as they work along the line. The same procedure may be followed to hang in the greater amount of excess on the lead line.

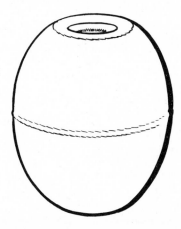

*Oval plastic purse seine cork, five inches long, four and one-half inches in diameter.*

## Seine Flotation

The spacing of corks on the cork line introduces something of a complicating factor in the hanging of webbing to the line because, essentially, the tie spacing is dictated by the spacing of corks. The hanging of purse seine webbing fortunately is not the matter of precision that the hanging of gillnet webbing must be. It is mostly a matter of knotting on the webbing where it may be most conveniently and securely done. No two seines are hung at exactly the same ratio along the cork line unless they are designed by one man out of identical materials assembled to identical specifications.

Corking a purse seine is based on a three to one buoyancy ratio, that is 300 pounds of flotation for each 100 pounds of dead weight. This would seem to mean 300 pounds of cork for each 100 pounds of seine. That, however, is not exactly the case because that ratio actually is based on buoyancy versus dead weight. Much of the seine itself tends to float and adds significantly to the buoyancy of the seine. A length of polypropylene line

pushed beneath the surface of the water pops back to the surface as soon as it is released. To a lesser degree, nylon acts similarly. Actually, almost the entire "dead weight" to be considered is that of the lead and rings.

A rather general rule calls for the use of from 55 to 60 corks to each 10 fathoms of cork line. But this rule is so subject to exceptions that it amounts to no rule at all. There must be considered the size and the type of cork, the condition of the corks (if not new), the buoyancy factor of the corks themselves (their own ratio of dead weight to flotation value) and, quite obviously, the weight of lead and rings. Other factors must be considered too, especially the type of seine, the conditions under which it must usually be expected to fish and, above all, perhaps, the opinion and experience of the man who will pick up the tab for the seine.

The actual spacing of corks on the line is another variable to be reckoned with, again a variable based upon the corks themselves and the opinion of the seine designer or its owner. Corks may be spaced singly and evenly; they may be spaced haphazardly (as they are on many older seines); they may be spaced by twos and threes and even fours. Most older corks, including those actually made of the bark of the cork oak, are cylindrical (as are some of plastic materials) with average dimensions of four inches in diameter by three inches in length. It is this genuine cork that may be found snugged together because such corks do not have, individually, the buoyancy of newer types of cork materials and two or three of the older must do the job of one of the newer. Nevertheless, too many cannot be placed side by side because eight or nine

*"Square" purse seine cork, four inches by three inches. This cork simulates shape of old-style floats of true cork but it, too, is plastic.*

inches is about the practical maximum distance for purse seine hangings in the interest of strength. It is not usual to hang meshes between corks placed too closely together although it may be done occasionally out of necessity.

Obviously then, purse seine corking is not an exact science and not even a third generation computer can make it so. There are still some elements of human experience, still some elements of the interplay of water and corks and webbing that cannot yet be programmed into a computer. Nor is much else of net construction a science. Net making is, more than anything, an art form, one ever subject to experimentation, a form as varied in effect as Beethoven  and the Beatles.

It was mentioned previously that all strips of webbing, both bunt and body, usually are laced or sewn into a whole before the body of webbing is hung to the cork and lead lines. The major exception to this rule, in net loft practice at least, is the sewing of the bunt and lead lines strips to the bunt end breast line in order to "close off" that end of the seine. This relatively simple procedure becomes somewhat more complicated if the uncommon practice of extending the lead line to form the breast line is followed merely because of the presence of an unwieldy length of line in the work area. It is a simpler matter to use the shorter length of breast line proper, no matter what its material may be. Polypropylene would seem to be the best because of its contribution to buoyancy although this virtue is outweighed, some men believe, by the slower rate of seine fall on the set.

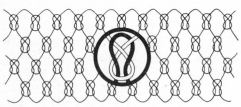

W. L. Scofield

*Strip of web with knots loosened. The enlargement shows weaver's knot. A vertical pull across the strip would run "with the knots," the direction of maximum strength.*

In any event, the breast line is measured exactly to the stretch measure of the bunt webbing and the webbing is sewn to the line mesh by mesh with a knot on every mesh. The 25-mesh lead line strip is sewn to the breast line in the same manner. This mesh by mesh sewing is used to insure all possible strength in this critical section of the seine. Then the webmen work along the bunt and lead line strips for the first 10-fathom stretch (the end of the bunt), sewing mesh for mesh. This procedure uses the heaviest hanging twine required on the seine with No. 42 to No. 60 being commonest. These heavy twines may also be used for hanging to the lead line because that line and the bunt itself experience stresses that ordinarily do not affect the rest of the seine. However, No. 30 and No. 36 are more common here.

With the strips secured to the breast line, the joining of the bunt strips begins mesh for mesh, each mesh securely knotted to its opposite number rather than loosely laced as is the common practice in the balance of the webbing. This procedure is as simple as any involved in the construction of a seine. But matters get somewhat more complicated when the end of the bunt section, the first 10-fathom stretch, is reached.

This second 10-fathom stretch, the first stretch of the body of the seine, presents the webmen with what may appear as an apparently-unsolvable problem. Here, they must deal with a cork line strip of four-inch mesh running 25 meshes deep; two 100-mesh strips of four-inch mesh to be divided by a taper of four-inch webbing beginning one mesh deep at the junction of bunt and body and deepening to 50 meshes through the rest of the length of the seine. Here, for the first time in the building of this seine, the art of picking up meshes must be put into play.

The webmen must reconcile, at the junction of bunt and body, the 200-mesh-deep bunt and the body beginning at 226 meshes deep. It is at this point that the cork line strip is sewn into the bunt. This does not allow for the lead line strip; this strip will be sewn to the bunt and laced to the body and does not figure in this matter. This strip, following the taper, rises to the bunt and breast line at a shallow angle, the virtue of the taper in the bunt end to allow fishing close to the beach.

The body of the seine is sewn to the bunt with a two-mesh pickup of body webbing at every eighth mesh of the bunt to allow for the extra 25 meshes of the cork line strip and the single beginning mesh of the taper. This is the solution of the problem that so baffles the novice. Schematically, as shown in the diagram of this seine, there would appear to be a right angle of cork line strip at the bunt junction. But the pickup of the 25 meshes of the strip will blend it into the bunt in such a manner that the meeting could be seen only if the entire bunt were to be lain flat. In use, the cork line will form one long graceful flotation strip.

On this same stretch, the webmen must begin to deal also with the insertion of the taper. This, cut according to the formula, will be rounded from the actual cut of 432 meshes to the nearest 50 meshes or to 450 meshes, one fathom too long for the stretch. But the taper extends over two 10-fathom stretches. The extra 18 meshes-four inches stretch measure to each mesh — is too little to worry about in comparison to the 432 meshes of the 24 fathoms of webbing of the two stretches or the entire 4,320 meshes of the 240-fathom strip. In normal lacing procedure, the few extra meshes will blend themselves into the total. Care should be taken, however, that the beginning of the taper itself is carefully fitted into the webbing around it. Before lacing begins, the knots along the cut must be cleaned of twine ends to form full meshes and the entire taper must be selvaged. It is recommended that the most experienced and careful of the webmen be trusted with this task,

When this stretch has been completed, the most troublesome part of the assembly of the webbing has been accomplished. The lacing and hanging of the remaining webbing, five strips now and 300 meshes deep at its maximum, is a most routine affair, 240 fathoms of monotony that salmon seiner crewmen endure without pay before each season in order to assure themselves of a berth.

Nevertheless, a certain amount of seine work must be done before and during every season. Area regulations, especially in Alaska, require differing lengths and depths. There must be patching done, corks replaced, webbing sections taken out and new ones put in. There must be replacement of lines and rings lost to foul bottom; everything bad that can happen to a purse seine will happen to it.

Rarely, a vessel operator will hire his own crewmen or professional webmen to rehang his seine during the off-season. But this can be, an expensive affair and most vessel operators much prefer to have the work done for free before the season opens. And most operators prefer also to hang their own new seines on the rare occasions when a new seine is indicated. A seine new from rings to cork line, from bunt to towing end, is seldom seen. Most seines are a hodge-podge of old and new webbing, old and new corks, old and new lines...

While three or four men have been working on the bunt and body webbing, another man or two men may be occupied with more menial jobs. Quite often these men are the youngest or least experienced members of the crew, men to whom such jobs seem to fall naturally, a fact of life not only in the construction of seines but in many other human endeavors.

Corks must be put onto the cork line and, since this seine is a new one, there must be something like 1,450 of them placed by hand along the line. They need not necessarily be spaced properly at this time since such spacing is essentially a function of the hanging of the webbing to the cork line. When corks are being strung on the line, the men doing the job must remember that on seines to be used in Alaska, vari-colored corks must be inserted among them so that they may be spaced every 10 fathoms along the line when the webbing is hung to it. Alaska regulations require twin corks of a color differing from the color of the rest of the corks on the line. Red against white is a popular combination—that is, a pair of red corks every 10 fathoms among the white corks. In actual practice, however, many cork lines may be found wanting in this regard because of fading of colors as well as wear on the corks. All corks of a long-used cork line may be pretty much of the same color.

Another job that rates low in popularity with men putting together a new seine is the feeding of sleeve leads onto the lead line, especially if that line be of manila. Manila, with its tiny, sharp fibers, is unpleasant to work with anyway; it becomes even less pleasant when 3,600 leads must be placed on the 225-fathom lead line of this Icy Strait seine, be it nylon or manila. The usual ratio of leads is 12 to 16, weighing four ounces apiece, to each fathom of line, four pounds per fathom at the maximum count. This is a job that takes no skill but lots of time...it is one as little liked as any job in fishing.

Not all men building seines must deal with sleeve leads and their eccentricities. Since World War II, several manufacturers of line have introduced lead-cored line, usually in nylon, that completely eliminates the sleeve lead. The lead coring is pencil-thick and weighs from three to four pounds to the fathom. It has come to be used almost exclusively on gillnets, while owners of new seines or vessel operators requiring new lead lines have turned to it increasingly. The cored line is more expensive than the sleeve-leaded line but its inherent simplicity makes it easier to work with when the lead line strip is hung to it.

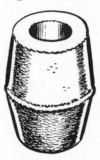

*Purse seine lead of five-ounce weight. This type of lead has been largely replaced on the Pacific Coast by cored lead line.*

Acceptance of this cored line was relatively slow because of the durability of older types of line and the understandable reluctance of vessel operators to part with a proved combination, especially when the sleeve-leaded line was still good enough to use. Some initial users said they believed the cored line tended to sink more slowly than sleeve-leaded lines despite the equal weight, fathom for fathom, of the newer line. This belief, voiced by many veterans of the seine fishery, may or may not have been correct and it appears to be based to some extent, at least, on the relative tendency of nylon line to float. No matter the correctness of this view, its exponents knew rightly that a fast-falling lead line will catch more fish than one that sinks slowly.

This is especially true in areas of fast tidal currents where fish may be moving rapidly because of the current and quite easily may pass under a slowly-falling lead line and out of the set. The necessity of fast-falling seines has been demonstrated repeatedly in the tropical tuna fishery where Americans have proved their predominance because of their devotion to fast-sinking seines. But even in this fishery, despite fast sets and fast-falling seines, speed boats and cherry bombs, water hauls are still something to be faced up to.

With corks on the cork line and leads in place on the lead line if cored line is not to be used, the next step is to hang the mass of webbing to these respective lines. The webmen turn again to the 10-fathom stretch, that convenient and unchanging unit of measure. (It might be well to emphasize here that this description of the hanging of a new seine is a schematic only, something of a "by-the-book" recital of putting all the components into a workable whole. This description would be approached most closely in the professional net loft. In the net yard, three vessel crews may approach the matter in three different ways with each of these ways agreeing only in principle with the "book" method. For example, it is common among West Coast salmon seiners to lace webbing strips together while, simultaneously, cork and lead line strips may undergo hanging to their respective lines and at the same moment, rings and ring bridles are being seized to the lead line—all of this in the same 10-fathom stretch. You're the skipper; you do it your way...)

It is here that the X-frames and horizontal frames that mark the stretch come most prominently into use. The horizontals indicate the exact length of the stretch over which the lines must be pulled taut; the X-frames support the lines at intervals along the stretch to reduce or eliminate sag in the lines while, at the same time, they hold the lines about waist high or somewhat more for the benefit of the men who must work along them. Other devices may be used too—the hooks and posts described earlier; aboard the traveling seine vessel where a strip must be inserted, pulled out or re-hung to a line, almost any jury rig can be used.

To repeat, the specifications of this seine call for 12 fathoms stretch measure of webbing for each 10 fathoms of cork line. These two additional fathoms must be hung into the line by picking up two meshes on every third hanging. The 12 fathoms may be measured exactly by the half, the quarter and the eighth in the manner previously described or it may be accommodated by the eye alone, the usual practice among experienced webmen for this short distance.

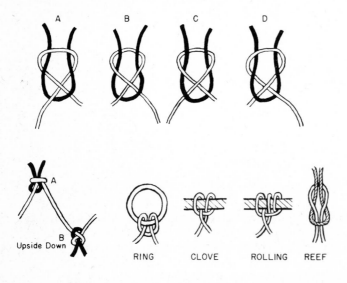

*Four aspects of the weaver's knot used in net manufacture and repair plus other knots used in webwork. A shows the weaver's knot as tied when working from the left; D, when working from right. C is to the right B to the left. A-B at lower left show the knot as tied upside down.*

The hanging of the webbing to the cork line begins on the line outside the first cork or set of corks, that is, outside the corks on either end of the seine. The first mesh is knotted to the cork line, then the needle is reeved through succeeding meshes until the location of the next hanging is reached, where again a knot is thrown in. As described earlier, the hanging distance may be almost entirely dependent upon the spacing of corks and/or the judgement of the webman. The necessity is that the webbing be hung to the cork line in its proper configuration, i.e, the fullness or "bagging" effect that is so pronounced in the gillnet with its extreme proportion of webbing to line, and to a lesser degree in the seine. No. 36 twine most commonly is used for

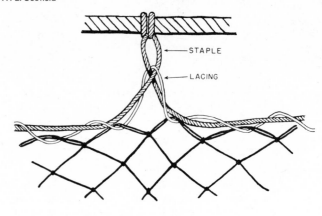

W. L. Scofield

*One method of hanging webbing to a cork line. The method, called "stapling" is not commonly used, although it provides strongest fastening. The lacing twine is shown as lighter strand.*

the hangings although heavier or lighter may be used according to the judgment of the seine owner. Proper strength is desirable but strength must be balanced delicately against the possibility of a parted cork line in a fouling situation.

Hanging to the lead line becomes a somewhat more complicated matter because of the greater amount of excess webbing to be accounted for because of the shorter line. This problem is negligible when hanging to the lead line of the drum seine where lead and cork lines may be of equal length or the lead line may be no more than one or two percent shorter than the cork line.

Hanging to the lead line also becomes a matter of somewhat more precision than hanging to the cork line of the average seine because hanging distances can (or should) be measured precisely. This seine calls for lead line hangings spaced either 4.5 or nine inches apart with a one-mesh pickup on the first spacing or a two-mesh pickup on the second spacing. In all probability, the nine-inch spacing would be chosen by most webmen because of the speed of hanging the webbing to the line. Here, in addition, the lead line is suspended in hangings from four to six inches below the webbing to allow greater flexibility on the haul. Twine heavier than No. 36 may be used because of the frequent encounters of the lead line with the bottom. But here again strength must be weighed against the possibility of a parted lead line. A seine can be fished for one haul, at least, with ripped webbing but it cannot be fished with a severed lead line.

While the lead line is being hung, ring bridles may be seized onto each stretch. Modern ring bridles usually are one fathom in length, cut out of three-eighths-inch polypropylene line with an eye spliced into each end. Round rings are secured at the midpoint of the bridle with a double half-hitch, then one eye splice is reeved back through the line at the ring to secure the hitches. Snap rings, a type that came increasingly into use during the late 1950's, slide freely on the bridle in most cases although they too may be secured in the same manner as the round ring. The virtue of the free ring is the added flexibility it gives to the lead line during the stress-laden pursing procedure. Pursing can be done with less strain on the lead line and on the rings since the rings can move easily with the direction of the stress.

The bridles are seized to the lead line with heavy twine, still with an eye, however, to the sacrifice of the bridle and ring rather than the lead line on rough bottom.

The design of this seine specifies rings on the breast lines. These rings are seized directly to the line without bridles. In common practice, however, many seines are built with no rings on the breast line in order to minimize the chance of fouling on the bottom, on the skiff or on the seiner itself during setting, towing and pursing.

The purse line itself, that most important element of a workable seine, plays no part in the actual construction of the seine since it is not run through the rings as the seine is being hung. It sees no action until the first set is made, when it feeds from the deck through the rings as the seine pays out.

The purse line is of two sizes of nylon. This seine requires two 125-fathom lengths of three-quarter-inch line and one section of seven-eighths-inch or one-inch line. This heavier section lies in the center of the purse line and is designed to support the weight of the rings and lead line when pursing is completed and the rings are hoisted aboard for the haul to begin. The line sections are connected by spring-loaded "figure eight" clips three and one-half inches long or by a swivel connection of heavy steel (see sketch). The figure eight clips often display a tendency to jam and thus slow down the easy separation of the line segments. The swivels do not jam and can be cleared with no delay so that pursing may get underway.

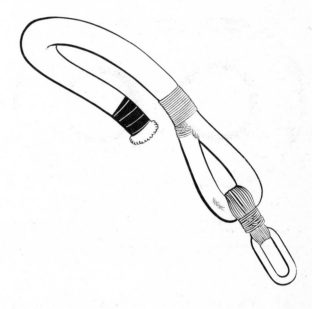

*Schematic configuration of running end of purse line with whipped backsplice and flatlink.*

Not all purse lines are of the pattern described here. Those on some seines may be of two sections only while others may have as many as four sections. Each of these latter types usually is of older line and the four-section purse line may merely indicate that the line has parted at one or more points at some time and has been reconnected with a clip.

A feature of the purse seine that must not be forgotten is the "fish bag," an appendix-like attachment to the bunt end of the seine used to lift fish aboard if there have not been enough in the haul to make brailing worthwhile or to dry up the net completely after brailing. In this sense, the fish bag functions exactly the same as the cod end of the trawl.

Like the bunt section, the fish bag is of three and one-half-inch mesh, cut 80 meshes deep, from No. 60 or No. 72 twine. It is sewn 40 meshes deep, mesh for mesh, just below the bunt end cork line, then doubled

W. L. Scofield

*Common method of lacing selvage edges of webbing strips. This can be used with webbing of different or same sizes.*

back upon itself for mesh to mesh sewing, thus to become cylindrical. The terminal is closed by a puckering string and clip arrangement like that of the cod end of the trawl and can be opened quickly as it and its load are swung aboard. Because of its position below the cork line, the contents of the bunt flows into it as the webbing of the bunt is hauled while, at the same time, cork line is secured along the rail of the seine skiff and the rail of the seiner to form something of a trough.

One more job must be done to finish off the seine. This is the backsplicing of "davit loops" into each end of the cork line. These loops are hung on steel cleats

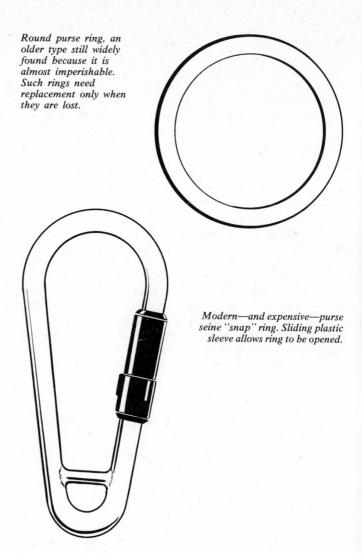

*Round purse ring, an older type still widely found because it is almost imperishable. Such rings need replacement only when they are lost.*

*Modern—and expensive—purse seine "snap" ring. Sliding plastic sleeve allows ring to be opened.*

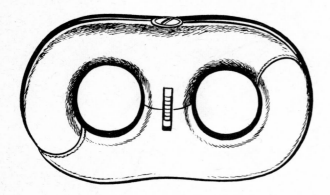

*Spring-loaded figure eight clip, designed to be used with flatlink of purse line above wherever connector is needed.*

on the pursing davit of the seine vessel before pursing begins. The loops and the cleats hold the ends of the cork line safely in position while the purse lines are brought aboard to be freed from their shackles near the davit loops before they are passed to the gypsies of the winch.

A new 300-fathom seine may be hung in the net loft in no more than two or three days with a complement of skilled webmen working on it. The hanging or rehanging of a seine by a vessel crew, working usually for no more than a free lunch from the skipper, proceeds at a more leisurely pace and can easily turn into a matter of two weeks or more, especially since many skippers do not feel it proper to push their unpaid men too rapidly. The "coffee break" can well extend into several hours.

But stretch by stretch, the work eventually is accomplished. In one manner or another, whether in professional net loft or on a shipside pier, there come together those bales of webbing, the reels of line and the myriad of hardware to form one of man's most efficient pieces of fishing gear.

The seine is big, awkward and ugly but it works.

## Of Men And Gear

The conventional salmon seiner using the Puretic Power Block customarily, but not invariably, works six men. The job can be done adequately but somewhat more slowly by five men. The splitting of the extra share among these five is an incentive to work longer and harder. The chief problem with five men only is the replacement of one or more because of accident or illness in remote areas such as Alaska's Inian Islands where extra hands are far away and hard to find. The drum seiner can get by easily with four men and, in emergency, with three.

These six usual crewmen are the skipper, who directs work procedures from the flying bridge controls (some skippers never leave the flying bridge during any stage of setting and hauling); the skiffman, who spends the entire set and haul period at various jobs requiring his skiff, and four deck crewmen. During the pursing of the seine, two of these work as winchmen, one to each running end of the purse line.

It is these four who, when pursing is completed, work the seine pile, and it is one of these who can be dispensed with if the skipper desires. A five-man crew with the skipper and skiffman committed, for the most part, to specific jobs, does cut down on the deck clean-up crew and the number of men available to pitch fish when off-loading.

It is customary for the vessel's cook to be excused from deck duties between hauls and from all fish handling so he may pursue his own job in the galley. The skiffman, however, often must help clean up after hauls although most are diligent at finding things to do around their skiffs until the deck is spic and span and it is safe to venture forward of the seine pile. On most boats the skiffman must pitch fish with the deckmen.

These observations are not absolutes. Customs vary between boats and a skipper may change his pattern of job assignments from one season to the next.

Together, the men, the seiner and its satellite skiff engage in a complex, sometimes spectacular and sometimes dangerous exercise in the pursuit and taking of fish. (Since the salmon seine fisheries of Washington, British Columbia and Alaska far exceed the southern fisheries in number of vessels and value of catch, this discussion will be confined to the salmon fishery although its principles, if not some of its exact techniques, are applicable to all seine fisheries. The principles of the salmon seine fishery have, with modifications, been adapted to most of the world's modern seine fisheries, specifically to those of Latin America.)

The seine skiff is both an invaluable tool and a threat to the well-being of its mother vessel and extreme care must be exercised in the handling and use of the skiff. When the seiner is running for comparatively short distances such as from one section of a fishing area to another with fish to be expected at any time, the skiff rides behind the seine vessel, snubbed up tightly to the stern, held in place, first of all, by its own painter, either of chain or twisted steel wire.

The skiff painter passes across the seine pile or the seine drum to a point where in turn, the painter shackle is secured into a mechanism called a "skiff release." This consists of a heavily-built pelican hook assembly (see sketch) welded on to the circumference of a steel frame and shield 10 inches in diameter. The hook itself is seven inches long with an eight-inch flatlink held in the bight of the hook. The painter is secured into this flatlink.

The tongue of the hook is held shut by a sliding lock and stout spring. A lanyard is secured to the lock. The base of the hook is welded into the squared end of an elongated U-bolt or flatlink-like device. A shackle or another flatlink fastens the release to a cable or heavy fiber line. This latter runs forward to the winch where it is tautened with a few turns of a gypsy and the running end then is secured to the winch bollard.

This succession of gear rides under extreme stress when the vessel is underway. The release, jerked open by a hard yank on the lanyard to drop the skiff, shows this stress when it suddenly is relieved. Men have been seriously injured when the release, weighing about 10 pounds, flies forward as the skiff is freed. Most seiner winches and the aft bulkheads of many seiner deckhouses show scars where the wild release has struck.

Handling of the skiff differs when the seiner runs for considerable distances as on the annual trek of many Puget Sound vessels to Southeastern Alaska. Over short runs, the seiners are unballasted except for what fish they may have aboard. But on the 1,000-mile trek from Seattle, for example, to the Inian Islands of Southeastern's Icy Strait, the seine, the Power Block, and all possible gear and perhaps some cargo are stowed in the hold for what beneficial effect they may offer. In this instance, the skiff rides flat on the fantail where it is secured by chain or cable tie-downs with its painter tightened by the winch and secured to the bollard for added stability.

Under normal weather and sea conditions, the skiff is no great problem as long as its fastenings are kept tight over this long journey, about 100 hours of running time for most vessels in late June or early July weather. It is during the fishing season itself that trouble may come along and the skiff begins to show its potential for acting up.

Seinermen, during the two, three or four days of closure during the week, prefer such bright spots as Juneau to the bleak Inians to pass their off time. The skiff, in this case, is hoisted aboard and settled onto the seine pile on the fantail. The same tie-downs and painter security are used but the seine pile does not afford the same firm surface that the decking itself did on the long run and skiffs often display an unpleasant tendency to shift position even on flat water.

In any event, heavy weather almost always calls for dropping the skiff to tow it. The seiner's center of gravity often enough is too high anyway; the added weight of the skiff riding high on the deck or the seine pile or angled against the drum in Canadian and Washington waters aggravates the situation considerably. Dropping the skiff under flat water conditions can be troublesome in itself because of its great weight; dropping it with wind and sea to complicate matters creates a downright dangerous situation that can be regarded as a common cause of accident among vessels and crewmen. Nevertheless, in the salmon fishery the seiner and the skiff must co-exist.

## Let 'Er Go!

The seine may be set in several ways, with any specific set a matter incumbent upon the skipper to decide after study of the run of the tide, the lie of the bottom and the horsepower of his vessel as weighed against tidal currents. The most used method of setting is called the "inside set" in which the skiff is dropped close upon the beach so it may hold its end of the seine close to the beach or turn right to tow parallel to the shore. Many skiffmen call this the "right

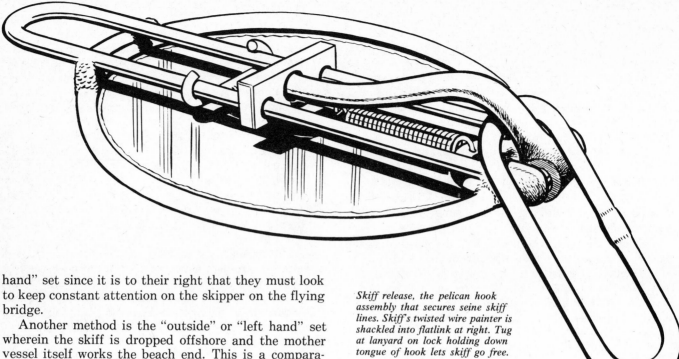

hand" set since it is to their right that they must look to keep constant attention on the skipper on the flying bridge.

Another method is the "outside" or "left hand" set wherein the skiff is dropped offshore and the mother vessel itself works the beach end. This is a comparatively uncommon use but one that is valuable in situations where it may counter unfavorable currents and a changed tide. The net can be set from its normal configuration for the outside set but all auxiliary gear such as purse blocks and ring bar must be moved to starboard positions from their customary port side posts to be used properly when the haul begins. Similarly, the net, as it comes aboard, is stacked on deck in the exact reverse of its normal configuration. Again, it may be set from that reverse position for return to a normal inside set but again all the auxiliary gear must be moved back to port for the haul. The whole affair is a nuisance to crewmen who must do the work; it is a matter of little concern to the skipper at his topside wheel station whose major interest is catching as many salmon as possible.

No purse seine can be fished successfully unless it is rigged properly. The basic assembly of the seine was completed with the tightening of the last knot when it first was hung. But certain small gear and lines lying outside the field of the webbing are vital to the proper function of the net. The best-hung seine will not take fish to its full ability without this auxiliary rigging in its exact sequence.

It is most important that there be no strain on the purse line during the set and tow in order that the lead line and its attendant purse rings may fall as quickly as possible to the full depth of the net. To this end, the purse line runs up to 10 percent longer in total length than the length of the cork line so that there is built-in flexibility to start with.

With the purse line possessing proper slack, modern seining techniques protect this flexibility by the use of two brass spring shackles (see sketch), called "Cana-

dian" releases after the country of their origin and manufacture. These are devices, roughly of a G-shape, about three by five inches in size, secured to each end of the cork line just ahead of the breast line junction at the davit loop. The purse line, running parallel to the breast line at each end of the seine, is set into the release by a flatlink. The line can be sprung from the release by an easy pull on a short lanyard when the tow has ended and the davit loops have been hung to their cleats at the purse blocks. During the tow, all strain thus falls along the cork line axis.

It has become increasingly apparent that there exist several alternatives for every method rigging fishing gear. So it is with the cork line—purse line—Canadian release complex. There is no standard; the only rule is that there must be maximum strength along the cork line and no strain whatsoever along the purse line during the tow. The commonest method of setting up the cork line to achieve this strength is to seize it to another line throughout its length. This line then assumes the tow line function. Or the cork line itself may be of heavier material than the line specified for the previously described Icy Strait seine and will work both as cork line and tow line.

In any event, the method of rigging the cork line determines the manner in which the Canadian release is set into the seine structure just ahead of the breast line/cork line junction. The simplest solution is to terminate the cork line assembly, whatever its rigging may be, at the Canadian release. Additional line then is needed ahead of the release for towing and handling. This line, usually called a "running line," is seized to a shackle and the shackle set into the purse line flatlink. When the release is sprung at the end of the tow, the joined lines come free for the beginning of the pursing cycle. (See sketch.)

With the seine and all its components set up to go to work, the skipper, when ready, orders the skiff dropped by any method that appeals to him—a whistle signal, a wave of the hand, a shouted (over the blast from the stack) "Let 'er go!"

The designated crewman tugs hard at the lanyard of the skiff release, standing aft of it a trifle if he is sufficiently acquainted with its eccentricities; the skiff drops free of the stern with a clatter startling to the stranger aboard, already half-deafened by the racket of a diesel engine that may range from 65 to upwards of 200 horsepower or more in an handful of extra-large, extra-powered skiffs designed primarily for use in certain passes of Southeastern Alaska and among the San Juan Islands of Washington.

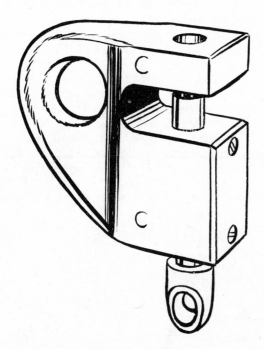

*Canadian release, vital gear item in purse line complex. Controlling lanyard runs from end of tongue at bottom of sketch.*

## Oh, To Be A Skiffman!

The skiffmen of the block seiner and the drum seiner find themselves immediately with problems of their own, each different and potentially dangerous. With all other factors equal, the block seiner skiffman has the better part of the deal although, haul for haul, he spends more time in his lonely skiff. The major possibility for accident, again with all other factors equal,

lies at the start of the set, at the moment when the skiff is dropped.

The block seiner skiffman, assuming his set is of the normal inside configuration, turns immediately to starboard after crawling forward to pick up his own painter and steers for the beach under full power. The chance of a net hangup in the seine pile are infinitesimal and, thus, so are his chances of broaching under the strain of an unyielding line secured to his towing bitt. The sole potential exception is that the purse line, running from its coils in the waist, may be fouled somewhere on deck. This is a most uncommon occurrence, at least to the extent that it might endanger the skiff, but it can take place. To minimize any such fouling some vessels use a wooden pole about two fathoms long lashed between the starboard shrouds and the after stay, the offside of the normal set. This pole is known variously as "purse bar" or "purse pole" and its function is to raise the purse line from the deck to reduce any chance of fouling. The line comes off its coil, passes up and over the pole from the offside and runs down to and through the rings as the net leaves the deck. Other vessel operators are content to let the purse line snake across the hatch cover on its journey through the rings to the bottom edge of the net with one deckman supposedly keeping tabs on it.

The skiffman of the drum seiner faces an ever-present danger on every set. Although the net comes aboard the drum through fairleads exactly as a fishing line is spread across a reel by a level-wind, the crush of cork line, webbing, lead line, rings and bridles and purse line on the seine drum lends itself easily to a backlash at any stage during the set. Here, instead of turning away full-bore as does the block seiner skiffman, the drum seiner skiffman backs off slowly from his mother boat; in essence he lets the seiner move away from him at its setting speed of two or three knots, a speed somewhat less than that generally used by block seiners. He watches carefully—and so does his skipper—for the backlash that could drag the seine skiff sideways far enough to cause it to broach and throw its crewman into the water, a deed uncomfortable at best and fatal at its worst for a man often heavily burdened with boots, warm clothing and rain gear. In most cases, the skiffman puts from 50 to 100 yards or even more between himself and the seiner before he begins his turn and steps up his power to tow.

One such dumping is described by a seiner skiffman of eight years' experience like this:

"The skipper said 'Hit it' and the guy on the deck pulled the pin. I had her in reverse with the wheel turning over just enough to give me some steerage way. I went forward to pull in the painter, all the time with an eye on the drum. Just as I got the painter in, I saw a hangup start toward the port side of the drum. I was about 10 to 15 yards off the stern then. Before I could get back to the wheel to head the bow square at the stern, the backlash was solid...the guys on deck were hollering at the skipper to back it off before I dumped. But the skiff had swung sideways just enough to get the full effect of the tight line and the forward movement of the seiner. I felt her starting to roll and I dived headfirst over the high side so that I wouldn't get hit by it on the roll or trapped under it.

"As soon as I came up, I saw I was clear of the skiff which was floating belly up by then and I started getting rid of everything I could. I got my boots off first (the fisherman's favored hip boots), then my rain parka, then my jacket. That's the one reason I wear rain parkas that zip down the front—you don't have to try to get them off over your head in the water.

"I was getting cold but I was treading water...I was downtide from the skiff so I just let her drift to me and I grabbed onto the skeg and let the drum haul us in. As soon as I got aboard I headed below and got dried off and into some nice warm clothes. By that time, they had the skiff righted...we headed to town then to get the skiff engine dried out and cleaned up. Out of the deal, the boat paid for new boots, a new jacket and a new rain parka for me...

"But we lost all the rest of that week's fishing—three good days." [1]

## The Seine At Work

Back to the block seiner set...the net ripples off the stern under the twin influences of the furiously churning skiff and the more massive forward movement of the seiner. Many fathoms of cork line appear at first glance to be a tangle that cannot be cleared as the corks first hit the water with webbing criss-crossed

___

1. Anonymous, 1973.

over them. But as the distance from the seiner increases, the weight of the sinking lead line and rings pulls the errant webbing with it and, as the set is completed, all to be seen is a line of cork up to 300 fathoms long.

But before the last few fathoms of the seine are off the deck and the main tow line becomes taut, additional security in the form of a second or auxiliary tow line is added. This comes from a hook on a double block hung from a shackle and flatlink on the boom about one-quarter of its length from the tip. The hook slips into a ring or flatlink seized into the tow line at a point where it rides from four to six fathoms off the stern during the tow. In actuality, this secondary line bears the main stress of the tow, especially on the drum seiner where the drum—except on a few specially-built vessels—cannot pivot. This inflexibility makes the second line mandatory on the drummer while the block seiner could do without the secondary protection if need be. Nevertheless, the added protection is handy since the seine still can be controlled if either line should part inboard of the junction. If it should part between the junction and the breast line, it can easily be picked up, spliced, and the tow continued. The main tow line of the drummer is secured by many turns around the drum; the main tow line of the block seiner is the extension of the cork line from its junction with the breast line, looped several times between the horns of the winch bollard and double-halfhitched for security.

With his skiff down and towing, with all his rigging working as it should, the seiner skipper finds himself faced many times with a variety of choices concerning his next move. Customarily, if his set be the normal, inside type, he runs, while completing the set on a fairly straight course off the beach. But his heading may vary somewhat because of tide, because of

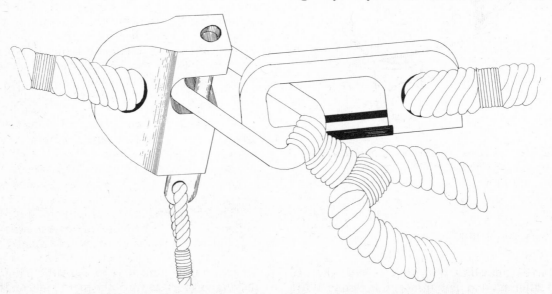

Warren Ernst

*Schematic configuration of purse line and cork line junction at towing or skiff end of the purse line. At left is Canadian release secured to cork line. Flatlink, center, is seized to purse line. Shackle in flatlink is connected to running line by which purse line is controlled until it is freed when pursing begins.*

wind direction and velocity or because of bottom conditions. But any deviation from that short setting run will be uptide. After the net is off and fishing, his course invariably will be an arc in the direction of the running tide.

It is during this period of the seine operation that the vessel operator "holds open" on his set, meaning that he allows his net to lie open to fish as fruitfully as possible over a length of time that seems proper to him. Preferably, the ideal open net would take the form of a rather flat or shallow U with the skiff holding its leg of the U as close to the beach as possible while the seiner would tow a somewhat longer leg offshore. This longer offshore leg would be predicated on the knowledge that salmon, moving along the beach on spawning runs, tend to turn away from it when faced with an obstruction. This is the principle that made the salmon trap a deadly taker of fish.

turns at fishing a favored spot. Such a one is the small Hansville ground at the north end of that state's Kitsap Peninsula, a headland marking the turn from the south into Admiralty Inlet. On Hansville and on others like it, open sets tend to be of shorter duration because each vessel takes its turn to set. A man who holds open longer than the man next in line feels is just may find himself "corked out" by that skipper, i.e., the next man up makes his set close uptide from the open seine, effectively cutting off the possibility of any more fish getting into the open net.

During the time the set is open, the skipper uses signals, usually a set of formalized arm movements, to direct his skiffman on the speeds he wishes used and the course to steer. Some vessels use Citizen's Band radio but even the best transmission is hard to read over the noise of the skiff engine. Headsets offset this somewhat although most skiffmen despise them.

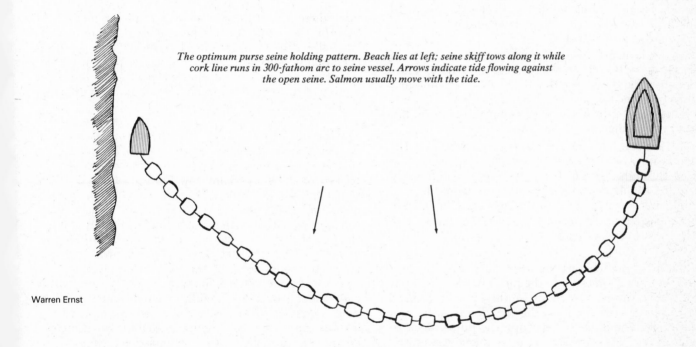

The optimum purse seine holding pattern. Beach lies at left; seine skiff tows along it while cork line runs in 300-fathom arc to seine vessel. Arrows indicate tide flowing against the open seine. Salmon usually move with the tide.

Warren Ernst

Other factors must be figured into the open set, however; the open set is subject to more variables than can easily be counted. Again, tide, weather and bottom must be taken into account as well as the presence of other vessels working the same area. Most sets wind up either in a deep U form or in a fishhook shape with the seiner representing the barb of the hook, the cork line the shank, with the skiff as the eye. A strongly-running tide may be accounted the most important factor in the shaping of the open set.

One human variable must be allowed for in some certain areas where many vessels work a small ground. This is particularly true in some "inside" waters of Washington where up to six or eight boats may take

When the skipper decides to close his set, he so signals the skiffman—manually, by whistle or, again, by radio. (Young skiffmen claim that skippers using arm signals stand purposely in front of the radar antenna complex or against the stack where it is virtually impossible to read the signals.) The skiff and the seiner turn toward each other immediately with the skiff at full ahead, the seiner under a lesser degree of power, since it is preferred to make the haul in water deeper than usually may be found inshore.

As the skiffman comes alongside the seiner, under the port bow if the normal set has been used, he backs his speed down to bare steerage way with his midships towing bitt a yard or so aft of the pursuing davit. The pursing davit is set just forward of midships; it and its opposite number are in a thwartships line with the winch. When not in use, the davit runs parallel to the rail; when it is to be used, it is turned 90 degrees and locked facing outward, away from the winch. This quarter-turn brings the purse blocks into working position to handle the lines feeding into them and running from them to the winch gypsies.

While the skiff is jockeying into position to pass off its lines, the seiner itself comes dead in the water. This is a crucial point in the haul cycle at any time, more so when wind and water conditions are adverse to the point that they can affect the operation unfavorably. Since all tension is off the tow lines now, the seine itself may display a remarkable affinity for the vessel and its underwater gear such as rudder and wheel. In tight passages with strong currents, the seine may become wrapped around the seiner, as the vessel turns in the swirls, before any move can be made to halt it. Above all things, the vessel cannot get underway with this mass of webbing around it; it falls to the skiffman and his skiff to attempt to clear the clutter as soon as he has been relieved of his lines.

Two movements get underway simultaneously as soon as the skiff and the seiner close up. One or more deckmen work the seiner tow lines, the main line and the line from the boom, as well as the purse line when it comes to hand in the sequence of operation. Another man works similarly with the skiff lines.

With the tow lines slacked off, the line secured to the winch is freed and moved to the aft purse block and reeved through it, then to the aft gypsy of the winch where it is hauled to the point that the boom line and flatlink are within reach from the deck. The double block is cleared and immediately secured to its shackle on the mast because the heavy block, swinging freely, can be dangerous.

Hauling of the tow line continues until the Canadian release complex comes up to the purse block. The davit loop is then hung to its cleat at the aft of the davit below the purse block. With the cork line now secure, the running line is passed through the purse block to the aft gypsy. Two or three turns are taken around the gypsy to secure the line. The release then is sprung at the davit and hauling continues until the flatlink and shackle connecting that line and the purse line have cleared the gypsy. The tow line is cleared and coiled down—or should be—and put aside from the working area in the waist while hauling of this end of the purse line carries on.

While this winch work is underway, a couple of fathoms of cork line running from the davit loop are pulled aft to the ring bar and secured along the rail until the hauling of the net itself is to begin. Positioning of this small part of the cork line along the rail frees the davit loop of some of the tension that falls on it during pursing. The same procedure is carried out forward from the other davit loop. Hauling can begin at any time after

pursing starts although ordinarily it does not begin until pursing is well underway. It can be delayed until the purse is complete and the rings have been brought aboard. But occasions may arise when it is necessary to get the seine aboard as quickly as possible and normal procedures must be forgotten.

As work proceeds on this end of the seine, similar activity has been underway at the skiff. The deckman assigned to the skiff takes the tow line from the skiffman and reeves it through the forward purse block to the forward winch gypsy. The line is hauled until the davit loop can be hung; the release is sprung, the lines are cleared and pursing of this end of the purse line begins. At the same time, the deckman manhandles cork line along the rail as far forward as the anchor windlass.

This work has freed the skiff for its second job, that of keeping the seiner out of her net during the haul. The skiffman takes his craft aft around the stern under the tow line if there still is room for his towing bitt to get under it. Otherwise he "jumps" the cork line in the other direction, as close to the bow as he can manage it, by riding over the corks, hoping as he does so that the line will not be fouled in the skiff wheel. This direction is chosen because, if there is fouling, 99 percent-plus of the seine may still be worked normally. The damaged section of the cork line can be worked by hand after the major operation is finished.

On the offside of the seiner, the skiffman prepares to tow the seiner to keep her out of the seine as the haul begins. Some skiffs carry a tow line which is passed off to the vessel; on others, the tow line stands and is passed to the skiff. Some tow lines are bridled at the seiner end, some are not. There is no rule. The job is a most important one since it allows, or should allow, working the seine without any fouling on any part of the vessel. Web or cork line fouling on the rudder assembly or in the wheel is not only a nuisance; it is a threat to the integrity of the seine (where enough routine patching must be done anyway); it conceivably can render the seiner unable to move under her own power; it can allow the escape of an entire catch if there are vast gaps after the fouling is cleared. Nevertheless, despite the utmost caution exercised by skipper and skiffman, such fouling does occur although often it can quickly be cleared with the use of long-handled plungers, looking for all the world like an oversized version of the "plumber's friend" used to clear toilets.

These plungers, incidentally, serve another purpose; they are used splashingly to frighten fish away from the gap between the breast lines of the seine before the purse is completed. Their action creates, in effect, the bubble curtain that has been found experimentally useful in controlling certain species of small, densely-schooling fish such as the Maine "sardine." When not in use, they are secured in the shrouds.

Pursing is a delicate and sometimes dangerous procedure because of the stresses now bearing on the purse line. After its long period of slackness, it now carries

Robert Browning

*Deck scene on drum purse seiner. Crewman holds the single purse line, running from pursing davit (out of picture at right) as the haul is momentarily stopped. Fish were gilled in webbing and were picked out before they reached the drum.*

the entire weight of the rings and lead line. Winching of any kind is an art in itself and it is especially an art on the relatively crude winches of the average salmon seiner. Veteran seiner winchmen can be noted by the number of missing fingers or portions thereof.

Two to three turns of the line are thrown around each gypsy; as pursing progresses, the winchmen must not only tend their lines on the winch, alert to any sudden order to slack off for any one of a variety of reasons, but they must, with one hand, coil the line in the waist behind them so that it may be ready to run freely on the next set. As the purse line, usually but not invariably in three sections with the short section having the largest diameter, comes aboard, the connectors are cleared so that the center section may be used to help hoist the rings without unnecessary interference from excess line.

As pursing proceeds, the purse rings begin, one by one, to come to the surface and bunch on the port beam aft of the pursing davit. When all purse line is aboard except that short strip of the center section carrying the rings, the purse is completed; the seine is sealed off to escape by the fish. (This does not apply to sharks, seals or sea lions who sometimes are trapped within the seine; usually, they can go through the webbing or across the cork line with ease.) Now the rings must be brought aboard and here again the invaluable double block and its tackle come into play.

## Caution—Men At Work

There are two methods whereby the block seiner may lift its rings. (The drum seiner has no problem in this regard.) In either method, the rings are lifted much as the loaded cod end of a trawl is strapped aboard. On the block seiner, a single length of chain may be passed around the bridles just below the ring cluster. Each end of the strap has a ring or shackle in it. The hook on the double block is passed through these and the whole weighty mass is lifted aboard by the winch and the rings lain on the hatch cover.

In the second method, two shorter lengths of chain, each with ring or shackle in one end, are passed around the purse line on each side of the ring cluster. The running end of each strap is passed through the ring; the running ends join over the ring cluster and the last links are threaded onto the double block hook. In either case, the procedure can prove hazardous because of the considerable weight of the ring cluster and the sometimes advanced age of the purse line or the double block tackle.

There is another option at this point for the block seiner skipper. With the rings on the hatch cover, it is necessary that they clear in turn as the haul begins to go through the Power Block and down to the ring bar. Some skippers, to achieve this orderliness, pass a steel bar through the rings so they will be pulled off in their proper sequence. If this device is used, it is necessary that the bar be withdrawn by degrees as the rings leave it in order that the whole affair may not be pulled off the hatch. This is not a common practice and most skippers take it upon themselves to tend the ring cluster often enough to prevent ring and bridle fouling under the pull of the Power Block.

To this point, the block may or may not have been put into use, depending upon the skipper's wishes and the circumstances of this haul. But now, with the purse tight up and the rings aboard, hauling can be no longer delayed (except by mechanical failure). To begin the haul, the aft davit loop, that at the junction of the forward breast line and the cork line, is freed and secured to a messenger line left dangling through the sheave of the Power Block after the set began. By this line, the seine is pulled up to the sheave, now turning under its hydraulic power.

The sheave grabs at the first straggling bit of webbing and cork line and carries it through and down to the fantail where the seine is to be piled. Four men normally take part in this procedure (this is the one operation wherein the cook spends more than a minimal amount of time on deck work). If the set has been the normal affair, one man stacks cork line in ragged coils to starboard. This is the hardest work of the seine piling procedure. To port of him, two men pile webbing, grabbing at it in great handsful to lay it as evenly as possible over its allotted area. The fourth man, working at the ring bar, puts down lead line as cleanly as he can in a limited space while, at the same time, he pushes each ring onto the ring bar as it comes down to him.

The block seine haul is a noisy and sometimes oft-interrupted procedure. There is the ever-present stack blast, the screech of corks squeezed in the sheave and shouted commands to stop the whole affair so small rips may be patched, gilled fish removed from the webbing or debris plucked from it to be thrown overboard outside the cork line. Throwing trash back inside the cork line is tantamount to trying to empty a garbage bucket to windward. These frequent starts and stops are easily handled because almost every block seiner of the 1970's carries a remote control for the block. The man in charge of the deck, normally the skipper (although some skippers never leave the topside station during the set and haul except in case of extreme emergency), usually carries the control suspended from a short loop around his neck so that both hands may be left free to deal with small jobs as they pop up. They pop up with great frequency...

Fathom by fathom, delay by delay, the seine is dried up...if the catch has been heavy or only fairly heavy, the presence of salmon within the confines of the webbing is quickly apparent. Their witless alarm is easy to see; there are jumpers and of these, a few lucky ones may make it over the cork line. Others fin at the surface, disturbed but not yet spooked by the ever-closing walls of webbing. These are money hauls; a veteran fisherman, after a few minutes' study of these surface indica-

tors, can come pretty close to guessing the tonnage of salmon in the catch.

But there are those others, the hauls that occur with increasing frequency—except for such seasons as the cycle years of the pinks in Southeastern Alaska and the Fraser River sockeye in Washington and lower British Columbia—where the first indication of fish comes when the haul is approaching completion. By this time, the salmon are thoroughly panicked; they dart despairingly from one panel of the webbing to another; from topside one may see for the first time and at some length, if the sun lies right and the water is clear, how perfectly the salmon fits into its medium...it is easy to guess wrong as to the salmon's speed through the water but a single flick of the broad tail seems enough to send the creature across the short distance he can move. And with every turn of the Power Block sheave, that distance grows less.

The final steps of the haul depend chiefly on the number of salmon in the seine. There have been times beyond count in years of great abundance when a single haul was enough to weigh down a seiner to the point where her scuppers come awash and another vessel had to be called in to take aboard what the first boat could not handle. And even in this latter third of the century, there still occurs, on widely infrequent occasions, a bonanza haul where one boat cannot handle the whole catch. Every bar from Seattle to Sand Point where fishermen gather hears great tales of great catches and in those bars there are more fish caught than most of the story-spinners ever have seen.

By the time the net has been dried up down to the bunt, that last 10-fathom stretch of the seine, there is no question as to procedure. If the fish are few enough, their few pounds or few hundred pounds can be hoisted aboard in the fish bag by means of the Power Block and dropped directly into the hold when the puckering string on the fish bag terminal is pulled.

But if there are more fish than the bag can handle easily, they must be brailed from the bunt. The brailer is an over-size dip net with a handle about eight feet long, a hoop 42 to 48 inches in diameter, with a bag that may be up to a fathom in depth. At the junction of handle and hoop, there is a ring wherein fits the hook of a single block (not the double block, this time). Powered by a winch gypsy, the brailer scoops salmon from the bunt by many hundredweight and deposits them in the hold through the terminal puckering string opening. Brailer loads by the dozen on a single haul...something to be hoped for but not too often to be seen in many seasons on many grounds.

Again that other valuable tool, the skiffman, goes into action. By the time the bunt is tight up, he frees his offside tow line and takes his craft around to the scene of action. His chief function here is to aid in the working of the bunt. The bunt is dried up to the point where there is just room enough for the mass of fish in it—assuming a catch of that proportion—and their weight is enough to sink the cork line. To avoid this, the skiffman pulls cork line by the fathom over his own rail to form a pocket from which the fish cannot escape and from which they can be brailed. If the brailer is to be used, the skiffman helps maneuver it into and out of the water and if there is enough brailing to be done, he may be joined in the skiff by another crewman to lend a hand. Even though the fish be few, the skiffman still takes on cork line to head off any last minute attempts at escape. Similarly, bunt cork line may be hand-hauled aboard the seiner and temporarily secured along the rail for the same reason, usually, however, only if there is a decent enough catch to be handled.

When the last fish is aboard, the final few fathoms of the bunt are run through the block and stacked; the messenger line is left in the block; the purse lines are set up again for the next set; accumulated debris is tossed over the side; jellyfish are swept or hosed through the scuppers and the decks cleaned down; the skiff is snubbed up again with its big release all set for the next time the skipper indicates:

"Let 'er go!"

If nothing at all has gone wrong, the drum seiner has used about 20 minutes to haul her seine after closing while the block seiner has worked 40 minutes and up.

## Drum Us Up A Fish

Among the American seine fleet, the drum seiner is by far in the minority. Since the drum cannot be used in Alaska, American use of it is confined to the seiner waters of Washington although it can be used anywhere in British Columbia, its place of birth. The drum seiner, indeed, may be a vanishing species among Americans because, to many fishermen, its costly installation with its complex hydraulic system, simply is not worth the money for a limited fishery.

In addition, most drummers are conventional seiners converted, sometimes clumsily, to take the drum. Only a few, perhaps no more than a dozen American boats, if that many, were built primarily as drum vessels with turntables that could move the drum to face to port or starboard, as the set went, to follow the stress

Robert Browning

*Drum seine at work. Cork line and other seine components come through fairleads at stern. Engineer at left controls the drum and movement of the fairleads. Haul is almost complete and crewmen at right tend purse ring "hairpin" to slip off rings as lead line comes aboard.*

of the tow and haul. But conversions or originals, they are efficient and fast workers and any good drum skipper, free to set as often as he wishes, can easily outdo the hardest working block seiner by about two to one any time there are enough fish around to make the effort worthwhile.

Something of the drum seiner operation has been described, particularly in reference to the use of the skiff and the tow lines. As indicated, the set, particularly by a vessel moving away too fast from its skiff, is the weak point in the entire drum operation. Most certainly, it is the weak point in the opinion of the drum seiner skiffman.

But possible backlashes aside, the drum seiner skipper has control over his setting procedure that the block seiner lacks. One man, running push-button controls with a quickly available braking system, can control the set from start to beginning of the tow. If necessary, he can leave his control station, usually at the starboard flange of the drum, to hook in the auxiliary tow line.

The drummer has no particular virtue during the period the set is held open but its supreme efficiency shows the moment the haul starts. For one thing, all the seine from cork line to lead line is drummed directly aboard with no recourse to any other gear except for gypsy head pursing of perhaps one-sixth of the purse line. Nor is this pursing the meticulous pursing of the block seiner with its neat coils of line in the waist. As this purse line comes aboard, it is tossed aft onto the hatch cover where it lies unheeded until the drum gobbles the seine up to the bunt and takes this fraction of the purse line with it.

Too, all but a few of the rings and their bridles go directly onto the drum with none of the intermediate handling needed on the block seiner. Those rings that do not go directly onto the drum are those on the bunt and just ahead of it.

The skiff passes off its end of the purse line and tow line/cork line just as it does with the block seiner. The lines go through the purse block in the same manner and a section of cork line is tied off along the rail just as it is done on the block seiner. Pursing by the drum and the winch begin simultaneously but the winchman is soon done with that task.

It is in the handling of the rings that do not go onto the drum that introduces a danger factor into drum seining that does not appear in quite the same form on the block seiner. Those rings that must be handled independently amount, on the average, to about three dozen, many of them with half-fathom bridles toward the bunt rather than the standard one-fathom bridle found all along the lead line of the conventional seine.

These rings bunch up in order on the beam just aft of the purse davit and they must be brought aboard to await their turn to go onto the drum. This is done by a device called either "hair pin" or "clothes pin" and of the two, the former is more descriptive because the affair does resemble a giant hair pin.

When the hair pin is needed, here again comes the single block with its hook. The hook fits into a ring at the junction of two chains of equal length secured to the upper leg of the hair pin. The lower leg of the pin is run through the rings snugged up on the purse line and the whole mass is lifted from the water by the winchman.

On flat water, the ring-loaded hair pin is no special problem. But in rough water, the pin, swinging freely from the boom, subject to all the forces that work on the vessel, can resemble a wrecking ball. Men have been hurt or hurled overboard when struck by the pin and its rings. Ordinarily, two men work it—one to guide it and hold it as steadily as possible during the final phase of the haul while the other slips rings off the pin as the lead line comes up through the fair leads to go onto the drum. With half-fathom bridles, it behooves this crewman to look sharp and work fast.

This hazard does not exist on some drum seiners. A few use a fixed ring bar like that of the block seiner while another minority uses a hair pin secured along the rail during the time the pursing procedure is underway.

Hair pin or no hair pin, the drum seiner cannot be faulted. It is as foolproof a method of fishing as any practiced anywhere in the world. About the only medium that can equal it for sheer efficiency is a hand grenade in a front yard fish pond.

## Gillnets

The gillnet is a productive tool in marine and freshwater fisheries the world over. In the chronology of net gear it must follow closely behind the beach seine, the ancestor of all nets, and, of course, well ahead—millennia ahead—of the trawl and the more complex forms of the seine. The gillnet would appear to be a logical evolutionary development of the simple beach seine when fishermen observed that closing off the course of a stream should be more productive than setting a seine blindly or on the fleeting signs of fish. And certainly, the weaving of a net was infinitely less labor, even with primitive tools and primitive materials, than the construction of a trap or weir or dam to corral the quarry. The gillnet, like all other nets and all other forms of gear, came into being at differing times in differing places as ancient man learned the techniques of fishing and it, the beach seine and the hook were monumental developments in the history of gear. With their coming into being, the pattern of fishing was set fairly well on the course on which it still continues today.

The gillnet ranks behind only the purse seine and the trawl in tonnage of fish taken wherever it is worked. Western European fishermen, particularly those of the United Kingdom, shoot their gillnets in lengths measured in miles, trailing upwind of their ocean-going drifters. The Japanese fish gillnets as long or longer in their ill-conceived high seas salmon fishery in the North Pacific. In American and Canadian West Coast

*Puget Sound gillnetter at work. Power reel hauls net through fairleads on stern. The boat, diesel powered, is of an old and fast-disappearing type.*

fisheries it is one of the three major gears used for salmon, the gillnet's principal fishery in the west. It is found in certain other minor fisheries, though, with smelt and shad of the Columbia being among its targets.

Superficially, the gillnet resembles the ancestral seine, but it fishes on an entirely different principle and it may be fished in three or four ways, with variations on these themes, while the purse seine, for example, may be fished in one way only. The gillnet can be fished at the surface, in midwater or on the bottom; it can be drifted or it can be anchored. But no matter how the gillnet may be fished, it can be equally as mulish as the purse seine, although its lesser weight and depth allow these obstinacies, in most cases, to be more easily overcome than those of the seine.

The Scots and the Japanese need fair-size ships and big crews to work their great strings of nets, but the salmon gillnet of the West Coast may be fished easily by one man (as it often is) or by a man and his wife, a common duo in the fishery, or by a man and his young son or young brother or kid nephew, although under

all but the most adverse conditions a single man and his power gear are quite enough to handle the net.

Alaska's Bristol Bay is the only United States fishing area where two experienced men usually crew full-size gillnet vessels. (There is a fleet in the Bay of boats from 16 to 20 feet or so, outboard-powered, that accounts for a good percentage of the catch; the "crew" of one of these may be only a small Aleut or Eskimo boy.) This Bristol Bay fishery is so intense at its peak during years of medium and heavy runs that only the strongest or most stubborn of men can fish a standard-size Bristol Bay boat alone at a time when he may pick up three or four thousand salmon on a short drift.

Bristol Bay gillnetters haul their nets before they pick them because this permits a faster picking operation than that possible with the usual gillnet power reel, and it takes both a skipper and his "boatpuller" to manhandle fish-heavy nets across the relatively ineffective power roller mounted above the transom. Picking salmon from such a haul is demanding work to be done in a hurry, but even a hurry takes too much time in that fishery. The power reel, as distinguished from the power roller, is entirely satisfactory for most gillnet fisheries but it does not allow the extra-fast picking of fish required on Bristol Bay.

Like the purse seine, the gillnet has a cork line and a lead line, both of them much lighter than those of the seine, but, since it need not be pursed, it lacks the ring and purse line gear of the seine. (Some gillnets in the Southeastern United States are roundhauled on schools of fish such as mackerel but the fish are taken by gilling in the web, not by pursing the net.) The gillnet's wall of webbing hangs fairly vertically in the water between corks and leads but it tends to bulge under current effect because of the excess of webbing hung into it. Much slack is built into the net because the fish swimming into a taut section of webbing tend to bounce away from it rather than entangle themselves in it. The gillnet fishes selectively according to the size of its mesh.

Circular 109 of the former Bureau of Commercial Fisheries, "Commercial Fishing Gear of the United States," lists eight nets that are lumped under the general term "gillnet." Of these eight, only two are of concern to West Coast fishermen. These are the drift net in its several forms and the set or anchor gillnet. The drift net is by far in the majority although the set net is widely used in Alaska. The set net is banned in British Columbia, Washington and Oregon except for the Indian fisheries of the Columbia River on treaty waters above Bonneville Dam.

The gillnet is as rigidly controlled as the purse seine. Washington specifies a maximum length of 300 fathoms in its inland waters where the gillnet is most often seen. The state sets no depth limits but the Puget Sound gillnet has come to be 120 meshes deep with the minimum mesh size set at five inches. Most nets are 5-1/8-inch mesh, however, except for certain larger sizes prescribed for varying times of the season and varying areas.

Alaska, as with the seine, sets regulations tailored to specific areas and to districts within the major areas. Lengths, depths and mesh sizes are spelled out minutely. The length of Bristol Bay red salmon gillnets is calculated on the aggregate amount of gear registered each year for this critical fishery. These maximum lengths may vary from year to year with 150 fathoms being the average maximum with a possible minimum of 50 or 75 fathoms. The maximum gillnet depth on Bristol Bay is 28 meshes but the actual depth may vary because of differing mesh sizes for districts within the general area. Set net length is only 50 fathoms but one fisherman may operate two or more set nets.

Southeastern Alaska gillnet rules are spelled out even more fully. Allowable maximum lengths run up to 300 fathoms with a minimum of 50 fathoms. Depths run to 60 meshes for eight-inch mesh with a maximum of 40 meshes depth for mesh more than eight inches. Set gillnets in Southeastern are limited to 50 fathoms. All gillnets in Alaska must be marked every 25 fathoms

with a single cork of a color contrasting with the dominant color of the cork line. Set gillnets are prohibited in Washington and British Columbia for salmon but account for more than half of Alaska gillnet licensings. (On the Columbia River, Indians may use set nets for salmon in their traditional sites above Bonneville Dam.)

British Columbia limits gillnets to 200 fathoms of length with a maximum depth of 60 meshes. Mesh sizes run from five inches to eight inches. Oregon sets Columbia River length maximums at 1,500 feet with mesh sizes running from four and a half inches to seven and a half inches, the same as those prescribed by Washington for the river. Depths of all these nets may vary fractionally because of the manner of hanging. Some set nets of five-inch mesh or smaller are sometimes used for shad.

All regulatory agencies concerned with gillnets insist, as noted, that gillnets be set "substantially in a straight line," that there be no circle or hook setting to prevent roundhauling on schools of fish. This applies equally to set and drift net but the drift net that may have been set originally in a straight line quite easily assumes any form wind and current impose. In the night with the drift once begun, few gillnet fishermen trouble themselves to tow their nets enough to put them back into a "substantially straight line."

The Alaska set net is anchored on the beach—none are allowed within a mile of a river mouth—with the offshore end secured to a variety of anchors and buoys. (See sketch.) There is really no standard pattern for setting up a net since most must be adjusted to the conditions of a particular site. The accompanying diagram is more elaborate than most set net configurations since it was designed to cope with conditions on the exposed west side of Kodiak Island. Others are as simple as a single stake on the beach with a mid-line buoy and two anchors and two buoys in a hook at the offshore end. The anchor itself may be as rudimentary as a 55-gallon oil drum with wooden cross pieces thrust at right angles through it to keep it from rolling with the current, the drum being filled with rocks to hold it down.

Spawning migrants tend to keep close to shore as they work their way toward their natal stream and they blunder into the set gillnet much as the trap of an earlier day took its fish close to the beach. But the similarity ends there. As in all gillnet fisheries, the set net is subject to stringent rules on length, mesh size and distance between locations. Seals and sea lions are particularly fond of set nets since they furnish a bountiful supply of food with a minimum of effort and one of the tasks of the set net fisherman is to persuade these creatures to stay clear of his nets. Gun fire is the most persuasive of the methods used.

Some set net fishermen pick their nets only at the bottom of the ebb tide. Others patrol the nets constantly, picking as they go to ward off or minimize damage from predators. Some net locations consistently show good earnings while during such peak years as 1965

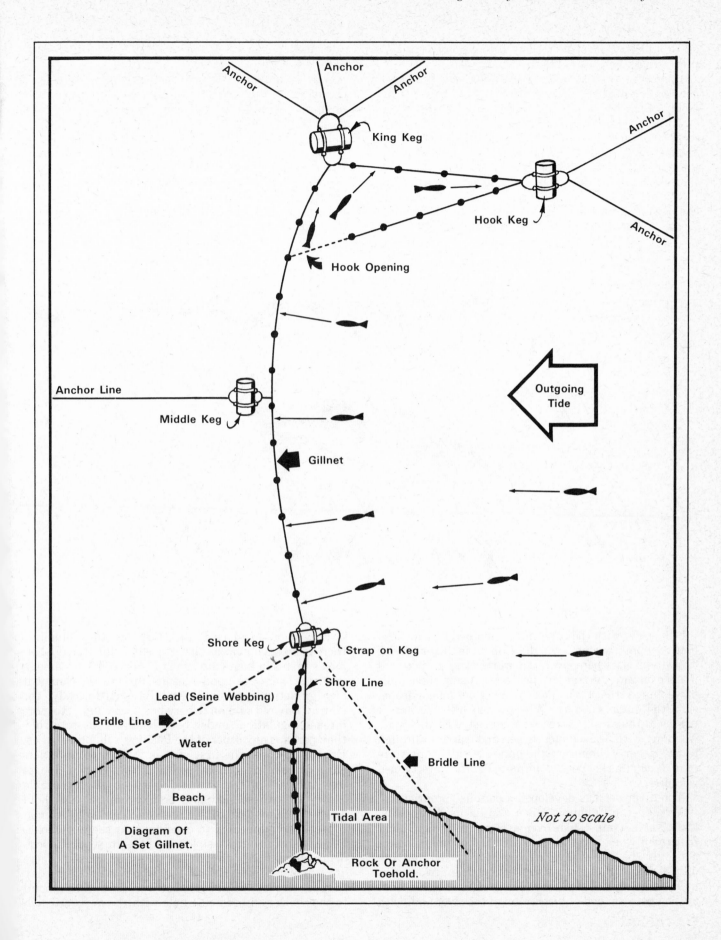

Anchor

Anchor

Anchor

King Keg

Anchor

Anchor

Hook Keg

Hook Opening

Anchor Line

Outgoing Tide

Middle Keg

Gillnet

Shore Keg

Strap on Keg

Shore Line

Lead (Seine Webbing)

Bridle Line

Water

Bridle Line

Beach

Tidal Area

*Not to scale*

Diagram Of
A Set Gillnet.

Rock Or Anchor
Toehold.

on Bristol Bay, most nets showed some profit. In that year, one of the best of the century on Bristol Bay, two brothers, one 14, the other 12, took 4,000 reds from a set net location north of Pederson Point, itself north of the mouth of the Naknek River. Not all nets were so profitable that year and the return from some every year is in direct ratio to the amount of energy the site operator displays in fishing it. Net locations, at least good net locations, are a valuable property and some sites have been held in the same family for several generations.

The drift gillnet, a mobile and deadly taker of fish, has been modified into the floater net with its cork line at the surface and the lead line hanging freely, and

good luck, reaches the end of it with a net full of salmon and no web to patch.

The Columbia River gillnet fishery, the only legal commercial salmon fishery on the river, also has given birth to the small gillnet boat called the bowpicker, described in detail in the Fishing Vessels section. Most gillnet boats work their nets from the stern but the bowpicker hauls its net from the bow. The Columbia River bowpicker has strayed into other gillnet fisheries in small numbers because of its comparatively low cost, but it has not gotten popular in salt-water areas where fishermen want better accommodations than those of the spartan Columbia River boat.

W. L. Scofield

the diver net with the cork line submerged and the lead line sweeping bottom. On the Columbia River at one time (and elsewhere over the world) the gillnet was made over into the trammel net, a gear using three walls of webbing rather than one. The inner net hangs deeper than the outer webbing. When a fish hits the net, it passes through the outer webbing, strikes the inner webbing with its smaller mesh and carries through to the larger webbing on the opposite side, thus trapping itself in the pocket formed by the intertwined webbing.

The diver net has developed a curious "proprietary" fishery on the Columbia River. To be used properly, the diver net must fish over a clean bottom. Gillnet fishermen work together to clear the bottom of snags and over the years have arrived at exclusive but unofficial drift rights along river sections from two to five miles long. In practice, the fisherman starts his drift at the upstream end of his section of the river and, with

Except on Alaska's Bristol Bay and the Columbia, there is no such boat as the "typical" gillnetter. Fishermen's tastes in boats are as varied as their likes among women. There are good reasons for the development of a gillnetter type on Bristol Bay. Historically, the stormy Bay red salmon fishery has been subject to more intense regulation, under both federal and state rule, than perhaps any other of the Pacific's salmon fisheries. Bristol Bay was the first area in Alaska to have nets chased out of river mouths. Until 1952, only sailboats were allowed in the gillnet fishery on the debatable theory that these slow and cranky craft helped hold down the salmon catch. No one concerned with salmon conservation appeared to consider that these sailboats helped tally the greatest loss of lives in separate accidents of any fishery in the Northeastern Pacific. There are still men fishing Bristol Bay who remember the day 26 men were lost when a northerly breeze backed around to the south and became a gale with the awkward sailboats on a lee shore. That was July 5, 1948, the "Bloody Fifth of July."

The gillnet is not regarded with affection by purse seiners in areas where it and the purse seine must compete for the available or at least allowable catch of salmon. In places like Bristol Bay where trolling is the only other method used for taking salmon, there is no friction. But on Puget Sound, however, where there is too much gear anyway, the misalliance between seiner and gillnetter has, at times, approached a state of guerilla warfare. Time after time, the friction between the two factions has come to the floor of the legislature of that state and initiative measures have reached the ballot on occasions as the competitors sought by politics to cripple each other. None of these propositions before the voters have been so nakedly worded but the underlying intent was obvious to any person even remotely familiar with the beleagured Puget Sound salmon fishery.

The object of the drift gillnet is, of course, to catch fish. To this end, the salt-water gillnetter fishes where experience tells him he is most likely to encounter fish. This means he usually fishes close to the beach to intercept salmon bound on their spawning runs or over grounds where feeder salmon are known to school. Most salmon taken by the gillnet, however, are that season's spawners. It is the troller who takes many immature salmon.

The gillnet is viewed with something less than affection too by some fisheries scientists who regard it as a relatively inefficient method of taking fish because of the high rate of "dropout" from the net. Fish well-gilled by the net die almost immediately by drowning. Many of them fall from the net when it is being hauled or worked by wave and current action and some researchers put the figure as high as 15 to 20 percent of the potential catch. It has been estimated that the Japanese high seas nets suffer as much as a 40 percent dropout rate. It is believed too that a substantial number of fish that do make contact with the net but subsequently escape it may die before they can spawn because of their vulnerability to bacteria and parasites through breaking or scraping of the skin. Despite these objections, only the troll boat rivals the gillnet boat in numbers and the gillnet accounts for about one-third of all salmon landings wherever it is fished.

## The Construction Of Gillnets

The gillnet is a creation somewhat simpler in design and form than the usually-big and ponderous purse seine. It is lighter, put together of materials considerably more fragile than those of the salt-water seine. It has fewer parts and it is easier to meld those parts into a coherent whole. It is easier to hang than the seine and often enough it is hung by one man, he who will fish it.

Gillnets are built of the same synthetic materials found in all modern nets. With the introduction of nylon and other synthetics after World War II, gillnet fishermen quickly learned that clear monofilament nylon was a deadly net material because fish often appeared unable to see it even in clear water in daylight. Fisheries officials learned this just as quickly as did the fishermen and the use of monofilament was banned.

Net makers, particularly the Japanese, began immediately to explore ways of getting around the prohibition and began to market webbing made of twisted white nylon fibers that approached the monofilament in effect. This webbing is not as hard (presumably) for salmon to see but it is a webbing that clearly violates the spirit of the ban on monofilament. Despite its efficiency, most North American gillnet fishermen limit themselves to various shades of green to conform to water and light conditions.

The Japanese themselves use monofilament in at least some of their high seas salmon gillnets in the North Pacific. Bristol Bay reds have been caught in American gillnets, still carrying with them necklaces of flat, ribbon-like monofilament, a lethal type because the flat webbing holds fish more securely after gilling than the round webbing. Perhaps the one thing to be said in favor of this material is that its use may reduce the incidence of dropout from the miles-long Japanese nets with the consequent savings, in the aggregate, of hundreds of thousands of pounds of salmon.

The gillnet is hung much more fully than the purse seine. The Icy Strait seine used previously as a model requires 12 fathoms of web for each 10 fathoms of cork line or 300 fathoms of web for the 250 fathoms of cork. For the most gillnet construction, designers use as a standard 220 fathoms of webbing for each 100 fathoms of cork line, or 660 fathoms for the Puget Sound gillnet built to the 300-fathom maximum.

Purse seine specifications, except for the rectangular drum seine, call for a lead line about 10 percent shorter than the cork line. The gillnet, on the contrary, is built with a lead line running from eight to ten percent longer than the cork line in order to offset the stresses set up by power reel hauling. The weight of the gillnet vessel itself also must be taken into account in determining the percentage of hangout on the lead line as well as the strength of the line materials in lead and cork lines. The lead line hangout and the weight of line materials may be one set of specifications for a light 28-foot Columbia River boat and something else again for the 40-foot offshore troller turned inside gillnetter for a part of the season. Wind, for one thing, influences the rate of drift of the towing boat and the small boat will be less affected by wind. But the bigger craft, with its greater square footage exposed to the same breeze, must be more influenced by it with a consequently greater strain on the cork and lead lines. A happening sometimes seen is excessive stretch of the cork line under such stress. When this takes place, the lead line may roll over the cork line with a resulting mare's nest that may take hours or even a couple of days to un-snarl.

To construct a Puget Sound gillnet of the maximum length and depth requires less area, less manpower and far less money than that needed for the full-size purse seine, up to 300 fathoms long and 350 meshes deep. A gillnet of maximum size with all new materials can be put together in a professional loft for about one-fifth or less of the cost of a new purse seine with highly skilled men doing the work.

Material needed for a 300-fathom Puget Sound gillnet 120 meshes deep includes:

About 325 fathoms of seven-sixteenths-inch braided polypropylene line for the cork line. (The extra line is needed to allow sufficient turns on the reel for security as well as enough free length to allow the net to hang well clear of the gillnet vessel on the drift.)

A minimum of 330 fathoms of three-eighths-inch core-leaded line, weighing 85 to 95 pounds to each 100 fathoms, for the lead line. (This allows for the extra length of the lead line.)

About 775 corks, five and one-fourth inches long by three inches in width, to provide flotation with these corks having a dead weight of 2 ounces each and a buoyancy factor of 16 ounces. These will be spaced at a ratio of 2.5 corks per fathom of cork line for a total of 750 but both ends of the net require extra corking which will use the extra 25 floats or most of them. The 25-fathom markers will be red.

Sufficient hanging twine for the cork line, No. 15 or No. 18 nylon.

Lead line twine, the twine for hanging webbing to the lead line, a sufficient quantity of No. 12 or No. 15 braided nylon.

Sufficient breast line material to secure 120 meshes at each end of the net, either a double lacing of No. 18 nylon hanging twine or single quarter-inch nylon or manila.

Exactly 660 fathoms of nylon webbing, five and one-eighth-inch stretch measure, of 210d/15 material, with a breaking strength of about 48 pounds, weighing about 38 pounds, measured by the usual standard of 100 meshes by 100 fathoms. The 120-mesh depth will weigh somewhat more but it is a factor of little importance.

This net can be a single 300-fathom strip and a minority of nets are built like that. But most fishermen prefer to break their net down into 50- or 100-fathom "shackles" or "shots." This net will be built of three 100-fathom shackles. There is a most valid reason for this subdivision of the webbing and lines. Gillnets are highly susceptible to damage but with the net in three or six shackles, a damaged shackle or shackles can be easily taken out of the net. The shackles are laced loosely together and it is only a matter of an hour or a bit more for a fast man with the needle to rip out lacing and lace two undamaged sections together.

At the junction of the cork line of each shackle, many gillnets are supplied with "knot protectors" to secure the line segments without joining them physically. The knot protector, about five inches long and three inches through, is a hollow, plastic, cork-like device with a hole in each end. It is threaded for easy takedown. The running end of each segment of cork line is inserted into the protector body and a knot of sufficient size to keep the line from slipping away is tied inside the protector.

The knot protector is not used on the lead line. The line segments can be joined by a sheet bend or other knot or by one of several type of patent connectors. The one commonly supplied by net lofts in the Pacific Northwest is a three and one-half-inch version of the big snap ring used on purse seines. This snap may be used singly or by twos with the line ends secured through eye splices or seizings.

Gillnet hanging in Pacific Northwest net lofts is done mostly at the hanging bench. The hanging bench is a simple, low bench equipped with movable arms by which hanging distances for cork and lead lines may be measured right to the turn of the hanging twine. One version of the hanging bench has a horizontal threaded steel rod opposed to a vertical rod for measuring hangings for the cork line as well as locating accurately the spacing of the corks. The other version has, as its principal feature, a retractable wooden arm by which hanging distances on the lead line may be set. The hanging bench appears to have originated in Norway but came into the net lofts of Washington State from Cordova, Alaska, where it seems to have been put to use as early as the first gillnet fishery for salmon in the territory. In addition to its exact measurements, it provides a comfortable working site for net loft men, many of whom are well past middle age with infirmities that make it hard for them to stand for the long periods the work requires. It was for this humanitarian reason as well as for the increased work load the netmen could take that impelled certain net lofts to put the hanging bench to use in the early 1930's.

As for the rest of the gillnet area of any net loft, the decking is smoothly painted and usually waxed in deference to the comparative fragility of the gillnet webbing. On the decking, every needed measurement has been reduced to fathoms, feet and inches while from the overhead are suspended the hooks needed for supporting lines as needed. The accents of Scandinavia are dominant in Pacific Northwest net lofts.

The gillnets put together in these lofts usually are those nets ordered by fishermen from areas remote enough to make it impractical or too expensive for the fishermen himself to appear in person to watch over the hanging of his new net. For the most part, gillnet fishermen prefer to hang their nets themselves because it saves a considerable part of the total cost of a new net. (Most gillnets are like most seines—an assembly of new and used line and webbing. The net is new just once. From then on, replacement parts are put into the net when they are needed.)

Many gillnets are rehung every spring in a basement or in a backyard because the work requires comparatively little area and it can be done as a man feels like doing it. Much net work is done, however, in areas like the two-acre net yard operated by the Port of Seattle, Washington, at its big Fishermen's Terminal just off the Lake Washington Ship Canal. This is a fresh-water area offering moorage for about 600 fishing vessels of all sizes and specialities.

The hanging bench can be seen in use here, especially among older fishermen. In other cases, gillnet men secure their lines on posts or X-frames and hang to them along a five- or 10-fathom stretch just as the seine webman does. The commonest method, however, is to work the net in the flat, using neither hanging bench nor supports. In this method, the net is laid out in a shape roughly a square of the total depth of the net. This requires an area of about 40 feet on each side or somewhat more or somewhat less according to the available space and the preference of the gillnet man. Since the fisherman knows the distance apart of his hangings it is a simple matter for an experienced man to work his lines past him from start to finish.

One of the virtues of working in the flat is that the method offers the novice at least a glimmering of the ultimate shape of the net, something not afforded by the hanging bench. The working surface, again, should

be as smooth as possible to minimize possible damage to the webbing.

The hanging of a new net is a rather more complicated job than the re-hanging of a net that has seen use. The elder net may need nothing more than the mending of rips or the insertion of a new shackle or a new section of cork or lead line or the replacement of sections of hanging twine. The new net exists only as reels of line, cases of corks and bundles of webbing. It is up to the fisherman to transform these components into a properly balanced and secured net.

With the work area selected and the necessary materials at hand, the first step should be to run corks along the cork line of each shackle. These need not be spaced at this time; they can be pushed to the end of the line or in that direction to be picked up one by one as the webbing is hung. The spacing of corks on the gillnet cork line is done more precisely than the spacing of the corks on most seines, most especially on those seines that have seen much work.

With the use of leaded line, there is no problem of hours of work slipping hard-to-handle sleeve leads along the lead line. Some fishermen still prefer the

Port of Seattle

*Fishermen's Terminal, Seattle, Washington. The facility provides berths for more than 600 fishing vessels of all types and sizes. It lies on the fresh water of the Lake Washington Ship Canal, inside the locks leading to Puget Sound.*

leads to the cored line, however, with the belief that it sinks more rapidly than the cored line.

At the junction of the breast line of the first shackle and the cork line, extra corks may be placed on the line outside the webbing. The number is up to the fisherman, enough, in his opinion, to give this section of the line a bit of needed extra buoyancy. These are seized individually to the line on 12-inch centers. Two, three, four or more may be used. Or none at all...

When a new seine is being hung, the bunt end breastline is hung to the webbing, sewn mesh by mesh, as first step in the webwork even before hanging to the cork line begins. The towing or seiner end of the net has its breastline hung as the last step in the webwork. Similarly, the float or buoy end breastline is sewn to the webbing, mesh by mesh too, as the usual first step in hanging a new gillnet with the towing end breastline hung last. These are the usual first and last steps in webwork proper but again, as in most matters dealing with nets, procedure often is a matter of every man to his own opinion.

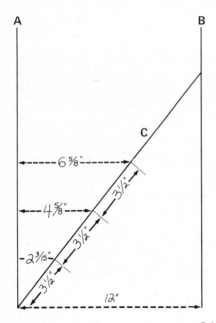

Robert Browning

*One method of calculating hanging intervals for gillnets (and purse seines). This method is based on same ratio of mesh size to fathoms of webbing but is the most cumbersome of the three.*

The usual second step in the gillnet webwork is to hang the first three meshes of the strip under the first cork of the cork line proper, not under any cork that may lie outside the breastline junction. The three meshes are to be picked up on the needle and knotted to the hanging beneath the cork. It is at this moment that the spacing of the web hangings become matters of importance because the balance of the net and its performance in the water depend on these complexly interrelated factors.

There are at least three methods to figure the hanging distances for the webbing. One, and the easiest, is a patented slide rule that allows determination of hanging spacings in inches as well as mesh pickup from a formula based on mesh size and the ratio of webbing to each 100 fathoms of cork line. The second, just as accurate but a bit slower, is the use of two vertical lines and a diagonal which uses mesh pickup and webbing-cork line ratio to arrive at the hanging in inches (See diagram.) The third is a chart similar to a highway mileage-destination chart which shows hanging spacings in inches according to mesh size and web-cork line ratio.

By use of any of these methods, it is easy to determine that this particular net, hung at the ratio of 220 fathoms of five and one-eighth-inch mesh to each 100 fathoms of cork line, requires hangings spaced at seven inches with a pickup of three meshes on every hanging. To carry the exercise further, the largest mesh used for salmon in Washington, eight and one-fourth inches, would be hung on seven and one-half-inch spacings with a two-mesh pickup at the 220 by 100 ratio. At the extreme ratio of web to cork line, 300 fathoms by 100, assuming five and one-eighth-inch mesh, hangings would be at six and seven-eighth inches with a four-mesh pickup.

By the book, the hanging interval must be precisely measured. This may be done most easily by using the cork line arms of the hanging bench. A needle of the exact length may be used for the web work, too, or a stick cut to measure can do the job. Not all men work this closely to the rule, however, choosing to measure by eye and some become remarkably skillful at approximating the distance between hangings. But the eye, no matter how well-trained, cannot do the job with the necessary precision and these men are a tiny minority. What works well enough with the purse seine does not work with the delicate gillnet. Purists insist that the hanging intervals take into account the minute turns of the twine in the knot that binds the web to the cork line. It is doubtful that many men working outside the net loft follow the book to this extreme, however.

Corks must be spaced as precisely as the web hangings. This net uses 2.5 corks to each fathom of the cork line, a ratio that spaces them on 28.8-inch centers. It is a generally-observed rule that corks must be secured individually to the cork line when hangings are spaced more than five and one-half inches apart. This procedure must be observed on this cork line with its seven-inch hangings with a cork being secured to the cork line after every fourth hanging. This may be done by throwing a clove hitch or double or triple half hitch at the left end of the cork, then passing the twine to the opposite end where it again is knotted, with this done until the twine has been passed across the cork at least three times. (Obviously, if the cork lashings are a part of the web hanging, the final pass across the cork must be knotted at the end from which hanging is proceeding.) The hanging bench with its use of exact distances allows the most precise spacing of corks. Nets hung outside net lofts without use of the hanging bench often do not have their corks spaced as exactly as they should be with the fisherman willing to accept an error of an inch or more. It will be noted that the gillnet is not as heavily corked as the purse seine because the cork line does not have to deal with the great weight of the seine. In addition, excessive corking produces a "bouncy" net in rough water conditions. This movement not only tends to frighten fish from the net but it also puts undue strain on the entire structure of the net.

*Gillnet cork, five and one-half by three and one-half inches.*

It is customary in professional net lofts for two men to hang each gillnet with one man using the hanging bench equipment with the cork line gear while the other hangs the lead line. When a net is hung by the fisherman, it is usual to hang the cork line first and finish up with the lead line. Thus, with this net, at least, the webbing is hung to the cork line until the end of the first shackle is reached. Some fishermen do not lace the shackles together, particularly on nets that will be fished in usually rough or fast water because of added flexibility for these conditions. But most shackles are laced loosely together with No. 18 nylon twine, usually doubled, or with the equivalent strength in manila or nylon quarter-inch or three-sixteenths-inch line. Lacing the shackles together may be held off until both cork and lead lines are hung.

Work proceeds shackle by shackle along the cork line, then, smoothly and monotonously—pick up three meshes, tie them off; pick up three more, tie them off, pick up and tie with a cork seized after every fourth hanging...

The lead line hanging is relatively simple because of the absence of complicating factors, especially when core-leaded line is used. But the lead line, running from eight to 10 percent longer than the cork line, requires hangings of different measure, in this case 7-3/4 inches. Here the lighter hanging twine, No. 12 or 15 braided nylon, is used in order that it may give, if necessary, under the strain put on the lead line during the haul. Parted hangings are preferable to a parted lead line. The lead line is hung from six to 18 inches below the webbing to allow also for that strain and additional flexibility is obtained by picking up only two meshes and letting the third lie on each hanging. In this way, the

gillnet actually can withstand more stress than is apparent from the lightness of its materials.

When the hanging of the webbing to the lead line is completed, the gillnet is essentially a finished product. There remains the sewing of the tow and float end breast lines (if this has not been done earlier) and the addition of the towing shackle to the cork line and a shackle for the light-bearing buoy or float at the off-end of the net. Complete, the net weighs at the most from 500 to 600 pounds.

There are as many options for the gillnet hanger as there are for the purse seine designer. The gillnet is the central element of the trammel net. The surface gillnet becomes the shallow diver net of the Columbia River when the ratio of cork to lead line is reversed. The lead line runs above 140 pounds to each 100 fathoms while small corks are spaced as much as a fathom apart to allow the net to submerge. Columbia River diver nets drift over the bottom and their webbing and other fixtures are heavier than the drift net just described in order to minimize damage from the hazards of the river bottom.

Some gillnets intended for certain Alaska waters are hung with a "weed line" suspended six inches or more below the cork line in order to escape as much as possible the web's contact with surface weed concentrations. The weed line itself may be of three-sixteenths or one-quarter inch hollow braid nylon. The webbing is hung to the weed line, obviously.

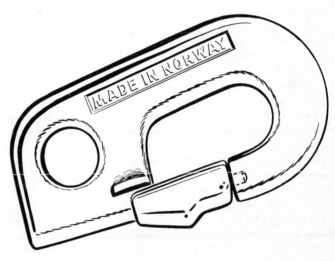

*Gillnet tow line shackle, a product of Norway, as is so much fishing gear used in the United States.*

A variant of this net uses a "bobber" cork line. This net is designed chiefly for rough water and is seldom seen outside British Columbia and Southeastern Alaska. The corks are seized to six- or eight-inch dropper lines with eye splices at the opposite end. The "cork" or tow line passes through these eyes. The webbing is hung to this line in a manner similar to the hanging of the weed line webbing.

The set gillnet differs in some detail from the drift net. The drift net is built with a maximum of "bend" in it and, by and large, it faces fewer hazards than the set net because the operator can often avoid deadheads and other drift. The set net, staked and anchored below the tide line, is subject to many hazards. Not the least of these is silt in many waters of Alaska, especially in Bristol Bay where such rivers as the Kvichak, the Nushagak and others feeding into the bay carry vast amounts of sediment. This silt, working in conjunction with tidal currents that often run like a mill race, is hard on every part of the net. These same currents carry floating debris even more destructive and the set net, consequently, is more heavily built than the drift net. Cork lines, for example, usually are of one-half inch polypropylene or even greater diameter instead of the seven-sixteenths favored for most drift nets. Lead lines run up to 120 pounds per 100 fathoms instead of the 75-85-pound weight used on drift nets. Cork and lead lines are of equal length, however, since the set net lead line does not have to meet the strain put on the drift net lead line during the haul. Mesh size follows the same regulations according to fishing area and district as those set forth for the drift net. Webbing usually is heavier than drift net webbing.

## Fishing The Gillnet

Setting the drift gillnet—the ideal set, not necessarily the usual one—is essentially the same procedure as setting the drum seine. The gillnetter, under normal conditions, drops his float rather close onto the beach and moves offshore in a straight line.

Instead of the heavy, highly-powered seine skiff of the purse seine vessel, the gillnetter has only a lighted float to buoy the off-end of his net, itself a creation of comparative fragility. Formerly, this float often was as simple as an inflated inner tube with a kerosene lantern secured to a piece of plywood lashed across its upper circumference. But this rude contraption has been almost entirely replaced by a shorter version of the pole and float used to mark the ends of longline strings. The modern float is a plastic pole about two inches through and six feet long with a weight at the bottom to hold it erect in the water and a battery-powered steady white light at its tip. Flotation comes from a water-proof pack of any one of several buoyant materials fixed to the pole. The gillnet vessel itself shows a steady red light at the masthead when it is fishing.

The power reel of the gillnetter is a diminutive version of the salmon seine drum and the trawl drum. It is hydraulically powered and can be controlled exactly as the bigger drums. It is most often found on the waters of Washington's big Puget Sound fishing district, through the length of British Columbia and in all of Alaska except Bristol Bay. This flexible device allows the gillnet fisherman complete control of his net at all stages of the set and haul.

To set his net, then, the fisherman first checks out his float, turns on its little light and places it at the transom where he can set it over the stern easily. Since the float is not heavy enough in itself to draw the net off the reel, the reel controls must be set for the desired tension. The float then is placed into the water with ample line between it at the net, and the fisherman, working at steering controls close by the reel, guides his boat off the beach while the net plays out behind it. Within a short while, except at extreme slack tide, current influence can be seen at work along the length of the net as the set is made. As the end of the cork line comes into sight on his reel, the fisherman prepares to brake his reel and bring his boat dead in the water. In order to avoid fouling, four to five fathoms of tow line must be allowed between the net and the boat, and this may not always be enough.

In calm water, with no wind, with currents not influencing the net unduly, the gillnet fisherman often finds himself with little to do other than drink coffee between hauls. But conditions are seldom so ideal and the fisherman may quickly be forced to haul before he had intended, especially if he is working among strong currents. It is on the drift that the gillnet displays its inherent ability to make a mess of things.

The purse seiner finds himself fully at the mercy of wind and current although he may be able to fend off these influences with his ability to set and haul his net quickly. The gillnet fisherman, however, finds himself even more bedeviled by the vagaries of nature since his method of fishing requires, in most areas under normal circumstances, that he hold his drift for a period of hours. This obviously cannot be true of the Columbia River or of narrow passes where currents run strongly at almost all stages of the tide.

But, if all goes as the fisherman would like to have it go, he can drift his net through most of one tide before he must haul and pick it. This will be found to be most true in areas of wide water where tidal currents are not constricted by the land and do not flow with the velocity they acquire in tight passages. Such grounds are, for example, the Kingston-Pilot Point waters along Washington's Kitsap Peninsula, the Salmon Bank area off that state's San Juan Island, British Columbia's Georgia Strait waters off the delta of the Fraser River, and in many of the great straits of Alaska. Bristol Bay drifts in poor years may be of several hours duration while in the years of heavy returns of red salmon, 30 minutes or less may be enough to pick up all the fish that can be handled.

To haul his net, the gillnet man again turns to his reel but again he is not hampered by the complex ring and purse line paraphernalia of the seine fisherman. The brake on the reel is released and, after assurance that his tow line is not fouled on the underwater structure of his vessel, the reel turns slowly ahead to retrieve the net. The net structure approaches the vessel over a smooth fender or roller and passes through fixed fairleads that feed it onto the drum. The fisherman must, in most cases, manually assist the smooth winding of the net onto the reel whereas the fairleads of the seine drum can be moved by the deckman at the drum controls.

Hauling the net can be a simple operation or one that may require considerable time. Fishermen using the gillnet reel pick their fish as they haul. Taking gilled fish from the net is an art in itself and untrained fingers sometimes are strongly tempted to resort to the knife. Salmon are bad enough to get out of the net but the gillnet does not catch only salmon; it takes everything that comes its way with the exception of those creatures small enough to slip through the meshes or big enough and strong enough to bull their way through it. Dogfish and other small sharks or rockfish with their dangerous spines can bring strong men close to something like tears.

Living things are not all the gillnet drift fisherman finds in his net. Kelp and other floating seaweeds are too common as is all the other flotsam and jetsam to be found in the water. The comparatively low-test twine of the gillnet parts much more easily than seine webbing and a man may spend most of a night with the needle patching up, at least, the biggest rips.

Fishing the drift gillnet probably is subject to more variables than the fishing of any other net, even the unpredictable purse seine with all its susceptibility to mishap. Not all sets are made in the manner of the one described. This set can be changed as easily as the purse seiner changes his customary right hand set to his left hand set. The gillnet fisherman marshals all his knowledge to put his boat and net into the water in the configuration in which it will be least subject to weather, current and the presence of other boats.

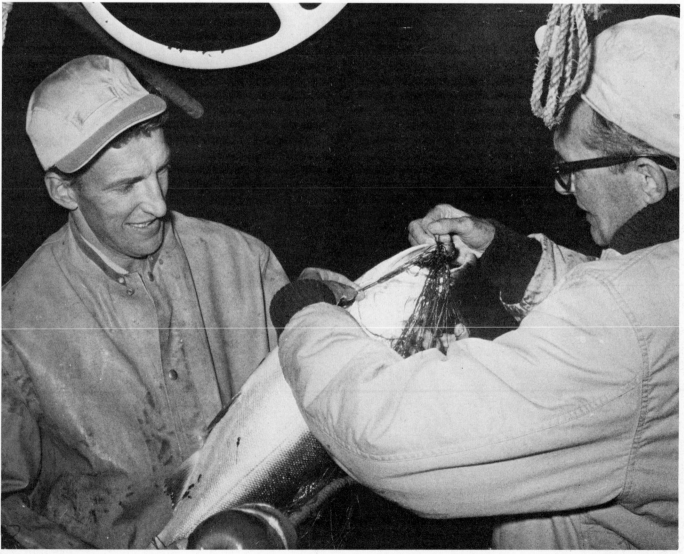

Seattle Post-Intelligencer

*Gillnetter picks silver salmon from webbing during early autumn season on mid-Puget Sound.*

Robert Browning

*Jackpot on Dock 8 . . . elderly man appears to have friend at Fishermen's Terminal, Seattle. The salmon he carries are gillnet-caught. The picture was taken during the autumn season on mid-Puget Sound where only gillnets and seines may fish. Trollers here have ended their season, but they can fish in coastal waters of the state.*

Washington Department of Fisheries

*Trolling pole frames troll vessel at work some 20 miles at sea off Destruction Island on upper coast of Washington.*

## The Troll Fishery

The third major method of taking salmon is by trolling. This fishery covers a larger area geographically than any other. Salmon trollers work from mid-California to Western Alaska with many of those out of Washington, Oregon and California ports turning to the offshore albacore fishery in July.

Trolling, comparatively, is the least-regulated of salmon fishing methods. With one small exception, there is no weekly closure off the Washington-Oregon coast, for example, such as that imposed on the inside gillnet and seine fisheries. Neither, however, is there any ocean net fishery for salmon permitted along this coastal area. Washington and Oregon do not allow commercial trolling on their inside waters such as the lower Columbia River or the waters east of Angeles Point on the Strait of Juan de Fuca. But with the exception of these relatively mild rules and some well-enforced size limits, the troller quite logically can describe himself as the freest among salmon fishermen.

The troller is mobile, his chief advantage; he can follow salmon where he wishes as long as the American

stays outside the Canadian three-mile limit and the Canadian stays outside the American three-mile limit. Many trollers from the lower Pacific Coast states begin their season in Alaska in April. They remain in the waters of that state until early June, then move south off Washington and Oregon when the coho season opens in mid-month. As soon as albacore begin to show off mid- or Southern California, many abandon the salmon fishery and go after the elusive tuna.

The troller can fish on several salmon species at the same time by using differing lures and changing trolling speeds and fishing depths. There is a wastage of fish in the troll fishery through the necessary gaffing of undersize salmon—"shakers"—to bring them aboard to get them off the hook.

The troll fishery almost exclusively supplies the fresh salmon market. Most trollermen treat their catch with care, cleaning them as soon as possible, icing them quickly and handling them gently.

Troll fishing for salmon is more of an art, without question, than any other form of taking them or any other fish of the West Coast except for the harpooners after swordfish. The grounds are bigger, the fish scattered. Much trolling is done "outside," out of the sheltered waters of the British Columbia coastline and Southeastern Alaska. All trolling in Washington, Oregon and California is on outside waters, sometimes at distances of 30 miles or more. Some Alaska trollers such as those fishing the west side of the Fairweather Ground or off Middleton Island in the Gulf of Alaska run even further offshore.

This far-out trolling is not especially a matter of choice since both the Gulf and the Pacific off the coast of the Lower 48 states are waters subject to drastic upset even in the summer months. But in those waters is where the salmon are and to take them, they must be followed. They are salmon, oftentimes immature feeders, that roam far and fast, fish as elusive as deer in the hunting season, fish not yet ready to follow the travel patterns of their elders, the spawning-bound migrants ready for their home streams.

The troll fisherman thus carries a burden not really shared by his fellows of the seine and gillnet fleets who, by regulation, fish inside and need mostly only to make their sets across the usual paths of migrating adult salmon and wait for the fish to come their way. Facing the troller is a lot of water that appears to bear no fish life whatsoever.

So the troller, with his whole ocean area to work, calls on his past experience, his intuition (or what he hopes is intuition but is mainly hope itself) as well as a healthy shot of good luck. He picks out his lures and rigs them with what he hopes is salmon sense and puts out his lines in what he believes to be salmon water. There are areas where salmon are known to concentrate at one time or another, drawn there by supplies of feed or the first urge to school up and head for home. But the salmon may not appear there on schedule. They may not appear at all that season. The salmon is a perverse creature and what may appeal to him one morning may be completely rejected that afternoon.

Neither the salmon nor any other fish reacts consciously to a situation. Fish are creatures of instinct, of reflexes triggered to turn to or away from a flash of light, a shadow, a movement in the water. It is the direction of the turn over the sum of the season that makes or breaks the troller's year. If his spoons or his herring or his hoochies are cleverly enough shaped and fished under conditions of light that stimulate a salmon to strike, well enough. If not, the lines can be dragged through the water by the week or the month with no great mortality among the fish.

There is a theory among some trollermen (perhaps hypothesis is the better word) that vessel noise—engine, wheel, gear, the thud of objects dropped on deck or in hold—translated into vibrations passing through the water may attract or repel salmon. This thinking quite probably is valid even though no one has made a study of it with the objectivity and the knowledge and equipment to check it out. But every man who ever has trolled knows fishermen who catch salmon when almost everyone else is willing to bet his last ton of ice there are no salmon for miles around.

Does the boat and its tuning mean the difference between the highliner and the run-of-the fleet fisherman?

That there is a distinction among fishermen is demonstrable. There are seiners who consistently make big hauls right after the men ahead of them have hauled nothing but jellyfish and salt water. There are gillnetters whose catch records over the years seem to show there are men who were born with the ability to "think" like a salmon or who, at least, can come pretty close to knowing—subconsciously, perhaps—what salmon are likely to do under the stimuli of varying conditions.

Trollermen who like the theory that noise of certain types drives salmon away from troll boats (no one has yet pinpointed the noise that would attract them) believe noises generated by the power train are most to blame. To them, this means that an engine properly seated with the shaft in exactly true line with the perfectly-balanced wheel is at least a part of the answer to that ever-new question:

"Where the hell are the fish?"

If this noise theory be true, there is reason to believe too that troll and stabilizer lines under certain wind and water conditions may contribute to the unease of salmon as readily as the worn shaft bearing or the out-of-kilter blade on the wheel. These lines, especially the stabilizer lines, growing taut under the stress induced by pitch and roll, whine like rigging in the wind, enough so at times to send chills down a man's back.

Why should they not have much the same effect on the ultra-sensitive salmon?

The trolling line's inability to distinguish among immature and mature salmon has brought to the fishery an increasing measure of criticism from some biologists and fisheries management men and, naturally enough, from sports fishermen. These critics contend that the offshore troll fishery takes too many "feeders and shakers," fish that may be one or two or even three years away from spawning. They contend further that such fishing is biologically unwise and that the troll fishery as presently constituted should be discouraged by any legal means. This belief was put down in writing in early 1970 by the Washington State Department of Fisheries which controls a substantial chunk of the Pacific salmon fishery. This "suggested" policy, contained in an over-all appraisal of fisheries responsibilities, said concerning the troll fishery:

"From a biological standpoint, it is preferred to permit commercial fishing as the spawning runs begin or as they (the spawners) separate from the vast ocean mixing areas so that control can be maintained over the harvest of specific stocks and optimum escapements to each stock can be more reasonably assured. This cannot now be accomplished as effectively as desired because of the intensive ocean fishery.

"An ocean sport fishery, even though it has the disadvantage of fishing mixed stocks, is socially and economically desirable...There is no parallel for a commercial troll fishery in the ocean. Compared with a net fishery which fishes on maturing runs, it has a number of disadvantages. This fishery operates on mixed stocks, allowing less precision in management, and harvests immature fish, consequently providing less total poundage from each stock than if the fish are harvested when mature.

"Furthermore, an ocean troll fishery is directly competitive with the ocean sport fishery as it generally harvests its fish before they enter the sport fishery whereas a net fishery on less active feeding fish is not so apt to be competitive. In many cases, the fish runs have already passed through the major sports fisheries prior to entering the net fisheries.

"A growing, relatively unrestricted troll fishery can only result in a greater restriction of other fisheries, particularly the inside net fisheries on maturing runs, in order that adequate spawning escapement be maintained. Allowing the ocean troll fishery to grow relatively unrestricted is, therefore, a very real form of discrimination against the net fisheries which can only fish on the remaining surplus. The troll fishery does have the desirable feature of providing fresh, high-quality salmon to the market at moderate cost over a long period of time. However, this advantage is not great enough that it would be economically desirable to allow continued, relatively unrestricted growth. The department must strive to seek that balance of catch among the ocean troll, inside nets and sport fisheries which provides the highest economics and esthetic return..."

Predictably, the trollers cried "Foul!" even as the coastal troll fleet continued to grow. But the trollers (and all commercials) could not help looking over their shoulders toward British Columbia where the Canadian Department of Fisheries began in 1968 to enforce a set of stiff rules regulating size of the provincial salmon fleet and entry into the fishery. This program had as its goal reduction of the fleet by half by eliminating part-time and unproductive fishermen. The situation in British Columbia is considerably different, however, than that of the four American states concerned with the Pacific salmon fishery in that the marine fisheries of Canada are regulated by the federal government and not by the several provinces. Thus, it was comparatively easy for the Canadians, working from far-away Ottawa, to impose the unwelcome limits on the B.C. fleet. Any similar action by the Pacific Coast states must be approved by the respective legislatures and, to be truly effective, all four states must act together to achieve proper policing of the salmon fishery or any other marine fishery. The chance of unilateral action by any state is remote and the chance of joint action is even more remote in view of the variety of economic interests and consequent pressures that are involved. But concern with conservation of resources and preservation of the environment is growing over all of North America and the limitation made so effective in British

Columbia should be kept in mind by all who have any interest whatsoever in the salmon fishery. An event that appears improbable in one year may be reality 10 years later. A reduction in all salmon gear, not merely troll gear, has been sought for 30 years in Washington and Alaska. Some of this reduction has been advocated by fishermen who would like to see some other gear be limited but not their gear. But much of the support for some kind of limitation has come from men like Dr. William F. Royce, the eminent, former University of Washington fisheries scientist whose concern is genuine and not selfish. Eventually, such views as his seem likely to prevail.

## Assembly Of Troll Gear

The assembly of salmon trolling gear used in the various fisheries of the Northeastern Pacific is not the fine art that distinguishes the design and hanging of such complex nets as the otter trawl and the purse seine. It is, primarily, the assembly in proper order of standard components available from all commercial fishing gear supply houses.

Trolling gear is "standard" in that every troll boat is rigged in essentially the same manner. But there are many manufacturers, domestic and foreign—specifically Japan and Canada (and Norway for hooks)—whose versions of a piece of standard gear may vary in considerable degree. It then becomes incumbent upon the individual fisherman to determine for himself what of this variety of gear items is best suited to him.

Here his decision must be weighed against his experience, the area or areas where he customarily works and the age and deck layout of his boat. The kelper working close among the coastal reefs has needs quite different from those of the "trip boat" fisherman who may be offshore for a week to four weeks or even more in search of salmon, albacore or one of the other species such as those sought in the Southern California fisheries.

The shape of modern salmon gear was pretty well set by World War II and the major difference in the years since has been the appearance of more durable materials and somewhat more sophisticated lures. Typical of the former is stainless steel fishing line of great strength (600- to 900-pound test) and ability to withstand the rigors of the troll fishery; typical of the latter is the increasingly more efficient flasher-type lure that does not catch fish by itself but attracts fish to it through its action. But still high on the list of favored inducements for salmon is the lowly and smelly herring, usually teamed with the flasher. Various spoons are used too, as well as a rubber or plastic device resembling a squid. This lure is known to the salmon troller as a "hoochie," because of its alleged similarity of action to that of the fabled hootchy-kootchy dancer who once, reportedly, was an integral ingredient of many American stag parties. (The albacore troller calls a related lure a "jig." The vernacular name "jig boat," used for an albacore troll boat, springs from this term. The word came originally from the feathered jig of the bait boat. The albacore jig consists of a chrome-plated lead head with vari-colored feathers streaming from it. White is the basic color with red, yellow or green used as contrast.)

Aside from the troller's poles, the main elements of his gear are several kinds of lines in addition to the fishing lines themselves; plastic floats to buoy and steady outside lines; lead or cast iron sinkers of several weights to keep his lines at a reasonable angle to the vertical as his boat moves through the water; a complement of small gear such as rings and clamps and springs and snubbers, all with their rightful and vital part in the gear assembly and, finally, his line haulers, the

gurdies without which modern trolling would not be possible.

The gurdy originated in the California salmon troll fishery and became standard equipment in two or three forms by the end of the 1940's. It sometimes is used in the albacore fishery by vessels taking advantage of the presence of albacore in waters they usually work chiefly for salmon. But it is not really a practical use for albacore because of the relatively slow hauling speed. The gurdy is indispensable. Each and every element in the gear assembly has its use but some pieces could be eliminated with the loss of nothing much but convenience and ease of operation. But any man who has had to haul a salmon line by hand with a 40-pound sinker at the end of it because of a gurdy malfunction will sign

*Troll gear: 1. Flasher for salmon.*
*2. Albacore jig.*
*3. Streamer fly.*

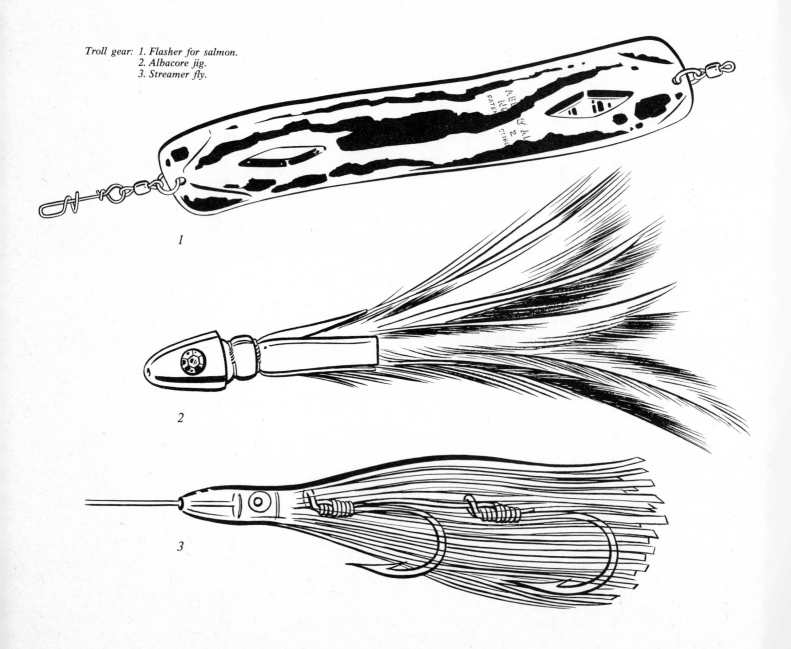

almost any kind of testimonial to the gurdy's contribution to trolling.

Many small inshore trollers use hand gurdies. With the light gear and short lines common to these craft, the hand gurdy works just as efficiently as the powered gurdy of the bigger vessel. Some of the smallest and lightest craft in this inshore fleet use outboard engines; hydraulic power takeoffs designed for the outboard have found increasing use among these small boats.

Gurdy assemblies may be powered by a takeoff from the main engine, by hydraulics or by electricity. Either of these latter is by far the preferable although hydraulic power is preferred by most men and is the system used on almost all new boats and on boats that have been remodeled. Most older trollers use mechanical power. This is often prone to breakdown while some installations, with exposed shafts and gearing, are a hazard in themselves.

The troller, among all the world's fishing vessels, stands out no matter where it is seen—fishing, running or tied up. It cannot be mistaken for anything else. Its distinctive characteristics are the tall poles from which the gear is rigged.

The poles may be likened to spreader bars. They allow the lines to be separated enough so they may fish without interfering with each other. The main poles are set amidships and abeam of the mast. There is a rough rule that says the main poles should be close to the length of the boat itself. A small boat, however, may have main poles a bit longer than itself while a big boat may have poles rather shorter.

The main poles are secured in brackets on the rails and, when in upright position, rest in the forks of a cross tree below the masthead. Some vessels use bow poles and these may lie aft when secured or else stand upright in their brackets with the aid of guy lines. The brackets for the bow poles are set into the foredeck rather than the rail. The poles themselves are of spruce, cedar or fir although aluminum poles appear increasingly in the troll fleet.

Poles are of several types, each of them increasingly costly. Wood poles are most common with several types of wood in use, among them cedar, fir, hemlock and spruce. Cedar is preferable among the woods because of its resistance to decay. Many poles are cut to size and roughly finished by the trollerman himself. But poles that have been machined are not uncommon and they do add somewhat to the appearance of a class of vessels that bears too much of a degree of shabbiness as it is. Replacement of a damaged wood pole usually is an easy affair, a matter of going ashore in some safe anchorage, selecting the proper tree, felling it, limbing it and shaping it to proper length and bracket measure. The green pole will not be as sturdy as a pole well dried but the new pole means the boat can go back to work again.

Metal poles and a combination of wood and metal have appeared increasingly since World War II, that upheaval that changed the direction of so many lives, so many endeavors and so many technologies. In the post-war years, advances in aluminum production and fabrication stemming from the war effort resulted in increasing abundance and comparatively modest cost of the metal. As a result, favorable strength-to-weight ratio made it an increasing favorite of builders in metal and by the middle 1950's its use for pleasure and work craft began to grow.

Gillnetters and seine skiffs were the first fishing-related boats to be built from it on the West Coast while, at the same time, trollermen became conscious of its virtues as material for their poles. With one exception, aluminum is the best material presently in use for poles. It is light, strong, resilient and virtually immune to the effects of salt water. It is used either for the full-length pole (hollow, obviously) or as a sleeve for the lower 15 to 20 feet or so of the pole. This use protects the wood against potential damage from stabilizer drag and other wear and provides, as well, a much stronger seating for the stabilizer fastenings and adds security at the bracket. Nor does aluminum, a non-magnetic metal, add to the compass deviation inherent in all watercraft with the smallest touch of metal about them. The cost of aluminum may not justify its installation on the day troller or on boats of marginal income or value. But its use seems clearly indicated on new vessels and on all boats with a consistent record of profit. And certainly, aluminum is the answer for albacore men who may be working a hundred or so miles offshore where replacement wood poles are few and hard to find. Many albacore boats with wood poles carry a spare as insurance, anyway.

An occasional troller may be seen with stainless steel poles, either full-length hollow poles of this semi-precious metal or sleeves like those of aluminum. Stainless steel has all the virtues of aluminum; it is beautifully shiny, beautifully strong, equally flexible and even more immune to the kiss of death of sea water. It also is so expensive that it has no practical value that aluminum does not possess too. For a seriously working troller, aluminum does all that stainless steel does. For a troller meant chiefly as a summertime plaything or as a tax loophole, stainless steel does add to the good looks and to the tax write-off. In an endeavor of this devious nature the ultimate absurdity would be a vessel all of stainless steel from keel to masthead...

The entire assembly of trolling gear from pole bracket to fishing leader is designed to withstand shock. Unfortunately, the only kind of shock welcome to the fisherman is too rarely felt—that of an outsize, salable fish such as a 60- or 70-pound king salmon. The more common types of shock are those that mean damage of lesser or greater degree unless their energy can somehow be dissipated. A stabilizer or a trolling line snags a deadhead; a 50-pound sinker hangs up on foul bottom; vessels crossing courses also cross lines...there is a considerable list of possibilities.

Along the train of gear there are fail-safes built into the assembly to minimize damage. To avoid major loss, the trollerman is prepared to accept minimal loss and a certain amount of work getting his gear back into fishing shape. For instance, there is no question about trading a rawhide thong and sinker for a trolling pole. The thong breaks; the sinker is lost; the affected pole scarcely bows under the slight stress. Sinkers are more expendable than pole and costly troll line.

Here are some pertinent definitions concerning salmon trolling gear:

Tag line—Lines from the trolling pole to the main lines, intended to allow a means of separating the main lines in the water and to bear their weight when they are fishing. The length of the tag line is the distance from its attachment at the pole to the rail at each side of the trolling pit. The tag line may be of manila, braided nylon or stainless steel.

Main line—The fishing lines themselves, high-test twisted stainless steel wire. The number of lines runs from four to six. They are secured to the tag lines when fishing by one of several methods.

Gurdy—Powered spools or reels in sets of two, three or four. Each gurdy spool contains and works one main line. Each spool is controlled by a clutch, brake and reverse gear.

Shock or shock absorber—A spring or length of elastic material introduced at one or more positions in the pole-tag line-main line complex to absorb some of the sudden strain on pole and lines when a heavy fish strikes or when the gear hangs up on the bottom. The shocks can be at the pole, at the trolling davit or at both.

Breaker line—A light cord, strip of rawhide, nylon or any other material of less strength than the main line or the tag line, set into the line assembly at the end of the tag line and between the main line and the sinker. Its purpose is to take the stress when the sinker fouls on a bottom hazard and by parting to save the main line and the pole.

Stops—Devices of brass, somewhat resembling elongated beads, about one-half inch or less in length, crimped onto the main line about four inches apart to prevent leader connectors from sliding along the line. The sets of stops usually are two and one-half fathoms apart on the line.

Bumper or snubber—A rubber device about 12 inches long and about one-third of an inch in diameter with rings and snap swivels firmly set into each end, used to dampen shock when a heavy fish hits. It is put into the leader assembly between the leader and the connector securing it to the main line between stops.

## From The Trolling Pit

There is a singular satisfaction in salmon troll fishing for the man who follows this fishery. For most purposes, he is king of all he surveys from his trolling pit. He may not make a fortune at his sometimes-arduous job but he has a greater degree of independence than any other West Coast Fisherman except the gillnetter. Low fish prices stir him to militancy and most salmon seasons begin with a strike; with a price agreed, he proceeds to forget the whole matter. He has an ocean that concerns him more. The kelper or the day boat fisherman may not worry about the sea too much but the trip fisherman, the man who spends much rough time offshore, knows he is dealing with a monster.

Often, he must face this menace himself because in many instances he is a man alone. If he does have crew, preferably the crewman, the "boatpuller," will be a kid relative who will work, knowingly or not, for less than the standard 12.5 percent share. Or perhaps the boatpuller is the fisherman's wife who will work for nothing as far as a cash share is concerned but who, nevertheless, usually carries the checkbook. An additional hand on a trip boat, wife or whomever, benefits the fisherman exceedingly in the galley. Most men fishing alone do not feed themselves properly; the exceptions are the men who seem to spend more time in the deck house than in the trolling pit. Decent cooking takes a certain amount of valuable time and a man who has a boat to handle, lines to run and fish to clean and ice seldom finds the needed time. It is simpler to open a can and, perhaps, eat the contents cold in the trolling pit while tending to business.

(One simple but nutritious dish, not acceptable probably to low-fat, low-cholesterol dieticians, consists of white beans, salt pork and onions simmered for eight hours or more on a slow stove. This dish lends itself to all kinds of condiments while split peas, yellow or green, can be substituted for the beans. The pot should be good for about three days of eating if it is kept hot enough or cold enough to inhibit bacterial action. Reefer facilities on many boats are limited or non-existent but that fact makes little difference to the fisherman—not while he has a hold with several tons of ice in it.)

The focal point of any salmon troll boat is the trolling pit—call it cockpit, fishing pit, the "hole." It all comes out equally—it is a cockpit large enough to allow the fisherman to work without exposing himself to falling overboard. There are some troller-gillnetter vessels that do not have a trolling pit as it usually is defined but the transom is high enough to offer most of the protection that the fisherman finds in a standard pit. From the trolling pit, the fisherman has control over the major functions of his craft—his engine, his steering, his lines and his depthfinder. These are remote controls that eliminate the need for a second person in the wheelhouse to steer and to monitor depths. They are the gear that does allow a lone man to fish safely enough and successfully.

The depthfinder is an instrument of greatest importance to the troll fisherman because much of the time he works on fairly shallow sounding where fish are known to migrate or congregate to feed. Some troll boats have a depthfinder mounted on an instrument

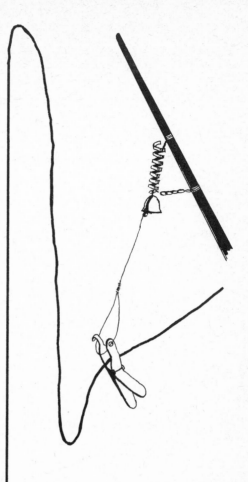

*This is a schematic configuration of salmon troll gear from pole to sinker. Essential elements of the assembly include the pole itself; coiled steel spring; a small bell on spring to alert troller to a strike; tag line and main or fishing line meeting at "clothespin." From the heavy-test steel line extends one "spread"—snap connector, rubber snubber, two- or three-fathom leader, flasher and whole herring. Line terminates in lead cannonball sinker.*

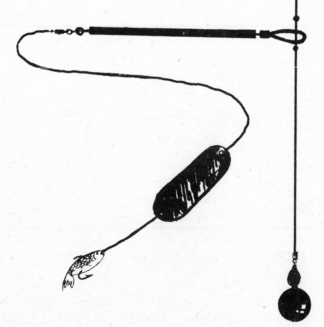

Warren Ernst

panel forward and within reach of the pit; others have the finder hanging on a movable panel on a bulkhead in the house whereby it may be swung to face astern so the fisherman can keep an eye on it. All these remotes are flashers although many vessels carry recording units in the wheelhouse. Proper interpretation of read-outs of these latter requires skilled instruction. Western dragger skippers have become experts in this art.

In addition to the above gear, the man in the pit also has handy the other tools he must have to do his job right. These include the sharpest of knives, stones for honing them to that edge; gaff, club, hooks, leader, spoons, hoochies, herring, and always gloves. The only practical glove for the troller is the cumbersome rubber type that comes well above his wrists, provided on the palms and fingers with a roughened surface to enable him to clutch fish handily and clean them rapidly and easily.

Leather or cloth gloves just won't do; within a few hours, washing or no washing, they develop an odor that must be cousin to that emanating from the slave ships of the Triangle Run so long ago. Such gloves begin to stink to high heaven because fish blood is a substance that begins to decompose even more rapidly than animal blood. Clothing—the below-the-elbows portions of jackets or rain gear—begins to smell too in short order because it is almost impossible to keep from some contact with blood, slime or viscera.

The well-furnished trolling pit comes with some form of overhead rain protection for the man working the pit. This can be a skiff—and it often is—secured to the lowered boom and the trolling block pipe davit or it may be something as elegant as plastic or plywood similarly secured. These devices usually protect more against rain than against sun because rain in the latitudes above Northern California is more common than sun at sea during the summertime. Unless it happens to be fog...

Although the trolling pit is the heart of the troll vessel it, in itself, poses something of a danger to fisherman and boat. It should be self-bailing because water coming aboard with the pit uncovered, whether by carelessness or accident, is an obvious threat to stability. For further safety there should be a step-up at port and starboard with a firm hand grip assured above because trolling pit areas can get slippery quickly. No matter what, trolling pits in the northern latitudes become cold and dank places of employment in a short while. Good boots are as important to the troll-erman as to any other fisherman although most do not favor the hip boots used universally by draggers and seiners.

Size of the pit is in direct ratio to the size of the vessel. A 40-footer should have a pit running from five to six feet athwartships and about three feet or a bit more fore and aft. Ideally, the aft edge or its coaming would be about 24 to 36 inches from the lip of the transom although with a double-ender this distance

would be more. The port and starboard coamings should allow about eight to 12 inches of work area between it and the rail. There cannot be too much more on a salmon troller since the fisherman must lean over port or starboard rail to gaff fish as the hauled line brings the leader within reach. An extra-wild or extra-large fish presents more than a mild challenge in use of the gaff.

Gurdy assemblies normally are positioned equidistant from the center line of the vessel within reasonable reach of the man in the pit. This should average out at about 12 to 18 inches from the pit coaming since the control levers must be quickly within reach. Instant response from a gurdy spool may mean the difference between damage and no damage in event of a potential hangup on the bottom. This does not mean a full hang-up but it does mean that if a sinker begins to skip along the bottom, a quick lift of the line usually will clear it before it truly is fouled. The sinker is expendable but if it can be saved, there is no point in allowing it to be expended.

The nature of the gurdy power can determine the deck layout somewhat. A mechanical power takeoff from the main engine usually must run along the deck, ordinarily to port of the hatch. This power assembly takes up room because of the gear and shaft housing involved. The power takeoff requires a shift control assembly between the sets of spools or at one side that takes up valuable deck space. The hoses and wiring for hydraulic or electric gurdies can be run under the deck and up to the gurdies with a consequent saving of limited space. The controls are small. Hydraulic fluid, incidentally, is one of the slipperiest substances known to man and it should be removed from deck surfaces immediately after line failures. Its effect can be neutralized partially and temporarily by blanketing it generously with rock salt. Fish slime and offal is slippery as well as quickly smelly and its riddance is a matter of esthetics as well as one of safety. Some vessels do not have deck hoses for this latter job and scrubbing down may be hard work if the stuff is allowed to dry. Getting rid of fish scales can become a matter of doing it one by one if they dry too much.

## Slaughterhouse

This problem of cleanliness exists around every trolling pit and every checker area on every troll boat. The butchering of salmon can become an incredibly messy job over a sustained period of activity. Any sizeable salmon or other fish possesses a substantial quantity of blood and a weight of viscera in proportion to its size. The first cut of the knife exposes this mass (and attracts every seagull for miles around). An expert with the knife can extract much of it in a single motion and fling it over the side to the birds. But there is an irreducible minimum that stays aboard and the cleaning of a couple of dozen fish shows it. Scales, blood and bits of gut cling to the vee-trough commonly used for butchering salmon; this same material is to be found around all the pit area. Blood and scales abound in the checkers because the gaff wound lets loose a free flow of blood before coagulation halts it.

Thus, to do properly by the catch and by the boat, the conscientious trollerman needs, in essence, two washing systems, one for his fish and one for his boat. The fish washing system may be as simple as a couple of deck buckets on a small boat or a hose and scraper setup on more elaborately geared vessels. The deck scrub-down system should be built around a high-pressure hose, sturdy scrub brushes and an appropriate amount of elbow grease, abetted by a chlorine solution.

As for the washing of fish, it is difficult to put too much thought and effort into this procedure. All fish should be killed, gilled and gutted as soon as possible after coming aboard. Salmon should be killed by the gaff handle or a club while still on the hook. Salmon die more quickly under a sharp rap on the head than most fish, a virtue of a kind. Members of the salmon species should not be allowed to thrash around on deck or in the checkers to die by themselves. They lose their scales easily and consequently show darkened skin area which reduces their market value. Their flesh bruises easily too, a fact that becomes quite apparent to the consumer who may already be less than entranced with his purchase if he buys the whole fish with the gaff wound prominent in the head.

It should be added that any fish to be cleaned must be dead before the knife is taken to it because an apparently dead fish coming to life in the cleaning trough under the first touch of the knife can cost a man a deep cut, a whole finger or even the hand. These wounds are extremely vulnerable to infection, sometimes a long way from quick medical care. The fisherman should

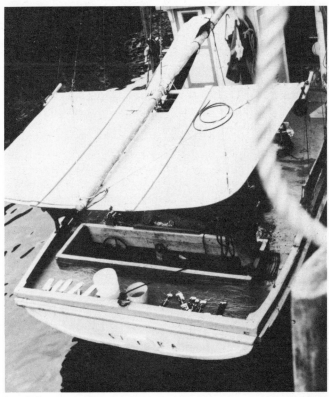

Dan Stair

*The troll pit becomes the trollerman's home away from home because, if he wishes to catch fish he must spend most of his waking hours in the pit. The pit of this Gulf of Alaska troller is typical of those among boats in this size range, 42 feet. The overhead protection is better than most of its kind, however.*

be particularly wary of the halibut, a tough fish that can go into a frenzy after hours of seeming death.

(The sooner a fish is killed after it is caught, the better are its keeping qualities in ice. The reasons are chemically complex and of no pertinence here; a corollary is that a fish killed with a minimum of struggle does not go into rigor mortis as quickly as one that has fought wildly over a period of minutes or hours. When rigor mortis does set in, the state lasts longer for the first fish than for the second. For the first fish, this late start and longer duration of rigor mortis has two advantages, or at least it does for the fisherman. For one, the still limp and pliable fish can be stowed away in a natural position rather than in the rigid, contorted shape of the fish subject to early rigor. For the second, again for complex reasons, the flesh of the first fish tends to stay in good shape longer. The flesh of net-caught fish almost invariably does not retain first-class condition as long as that of most troll-caught fish because of the usual troller habit of quick death, quick cleaning and quick icing. Trawl-caught fish may struggle for several hours against the net, then die of crushing in the cod end or of suffocation on deck. Salmon taken in the gillnet or the purse seine similarly lose quality more rapidly than the troll-caught fish for the

*Seiner crewmen at dawn ... first haul gets underway with the rising sun. Skipper controls hydraulic power with hand set. At stern, one man works, from left, the cork line; two in center pile web while man at right wrestles with recalcitrant lead line and purse rings.*

same reasons—struggle and drowning in the gillnet, struggle in the seine and suffocation on deck or in the hold. The sight and sound of a thousand salmon flopping themselves to death represents the sound of money, truly enough, but this symphony does nothing to improve the quality of the fish. Unfortunately, if seinermen were to have to put to death, gill and gut every fish coming aboard, there would be time for little else after one good haul.)

The careful washing of gilled and gutted salmon should be a matter of rote for the good trollerman, something to be performed as part of second nature. A fish clean inside and out lasts longer and better in the ice, all other factors being normal. Of particular importance is the removal of every trace of blood from the body and head cavities. The blood itself is an invitation to the bacteria that thrive even under the best of icing. Its decomposition imparts to the whole fish an unpleasant odor that it may carry with it all the way to the fish market. Wash water should, if at all practicable, carry with it a chlorine solution since this chemical has been proven to have a beneficial effect on the keeping qualities of fish.

The cleaning of fish is a problem the trollerman need face only after he has started to fish; before he is in shape to catch a single fish, there are others to confront him.

## Water Hazards

All manner of mishaps, some small and inconsequential, others big and potentially devastating, lie in wait for the trollerman (and all other fishermen), some even before he leaves port. These, other than fire, usually can be handled with little trouble by the experienced fisherman. It is those accidents offshore, all the way from Baja California to Western Alaska, especially if he fishes alone, that give fits to the fisherman. (As for the term "offshore," some of the great waterways of Southeastern Alaska—Chatham and Clarence Straits and Frederick Sound, for example—might just as well be classified as offshore. They are wide, mean and lonely, as inhospitable as the Washington coast.) The list of possibilities would appear endless but most can be headed off by the fisherman who senses and deals with a developing dangerous situation.

The major hazards the fisherman must face are fire, sudden illness (especially if he is alone); collision, loss of power or steering, storm and gear mishaps. Fire and illness need no explanation; gasoline explosion and the heart attack or stroke account for a certain number of men every season. Collision includes that with boats of his own size or thereabouts, with big shipping booming through night or fog and, most dangerous of all, collision with an unseen floating object as he travels under full power. Loss of power can mean many things, some repairable by the fisherman in a few minutes or hours, others like a broken shaft or lost wheel meaning

a tow at the very least. Storm needs no explanation. Gear mishaps, that is, accident involving elements of his fishing gear alone, are legion. Some the trollerman has in common with all fisherman; some are reserved for him alone.

Among them is one seen fairly often, especially on single-manned boats, and one, if it be carried far enough, that can mean potential loss of a boat. This is the "wild pole," a trolling pole getting out of control on the set or the haul. In this instance, for any one of several reasons, the pole misses its seating in the fork of the crosstree high on the mast and slams across the house. Big wooden poles are heavy and they are quite capable of smashing through a wooden house. Aluminum poles are lighter, length for length and diameter for diameter, but they can come down crashingly too. At the least, such an accident may mean the loss of several days of fishing. At the worst, it can let loose fire from a galley stove or cripple steering and navigational gear and, with these gone and storm coming, the possibility of a lost vessel is high.

The poles are stayed from four points while they are working but, when they are being lowered or raised, tension slacks off on the stays and the major control comes from the falls. Theoretically, the forestay should prevent the pole from going aft of the fork but this is more theory than fact. The backstay may be no more effective than the forestay while control from the falls is minimal and is least effective at the critical point in the pole's descent or rise.

This critical point appears when the pole has been raised or lowered to the zone where gravity has no real grip on it and the vessel is in the trough or in a confused seaway with no pattern to it. The danger is most acute when a pole is being raised and must be seated home in the crosstree fork. But a pole can get out of control when it is being lowered too although the incidence is less. It can happen, for example, when a starboard pole, in the zero gravity zone, is caught by a heavy roll to port. If all goes well, the pole will fall into the fork. If not, it can slash through rigging fore or aft of the fork and slice into the house. If it falls far enough aft, it may catch a fisherman at the pin rail. The pole can go wild too in the same area that is "zero," as far as gravity is concerned, in a roll situation or in a pitching situation where the pole may be whipped fore or aft as the boat goes. Here, the stays should exert more effect but there always is a chance that the pole may fall inboard. Here indeed, is the place for a mighty heave at the falls if the pole is close enough to the fork or the use of a pike pole, boat hook or other device to fend it off if it is tilting too far inboard. Two-man boats with problem poles often use one man behind a pike pole to hold the pole clear or shove it outboard far enough for gravity to take it down. The tendency of poles to go wide is most often seen on boats with wood poles too large for the size of the boat or its crosstree forks. The problem appears minimal on modern, tautly rigged vessels with aluminum poles. Many new trollers are constructed with a metal yoke secured to the mast and the poles in a thwartships position that eliminates the danger of the wild pole. But these boats are few.

Another problem to be endured by trollers is a season-long problem, one which can only be alleviated by the layoff between seasons. It is an occupational thing and one which afflicts all fishermen handling nets and lines. But of all of them, the trollerman suffers most—from tired and cut hands. When the troller runs his lines, he literally runs them through his fingers. This procedure is an inescapable fact of trolling life. His twisted steel line has a nasty habit of fraying, sometimes through carelessness of the man in the pit, sometimes and usually from outside causes. The fraying and consequent breaking of the strands results in a needle-sharp, almost invisible point at the break. These broken strands tear at even the toughest skin as the line slides through the hand. Line splices do the same thing but line splices can be seen coming and can be avoided. But the touch of the hand on the line cannot be avoided if the fisherman is to have a full awareness of the state of his lines as he sets or hauls them. Gloves cannot be used effectively any more than they can be used in so many other conditions in fishing where their efficient use would make work more comfortable.

The trollerman has another monkey on his back also, one that must be faced by all fisherman, that of getting his gear in the wheel. All vessels of the four major methods of fishing—dragging, seining, gillnetting and trolling—run into this one although the drum trawler probably is least exposed to it. It is a common occurrence in the salmon seine fishery where so much effort takes place in narrow passes or similar turbulent areas where a boat may seem to spin like a top on the haul. Seiner crewmen with a wet suit, tanks and some diving skill often find themselves in demand beyond their own vessel. A great amount of webbing in the wheel may mean a tow to the nearest boatyard or grid if there is no diver to cut it free.

The accident, if it occurs, is even more serious for the trollerman than for the seiner since the latter has its skiff to keep it off the rocks or a lee shore. But a troll boat, caught offshore with scores of yards of high-test steel line in the wheel, has no recourse other than whistling for help and hoping the weather holds good. Help may be close by or it may be far away. But without exception almost, the boat must be hauled out of the water or floated onto a grid to get the line out.

There are two prime causes for line in a troller wheel. One is a short turn where the lines not only foul each other but also the screw. One time in 100 there may be a valid reason for an emergency turn with its attendant risk; the rest of the time such a turn stems from inexperience, carelessness or panic. The second reason is more valid—loss of power. In this case, the fishing vessel, caught by wind and sea, with lines dangling uncontrolled in the water, may be hit by the same mechanics that fouled the screw in the first instance. In this latter case, however, with the wheel dead, the foul-ing may result in nothing more than the awkward job of clearing the lines blindly with the hope the effort does not add to the trouble. The biggest effort may be expended on cleaning up the mare's nest caused by the lines' witless fondness for each other. There is no affinity like that of one line for another under the worst possible circumstances...

## Gear Away!

Before the troll fisherman can set his lines, his poles must be lowered away. This is done by releasing the falls at the pinboard on the shrouds at port and starboard, the most common arrangement for this gear although some older and smaller boats may have their gear secured differently. The lines may run through one single block, two single blocks or a pair of double blocks. The double blocks make handling heavy poles easier for one man.

When the falls are clear, the pole is pushed clear of its fork on the cross tree as the boat, on its automatic pilot, steers a course hopefully calculated to minimize rolling. The rate of descent of the pole is controlled by the falls and the descent is checked when the pole reaches an angle approximating 40 to 50 degrees from the vertical. The pole usually lies about five degrees forward of a right angle to the center line of the craft. The falls are secured and the fore and aft stays are tightened by means of turnbuckles to prevent movement of the pole. Most troll boats, not all, also use an understay running from a ring bolt or chainplate under the pole bracket or from a point a bit aft of the bracket. This stay prevents vertical motion of the pole and commonly runs about one-third to one-half of its length to the stabilizer hanging. It cannot run beyond the stabilizer hanging because it then would interfere with movement of the vital stabilizer. This is not a standing stay; it must be cleared before the pole can be raised. Tautness is adjustable by means of a turnbuckle or by a simple take-up in chain links by a shackle or G-hook. The poles, when all rigging is secure, are stayed from four points—from below, from fore and aft and by the falls.

The stabilizers follow the poles in the order of setting. On well-geared vessels, the stabilizers are hung in brackets secured to the bulwark below the pole bracket or to the bulkhead of the deckhouse on each side. Modern stabilizers are of steel and lead, built in a delta or bat wing shape, and are suspended from a ring on the upper planing surface. There are several variations in shape and weight but most have a torpedo-shaped lead log welded to the bottom of the planing surface for ballast. They ride from five to 10 feet below the surface, a depth dependent upon the size of the vessel. They are suspended from the pole by twisted steel wire or chain. The lines are secured to the pole variously—by a steel collar and ring; by a wire seizing on some older rigs or by a ring bolt through a wood pole or a hollow metal pole.

For various reasons, such as masses of weed fouled on it, the fisherman may wish to retrieve his stabilizer while maintaining way with no loss of fishing time. The simplest way to do this is by using a grapnel heaved forward of the stabilizer wire. A more involved system, one that adds to an already complex system of lines on the poles, is a recovery or haul line running from the vicinity of the pole bracket at the rail to the stabilizer wire. At the wire, the line carries an eye or thimble to allow it to move freely up and down the wire with the motion of the vessel. The line also has enough slack to allow unhampered movement of the stabilizer.

In water of relative smoothness, each stabilizer can be launched after the pole is secure; in rough water, this may cause an excessive list while the second pole is being set. In this case, the stabilizers should go into the water as closely together in time as possible.

With poles and stabilizers down, actual setting of the lines begins. The trollerman works from the bottom up on his lines and from the outboard lines inboard to minimize chances of fouling. If lines are counted in order from the outer line of the port pole on the two-pole six-line boat, the lines would be Numbers One, Two and Three. From outboard on the starboard pole, they are Six, Five and Four. If a man begins setting with the port pole, Line One goes first, then Two, then Three; from starboard, Line Six sets first and is followed in order by Five and Four.

In most gearing up for salmon trolling, the inside and outside lines of each pole carry the most weight for a quite logical reason—to avoid line and vessel fouling. The inside lines, usually called "deep" lines, are vulnerable to fouling on the rudder and skeg or in the wheel during ill-advised turns or in strong crosscurrents. For this reason these lines are heavily sinkered so they will hang almost vertically in the water at trolling speed, in this manner being less susceptible to such mishap. Weights of from 40 to 65 pounds may be used, anything best calculated to avoid line contact with the underwater complex at the stern. This heavy weighting also minimizes risk of the inside lines fouling each other on turns or under current influence.

In like manner, the outboard lines are rigged with heavy sinkers, both that they may fish deeply and so they may pass forward of and below the middle line when they are hauled. The weights run about 10 pounds less than on the deep lines in consideration of strain on the pole tip, a matter of concern when wooden poles are used. The middle lines, the lines here numbered Two and Five, carry the least weight with sinkers in the general range of 20 to 35 pounds. These are the longest of the six lines but they fish at least depths so they will not interfere with hauling and setting the outside lines.

This manner of rigging is not an invariable rule. It does not apply to kelpers with their entirely different gearing nor does it apply specifically to four-line boats since these have only two lines to a pole to deal with and sinkers of nearly equal weight can be used on each line if the fisherman prefers. But the system is one used by most men working six-line boats.

The two outer lines on each pole of a six-line boat are "float" lines, lines buoyed by several shapes and sizes of float material to allow them to fish well behind the boat with greater lengths of line working than those used ordinarily by the inside lines. Torpedo-like styrofoam "logs" from three to four feet long and from 12 to 15 inches in diameter with as much as 300 pounds' buoyancy are used on well-geared vessels. The outside lines of four-line boats usually are float lines although they can be fished without the floats.

On the under surface of the float there is secured a line clamp assembly occupying about the middle third of the float. The main line is passed through these clamps and they are snapped closed to hold the line so it cannot slide through them. Floats are secured to the lines so that the center float rides farther aft than the outer float. Respective differences might average 125 and 75 feet behind the pole although distances can be more or less, depending on wind, current, wave action and the whim of the fisherman. The outermost float, as noted, is set closer to the pole so that line with its heavy sinker can be hauled without fouling the center line.

The first piece of gear the fisherman touches after his poles and stabilizers are down is the cannonball sinker. In the described rig, it is the 40- to 65-pound lead ball used on the No. One and No. Four lines. These sinkers, in sets of two, three or four, are set in concave holders of wood or metal bolted on the rail or in any other location handy to the trolling pit. The concavities are deep enough to take about half the circumference of the weight. Most of these holders are fashioned to the fisherman's own design although holders consisting of no more than two or three holes cut into a piece of plywood are sometimes seen on old boats. Lead is preferable but cast iron is cheaper. Each sinker has a ring and swivel set into it.

To the ring, the fisherman attaches his first main line, Line No. One, the port outboard line. The line is not connected directly to the sinker ring. Between line and sinker is set a short length of rawhide or other breaker material of comparatively light test to take the shock of a hangup on the bottom and, by parting, to save line and pole although the sinker will stay on the bottom. This breaker line carries swivels and connectors at each end. The end to go onto the troll line is connected to a twisted or back-spliced loop or to a ring. The rigging possibilities are many and two rigs exactly alike are seldom seen. In some sinker rigging, a coiled steel spring from six to eight inches long and about an inch in diameter is inserted between the breaker line and the main line to further reduce shock.

Whatever his rigging, the trollerman's next task is to put his leaders and lures or bait on his lines. The first leader assembly, known as a "spread," goes about a fathom or a fathom and a half above the sinker. Subsequent spreads are set along the line at intervals of

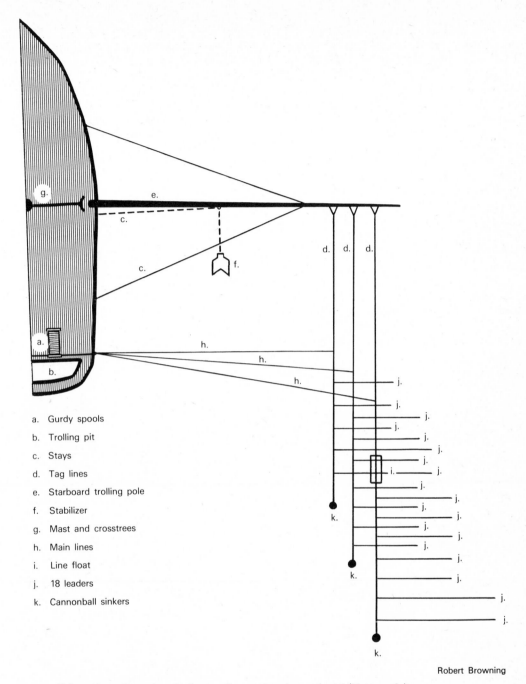

a. Gurdy spools

b. Trolling pit

c. Stays

d. Tag lines

e. Starboard trolling pole

f. Stabilizer

g. Mast and crosstrees

h. Main lines

i. Line float

j. 18 leaders

k. Cannonball sinkers

Robert Browning

*Schematic configuration of salmon troll gear, viewed at starboard. (Not to scale.)*

from two to three fathoms, depending upon the fisherman and the species being fished. When kings only are the target, these spreads may be set on the line at every other pair of stops while if silvers are the main chance, spreads go on every pair of stops. Spreads for kings may also go on the lower leg of each line with spreads aimed at silvers riding above them. In deep water, all lines can take 10 or more spreads if desired.

At the end of the previous day's fishing, the lines have been hauled and the sinkers set in their holders with the main lines running through the davit block or blocks, still secured to the sinkers. This saves time when a man is trying to get set for the morning bite just at first light. But the leaders are removed and stowed for the night, usually under burlap or other material wet down to keep the monofilament nylon pliable and easy to handle. (Too much of the time, the sea and/or rain do the job automatically.) The tag lines have been removed too and anchored to a heavy-gauge wire or bracket at the davit.

To set his first line, the fisherman fires up his gurdy power, whatever it may be—power takeoff, hydraulic or electric—and releases (gently) the brake on the appropriate spool as he lifts the sinker from its cradle and guides it carefully over the side. The spool for the No. One line ordinarily is the forward working spool in the assembly. A six-line boat may use a four-spool assembly on each side to have a spare if the need arises.

The line is lowered away until the first pair of stops is at hand. Between them, the fisherman snaps the connector of the first spread. Spread after spread follows until the desired number are in place and the line is lowered away to a depth at which the top spread works about a fathom beneath the surface. If 10 spreads are used, each two fathoms apart, the deepest spread on this line would fish at a depth of 120 feet if the line hung vertically in the water from the line float. But vessel movement keeps it at an angle above the vertical unless it comes under undue influence from currents. This depth can be varied considerably, however, by the place the fisherman chooses to clamp on his line float. He must keep in mind, however, that this line must be short enough not to foul the middle line on hauling or when turning. He next turns to his tag lines.

The tag lines, running from the pole to the main lines, are designed to bear the weight of the main line and its gear through the fishing cycle until the line must be hauled to pick fish or secure for the night. The tag line has gone through a lengthy evolution since its first appearance in the troll fishery before the turn of the century when trolling poles replaced hand lines or rods as standard gear. It has been known by a half-dozen other names but its function has not changed over the years.

In modern trolling, the tag line, except on the oldest of boats or among the most old-fashioned of fishermen, is connected to the main line in one of two ways. One method, the less common perhaps, involves use only of the porcelain ring or insulator through which the main line runs. The tag line ends in a wire snap or swivel and connector which snap into a ring and short length of line seized to the porcelain ring. The use of this snap arrangement does away with the clamp or clothespin favored by many fishermen.

The troller's clothespin does not much resemble the wooden or plastic domestic tool from which it takes its name although its action principle is the same. It is a plier-like device of nickel-silver or bronze alloy with jaws held shut by a powerful spring. The tag line is secured through a hole in the head of the top leg. The clamp is fastened to the main line by squeezing the grips to open the jaws, then releasing them when the line is fast in the jaws. The tag line then is run through a hook at the end of the lower leg to steady it. Once the tag line is fast to the main line, the line is freed from manual control and allowed to swing outboard until it steadies in the water behind the pole. The clothespin may be used for all lines or it may be used on the deep lines only, with the insulator used on float lines.

Line No. Two is set the same way except that the float rides farther aft. The deep line, No. Three, is rigged and set like its mates except for the absence of line floats. The starboard lines are set in the same order from outboard with the floats in the same pattern. A single man, working fast, uses a half-hour to get his poles down, his stabilizers launched and his lines rigged and set.

## Heads or Tails?

When the fisherman is selecting lures for his spreads, he faces a multiple or even myriad choice. What he chooses may be based on his experience, or hearsay or solely on good salesmanship. When he uses bait, he need not deal with these variables. He uses herring, either fresh or frozen, the whole fish, the plug-cut fish or the cut spinner strip.

Manufacturers of troll lures turn out an awesome array of plugs, spoons and hoochies as well as the flashers or dodgers used in conjunction with some of this gear and with herring. The flashers have proven themselves solid fish takers. White finishes sometimes are used for flashers but highest in favor is a smooth-finish chrome. Flashers of hammered chrome which, with its multi-facet surface reflects light farther and better, began to appear in the fishery after 1970.

Plugs seem to rate lowest in the order of use among lures and when they are used they are seen chiefly in

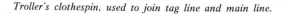

*Troller's clothespin, used to join tag line and main line.*

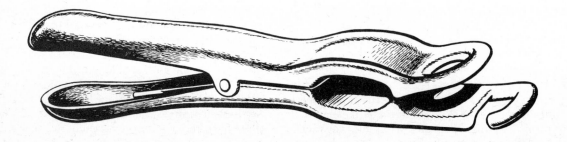

the king fishery. Before World War II, they were probably the most popular of artificial lures but post-war
refinement of spoons and the hoochie types displaced
them.

It has been widely observed in the troll fishery that
spoons may not properly be used behind flashers.
Nevertheless, a minority of iconoclasts does use spoons
with flashers and they do take fish with them. Similarly, another rule says that hoochies are not to be
fished alone but behind flashers but some men do use
them alone and they do take fish with them.

Another maxim says that whole herring should be
fished only behind flashers but many good trollers
fish them solo, rigged to wobble and weave through
the water. Often, herring and flasher spreads and her-
ring-only spreads are alternated on the same line.

The greatest difference of opinion among knowledgeable men, not necessarily working fishermen who
might not see far beyond their own patterns of activity,
lies in the field of preferred colors for spoons. There is
general agreement on the virtues of spoons with hammered metallic finishes (rather than painted) and on
spoon sizes, Six to Eight (roughly that length in inches)
for kings, Three through Five for silvers.

One consultant familiar with trolling, both from
the pit and from the business office, maintains spoons
for kings in order of effectiveness are 50/50 brass and
copper; 50/50 brass and chrome; gold/bronze, brass,
phosphorus/bronze and 50/50 cerise pearl. A second,
fisherman turned educator, maintains cerise pearl is
not acceptable for kings. Both agree, however, that
hoochie-type lures, in order of preference, are green,
blue, yellow, orange and red.

When coho season opens in mid-June, these aggressive feeders are taken mostly on spoons or on hoochies.
Opinion diverges mostly sharply here. The first consultant lists favorite colors and combinations for
silvers as chartreuse with red dots; chartreuse; cerise
with black dots; 50/50 brass and chrome; 50/50 cerise
and pearl; brass, green and orange, not in combination;
50/50 fire and pearl, and 50/50 blue and chrome.

The second dislikes almost everything on that list
and reduces it to four colors only—cerise, red, fire and
chartreuse. Both accept hoochies in orange, yellow,
red, blue and green. The first offers shiny streamer
flies behind flashers; the second discounts them entirely.

Trolling speed for salmon may be put variously at
from one to three knots. But speeds in figures such as
rpms mean little or nothing because of the eccentricities of wind, sea action and currents. A boat with its
wheel giving it a hypothetical speed of three knots while
running with a three-knot current will be making six
knots over the bottom but only three knots through
the water. This is the fishing speed of its lures despite
the total speed of the vessel. This boat moving at
three knots against that same current hypothetically is
standing still in relation to the bottom but its lures,

nevertheless, are traveling at three knots through
the water because of the three-knot current.

As a generality trollers fish slower and deeper for
kings, higher and faster for silvers and pinks. (The
pink has become a respected troll fish although as late
as the mid-1950's troll pinks were regarded mostly as
trash fish because market demand was almost non-
existent. Pinks bring lower prices than chinook or coho
as they do in all salmon fisheries. The sockeye too has
become a troll quarry with the discovery that the species is susceptible to a limited range of lures when it
switches from its plankton diet in June, July and August of its last year at sea to small bait fish and squid.
Sockeye take a red or reddish purple spoon or hoochie
behind a flasher moving at about one knot at depths
from 90 feet upward. Its value to the fisherman lies
between that of the silver and the large red king. Such
kings, in sizes suitable for mild cure, are the salmon
fisherman's prize catch.) Thus, an intimate knowledge
of currents and tidal changes is required of the troll
fisherman, more so in that respect than in any other
salmon fishery. This is particularly true where there
are river discharges as off the mouth of the Columbia
where the intrusion of river water compounds the ir-
regularity of ocean currents. A man who may be a highliner on his home grounds may be only a greenhorn a
couple of hundred miles away where the current patterns are a mystery to him.

## Fish On!

Most trollers hang tiny, tinny bells at the junction
of the pole spring assembly and the tag line in order
that the fisherman may be apprised of a strike. The
bell is particularly necessary on boats with long heavy
poles of wood or of aluminum because the strike of a
small fish may not jar the gear perceptibly. Or a man
at work in the cockpit, cleaning fish or sharpening
knives or hooks or tying leaders or doing any other of
his myriad chores might not even notice the strike of
a fair-size fish were it not for the bell.

The bell tinkles, often a small sound against the
sweep of the breeze through the rigging, the murmur
of the slow-speed exhaust from the stack, the working
of the water against the hull. The sound of the bell does
not mean necessarily that one spread has a fish on it;
weed drifting past may foul the line and bring the bell
to life. But weed usually can be seen; in any event it
sets up a tremor that sounds the bell continuously.
A hooked fish reacts spasmodically, in most cases, to
the sting in his jaw, fighting the hook for a few seconds
at a time, then seeming to rest briefly before renewing
the struggle. Some few fish may hit the lure, struggle
briefly, then swim along with the pull of the line with
little reaction from the gear.

Some men haul their lines on the strike; others prefer to let the fish "soak" for a span in the belief that
the hooked salmon fighting the gear may attract others
to the scene and to the hooks. Or they may let a big
one simply tire himself out on the gear. The swirling

salmon on the leader displays the same flashing characteristics as the dodger. This practice is popular but it does result in the death by drowning of "shakers," immature salmon too small to be taken. They weaken and die quickly. It also may result in scrap fish uselessly occupying spreads that might better be working for salmon. No matter...it has been noted repeatedly that the operation of fishing gear is pretty much of an "every man for himself" type of thing.

This is a fish—not weed; the fisherman smokes a cigarette during the soak and, halfway to the butt, another bell sounds. He now has a fish on his No. Four line, the inboard starboard line and another on Line One, the outboard port line. He decides to run Four first, the quicker and easier of the two to work.

The gurdy spool controlling this line is the one aft in the starboard assembly; he pushes forward the control lever to engage the clutch and the line begins to come aboard, up to the outer davit block, inboard to the block hanging above the gurdy and down to the spool. His left hand controls the speed of the haul; with his right he tests the line for strain, swearing as a broken strand of the wire cuts across his finger tips. But the haul goes smoothly, the steel line running effortlessly through the porcelain ring that connects it with the tag line, spreading itself evenly between the flanges across the width of the spool (on newer models; on older designs, the fisherman must "level wind" the line manually).

The tag line comes alongside within easy reach of the trolling pit. The fisherman unsnaps it if it is the type using the snap-on swivel arrangement with the ring or unclamps the clothespin that holds the main line in the other rigging method and secures it carefully to the wire or bracket on the davit. A tag line allowed to get away means a line out of business for the day unless the fisherman wants to pull his pole to regain it. It is too light and too short to be recovered with grapnel. Greenhorn trollers learn this lesson fast.

The first spread is in sight and barren. When the leader connector is within reach, the gurdy is stopped and carefully the connector is unsnapped and the leader brought aboard. These must be handled with attention because those with flashers or flasher combinations exert a heavy pull and careless or inexpert handling can lose them—$15 to $20 worth of gear gone forever. Greenhorn trollers learn this one fast too.

Some men, with ample working space around the cockpit, coil each leader down as it comes aboard. Others snap the connectors to a wire running athwartships between the pit and the transom and stream the leaders behind the boat. Both systems have their virtues; of the two, the coil-down seems preferable because a husky fish on a leader a couple of fathoms long can cross the trailing leaders and foul them mercilessly as he comes up to the boat. A following sea also can cause the trailing leaders to foul each other. But it does take speed and skill to coil leaders so they may be snapped on to run freely on the next set.

The third spread is the one with the fish, anything from a shaker to a good-sized chinook. Some fishermen, if they are sure the small fish is a shaker, skip the gaff and bring him aboard on the leader so the hook can be removed from his jaw and he can be returned to the sea with minimum shock and physical damage. Some men do not; the fish is gaffed, brought aboard, tried for size on a measuring scale along the fore coaming of the pit and tossed over the side if he fails to meet the minimum length requirement. The survival rate of shakers handled as gently as possible is an unknown quantity; the survival rate of those treated roughly is even more doubtful although salmon that obviously have been gaffed and returned to the water do show up in later troll catches and in inside net fishing areas.

The hooked fish is a silver of about 10 pounds who went after an orange hoochie wobbling its way enticingly through the water. As the leader connector comes up closely, it is disconnected and the fisherman grasps it as far down its length as he can, looping the nylon in his left hand as he leans over the rail from the pit and brings the head of the fish out of the water. With his right hand he wields the gaff. The essence is to gaff the fish cleanly in the head and bring him aboard in one sweeping motion.

But a hooked fish is unpredictable; he may submit docilely to his fate, especially in the case of small and tired fish. Or he may begin a rolling, thrashing action that can straighten the hook and free him. Or he may simply roll violently enough that the hook pulls free. A sharp rap on the snout with the backside of the gaff hook will cool such a struggle quickly because the fish stuns easily but if it is carelessly done, the clumsy swipe may knock him off the hook. A clean swipe with the gaff hook itself will get the fish; a clumsy swipe with the hook may knock him free also. A .22 caliber rifle may be used to kill a big fish with a single shot through the head before he is gaffed. But a fish gaffed properly can be swung aboard, killed with a single blow and the hook pried from his jaw with a twist of the gaff hook. Then he can be lain gently in the checker so that he does not bruise or lose scales.

That was Line Four...Line One is a bit different... longer and deeper with more spreads and two lines between it and the port rail. But the haul starts in the same manner with the line running through the insulator ring at the end of the tag line as the line shortens and passes forward of the No. Two line and beneath it so they do not foul. The tag line comes alongside and is disconnected. The weight of the line rests now on the davit blocks and the gurdy. The float comes within reach, the clamps are released and the float placed within reach of the pit.

The spreads come up and come off the line; the leaders are coiled and placed carefully aft of the pit. Spread after spread...the fifth, the sixth, the seventh, the eighth, the ninth. On the 10th, the bottom spread at the end of about 40 fathoms of line, is the quarry—a lingcod that will go 40 pounds headed, gilled and gutted

...at six or eight cents a pound, taken on a hook that had been fishing a whole herring. His value is about one-tenth that of a coho, about one-twentieth that of a chinook. But 40 pounds of lingcod buys a half-case of beer or six gallons or so of diesel oil or a restaurant meal the next time ashore.

The fisherman sighs, gaffs and hauls aboard the big-headed, ugly fish with its sweet, white flesh. A nickel is a nickel and they add up to dimes and dollars.

Lingcod, halibut and sablefish are the only "trash" fish the salmon troller bothers with. Rockfish and hake have no market value for the troll boat nor do any other of the many species that may take the bait or lure. Rockfish often are used by the fisherman himself—if he likes fish—because they can be filleted easily with his salmon knives and the flesh of most is the equal of that of lingcod. So it is with too-small halibut, the "baby chickens" the fish buyer cannot legally consider. A man who really likes fish can do a considerable amount of foraging from among his varied catch.

This is salmon trolling...no picnic certainly, but a fishery wherein each man may do as he pleases within the rather tenuous control regulations stemming from the four Pacific Coast states and the Province of British Columbia. Other than season openers and closers and the minimum size of fish, there is not much else to bother the trollerman—other than the sea itself and the obstinacy of his gear.

## Albacore Trolling

The albacore troll fishery centers around a colorful and fast-moving fish of the deep sea, often hard to find in the ocean expanses it prefers, a fish as contrary in its like and dislikes as any other fish taken on a troller lure. It is the only fishery of the Northeastern Pacific that involves a boat moving at good speed through the water if not over the bottom. The albacore must be fished at speeds between six and eight knots. The fish will strike a lure moving even faster, up to 10 or 11 knots, but somewhat slower speeds are mostly used because of the difficulty of hauling fish at the higher speeds. When albacore are plentiful and greedy and the weather is tolerable, the fishery can be an exciting and even pleasant affair, something that cannot be said of most fisheries most of the time.

As with all fisheries, weather is a sharply-limiting factor in this fishery. The albacore trollers are especially vulnerable to bad weather because of the far-offshore location of the fish during most runs. Eight or nine knots can be reckoned good speed for most trollers. In best sea conditions with top engine performance and light load it takes a nine-knot boat 10 hours or more to cover 100 miles. Often enough, this is too much time. It was this albacore fleet that was hit so hard on October 12, 1962, the day a singular maverick typhoon or what was left of it stormed in from the west, thousands of miles from where it should have been and caught several dozen albacore vessels still

at work 100 miles or so off the northern California-southern Oregon coast. There had been 10 or 12 hours of radio warnings but some boats with tons of albacore in their holds still were 40 or 50 miles off the coast when they were overwhelmed. About 20 men (no exact tally ever has been made) died in the peak winds of 120 knots, a steady blow for six or eight hours.

The search for albacore differs somewhat from that for troll salmon. Salmon seiners and gillnetters in their usual waters and trollers in British Columbia and Alaska (where they can fish many inside waters) as well as those of the lower coast stick to the long-known migration routes of the salmon or to specific areas where the fish may congregate to feed. This manner of fishing is somewhat analogous to the activities of the traffic policeman who is quite sure he knows the flow patterns of the bulk of the traffic he must watch. Outside salmon trollers must search more widely for their fish but these offshore salmon do stick to some kind of pattern during inshore migration and during the ocean feeding and growing phase of their life cycle. This accounts for the concentration of trollers off the mouth of the Columbia River, the Strait of Juan de Fuca and on the Fairweather Ground in the Gulf of Alaska north of Cape Spencer, for example.

But the pelagic albacore is another kettle of fish, if that expression may be excused. He ranges over great reaches of the sea with extreme mobility and becomes available to West Coast fishermen in the summer months with apparent unpredictability along that part of the Eastern Ocean from mid-Baja California to Vancouver Island or even, on rare occasions, to the north of the island. His lateral range extends from about the offshore limit of the salmon troller of California, Oregon and Washington, something like 20 to 30 miles, to an extreme of about 300 miles (for practicable fishing), running some 1,200 or 1,300 linear miles from south to north. This total area of about 400,000 square miles is more than one-third of the area of the Mediterranean Sea and is roughly equivalent to the Sea of Japan... in other words, a lot of water to explore blindly for fast-moving, widely-scattered schools of varying density.

In the early days of the fishery, an era that might be said to extend into the early 1960's, the search for albacore was carried on rather in the manner of a blind man trying to find an object in a strange room with no cane to guide him. Fishermen were familiar to an extent with the "normal" movements of the albacore in normal migration years...but there are, perhaps, fewer normal years in the albacore fishery than in any other fishery of the Eastern Ocean. Only scientists were aware to any extent of the rules that govern the movement of the albacore; this knowledge had not yet filtered down to the fishermen. Just the same, fishermen did find albacore in greater or lesser quantities almost every season, searching the sea under terms of their own rule-of-thumb knowledge. It was a fragmented effort, an empirical effort that worked—more or less. A lot of oil went to waste during the first half-century of the albacore troll and bait fishery.

A certain amount of oil still is expended fruitlessly every season on the search for albacore but no longer must the fishermen depend solely upon themselves and what experience has taught them. The appearance of the albacore is not quite as unpredictable as it would appear from casual observation. Science, government, higher education and fishermen have teamed to produce a fairly workable method of predicting when and where the albacore may be found and how many of them may be potentially available, if not amenable, to the lures of the fishermen. This cooperative effort has substantially reduced, for most fishermen, the long hours and the long days of search across thousands of square miles of often tumultuous ocean. As well as the consumption of oil....

Official agencies playing parts in this research, the ones with the proper tools for the job, are the National Marine Fisheries Service, working chiefly out of its Southwest Fisheries Center at La Jolla, California; the Division of Marine Resources of the University of

Washington at Seattle; Oregon State University at Corvallis, Oregon; the Navy Fleet Weather Center at Monterey, California; the Oregon Fish Commission; the California Department of Fish and Game, and the National Weather Service with an assist from the weather satellites launched by the National Aeronautics and Space Administration.

The NMFS, the California Department and Oregon State University would appear to carry the major burden of determining the migration and incidence of albacore from the time the runs first appear, whether to the south or the north of Guadalupe Island, until the time the last commercially available schools disappear into the far offshore waters on their journey into the Central Pacific and beyond.

Research activities by fishermen themselves include specifically the charter of commercial vessels by the American Fishermen's Research Foundation, a fishermen's group financed by a self-imposed donation of $10 a ton on all domestically-caught albacore landed on the Pacific Coast (as distinguished from import fish). During the first year of this work, 1971, some $200,000

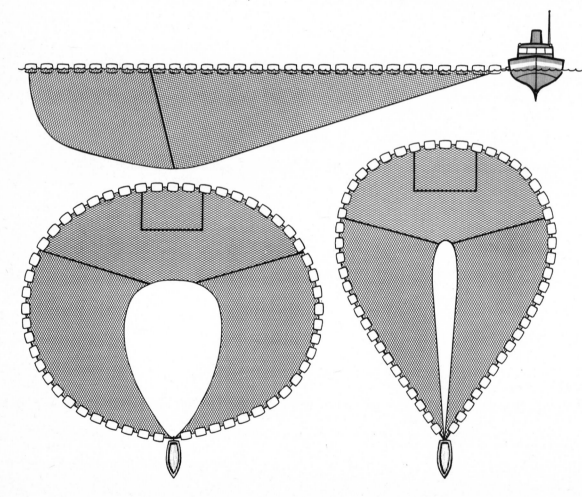

*The vanishing lampara net: Top and left, the gear fully set; right, half-hauled, with lead line closing to block escape of the catch. The lampara once out-numbered all other round-haul nets in the western fisheries.*

was raised by this method to finance the activities of six vessels following albacore further offshore and later into the year than they normally are observed off the Pacific Coast.

This surveillance extends in general from May until late October. Every means of detection and analysis is utilized—research cruises along the entire reach of the waters where albacore may be expected, coupled with accurate observations on oceanographic and atmospheric trends, the criteria that determine the distance and duration of the travels of the albacore.

This basic information, kept up to date by constant surveillance and test fishing, is extrapolated in forecasts of both short- and long-term usefulness in terms of potentially-productive fishing areas. In addition, each of these major agencies concerns itself with other, more mundane matters of interest to working fishermen such as changes in marketing and off-loading conditions. Individual fishermen, or many of them anyway, go along well with this official effort by reporting fishable concentrations of albacore. Not all of them, however...these are the men who pay little or no attention to the co-op research and report structure and its results and those who, if they do find albacore, keep it an affair of the utmost privacy. They regard this reticence as a necessity because of the increased size of the fleet.

This concentrated research and report effort has, in addition to its obvious and immediate benefits for the fishermen, a special long-term virtue in that it has added and continues to add to the rather sketchy knowledge of the albacore and his habits as well as contributing to the general knowledge of the oceanography of the Eastern Pacific.

## Same Game

Although salmon and albacore trolling embody the same principle, that most ancient one of enticing a fish with a hook camouflaged with something the fisherman hopes will attract his quarry, the application of the principle varies between the two fisheries. Both methods involve towing a number of hooks rigged in any one of a variety of ways through the water at differing speeds and depths. The salmon troller uses the most hooks and tows them at the lowest speeds; the albacore troller, as ingenious as he may be, fishes with one hook to a line and the number of hooks he uses depends, then, upon the number of lines he can rig and handle. A good figure would seem to be from 10 to 14 lines with 12, perhaps, as a schematic average. But an albacore trollerman's line rigging may vary from day to day according to his whim, the weather and the number of fish striking his lures. Even with a powered line hauler, one man finds handling 200 or 300 or 400 fish during the number of daylight hours he can work to be a considerable chore. Not that the idea of so much labor repels him else he would not be there; it is only that one man eventually comes to the point where his strength runs out. From a money standpoint, the one-man boat is desirable although there is a risk factor here that cannot be ignored. A two- or even three-man vessel can work harder through the same hours than the single-manned craft but, unless fishing is exceedingly good, the inevitable point of diminishing returns comes along. The husband and wife team, in addition to its obvious advantages for the troller or gillnetter, also keeps the money in the family.

The troller went into the albacore fishery long ago anyway because of these economics of the fishery; one man in a small boat can sustain himself and make a modest amount of money. Except for periods of natural fluctuation in albacore stocks, the fishery has been a good one to many men over the years and in all probability it will continue to be so although eventual regulation of some sort will be required. Such regulation of a pelagic stock with no national origin such as that of the various salmons will be even harder to arrange than the rules that to some extent regulate a part of the yellowfin tuna fishery of the Central Eastern Pacific.

The albacore has been described, in its periods of high speed travel, its common mode of movement except when feeding, as "coming out of the side of a swell like a flying fish about to take off. Those long pectorals make them look something like airplanes..."[1] On occasion, when seas are high enough and close enough together, the albacore plunges, in its mad course, from one to the next without ever touching the troughs until at last some factor—the presence of feed, a change in water temperature—causes it to slow its speed or change its course.

Without exception, the world of fishes is a carnivorous world and the marine fishes are hungrier and fiercer than many or most of their fresh-water relatives. The anchovy, for its size, is a terror to be compared in degree with the sharks. The albacore is as fierce as any marine species among the sea creatures that comprise its favored food. Perhaps tops on its gastronomic list is the squid, where it is available, whence stems the feathered jig. Boated albacore often start disgorging their stomach contents before the hook is shaken loose and during a day's fishing a good weight of squid will be vomited onto the decks and into the checkers, some of them still alive as the distressed albacore rids itself of them. But no feeding albacore will disdain anchovy, herring, smelt or other fish in those size ranges.

It is, then, to catch this fast-moving high seas creature that the albacore troller must rig his lines so differently from those of the salmon troller. Standard salmon gear does not haul fast enough. In essence, the albacore man streams his lines directly aft from the poles and controls them by tag or haul lines connected

1. Huswick, David J., interview, January 24, 1972.

to the main lines at about the point where they meet the water. By so doing, he dispenses with the clumsy (to him) gear of the salmon troller; he has no complexly powered gurdy assemblies, no davit blocks, no davit and no heavy sinkers to slow him down. In most instances he uses no lead at all on his lines except in those instances when water conditions require it to keep his lures from bouncing free of the water. One or two lines on each pole may be weighted with chain heavy enough to sink line and lure so that outside lines may be hauled over them without snagging.

The albacore troller uses either the single coil spring or the vee-form coil found on most salmon trollers, although older boats in both fisheries may settle for a double seizing ending in a double half-hitch to anchor their lines to their poles. The vee-form seems preferable because of the added flexibility it provides against sudden shock. The line is secured to the spring by a ring and swivel connection although some fishermen insert a rubber bumper between swivel and line.

Most salmon trollers content themselves with two or three springs to each pole (remembering that Alaska law limits troll lines in that state to four to a boat) but the specialized tuna troller may have up to eight springs to each pole, about as many as can be handled without all sorts of potential foulups involving lines and the stabilizers. This means 16 lines from his poles (a rarity, actually) but this figure does not take into account the other lines he may rig from various parts of his vessel.

From the spring to the lure, the line arrangement is as uncomplicated as any fishing gear to be found anywhere, as uncomplicated almost as a cane pole, a linen line, a cork bobber and a hook with a worm on it. At the end of the main line, if any weight is to be used, a small crescent lead, usually of one ounce, is attached by a single or double swivel to a loop tied into the line or, sometimes, put there by a painstaking backsplice. The leader, in turn, has, the same swivel arrangement between it and the lead while at the fishing end of the leader a similar swivel connects leader and lure. The swivels, common to all trolling, are needed to help keep a hooked and rolling fish from freeing himself while, at the same time, they allow free play for the catching gear and thus cut down on any tendency of the leader to break.

The outside lines on each pole are geared and set first with the succession running inboard to the boat. Salmon troll line floats are not used. Most fishermen put out their lines in the same succession set after set. While the lines are going out the boat travels a straight course to avoid any fouling of the lines. If fish are plentiful and feeding, the first lines out can have fish on them before the lines on the other side have been let go.

Here is the assembly of lines used by the *Astrid,* of Seattle, an arrangement dictated by years of experience in the tuna fishery.

*Astrid's* line arrangement may be regarded as typical of a vessel of her type, 46 feet long, diesel powered, galley below, a craft starting out each season in the salmon fishery, then switching to albacore when they appear. Although the *Astrid* hauls only by hand, there is no essential difference between her gearing and that of other, bigger vessels using hydraulic line pullers. No matter how it may be figured, there are only so many ways to gear up a troll boat for maximum efficiency.

The drawing shows the four-line hookup of the starboard pole, identical obviously to that of the port pole, as well as the starboard line of the two fished directly from the stern of the *Astrid,* 10 lines in all. They are numbered from one to five outboard from each stern line, in the terminology of the *Astrid.* But to each man, his own terminology...

All main lines are 350-pound test braided nylon. Leaders are one and one-half fathom 200-pound test monofilament. The haul or in-haul lines are 350-pound test braided nylon, secured to the main line by a ball bearing swivel. Because *Astrid* does fish salmon before the start of the albacore fishery, her salmon gear stays aboard and the haul lines are secured to her block davit although the salmon gear itself plays no part in the *Astrid's* tuna work. The stern lines are hung to rubber snubbers to eliminate chafing of the nylon and these latter are secured to eye screws mounted to the bulwark aft of the cockpit. On bigger vessels designed mostly for albacore, the haul lines are hung to a stay by various methods, any one of several ways that happens to appeal to the vessel operator. This specialized vessel (although it can convert to other fisheries) stands out because of the relative lack of gear clutter around the stern, the absence of a true cockpit and by the presence of one or more line haulers. (The haul line may also be called the tag line but it is not the tag line as used by the salmon troller and its purpose differs somewhat. The salmon boat tag line is designed to relieve the main line of undue shock and to hold it in its allotted position outboard of the fishing vessel.) All main lines except the stern lines are buffered by vee-form coiled spring shock absorbers.

Line No. One is six and one-half fathoms long and, with its twin, is the easiest of the lines to handle. Line No. Two, first on the pole, is six fathoms; Three is the longest at 18 fathoms; No. Four is only four fathoms and is weighted with some 18 inches of chain at the water to hang almost vertically as it works; No. Five is second in length at 11 fathoms. To each of these main line lengths must be added leader length.

It is the handling of these lines when fish are striking in good numbers with a bouncy sea running that albacore fishermen may be distinguished from greenhorns. Obviously, each line must be hauled so that it does not snag any other line (although this desirable situation is not always the real state of affairs) and this is the reason for the varying lengths and their positions on the poles.

Line hauling follows this pattern:

Line Five, fish on, passes OVER its short mate, Line Four, UNDER Line Three and OVER Line Two until the junction of the haul line and the main line can be grasped by the fisherman. The main line is then hauled and loosely coiled on deck or along a coaming of the trolling pit. (Not all albacore boats are built with pits, especially those of the Alaska-limit size of 58 feet and up.) The fish is shaken free of the hook and the line is returned to the water in reverse sequence.

Similarly, Line Four passes UNDER all lines inboard of it and the boat while Line Three passes OVER Line Two. The stern lines are hauled directly.

This is the intended procedure but what man intends is not necessarily what nature and fish may do with good intentions. In decent weather with fish strik-ing singly or with no more than two or three on simultaneously, no problems should arise other than those prompted by inexperience, carelessness or a fish big enough and husky enough to try to have his own way as he is hauled toward the boat. A fish this big is rare enough; troll-caught tuna run from nine pounds, the smallest accepted by most buyers, to around 25 pounds.

Fish hitting on all lines for an extended period (just what the fisherman wants, of course) may cause confusion with lines snagged in the water or fouling each other aboard the fishing boat. Bad weather adds to the mess with a rolling, pitching boat tossing in an element of danger, posing to the fisherman at times the choice of buttoning up or staying on the fish. Floating debris, especially masses of weed, do nothing to help any troller at any time while a turn made too short can foul lines,

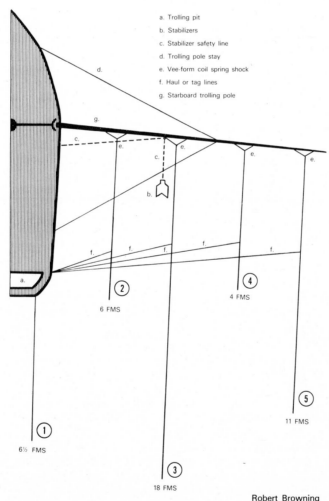

a. Trolling pit

b. Stabilizers

c. Stabilizer safety line

d. Trolling pole stay

e. Vee-form coil spring shock

f. Haul or tag lines

g. Starboard trolling pole

Robert Browning

*Schematic configuration of albacore troll gear, viewed at starboard. This rigging lacks the complexity of salmon troll gear. (Not to scale.)*

even to the extent of getting them into the skeg or the wheel. Power loss is not the gear problem it can be to the salmon troller who must depend on his main engine if he uses mechanical or hydraulic power for his gurdies although the albacore man may find himself with lines snagged on stabilizers or his stern underwater components before he can pull them.

As for stabilizers, they are the one piece of gear above all others that makes life bearable at sea for small boat fishermen in bumpy weather. They, combined with the sea anchor, offer at least a measure of steadiness to tired men on a boat just riding out the night for a few hours sleep or to a boat that must depend on them for the difference between safety and disaster.

But to a troll boat working, the stabilizers may present a problem, that of lines fouling on them for several reasons. On a turn or in seas of any considerable size, the stabilizers may swing fore and aft of their proper position or inboard or outboard of it in reaction to the irregular movement of the vessel with the possibility of snagging a line. To combat this, some albacore trollers secure a line of any preferred size of manila or nylon to the leading edge of each stabilizer and run them forward through the bow chocks to secure them to the bow deck cleat. On the other hand, some veteran fishermen do not use stabilizers when they are working tuna because of the turbulence they set up which is believed to spook the fish.

## To Catch A Fish

Taking albacore or any market fish for that matter is not as simple as scooping them out of a live tank in one swoop and calling the job done. The albacore's tank is an uncommonly big one and, when its surface area is measured against the estimated number of albacore, the odds on encountering even a single fish would appear incalculable. But, elusive as he is, the albacore does follow something of an accustomed path through the sea (although a wide and winding path it may be) and that he can be caught is attested by the million of pounds of albacore that have gone to the store in flat cans during the past three-quarters of a century.

Bait fishing, the elder method of taking albacore, still has its adherents, almost all of them from the California ports where the method was introduced more than 100 years ago. But the bait boat of this part of the 20th Century is not the belle of the fleet that she was until the late 1950's when the Power Block, nylon netting and the rise of the big seiners made a poor relation of her. These vessels run in the 60- to 80-foot range and work with minimum crews rather than the 20 to 30 hard-working, hard-fisted, hard-drinking men, mostly of Mediterranean descent, who made up the crews of the old clippers.

But the actual method of fishing remains essentially what it has been for a couple of generations, modified, of course, as all West Coast fisheries and others the world over have been by new materials. The traditional technique and still the one in use today in its general application has been authoritatively described:[1]

"Essentially, the bait method was to troll until fish were located. The boat then was stopped and live bait or ground up fish was thrown out as chum until a school was actively feeding about the boat. The fish were caught by means of a short bamboo jack pole, short line and a lure or baited barbless hook. The commonly used lure was a bone or metal head with bright eye and brightly colored feather streamers. Originally, the lure was called a squid but later was known as a striker and the short pole was called a jack pole. The live bait, striker or jack pole method of fishing was used by many commercial fishermen for many species of fish, including several of the tunas...

"The two methods, trolling and live bait fishing for albacore, barracuda, white seabass and other species are usually referred to as jigging and bait fishing (respectively). The latter term might better be 'live bait' because there are several variations in fishing methods and in the use of dead bait. Some boats carry a small tank of live bait for jack poling as well as salted bait for trolling. Trollers sometimes carry salted bait and circle the boat about a spot where a fish has struck a troll line. Salted bait or ground fish is then scattered to chum up a school. By contrast, it is not uncommon for an albacore bait boat, with a tank of live fish, to find bait fishing too slow. The crew then turns to trolling. The catch for that day is partly from live bait and a portion from trolling. Occasionally, the delivery by a bait boat is as high as 90 percent jigged albacore.

"Albacore are wary fish and frequently will make a slow pass at a lure, then turn aside. The fishermen call these 'cold fish.' At other times they may become excited and the 'hot' school will strike at anything. At times albacore may be seen deep under the boat but they will not rise to the surface to take chum or live bait thrown out by the fishermen. At such times, a live fish from the tank may be hooked through the back and lowered to deep water by a hand line. Formerly, hand lines often were used, one line to each man, with a hooked sardine or anchovy as bait an on special occasions when fish are deep and 'cold' hand lining is still practiced. In order to excite a school of albacore, the Japanese fishermen used a 'flicker' while throwing out live bait. This device, a tuft of feathers on the end of a light bamboo pole, was used to flick the water surface to resemble a surfacing school of sardines or anchovies. A tin cup on a light pole also was used to scatter drops of water on the surface (for the same purpose)."

---

1. Scofield, W.L., "Trolling Gear in California," California Department of Fish and Game, 1955.

## Troll Boat Leads Fleet

The economics of the albacore fishery are not especially kind to the bait boat during mediocre or poor seasons and for the reason cited previously, that one man in a small boat can make out, there are a score or more troll boats in the fishery for every bait boat to be found in it. Trollers or jig boats, call them what you will, take the bulk of the West Coast albacore catch.

No one, neither the world's top fisheries scientist nor the best troller, knows why fish strike at a specific artificial lure. It is assumed they do so in response to the feeding reflex but this assumption may be as questionable as the many theories concerning the homing reflex of salmon and other anadromous fishes. But, with use of baits such as herring for salmon, it is hard to sidestep the fact that the taste and/or smell of feed in the water must bring them to the bait and the hook.

But the albacore often is a cautious fish and the water in the vicinity of a fishing vessel may show signs of fish even though the fish show no sign whatsoever of interest in the lures being towed past them. Or they may turn, in the space of a minute or two, into ravenous creatures striking at anything that moves around them.

Albacore are described as schooling fish but usually the schools do not have the cohesiveness of salmon or mackerel schools and certainly not the closely-packed togetherness of such fish as herring and anchovy. What might called a "school" can be spread out over a considerable area of sea. The albacore troller may work back and forth for hours through what he knows to be albacore water without getting a single strike. One or two fish hitting closely together may be an indication of many more to come or they may mean that only one or two of a multitude are feeding.

As with salmon trollers, there is a body of thought among albacore trollers that there are boats best known as "fishy" boats, boats that seem, trip after trip, to catch more fish than other boats in the same area fished by men of the same experience as that of the man catching fish. There is no understanding of this phenomenon in either fishery but the most probable explanation seems to be that applicable to the salmon troll fishery — that either lack of power train noise or the setting up of vibrations attractive to the albacore draw them to the "fishy" boat.

An albacore troller may spend hours or days with no fish or a few fish, only, coming aboard. Then there may strike that happy hour when albacore seem to be waiting in line to take his jigs. The long dull hours with no fish are forgotten; fish on all lines mean hard work, most especially for the vessel hauling by hand. Getting a threshing 15- or 20-pound fish as vigorous as the albacore on deck from the end of a line 20 fathoms or more in length takes skill and strength. It is a job that, over the course of a good season, can put strong arms and shoulders on a man. And here is one of the few occasions in working with fishing gear that gloves can be used without impeding the job. Nylon line of the size used by most albacore men is hard on the hands at any time in the season and cotton gloves do a good job of protecting tender hands from the beginning of the season to its chilly end.

The power line hauler was developed on the gurdy principle to offset most of this struggle with the albacore. In this instance, the line hauler is an open vee-sheave of aluminum, chiefly, mounted on the stern rail. Some vessels carry two, each set closely to the port and starboard rail respectively. When a fish strikes, the deckman brings in the in-haul line manually until he can reach the main line. He throws this line into the sheave where friction grasps it and hauls it inboard until the fish can be lifted from the water and brought over the rail. The hook is freed quickly and inspected by careful fishermen before it is returned to the water; it may need a touch-up with a stone to insure its continued fish-taking capability.

So, just as the salmon gurdy, the halibut gurdy, the crab block and the Puretic block have taken much of the back-straining labor from their fisheries so has the line hauler done it for the albacore troll fishery. The haulers may be powered either by electricity or hydraulics. Controls are located beside the hauler near the rail. The rate of haul can be regulated precisely with 20 to 40 fathoms a minute about average.

It is an axiom among troll fishermen that a man does not catch fish by trolling away from them. Once fish have been taken, the albacore troller, like the salmon troller, makes a careful turn to go back through the same water. Feeding albacore may tend to follow a boat just as salmon will during certain phases of their feeding activity.

Albacore usually are left on deck or in checkers for a short time to bleed. In addition, their body temperature may be higher than that of most fish because of the warm waters which they frequent and, where possible, they are allowed to cool somewhat to reduce ice melt. This, however, like all good things, can be carried to an extreme and these small tuna should be washed down and gotten into the fish hold without undue delay.

But a one-or two-man boat in heavy fishing may (and always does) take advantage of the presence of striking fish and the catch may lie on deck or in the checkers for a considerable length of time. Albacore are not gilled and gutted as troll salmon are but are iced or frozen in the round. It is preferable that they be gotten into the ice before rigor mortis begins to affect them since they can be better stowed before the stiffness sets in.

Icing in the round means that the vessel without refrigeration may count on no more than 10 to 12 days for a trip after the first fish come aboard. This includes most trollers, especially those that primarily are salmon trollers. There are some big and newer boats, built specifically for albacore but convertible to other fisheries, with glaze-ice freezer systems that may stay at sea until they are fully loaded, their catch frozen solidly and safely for the cannery—where, during periods of big landings, they usually are frozen anyway. A few modern salmon trollers use this same system and they too can stay out until they are "plugged," a most desirable state of affairs for any fisherman.

The boat using ice only or ice with minimal refrigeration intended chiefly for slowing of its ice melt rate finds itself at a disadvantage in either the albacore or the salmon troll fisheries. Quality requirements make it mandatory that a trip be cut short, no matter the size of the load, when the fish first caught are still marketable by the time the cannery or the fish buyer looks at them. Too often, this may mean a losing trip for a vessel that has been at sea for a week or more with the consequent heavy consumption of fuel oil and other supplies. It is in this situation that the small salmon day boat or the kelper fishing from an outboard craft with sports gear as his equipment may appear to have something of an advantage over the trip boat many times during an average season because neither uses ice and if there are no fish (or at least no fish striking) they

can go back to their floats to wait until tomorrow. This apparent advantage, however, may not actually exist because their uniced fish often are of poorer quality and they cannot find, prospect or follow fish.

It seems easy to guess that the day of the ice boat, working without refrigeration of any kind, will end sooner than most trollerman like to think because of increasing demands for higher quality fish, while, at the same time, there comes the necessary and inevitable replacement of older boats by more efficient vessels engineered specifically to achieve that higher quality.

The albacore troll fleet is on the increase; the salmon fleet of all gears must necessarily find itself diminished as limited entry into that most historic of West Coast fisheries becomes a reality in the four Pacific Coast states as it has become a reality in British Columbia.

But, as long as the quarry of both fisheries exists in commercial quantities, there will be a troll fleet of some proportion working it from Baja California to Western Alaska—independent men in the most independent of fisheries of the Eastern Ocean.

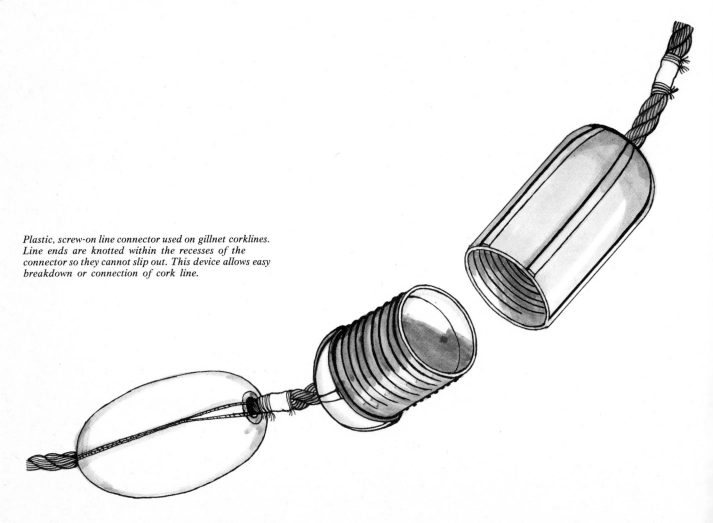

*Plastic, screw-on line connector used on gillnet corklines. Line ends are knotted within the recesses of the connector so they cannot slip out. This device allows easy breakdown or connection of cork line.*

Linda Rogers

*Halibut schooner* Zenith, *archetype of the durable longline vessel developed just after the turn of the century. Line chute sits on stern. The* Zenith *was lost off the Washington coast in a gale in January, 1972, after a career of almost 50 years.*

## Longlines And Their Use

The first man to use a fishhook must soon have learned that the more hooks he employed and the farther he spread them, within reason, the more likely he was to take fish in quantity. From this prehistoric endeavor, there developed through millennia the fishing gear known today in the Northeastern Pacific as the "longline"and by two or three other names elsewhere in the United States. But no matter what name may be used, each means almost exactly the same piece of gear—a long length of line or many lengths, with a series of hooks on leaders of several lengths secured to the main line at varying distances apart.

The longline used in the waters of interest here fishes for halibut and sablefish (pots have been adapted to this latter species) and lies directly on the bottom. Fishermen of other nations, especially the Japanese, use longlines for tuna, swordfish and shark. These lines are heavily buoyed and fish relatively close to the surface.

In the Northeastern Pacific, the longline is the only legal means for Americans and Canadians to take hali-

but, a rule that has been enforced by the International Pacific Halibut Commission over the repeated protests of the trawlers who would like, at least, too keep the halibut they take in their bottom-working nets as an incidental catch. The IPHC, with the whole-hearted and obvious support of the longliners, has been adamant in its refusal to allow the retention of dragger-caught halibut with the exception of experimental fishery tried during one season in the area south of the Columbia River where little halibut fishing is done anyway. The rules of the IPHC do not, of course, affect in the slightest the trawl fleets of the Asians working the halibut grounds of the Bering Sea and the Gulf of Alaska. But through the International North Pacific Fisheries Commission which has as its major function the control of North American salmon stocks, there have been treaties with Japan which have succeeded, to some extent, in keeping her vessels out of certain halibut "nursery" waters. Similar agreement also keeps the Asian draggers out of some major fishing areas during the halibut season because trawl and longline gear are no more compatible than gasoline and lighted matches.

The longline appeared on the Pacific Coast with the rise of the commercial halibut fishery after 1889. It first was used by the dory schooners and the term "skate," the primary unit of the longline, came in those days to be applied to the individual sections of line two men in a dory could handle, with a lot of hard work, during the heyday of the dory fishery for halibut. This period ended by World War I or a bit later because the appearance of the powered halibut "schooner" made both sailing vessels and the big steam dory vessels obsolete. The question became moot by 1935 in any event when dory fishing for halibut was outlawed by the IPHC. (The dory fishery for Pacific cod lasted until after World War II and its demise was largely a matter of economics. Trawlers could take cod faster, more cheaply and more efficiently than the dory gear and increasingly expensive manpower.)

Every fishing vessel is marked by the tools of her seasonal trade as many switch from one fishery to another, to any one that looks as if it might be more profitable than the one then engaged in. The tall poles mark the troller; the seine, the Power Block or drum and the heavy seine skiff mark the seine boat; the gallows, drum and otter boards mark the dragger; the pots and crab block mark the crabber.

The halibut vessel, whether it be the venerable power schooner or the western combination vessel, is as easily distinguished as any of these—as long as she is fishing halibut or sablefish. She can be re-geared with a minimum of time and money for trolling or dragging. But as long she is working with the longline, her tools give her away—the long bamboo poles with light and flag to mark each end of the string; by the baiting tent usually on the port quarter; by a dory nestling above the wheelhouse (on the schooner); by the roller and gurdy complex for hauling the gear, the line chute at the stern and, on some boats, the tubs that contain the skates of gear. The tubs are mostly found on the western combination boat geared for halibut fishing although they are sometimes found on the schooners.

To accept at face value a schematic drawing of a conventional longline rigging, the gear would seem to be the simplest of the major methods used for taking fish in the Northeastern Pacific. But matters are not always as they may appear and the longline is not as elementary as its bare outline might suggest to the unknowing.

The longline at work is subject to all the stresses of the seine and trawl in setting and hauling (the mere act of getting big fishing gear into and out of the water puts continual stress on every component and on the fishing vessel itself). The usual single string of 10 skates of 300 fathoms each reaches across 18,000 feet of bottom, just about three nautical miles, and there may be three or four or more of these strings side by side while the largest of tuna seines encompasses no more than six to eight acres of surface at its greatest extent. To boot, this oversize seine is no more than 60 fathoms or a bit more in depth while the halibut longline fishes down to 600 fathoms and below with 150 to 300 fathoms being worked routinely.

And all this great reach of line depends solely on the durability of the ground line (sometimes called main line because it is the main fishing line), made up of one-quarter-inch three-strand hand lay nylon with a breaking point, when new and undamaged, of no more than 1,800 pounds. A fragile reed, indeed...

But with this gear, a lot of sweat and little sleep during periods of fishing, the American and Canadian halibut fleet takes, in really good years, 50 million pounds or more of this excellent fish, working for the most part in the Gulf of Alaska and north and west in some of the world's nastiest waters. Every season, as in every salt-water fishery in every season, boats and men vanish with no bit of flotsam whatsoever to tell the tale.

Diagrammatically, halibut longline gear runs from the surface to the bottom from the upright pole to a plastic buoy (sometimes a keg on older rigs) thence to the bottom by a line from the buoy to an anchor. From

*Schematic configuration of longline gear. Line hauler or gurdy is a center, below, with line chute at right.*

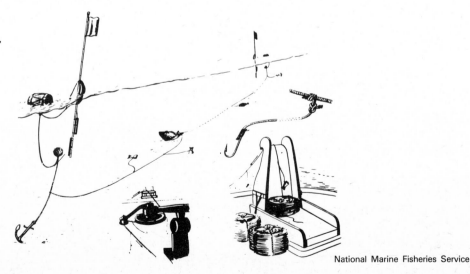

National Marine Fisheries Service

the anchor the ground line extends skate by skate to form the string. At the far end of the string is set a second anchor while above it lie the buoy line, the buoy and the pole and flag.

Along the ground line is secured the actual catching gear, a series of leaders and hooks spliced into the ground line at distances from 13 feet upwards. These are called "gangions" but the word is subject to a variety of pronunciations, often compounded by a Norwegian accent, ranging through "ganyon," the commonest, to "gangen," to "ganGEEyon," or simply "ganging line." No matter the pronunciation, the word means the same thing—something with which to catch fish.

Thirteen feet used to be the standard spread between gangions. In the 1960's, fishermen began to discover

"If the catch per unit of effort rose from year to year, the abundance of halibut was assumed to be increasing. If it fell, the abundance was assumed to be decreasing.

"In the mid-1960's, halibut fishermen, faced with decreasing abundance and more scattered fish, increased the hook spacing to 18 feet, then to 21 feet and some to 24 feet. The change allows fishermen to cover more area with the same number of hooks with no increase in bait cost or hauling time. The staff adjusted the standard skate in their calculations but assumed that hooks spaced 13 feet apart were as efficient as hooks spaced 18 or 21 feet apart.

"During the summer of 1971, the commission chartered the schooner *Chelsea* (of Seattle) and spent 30 days fishing with 12-foot, 18-foot, 21-foot and 24-foot

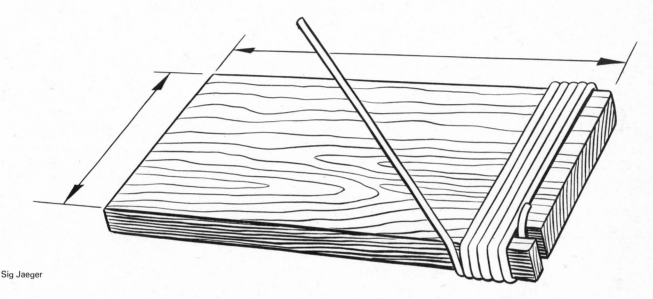

Sig Jaeger

*Becket board, section of plank used to cut gangions to uniform lengths. Width averages five inches while length varies.*

that widening the distance between spreads allowed fewer hooks to a string but, up to a certain and variable point, resulted in a bigger catch because more ground was being fished.

This seemed to result in an apparent over-assessment of halibut stocks by the International Pacific Halibut Commission. How this came about has been explained editorially in the monthly publication, *National Fisherman:*[1]

"Until the early 1960's, most halibut fishermen were using ground lines on which the gangions and hooks were spaced 13 feet apart. The (IPHC) staff developed a standard 'skate' which was composed of a given number of hooks and which was considered a 'single unit of effort.' They then based their calculations of (halibut) abundance on how many pounds of fish a standard skate would catch and called the result 'catch per unit of effort!' "

gear east of Kodiak...the intensive comparison of the four types of gear revealed that the catch per hook changes as the hook spacing changes and that with increased spacing, each hook becomes more efficient.

"Therefore, for the past several years, the staff's calculations of abundance have been overly-optimistic and 1972 (was) when the quotas were cut to compensate for overfishing in the late 1960's."

Hook spacing is a developing art rather than a science and longline skippers who like to experiment have spaced their hooks as far apart as 36 feet. Gear of 26 feet has become rather common in the halibut fishery although 36-inch gear is about the standard in the sablefish fishery. The sablefish is a vigorous, soft-

---

1. Philips, Richard, Pacific editor, *National Fisherman,* October, 1972.

mouthed fish and longline gear fishing for the species is hauled more frequently than halibut gear because of the tendency of the sablefish to free themselves from the hook.

The optimum spacing of hooks must be calculated on the density of fish and the availability of manpower. An experienced halibut fisherman, for example, can bait a skate of 12-foot gear in 21 minutes but it takes him only 13 minutes to bait a skate of 24-foot gear. With the growing tendency to reduce crews on longline vessels, the speed of baiting must obviously be taken into account. Bait cost also must be considered as must the speed of hauling.

Spacing has been explained like this:[1]

"...in halibut fishing, if 21-foot gear with 84 hooks to an 1,800-foot skate yields two 50-pound fish per skate, this is good. In the case of black cod (sablefish), using a three-foot hook spacing, with 590 hooks to the skate, 25 fish per skate, with an average weight of five to six pounds, would be very good.

"Data gathered from 1965 through 1967 by the schooner *Seattle* indicates that at fish densities of, say, 150 pounds per skate, doubling the hook spacing from 21 feet to 42 feet, thus reducing the hooks by half, cuts the number of fish caught by less than 20 percent.

"But the weight of the fish caught was reduced only by 10 to 12 percent because the increased hook spacing was apparently selective for larger fish. When fish densities were high, indications were that under such conditions the number of fish caught was more nearly proportional to the number of hooks per skate. A case in point was at St. Matthew Island (high in the Bering Sea) when the catch was at or near gear saturation, catching well over 1,000 pounds to the skate.

"Hook spacing in relation to the cost of labor means that with wide hook spacing, less manpower is needed to bait the gear in a given time period...than if the hooks were more closely spaced."

Although the longline does not have the complexity of the purse seine or the trawl, it is made up of materials of rigid specification, each of which must bear up under the stresses just described. Of all the materials built into the gear, the bamboo flagpole must be accounted the most fragile. It is 17 feet long and about one and one-half inches through at the butt and about one-half inch at the tip. At the tip of the pole, a light, flashing white when working, is lashed. (There have been questions brought up from time to time by the Coast Guard about the effect of these marker lights when gear is being fished close against the land at night because there is always the chance that mariners might mistake the pole light for a shoreside navigational aid.)

Modern light assemblies are powered by an electronic battery pack of four units, working on the same principle as that of the stroboscopic light used for photography and other purposes although its power is much less. Flying from the pole just below the light (the battery pack is below water on some poles) is a flag

running from 12 to 14 inches square. Dark red is preferred because of its high visibility although yellow and International Orange sometimes are used.

The pole rides with 12 feet of its length above the surface. It is buoyed by three plastic floats shaped much like a spool of thread with passages through them along the axis. The pole is passed through these. Each float is 10 inches long and about eight inches thick. The pole is steadied in the water by a 17-to 25-pound sash weight lashed to the butt.

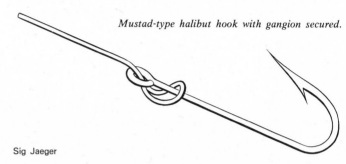

*Mustad-type halibut hook with gangion secured.*

Sig Jaeger

The pole assembly is only a means of letting the skipper know where the beginning of his string lies when he comes back to pick it up. The real work of supporting the gear leading down to the anchor is done by an inflatable plastic buoy, much the same as those used for crab pot markers. In its inflated configuration it is 60 inches in circumference and rather resembles a child's toy balloon with one end drawn to a point. The fisherman calls it a "bag."

Built into the buoy structure at the point, the end floating beneath the surface, is a tiny bridle from which dangles a 12-inch length of nylon of one-quarter-inch diameter. This short line handles two others, one a length of quarter-inch nylon between the buoy and the pole. At the pole, this line is clovehitched to a bridle straddling the float assembly. At the buoy end, this line is secured with a three-or four-turn hangman-type knot.

The second line is the first of two segments of buoy line, the line running to the anchor on the bottom. The upper end of the buoy line is secured to the short line by a clove hitch or becket bend. The first leg, also of one-quarter-inch nylon, uses one or more 50-fathom coils of line to its junction with the second leg, usually three-eighths-inch polypropylene, preferred these days because of its positive buoyancy. This leg may range from 20 fathoms upward in length, according to the preference of the skipper. The buoyancy of the polypropylene prevents chafing of the buoy line on the bottom in slack-water conditions.

At the end of this length of line, often enough going below 600 fathoms, will be found the anchor that holds

1. Jaeger, Capt. Sig, communication to the author, September 13, 1972. All following quoted material is from the same source. This entire section on longlines is based largely on Capt. Jaeger's methods.

the entire gear assembly in place. This anchor most often is of the kedge type and its weight ranges from 35 to 45 pounds. As with most anchor combinations, a length of heavy chain is shackled to the stock to hold the anchor flat against the bottom to minimize any chance of dragging.

At the junction of the chain and the polyprop line, there is a ring or shackle to which a three-eighths-inch nylon "slipshot" is secured by hitches. Its length varies and the ground line is secured to its off end. The word "slipshot" is found in no dictionary but is merely a line connecting the anchor complex and the ground line. Its length is determined, when used on the halibut schooner, by the length of the vessel from the break of the poop deck to the stern because the anchors are stowed near the break. In length it may run from 20 to 40 feet.

As noted, the standard skate is 300 fathoms long, composed of 72-thread nylon with a test of 1,800 pounds. It is called "32-pound gear" by the men who use it because it weighs 32 pounds to the skate. Its specifications are subjective as much as objective and they go like this:

"The ground line cannot be laid up too soft or too hard. In the first place, it will lie in a pile and cannot be easily handled and it is a mess to set out. If laid up too stiff, it will be kinky and difficult to handle. This matter of hardness of lay is important...and difficult to describe.

"A proper hardness means that when it is held in the hand horizontally, the line will form a gentle arc down-ward from the point of suspension. If too soft, it will hang straight down. If too hard, it will be difficult to form coils and will not lie down...it is springy, like wire.

"...an aspect of coiling gear to be set out—coils should not be perfectly round  but lain on each other like clover leaves. The idea is that each coil (turn of the line) should slightly cross the coil (beneath it) so it will not drop down inside...if it does so, this substantially increases the chances of a snarl when setting out."

Each skate has a two-foot loop backspliced into each end. These are used with the English knot to secure one to another as they pay out through the line chute during the set.

The gangion is tied to a becket reeved through the ground line. The becket line is from 12 to 14 inches long. One end of the line is doubled back to form a bight, the ends are "married" and passed through an open strand of ground line, going around it one turn and pulled under itself to form a hitch, then pulled tight. The becket holds in direct ratio to the strain upon it.

Increasingly seen, however, especially on modern vessels where various types of reels—such as the gill-net reel—are used to haul ground lines, are several types of snap gear which clip directly to the ground line and are removed as the line is hauled. Most of these clips are of corrosion-resistant stainless or galvanized steel and are entirely practical for mosquito fleet boats.

The gangion itself, 72-thread braided nylon, averages about 58 inches in length. The becket end of the gan-

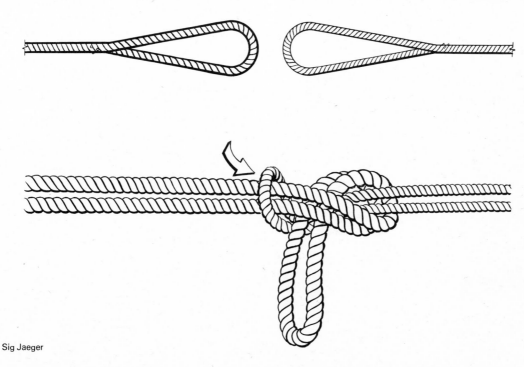

Sig Jaeger

*Loops in end of skate, top, and method, bottom, of using English knot to connect the loops.*

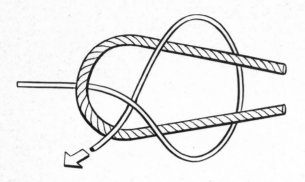
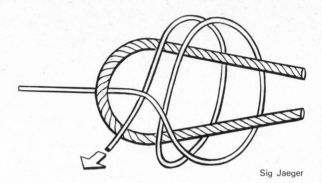

Sig Jaeger

*Method of securing gangion, small line, to becket loop with single, left, or double, right,
English knot.*

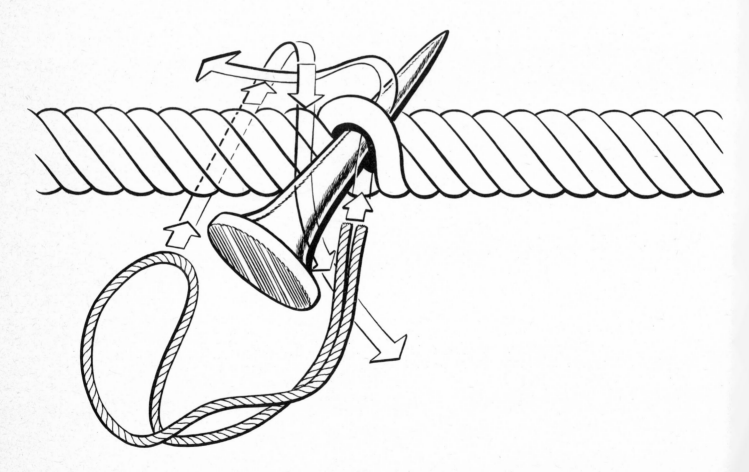

Sig Jaeger

*Method of reeving becket through ground line with fid opening strands. Running ends and
loop follow respective arrows.*

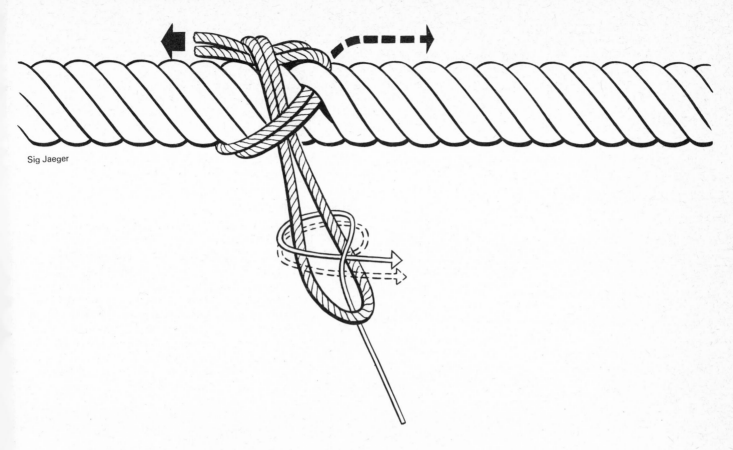

Sig Jaeger

*Completed knotting of becket and ground line with gangion secured to becket loop below.*

gion is secured to the becket loop with a double English knot while the hook is tied to the off end with any one of several kinds of knots, including the old and long-tried fishermens's bend. Any of several knots serve well here. Until the early 1960's, flat-shank hooks had traditionally been used in the halibut fishery. But monofilament nylon line was hard to bend to these hooks and was, besides, subject to sometimes-dangerous stretch if the hook fouled. The problem finally was solved with a switch to eyed hooks and braided nylon.

It should not be necessary to add that the preferred hook in the North Pacific halibut fishery is the famed Mustad hook of Norway. Sizes 12/0 or 13/0 are preferred for halibut while 6/0 to 8/0 are used for the smaller and softer-mouthed sablefish.

## Baiting The Gear

The setting of any commercial fishing gear is subject almost always to a certain amount of annoyance at the least and, at the worst, to massive snafus with corollary danger to crewmen. This is as true of the longline fishery as of any other fishery using massive gear. Even the salmon troller working a 36-foot boat can find himself in a bit of a bind when he is rigging and setting his lines.

The longline finds itself especially subject to snarls if something goes awry as the gear runs through the stern chute. The result may be a fine foulup involving one or more skates; if this should occur, the set cannot be halted to clean up the mess. The set keeps running and the fouling is cleared on the haul.

Although baiting machines have been used experimentally in the Pacific halibut fishery, tried and true hand baiting still does the work. Each hook on each gangion must be baited separately with the bait of the skipper's choice. Baits usually are mixed on the same string. Herring, octopus, gray cod and sablefish are among the usual baits. Octopus has attained much popularity since the 1960's because it stays fresh longer than herring and lasts better on the hook amid the rigors of the bottom than the soft-fleshed herring. There is incidental live baiting also that occurs when small bottom-prowling fish take the original bait and find themselves fast to the hook. Their struggles then attract halibut or other bigger fish who in turn are hooked. Bait cost per voyage for a deep sea vessel may run from $3,000 to $6,000. Most octopus is imported from Japan and South Korea.

Baiting takes place on a table or bench in the baiting shed, hook by hook, skate by skate. As each hook is baited and the ground line coiled down, the baited hook is allowed to fall naturally within the circumference of the coil. This operation is done carefully; if a skate fouls on the set, this is where the fault may lie.

The skate is baited and coiled on a small, square piece of canvas with a grommet and eye arrangement and short ties. A canvas top of the same size also is used to protect the lie of the line, gangions and hooks when the gear is moved. It is described:

"Baited skates of gear must have bottoms and tops ...squares of canvas about 18 inches square. The top needs nothing on it and simply covers the baited skate so nothing will drop in. It also helps hold the hooks on place. If they slide about, that makes snarls.

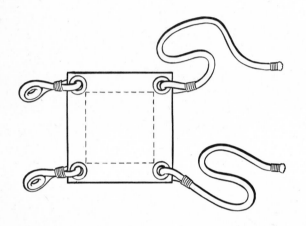

*Canvas protector for baited skate.*

"The skate bottom has eyes or grommets in all four corners. Two grommets have short lines spliced in with thimbles at the end. The other grommets have 'skate legs' spliced in, pennants long enough to be crossed over the skate, run through the thimbles and cinched, then tied. A skate should be tied tightly to prevent movement of the bait within.

"When baiting, place the coils on the skate bottom ...as each hook is baited, drop it into the center of the

*Canvas protector folded and knotted over baited skate.*

coil and distribute each successive bait evenly within the coil. But do not disturb the bait again after it has been placed. The bait should rise in a pile about equal to the height of the (coiled) skate.

"Once baited, lay a skate top on, then pass a skate leg diagonally across, then through the thimble, tighten it and take a hitch. Pass the other leg over diagonally (with a round hitch at the crossing of the first leg) then pass through the thimble, cinch up and tie the end on top. Use hitches rather than knots because they must be untied fast when setting.

## Setting The Gear

"In preparing a skate for setting, lift it onto the chute table. Untie one leg and with that leg directly aft (or the leg with the thimble in it), grasp it behind the skate with the right hand. With the left hand, grasp the still-tied remaining leg and lift the skate off the table sharply enough that the other leg may be pulled under the skate by the right hand. The skate bottom then should be folded under itself like a folded diaper.

"Untie the other leg. Hold the left hand against the nearest side of the skate and with the right hand grasp the skate leg nearest you...give it a sharp tug and the bottom comes from beneath the skate without disturbing the bait.

"Be certain that the bottom end (running end) is free, ready to tie onto the top (running) end of the next skate. As the last coil of the previous skate goes out the chute, push in the next skate. (Each end of the skate has a long eye or loop backspliced into it; they are tied together with the English knot.) When pushing a skate into the chute spread the hands at the sides of the skate so the thrust is distributed without disturbing the coils.

"If the chute table has been wet down, the skates initially slide easily and after that the slime from the bait furnishes its own lubrication.

As for the line chute itself:

"With the exception of snap-on gear, where the ground line comes off a reel and over the stern, skate and tub gear are set from the chute on the stern. (See sketch.) The chute is built of metal or is lined with metal so hooks will not hang up in it. The bottom of the chute must be wide enough to accommodate the skate comfortably. The neck at the top of the chute should center over the coil of the gear so that the gear is pulled up vertically from the center when setting.

"The chute must be as close to the stern as possible so that baited hooks won't drop on deck or hang up on the rail as they go out. The speed of setting is limited only by how well the bait can withstand the pull set up by going through the chute, yet remain on the hook.

"In the case of tub gear, the bottom of the chute should be wide enough to accommodate two tubs side by side because the tub emptied of gear must be removed and replaced with a full one while the gear is being set."

Two men customarily set the gear, one to handle flagpole, buoy and anchor, the other the skates. The skipper of the halibut schooner usually cannot see, from his place at the wheel, all of the procedure because his view astern is blocked by baiting tent and other impedimenta. The skipper of the western combination vessel fishing halibut can keep an eye on the entire operation from his topside wheel just as the skipper of the seiner controls his set from the same station.

The halibut schooner set begins with the flag and buoy positioned on the port side of the vessel slightly ahead of the wheel house. Setting speed is seven or eight knots. On signal from the skipper, the flag pole and buoy are dropped overboard with the first coil of buoy line following. As this leg of the line begins to straighten itself, the balance of the buoy line goes over. When all the buoy line is clear of the vessel so that the anchor will not foul in it, that essential piece of gear is dropped over.

The longline is more susceptible, perhaps, to mishap on setting than any other gear with the possible exception of the drum purse seine. The skates themselves, made complex with their gangions and hooks, would appear to be a standing invitation to fouling. But before the skates have a chance at it, the buoy line complex presents problems of its own:

"...the length of buoy line is in excess of the depth... to give enough scope to insure that the anchor hangs up. A rule-of-thumb for buoy line scope is about 25 percent more than the depth, i.e., in 400 fathoms, 500 fathoms of buoy line should be adequate.

"The first end of the set is started by throwing off the flag and bag (buoy) and a 25-fathom shot of buoy line. Each coil of buoy line is thrown out in turn, keeping in mind that the side of the coil **toward** the anchor should be down; as the anchor sinks, it will pull buoy line from the **bottom** of the coil in the water.

"If this procedure is reversed, pulling buoy line from the top may cause snarls. Setting buoy line is an exacting task since snarls large enough to use up the scope will mean the sinking of the flag and bag.

"...while the set is in progress, the buoy line is being readied for the end (of the string) in the same manner as the beginning. It still holds that buoy line coils thrown out must have the anchor side down.

"The slipshot is passed aft and tied nearby, ready for the last skate...the slipshot must be tied to the bottom end of this final skate and as this nears, the boat should be slowed and the propeller stopped as the last end of the skate goes over.

"This usually is thrown over by the man at the chute, casting the slipshot over the chute, while, simultaneously, the man in the waist throws the anchor with one hand and drops a coil of gear (buoy line) with it, then casts off the rest of the buoy line and the flag.

"Arranging the buoy line for each end requires thinking through the entire process by the man on the job. He must review the exact way it goes out, whether first end or last end. The most problems are found here because the procedure is not thought through thoroughly. **The flag goes out first on the first end; the anchor goes out first on the last end.**

"Each coil of buoy line should be cast well clear of the boat in small, easy-to-handle coils. A good technique is to throw the buoy line so the coils spray out rather than lie in a concentrated coil in the water.

"Under no condition should the line handler get between the buoy line and the rail. The gear must always be kept between him and the rail. It is a good idea to keep a sharp knife handy at the rail. A knife should also be kept handy at the chute. Some men also keep a broomstick at hand to clear a hook that flips out of the coil before it can pick up and snarl several coils."

As the anchor sinks, the tension it sets up immediately begins to pull the running end of the first skate of gear from its position by the stern chute up and over and through the guards of the chute. At this time two skates have been tied together and others are moved into position and secured to insure the set running continuously. Some vessels use patent snaps of stainless or galvanized steel. Some of these have breaking points of more than 4,000 pounds, far exceeding that of the ground line. Nevertheless, most deep sea skippers still prefer the older method of connecting their skates.

Before the last skate has begun to pass over the stern chute, the deckmen have set up the second pole, float, buoy lines and anchor so these may go into the water in their exact order without putting any strain on the ground line. In some unusually long strings, a flag pole-buoy-anchor combination may be inserted along about the middle of the string to keep it in position. More commonly used, however, is a short length of heavy chain, usually about four feet long, secured between each skate or between every second skate to offset movement on the bottom. The addition of the chain, although it may be necessary, would appear to complicate an already-exacting setting and hauling procedure.

Longline gear sinks slowly despite whatever may be added to it in the way of extra weight. Since the nylon of the main line has a slight negative buoyancy, it is estimated that it takes about 20 minutes for the entire line to settle to the bottom, with the time dependent obviously on depth.

Although not all men follow the practice, some skippers make their sets along loran lines, especially and obviously when working out of sight of land. The loran line is not a straight line although its arc depends on distance and direction of the master and slave stations. A 10-skate string may be laid out along a line with as much as two degrees of arc in it. Although this slight bend in the line makes no difference at all to the halibut, it is essential to follow the primary line if the gear is to be kept parallel, so that all sets fish equally well by covering the ground systematically.

The use of loran lines for setting has at least three virtues: the loran line lets the skipper know exactly where his first string lies on the bottom; it allows him to set his other strings in an exact relationship to the first, and it allows him to return, especially with the use of Loran C, through rotten visibility to the marker at the beginning of the first string to start his haul. In poor visibility, good radar is an additional valuable tool in locating flags with radar reflectors.

Strings are not spaced randomly. The spacing is called "berthing" by longliners and is based on a pre-calculated distance between strings. This is the distance by which the skipper believes he can fish much ground productively without leaving too much untouched area between strings. In essence, this is the same principle on which the fishermen began in the mid-60's to space their hooks further apart on the ground line. A distance between strings of about 3,000 feet or more might be judged a good average.

By the time the halibut fishery had fully come of age, the years in the first third of this century when the power schooner had replaced the sailing schooner and the steamship and their dories, berthing had become something of a science. Berthing was reckoned by minutes—a "four-minute berth, a six-minute berth" or whatever.

This was the distance the vessel could move across the bottom in those time periods and this was the distance at which the strings were set from each other. The six-minute berth was regarded as a standard, more or less, because, under normal conditions, this was about a nautical mile, the ground that could be covered in a six-minute run.

These time elements were used because "strings of gear should be fished methodically, berthed parallel to each other at the optimum distance apart. The width of the berth should be governed by the most efficient distance at which all or most of the fish are (being) caught in that area."

Unfortunately, setting out parallel often was more desired than accomplished. It could be done with a fair degree of accuracy if a vessel were working closely enough to the land in clear weather to set on bearings. But offshore, dead reckoning had to be used in conditions of bad visibility after the first string was down. Even the best of navigators may find himself in error (with errors compounded) when using dead reckoning. This dead reckoning was reduced to the timed run on a heading hopefully at a 90-degree angle to the first string and each subsequent string. The four-minute berth was preferable to most men but too often it resulted in strings drifting across each other and the six-minute berth was most used for that reason.

Introduction of loran and its quick adoption by longliners (and most other offshore fishing vessels) brought some berthing down to less than three minutes in water where the gear could reach bottom without undue drift. This gives a berthing distance of about 2,700 feet although most men feel that distance is not enough.

However, the tight berthing did produce, for highly-competent skippers, a better catch, up to 20 percent,

from areas previously fished on dead reckoning and six-minute berths.

Nor are strings left in the water for random periods. The average "soak" might be put at about 24 hours unless weather interferes with pickup or it might be less if it is apparent that fish are abundant and that hauling lines, picking fish and resetting immediately may pay off. Gear being used for prospect fishing, i. e., gear searching for fish, may be left in the water for less than six hours, then hauled so the vessel can continue its search elsewhere.

The longline is vulnerable to several types of accident as it works away deep beneath the surface. Bottom scouring by currents can part the line as can chafing against bottom obstructions. But the longline skipper has a pretty fair piece of gear to help him pick up his parted ground line. This is a drag, a 20-foot length or more of three-quarter-inch chain shackled to enough line to find bottom. To each third link of the chain are welded two hooks that resemble, as much as anything, an oversize staple with the points ground sharp. Since the skipper, under most circumstances, has a pretty good idea where his ground line is anyway, especially if he used loran to set, he zigzags his vessel along the suspected location of the line until the heavy chain and its hooks foul it and claw onto it. It then becomes a simple job of hauling the line aboard, breaking it down into skates and starting all over again.

The recovery operation is detailed like this:

"Recovering lost gear...is a necessary competence in any longline operation. In depths up to 350 to 500 fathoms, this can be done with appropriate gear and technique. (A lost string of heavy sablefish pots has been brought up from 420 fathoms.)

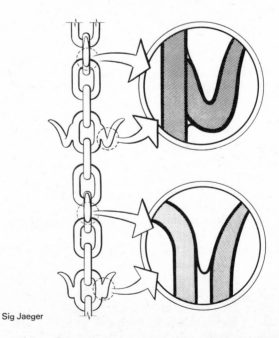

Sig Jaeger

*Longline skipper's ace in the hole—chain and tooth drag gear for retrieving lost gear.*

"The location of the lost string must be determined as precisely as possible. This positioning must be determined each time a string is set by loran or Decca readings; relative bearings on adjoining gear or landmarks; known configuration of the bottom, where changes are abrupt and records have been kept; running time and direction from a marker; any combination of two or more of the above.

"Additionally, a marker buoy is needed, consisting of an anchor, sufficient buoy line for the depth, and a bag and flag that can be seen a half-mile or so.

"Once the location has been established...the marker is dropped anchor first. When the buoy line goes slack, with the anchor on the bottom, as little scope as possible is left to the bag so that it will not be too far off the mark. But there should be enough scope that the tide or the wind will not make the flag and bag pull the anchor out of position. The flag will also indicate the tide strength and its direction.

"The drag teeth (see sketch) should form a sharp vee with the shank but with no sharp corners to chafe the recovered gear. The peak of the vee should be round and smooth and about one-third the diameter of the gear to be recovered. The number of drag teeth should be six or seven pair. If using a grapnel-type drag, a length of chain in the eye of the shank helps keep the drag parallel to the bottom.

"About 50 or 60 fathoms from the drag, a weight such as a chain or a steel bar is needed to help keep the drag on the bottom by absorbing the surges transmitted down the drag line by the boat's heaving in the sea.

"The drag line itself should be at least as strong as the gear being recovered, preferably a size larger because of the heavy lift. It should have negative buoyancy, not the positive buoyancy of polypropylene, for example. Nylon sinks and is alright...

"The extra weights, such as chain or bar, should be tied to the drag line by a pennant small enough that the knot will pass around the hauling sheave of the gurdy or be easily removable when hauled on deck.

"Only experimenting will tell how far uptide or upwind the drag should be dropped so it will reach the bottom and can be towed across the lost gear. Several attempts may be necessary.

"A watchful eye must be kept on the time used in trying to recover lost gear. If the lost time could be more financially rewarding by hauling the balance of the gear rather than trying to recover the value of a lost string, obviously the former should govern."

## Hauling The Longline

The skipper of the longline vessel starting to haul gear finds many variables he must take into account and he is powerless to control some of them. He must consider direction and velocity of tide and wind, which he can do nothing about whatsoever. He does have discretion, though, in deciding from which end to haul and which string to haul first. Wind and tide will decide for him, if they are strong, from which end to proceed. Too much wind may drift the boat over the gear faster than it can be hauled. If he is at the beginning of a trip, he first may haul a string in new territory to assess prospects there. Many times he may haul two or three strings before resetting any since setting in that manner saves much time. If a flag and buoy are gone or the line has parted he may have no choice about which end to haul. He must take the end with the marker gear still intact.

Faced, thus, with several choices, the longline skipper must decide quickly which course he must follow. Most men prefer to start the haul from the end of the string lying downwind and/or downtide. In this haul, the vessel is more easily kept in line with the lie of the string and can be "kicked" ahead to follow the ground line as it comes aboard. Here, hauling speed and strength of wind and tidal drift must be carefully calculated to avoid too much stress on the fragile ground line.

Some men of special skill prefer to haul with the tide. This procedure involves an equally careful calculation of tidal set and wind velocity and direction. In this manner of hauling, the estimated speed of the tidal current must be deducted from the usual hauling speed and the speed of the boat reduced accordingly so that the gear may come aboard as smoothly and evenly as possible.

In addition, this haul involves a careful juggling of vessel speed and handling to hold the vessel in line and to keep it from running over the gear. In deep water, however, tidal currents generally are of such a broad sweep (as contrasted to the velocities of inshore ebb and flood) that they often actually have little effect and wind may be the major complication. In a good tide run, small vessels hauling with the tide may be moved along the string by the pull exerted on the gear by the gurdy as it brings the line aboard and the engine need not be used.

Crosscurrent and crosswind, singly or together, bring with them their own problems. Either or both can give a skipper fits as he tries to keep his vessel on a heading that will hold it along and above the string so that stress on the ground line can be held to a minimum.

Any person who has handled or attempted to handle a single-screw vessel of comparatively deep draft (seven to 12 feet or so) in a confined area easily can understand the problems confronting the longline skipper as he tries to stay with his gear in any of the described situations.

The vessel itself and the bottom from which it is hauling the gear present hazards too:

"When hauling gear, especially in deep water with its increased tension (on the ground line) the vessel should guide on the gear as it comes up to the roller (on the starboard rail) so the gear may be kept as nearly vertical as possible.

"If even a slight angle develops in the direction of hauling, the strain is multiplied greatly. Care should also be exercised that the ground line does not chafe against the bilges. If the gear does chafe against the vessel's bottom, it may be standing at all (any) angles to the bottom without the operator knowing in which direction because the angle between the roller and bilge remains constant and won't indicate if the angle of the gear changes. The only clue is that the tension on the gear increases as it goes farther and farther under the boat.

"In event the gear hangs up on the bottom, hauling should be slowed or stopped, and the boat brought up the gear so it is vertical. Sometimes the gear can be broken loose by moving gently in the direction of tide or wind, then coming back to the vertical, keeping tension on the line. In a chop or heavy sea, there is a substantial likelihood of the gear parting.

"With nylon ground line, which has considerable stretch, the hangup situation can be dangerous to the crew if it is attempted to throw the gear off the gurdy (while under tension). If this is attempted, it should be made certain that a number of hooks are removed so that when the gear flies out of the sheave, no hooks are flying around.

"An alternative in a hangup is to haul the gear tightly, then lash a bag to it close to the water, cut the ground line and start hauling from the other end. It must be assured, when the gear is out, that enough is left above the bag that it may be easily led over the roller and around the gurdy to resume hauling.

"When hauling from the other end toward the hangup, it may let go by itself. If not, by hauling to the bag and taking that end aboard also, both ends can be hauled simultaneously until the gear comes free or parts.

"In a hangup situation, the deeper the gear lies, the more slowly does the increased strain of hauling distribute itself evenly toward the bottom. Increase the strain only a foot or so at a time, wait a minute or so, then take up more. It is a good idea also to heave as the boat goes down in the trough between swells, then slack off on the uplift."

But if all goes right, if wind and tide are as the skipper desires them, he is free to start his haul at the preferred end. The operation begins when the crewmen

National Fisherman

*Stern ramp trawler making haul. This trawl method, common on the Atlantic coast, has been slow to come to Americans in the Pacific although Asians fishing the Eastern Pacific have converted to it almost entirely because of its simplicity and safety.*

Pacific Fisherman

*The unfriendly shores of the Eastern Ocean.*

lift aboard the flagpole and buoy assembly. These immediately are freed from the buoy line and stowed away until they are to be reset. The buoy line passes between the fairleads of the roller secured to the starboard rail and the line is led through the sheaves of the gurdy which takes much of the hard work out of this fishery as similar gear has done in others. The gurdy complex sits about midships along the center line of the vessel. The sheave rotates horizontally in contrast to the vertical movement of the salmon gurdy, the crab block and the albacore gurdy.

When the gear starts aboard, the running end is passed halfway around the line sheave and, in turn is passed off to a smaller sheave working at the same speed and at the same level as the main sheave. This small sheave is called a fairlead and from it the man coiling line takes the gear to work it. From above, the line running from the large to the small sheave assumes an S-shape. The gurdy controls the pace of the haul, fast or slow, according to the calculation of the man at the roller. Power comes from the main engine, either by a mechanical power take-off from the main engine or by a hydraulic system. Hydraulic systems have largely replaced the mechanical power of older vessels of the longline fishery and they or electric power are standard equipment on all modern or remodeled vessels in all fisheries of the Northeastern Pacific.

There are two fail-safes built into the gurdy. One is a stop control for the rollerman so that if he becomes

hooked up or otherwise entangled in the gear, he can cut power instantly. The skipper also can do this from the wheelhouse. The second is a pawl in the gear casing that becomes activated when the gurdy stops so that the gear cannot backhaul despite any strain on it.

On the edge of the sheave facing the gurdy fairlead is a metal wedge inserted into the vee of the sheave which is called a "finger." It is used to clear the line from the sheave if excessive strain jams it there.

Most modern gurdy assemblies carry a second sheave mounted atop the gurdy, one with a diameter greater than that of the main sheave. This sheave rotates at the same speed as the line sheave but it is used to haul buoy lines because its greater diameter picks them up faster than the line sheave.

The holler on deck is "Keep that damn gurdy going...!" because at any time the gurdy is idle, the boat is not fishing and a gurdy going fullbore is the one piece of gear that gets the catch aboard.

As described, two men are involved in the actual hauling procedure when the line starts coming aboard, he who coils it down as it comes off the gurdy while the second tends the line at the roller. At the same time, he keeps a running check on the line with one hand— much as the salmon troller uses his sense of touch on his lines when hauling them to estimate the strain. It is the skipper's job to watch the direction the line takes so he may con his vessel properly along it. Positioning and moving the vessel along the line just right minimizes stress on the line with a consequent reduction in damage caused by undue strain on the gear.

When the anchor comes aboard, it is released from the ground line after that line is safely secured by a couple of turns around the sheave and it too is stowed until it is again needed. With these preliminaries completed, the skates start coming aboard with their fish— if any—and here is where the man working at the roller must display strength and great skill. Halibut may run up to 400 pounds or even more (although specimens of that size, always females, are seldom seen in this era). But fish of 100 pounds and up are common. In addition, the halibut is a strong and vigorously resisting fish which must be handled with care since a swipe of a big specimen's tail can shatter the bones of a man's leg.

The fish must be gaffed in the head as soon as the gaff can reach it and dragged over the rail in one sweeping motion by the man tending the roller. Fish gaffed in the body are sold as seconds and a good man with the gaff insures a higher percentage of Number One fish. This job often must be done with a heavy sea running and the fisherman not only must contend with the recalcitrant halibut but with the dangers posed by a rolling and pitching vessel. The gaff man also faces another danger. A hook may straighten or be pulled free at the roller and the flying hook instantly becomes a deadly weapon. All nylon has a degree of elasticity and not even braided nylon is free of this tendency. It is the recoil of this elasticity that makes the hook a danger.

Skate by skate, fish by fish, the ground line comes aboard. As the fish are brought in, they are deposited in checkers until they can be cleaned. As the skates come free of the gurdy, they are released from each other and coiled on their canvas carriers. The baiter takes the coils aft to the baiting tent where bad hooks are replaced and each hook is rebaited. Worn sections of the line are cut out and the ends retied.

In his "spare time," the baiter may also be the man who cleans the catch or a part of it, at least. On the dressing table by the hatch cover, the fish are gilled, gutted, scraped clean and placed in a holding checker until they can be iced. (Recommendations for handling and icing of halibut as outlined by the National Marine Fisheries Service may be found in detail in Appendix IV, Part 1.)

In general, each fish is passed into the hold tail first to minimize bruising. The visceral cavity, the "poke," is filled with ice. Halibut are not shelf iced but instead are placed in pens for storage, all of them white side up except for the top layer of fish. It has been learned over the years that full icing of the poke is sufficient to bring down and maintain proper internal temperature. Ice is not placed between the layers because ice tends to cause skin damage which results in less desirable quality. But ice is placed around the edges of each layer, between the mass of fish and the hull and atop the last layer. Most deep sea vessels now carry refrigeration systems which reduce excessive ice melt and hold the catch at 32 degrees F. or slightly above.

The last skate brings up with it the slipshot and the anchor. This leaves the buoy line slack in the water and it is hauled at high speed. The buoy and the flag pole are the last elements of the string to be retrieved.

With the entire string aboard and broken down into its component parts, the skipper again has the chance to exercise the options open to him—to reset this one or haul a couple more or...

Pacific Fisherman

*Wood-hulled, gasoline-powered gillnet boat built specifically for work on Bristol Bay. The power roller, as on stern of this boat, is preferred to the power reel for the high-speed Bay fishery because it allows faster hauling of nets.*

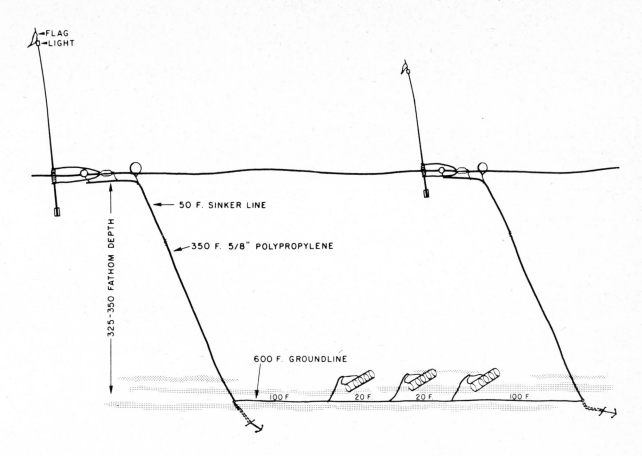

*Standard longline gear modified to use pots for sablefish, crab and other species.*
*This gearing has promise for a flourishing fishery.*

## Pot Fishing Gear

The use of pots to catch some species of fish and all manner of crustaceans was a fairly early development in the fisherman's never-ending quest for more efficient means of taking his quarry. Without doubt, the pot appeared through the world's fishing areas ages before the first history was written down. No one can know for sure what these first pots looked like but it is certain enough, from examination of pots still in use among the world's vanishing primitive peoples, that no significant change has taken place in general design although the use of increasingly sophisticated materials over the centuries has made them bigger and stronger and better catchers of the creatures they are designed to trap.

The first pots must have been laboriously built of the tough flexible branches of trees like the willow, the members bound together by willow withes or raw hide or liana or flax or papyrus or whatever the fisherman had at hand when the work was being done. Wood, in more polished configuration, has not yet disappeared entirely from pot gear and some still may be found working in a small way in the fisheries of the Northeastern Pacific. Although crab fishermen of the

area long ago graduated to metal for their pots, much shrimp fishing still is carried on with pots built of cedar slats and a fine-mesh wire screen. These are inexpensive, easy to build without welding and electrolytic damage is not a matter of much concern to the men who work them.

The great crab fisheries worked by the West Coast fishermen of the United States and Canada are carried on solely with pots because of the comparative simplicity of regulating conservation measures through them. Foreign fishermen working king crab off the northwest coast still use tangle nets, to some degree, a quite efficient device for taking crabs in large numbers, although they will be phased out in time. But the tangle net cannot distinguish among crabs and it takes juveniles with adults, ineligible females with eligible males. The net injures crab and it makes the recovery and return to the water of these ineligible crabs an arduous and unrewarding task, one seemingly little bothered with by some foreign fishermen who concern themselves only with market crab and consign the others to whatever fate may await them back on the bottom. The king crab pot takes these unusable crab too but it does not hurt them except as there may be some crushing when a heavily-loaded pot is hauled aboard, and it is a simple matter to sort out the sub-legals and females when a pot is emptied.

The American king crab fishery was not always a pot fishery. In its Stone Age days before and right after World War II, any gear went and it was not until 1954 when the fishery was in its early maturity that tangle nets and trawls were outlawed and pots made the only legal gear. But pots had been the major gear for several years by then after considerable sweat and money had gone into development of a fairly standard large pot for the large crab.

By contrast, the coastal Dungeness crab fishery of Washington, Oregon and California is an affair of antiquity. Fisheries scientist A. T. Pruter wrote that the first commercial crab landings in Washington and Oregon occurred in the late 1880's. A major fishery did not shape up, however, until the 1920's were well along, this coincidental with development of canning and freezing methods that made crab safe for consumption in areas more than a few miles away from the piers where they were landed.

Dungeness fishermen long ago settled on a uniform or nearly uniform pot design, a round, squatty contraption of size and strength consistent with the size of the smallish Dungeness crab. But a few minutes of research along the piers of any crab port will disclose a variety of construction within this uniformity, a variety based on age of the pot as much as anything. A single boat may fish pots ranging from the skipper's earliest years to others of design and materials so new that the pot does not yet appear in general literature on the fishery. The shrimp pot fishery contributes a quantity of large or prawn-type shrimp to the fresh market wherever these are to be found in suitable concentrations along the coast. But in the main, the shrimp pot fishery may best be characterized as a subsistence fishery, one on a par with the one-mule, 10-acre hill farms of Appalachia. A good part of the catch goes for home and neighborhood use, especially along the remote coasts of British Columbia and Alaska where shrimp are easy enough to get to but markets are hard to come by.

This is not the case, however, in those areas like Washington's Puget Sound where there are multiple markets and multiplying demand but few big shrimp available from handy waters. There has been a small-scale commercial fishery in parts of Puget Sound for a couple of generations but it has declined badly since 1964. In that year, some 70,000 pounds of big shrimp, chiefly the spot shrimp, *Pandalus platyceros*, were landed for market. These shrimp ran from 10 to 17 to the pound, heads-on. By 1969, landings were down to less than 20,000 pounds and a BCF exploration of other areas on the Sound found no new grounds where the big shrimp were available in commercial quantities.

The situation appears to be more favorable in parts of Southeastern Alaska where the spot shrimp abounds. Fishing has not yet dented the populations as it apparently has done, at least in part, in Puget Sound.

(Pollution and other factors may be even more responsible for the decline in abundance.) The BCF describes the shrimp as being of large size and superior quality, available the year around, "ideal for an off-season small boat fishery." Exploratory fishing with various types of trawls around Ketchikan have shown that the pot is still the best method for taking these big shrimp because of the foul bottom they prefer.

## The Pot Fishery For Sablefish

The concept of the pot as a means of catching certain fin-fishes is as old as the concept of the pot itself as an efficient piece of gear in specific situation. In North America, however, the pot in its various forms has been used almost solely to take such desirable crustaceans as the lobster, some of the larger shrimp. Many of these pots have been found to be attractive to some species of fish although no commercial fishery worthy of note developed from casual observation of the phenomenon. A subsistence fishery has been built up around this use of fish from pots that is as old as the first pot use in the waters of the continent and generations of fisher families have grown and thrived on this incidental catch.

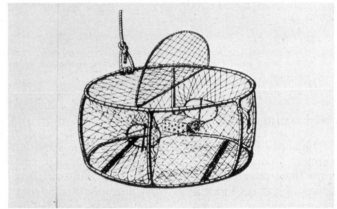

National Marine Fisheries Service

*Dungeness crab pot.*

With the development of the king crab fishery of the Northeastern Pacific after World War II, it was seen quickly that the large pots that began to come into use after 1950 sometimes contained as many fish by weight as they did crab. Among these fish was the sablefish, *Anopoploma fimbria*, a toothsome denizen of deep water from mid-California to the Bering Sea. The species is commonly called black cod by fishermen and usually comes into the market under that name although it is not a cod and the shading of its back may range from green to charcoal. It also is known variously by a number of other equally-inaccurate names such as coal fish, butterfish, coal cod, blue cod and candlefish, appellations also applied indiscriminantly to many other fishes in many other places. It may range up to 40 inches in length and 20 pounds in weight although specimens of this size are rare. Sablefish and the petrale sole bring the highest prices of all his catch to the drag fisherman.

Traditionally, sablefish have been taken in trawls and on longlines. The longline catch comes both as an incidental catch on gear intended for halibut and as the primary species in a specialized fishery conducted chiefly along the Oregon coast during the autumn months. Halibut fishermen are not especially pleased with any high incidence of sablefish on their gear since its best price, a good one to the dragger, may be only a third or even less of the price of top-grade halibut.

Both of these older methods of taking sablefish have several failings. Predators such as sharks, seals, and sea lions, deepwater crabs and some species of bottomfish consider fish on longline gear as dinner on the table. The mutilation rate may often be high or even devastating as far as the fisherman is concerned since fish heads sans bodies on his hooks are worth something less than nothing to him.

The trawl fishery, with its codends sometimes heavy with thousands of pounds of fish of various species, may turn out crushed sablefish of little or no value or sablefish rendered almost equally worthless by punctures from the spines of such heavy-finned creatures as the rockfishes or lingcod. Drag fishermen, if the catch is not too heavy and the time is available, usually gill and gut sablefish, almost the only example of dressing fish in the trawl fishery of the Northeastern Pacific. (The wings of edible skates are trimmed off and held for the market.)

Fishermen and personnel of the present National Marine Fisheries Service and the Alaska Department of Fish and Game soon became aware that sablefish found in crab pots that were not too loaded with crab and thus not crushed on the haul were healthy, strong and un-mutilated, good marketable fish if there happened to be enough of them in a string of pots. Crab fishermen did not fool around with them because sablefish were not their quarry. But their presence and condition were not forgotten.

In the middle 1960's, the potential of a pot fishery for sablefish became an object of interest to the old Bureau of Commercial Fisheries Exploratory Fishing and Gear Research Base in Seattle. Over the years, through a myriad of trials and errors and minor successes, the study has resulted in development of a pot fishery for sablefish that has passed the experimental stage but has not yet approached the status of a full-ahead commercial fishery for every man who might be interested in it.

The initial attempts at turning the sablefish pot fishery into a viable commercial fishery began as a simple pot fishery in the sense that the crab fisheries per se are pot fisheries. But the end result showed that a combination of the historic longline fishery and the pure pot fishery offers advantages to the fisherman that neither of these older methods can offer in itself.

The first experimental fishing on any real scale used modified king crab pots. These were standard pots altered to six feet by eight feet by three feet with nine-inch webbing and two tunnel entrances, one at each end. This modification caught fish but it was not particularly fruitful nor had it really been expected to be so. It was a trial shot, a first attempt at standardizing the haphazard observation that sometimes, in some circumstances, sablefish did get into crab pots with no apparent difficulty—and they could not get out!

These first pots then were modified again until they came eventually to have three-inch mesh and four tunnels with eight-inch entrances on the tunnels. These were the first pots that could reasonably be called successful. They were, remember, in all respects except their modifications, king crab pots with the standard buoy system and were set and hauled as are all king crab pots.

The BCF did not have the time, the men or the vessels to put into this single experiment on the scale it deserved. The next step was to charter a vessel to carry on a continuing testing program and here the BCF made one of the better deals of its history. The charter went to the halibut schooner *Seattle,* an archetype of the classic West Coast schooner, a vessel owned and skippered by Capt. Sig Jaeger, an experienced and experimentally-inclined veteran of the longline fishery, a man who was not only a good fisherman but also one quite capable of writing about fishing and talking about it among men of considerable academic background.

The charter called for Jaeger and the *Seattle* to simulate a 12-day commercial trip. The only added gear included a block davit, a crab block and 23 complete sets of pot gear supplied by the BCF. First results of this fishing were "encouraging" and Jaeger offered to continue experimental pot fishing without charter. This offer resulted in a one-year agreement whereby Jaeger and the *Seattle* would conduct cooperative research with NMFS to furnish deck gear and pots in return for Jaeger's increasing knowledge of the fishery.

Through the winter of 1969-70, the *Seattle* labored away off the Washington coast, fishing a variety of pots during miserable weather that resulted in a high pot loss (as well as eight others that vanished while the *Seattle* lay in Fishermen's Terminal in Seattle). The result of her efforts during that time showed that modified crab pots measuring six feet by eight feet by 32 inches were most productive.

During this time, the *Seattle* broke even or less on her efforts and Jaeger noted in a subsequent newspaper story that only faith in the eventual success of the fishery sustained him and his crew through this difficult period.

While Jaeger carried on, the NMFS research team turned to a study of a hybrid system, one using newly designed pots combined with the longline. The first work on this phase of the research was done aboard the NMFS' veteran research vessel *John N. Cobb.*

The *Cobb* fished both cylindrical and rectangular pots with the most success achieved with a cylindrical pot with a double-tunnel entrance on one end. It took just two weeks of experimental fishing to demonstrate that the new system had a good potential for the development of a commercially-viable fishery.

Here again, the *Seattle* and her determined skipper went into action. The *Seattle* took on 10 pots to set on a longline while, at the same time, she continued to fish her modified crab pots. Concurrently, the *M/V Seaview*, Konrad Uri skipper, took on 80 of the double-tunnel cylindrical pots to fish four strings of 20 each in an experiment to determine if sablefish could be taken with them and delivered alive to the market.

Despite a multitude of problems, the *Seaview* showed that sablefish in commercial quantities could be brought to the processor although no commercial follow-up has taken place.

The *Seattle,* through calm and storm, through poor catches and good catches, her crewmen sustained by her strong-willed skipper, stayed in the fishery for more than a year although no one made much more than expenses. (Jaeger, a dominant figure in the North Pacific longline fishery, once played a figurative Captain Ahab in a television documentary shown in the late 1960's. The 60-minute film depicted the *Seattle* and Jaeger as latter-day counterparts of the whaler *Pequod* and Captain Ahab in a long and largely fruitless—until the last moment—search for halibut in the Gulf of Alaska. The film was one of a series wherein modern-day mariners were likened to real and fictional characters of the sea in days past. For some reason, however, no one ever was asked to re-enact the careers of Bluebeard or Captain Kidd.)

The most practical pot used by the *Seattle* was the standing cylindrical pot eight feet by three feet. The longitudinal members or stringers and the rings are of one-half-inch soft steel rod welded at all points of contact. This skeleton supports a webbing of 11-gauge two- by four-inch mesh of galvanized wire. The pots are in two four-foot sections, hog-ringed together and to the steel support rings. The tunnels are built of two and one-half-inch No. 21 nylon, the first leading into the second to form a natural fyke. Thus no triggers and gates are required. Each tunnel is cut two feet long since shallower tunnels do not appear to fish as well as the deeper ones. A semi-circular bait and fish removal door, half the diameter of the pot, is built into the opposite end. The longline gear itself is essentially that of the halibut fishery and is described in connection with the fishery.

If a modest or a large-scale pot fishery for sablefish is to develop on the West Coast, these pots may be replaced by folding pots that have been proved practical in experimental fishing conducted by NMFS itself. The collapsible pots are eight feet long and 34 inches square with the same basic metal frame as the standing pots. The mesh is two inches square to allow escape of small fish.

Longitudinal and end sections are built individually, then the sections are joined by coiled springs crimped to keep them in place. These springs, acting as hinges, allow the pots to collapse and be stacked as high as necessary. It is estimated a fishing vessel in the 60-

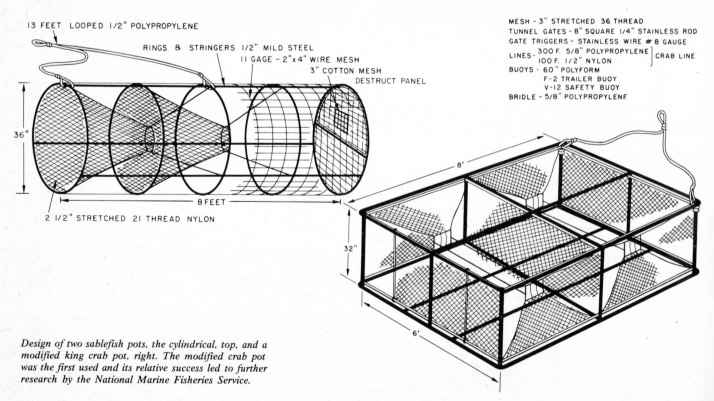

13 FEET  LOOPED 1/2" POLYPROPYLENE

RINGS 8 STRINGERS 1/2" MILD STEEL
11 GAGE – 2"x 4" WIRE MESH
3" COTTON MESH
DESTRUCT PANEL

36"

8 FEET

2 1/2" STRETCHED 21 THREAD NYLON

MESH – 3" STRETCHED  36 THREAD
TUNNEL GATES – 8" SQUARE 1/4" STAINLESS ROD
GATE TRIGGERS – STAINLESS WIRE # 8 GAUGE
LINES – 300 F. 5/8" POLYPROPYLENE
          100 F. 1/2" NYLON            CRAB LINE
BUOYS – 60" POLYFORM
          F-2 TRAILER BUOY
          V-12 SAFETY BUOY
BRIDLE – 5/8" POLYPROPYLENE

8'

32"

6'

*Design of two sablefish pots, the cylindrical, top, and a modified king crab pot, right. The modified crab pot was the first used and its relative success led to further research by the National Marine Fisheries Service.*

to 70-foot class could stack up to 200 folding pots on her decks.

The advantages of collapsible pots are obvious in that a vessel of the above size can load, at one time, so many more of them than of the standing type. Many more strings can be fished with less travel between port and fishing ground and the last strings put down can be kept fishing while the catcher boat returns to port to unload. Thus a virtually-continuous fishery can be carried on.

An essential feature of all the pots used after the initial experiments is a "destruct" panel to eliminate the fishing of pots after they have been lost. The NMFS recommends a panel of cotton webbing about nine inches square. Cotton deteriorates relatively rapidly in salt water and a 21-thread webbing should rot away in less than five months.

The effect of fishing by lost pots is described:[1]

"On one occasion, nine king crab-style pots which had been lost for about six weeks were recovered and found to contain numerous skeletons and black cod skins in addition to up to 24 live black cod per pot. During fishing trials, it is quite common to find dead black cod in pots and probably most other pot gear will continue to fish with dead fish serving as bait to attract more fish which eventually die to attract more fish, ad infinitum, until the pot deteriorates to the point where the fish can escape. This process…is slow and could take years, depending upon the materials employed…The consequence of large numbers of "ghost" pots fishing in an unregulated manner…could be catastrophic to the black cod fishery…"

So goes the conception, gestation and birth of a new fishery.

## The Dungeness Fishery At Work

The Dungeness crab fishery is a fishery of some peril to its men, a fishery of irritating fluctuation in abundance and quality of crab but a fishery that in some years comes up fully to the greatest expectations of fishermen. Just such a cyclic season was that of 1968-69 along the coast from Northern California to the Strait of Juan de Fuca where the crab were plentiful, big and of solid meat. That was the season that highline fishermen bought big cars, built new homes or remodeled the old ones and took vacations in exotic places in the off-season. Not many of them, unfortunately, invested in new boats or in major work on the old faithfuls that had brought them the big catch. Such seasons should come along about once every five years. Similar seasons back to back do not seem to occur.

That season, the natural factors that influence crab abundance were working all for the crab. Weather, in the main, was good for the fishermen. There were not too many extended periods of bad weather when the harbor or river mouth bars could not be crossed to work the pot strings. These bars, those of Grays Harbor or the mouth of the Columbia River or Tillamook Bay and all the other shelters along the coast, are the greatest hazard to crab fishermen of the lower West Coast states because theirs is a winter fishery, carried on, with minor exceptions, in the open sea on a lee shore. This situation can be unpleasant enough at times during the summer months and it can be many times more unpleasant during the winter months when the bars may break so badly not even the biggest of ships can cross them. In Western Alaska, winter weather is the thing that makes that area's Dungeness fishery a summer affair because the small Dungeness boats must fish in shallower and more dangerous water than the usually-bigger vessels of the king crab fleet.

The Alaska Dungeness fishery, more particularly the fishery landing its catches at Kodiak, the nation's top crab port, is a fruitful one and one that grows each year. One reason for this growth is the chance to use vessels that are not big enough and stable enough for the wintertime crab fisheries such as king crab fishery. The Kodiak fishery has been described thus:[2]

"The principle gear in the Dungeness crab fishery is a round pot. The pots are 42, 48 and 60 inches in diameter. They are constructed of three-quarter-inch round steel stock with two pieces of 1-1/2-inch stock welded to the bottom of the pot for ballast. The pot frame is wrapped with rubber strips cut from inner tubes. Then it is covered with stainless steel wire woven in a four-inch stretch mesh. The rubber insulator between the stainless steel mesh and the steel of the pot frame prevents disintegration by electrolysis. Each pot contains a four-inch ring and two eight-inch by four-inch oval tunnels with triggers that close the tunnels so the large crabs cannot escape. A small ring, generally welded to one pot-frame member near the top of the pot, allows sub-legal crab to escape. The crabs are removed and the bait cans changed through a door on the top of the pot. The door is made of a stainless steel rod one-quarter to three-eighths of an inch in diameter. The door is hinged at each end and locked in the closed position with rubber straps and hooks.

"The pots are baited with razor clams, squid or herring kept frozen aboard the boat and thawed just before use. Razor clams are crushed before being put into the bait can; American squid are used whole; the larger Japanese squid are cut into five to 10 pieces while herring are cut into pieces one or two inches long. The bait is held in stainless steel louvered bait cans seven inches in diameter and four inches deep. They have hinged tops and are attached inside each pot with stainless steel hooks and rubber straps.

---

1. Hipkins, Fred W. and Alan J. Beardsley, "Development of a Pot System for Harvesting Black Cod," National Marine Fisheries Service, Seattle, Washington, 1970.

2. Meyer, Robert M., "The Dungeness Crab Fishery Around Kodiak," *Commercial Fisheries Review,* September, 1968.

"Each pot has a polypropylene or similar line with a plastic foam buoy attached. The line is 10 to 20 fathoms long, depending on the depth to be fished. The buoy is 18 inches long by four inches in diameter and is tapered at the bottom to reduce the chances of fouling by kelp. The buoys and lines must be dipped periodically in a chlorine solution to remove fouling organisms, primarily algae and hydrozoans.

"Just before the gear is set, the skipper selects a course and sets the vessel's autopilot. It is important that the boat be maintained on a straight course to facilitate the recovery of pots in rough or foggy weather. As the setting of gear begins, the buoy line and buoy of the first pot and a colored float used to mark the beginning of the string are trailed behind the boat. At word from the skipper, the first pot is pushed overboard. The next pot and its line and buoy are carried to the rail. The process is repeated until the last pot of the string, also marked by a colored float, is set. A string may contain from 30 to 60 pots. Each boat fishes several strings. These usually are set parallel to each other and approximately parallel to the beach...

"An efficient crew can lift and reset over 300 pots in a 10-hour day. In good weather, two men can pick and reset more than 60 pots an hour. To reach this rate, an assembly line approach must be used in handling the gear. The boat is brought alongside the buoy to be retrieved. The line is brought on board with a boat hook. It is put in a hydraulic power block mounted on the end of a boom. The boom is lowered so the line can be set in the block and then it is raised so the pot can be swung aboard the boat and emptied into the sorting box. One man hauls the pot by keeping tension on the line while the other man fills a bait can and sorts the catch of the previous pot.

"After the pot is aboard, the catch is removed and the bait changed. Dungeness crabs are put into the sea water tank and fish, octopus and king crab are thrown overboard. The pot is pushed overboard about 75 feet before the next buoy is reached. The boat does not stop at each pot; the crewmen must haul the pots while the boat is underway at a speed of about two knots. Because the boats usually work into the wind, they must, on reaching the end of a string, run to the opposite end of the next string."

Coffee break on the run back...

This description of Dungeness fishing is the description of the classic set and haul, a procedure going as the book says it should go. But this picture necessarily is a schematic, not the usual set and haul where a myriad of annoyances may be at work. Murphy's Law says that if anything possibly can go wrong, it will. Especially when fishing. Weather goes sour no matter what the forecast says, gear breaks down, lines foul, the bait spoils, the skipper has a grouch, the pots get sanded in, the bar is beginning to slop, and all the crab have moved elsewhere...and sometimes a man gets his foot in the bight of a plunging pot line.

Alaska Dungeness fishermen, at least those of the Kodiak fishery, favor bigger vessels than those commonly used by crabbers of the three contiguous coastal states. The Alaskans like the western seiner or multipurpose vessel or the old workhouse power scow, a stable, easy-to-work-from platform for their fishery. Crabbing off Washington, Oregon and California seems to be the province of a rather rag-tag and bob-tail collection of vessels. The seiner is scattered through the fishery but much work is done by trollers converted from their summer job and by vessels smaller than the limit seiner built especially for the fishery or converted to it. Altogether, the Dungeness fleet presents something less than an imposing picture of vessel beauty although it is, in the main, a fairly husky and seaworthy fleet.

## The Fish Wheel

No where else in all of North America except on some river systems of Alaska and Yukon Territory is the fish wheel found. The fish wheel in these areas where its use is permitted is a fish-catching device operated by natives of the country and it is mostly used for subsistence fishing.

It consists essentially, according to the old BCF description, of "a series of lift nets attached to a circular frame operated by the current of a river. As a fish swims near the wheel, it is scooped up, slides toward the axle as the wheel turns and is then deposited in a box or on a scow." The following description of the fish wheel describes the Eskimo- and Indian-built wheels found along the Yukon and Kuskokwim rivers of Alaska:[1]

"To make a fish wheel is a task which requires abundant skill and an even greater amount of patience, referring to the true fish wheel made of native materials...These wheels are made of logs, spruce poles, wooden pegs, and a host of ingenious innovations passed down from father to son. Surprisingly, the fish wheel was not invented in Alaska. Forerunners of Alaskan wheels were used on the Sacramento River in California, and later on the Columbia River in the State of Washington. The name of the builder of the first wheel used in Alaska is apparently unrecorded. He was a white man, however, and the first wheel was believed to have been put into operation on the Tanana River in the year 1904. Within a relatively short time, Alaskan natives on the Yukon River adopted the device and were using it extensively. Fish wheels came to the Kuskokwim River in 1914, and were used there by white prospectors near the village of Georgetown. Since these early beginnings, salmon have been harvested with wheels on virtually all of the major Alaskan watersheds.

---

1. Greiner, J.D., "The Fish Wheel," *ALASKA*® magazine, September, 1970.

"Few tools are needed in the actual construction of a fish wheel. A sharp axe, a large wood chisel, a sledge hammer, an ordinary carpenter's hammer, and a large ship auger, in addition to staples, nails, wire, poles and boards seem to represent the complete list.

"A fish wheel actually consists of two major parts: the base, constructed of large logs much like a raft, and the actual wheel, which consists of two scoop-like baskets of wire mesh mounted on opposite sides of the axle. The base is usually constructed first. Spruce logs are cut early enough before they are needed to dry properly. These logs must be only partially dry so that they float well, but not bone dry, for if they are, they will rot when placed in the water for long periods of time. The base logs are peeled, for if bark were left on, rotting would also result. The logs used in the base are cut in 24- to 30-foot lengths and are usually about 16 to 20 inches in diameter. Next, stanchions are constructed. The stanchions are vertical cradles in which the axle is mounted. Each stanchion is fitted with a block which fits the curvature of the axle end. These blocks are adjustable for height, and are held in place with a heavy wooden pin, which slips through one of the several pairs of large holes in the stanchion sides.

"The wheel itself is a marvel of ingenuity, and consists of a rather complex frame of peeled spruce poles. These poles, about three inches in diameter, are gathered, peeled, and used in their green condition, for they tend to split less, and once in the water, seem to last longer than dry wood. The baskets are mounted on opposite sides of the axle and face in opposite directions. Heavy galvanized screening of two-inch mesh seems to be preferred for covering these baskets. Each basket is provided with a slope board at its axle end, so that a fish, which slides downward toward the axle as the basket approaches its vertical point, will continue to slide toward the fishbox at the side of the wheel. The wheel must be balanced so that when it is operating, it turns uniformly, with two paddle boards, the length and width of which can be varied, depending upon the demands of the river current. These boards catch the current as the baskets reach the horizontal position. Since at this point both baskets are free of the water, they serve to keep the wheel turning until the leading basket reenters the water. Thus, a fish wheel becomes almost a perpetual motion machine, with no operating expenses save the replacement of the odd brace or board which might become damaged as the wheel turns continuously through the long summer days. Even these minor repairs are kept to a minimum by wrapping joints in the bracing with soft wire and providing a bearing surface on the axle ends, by wrapping them also with wire. Some builders prefer to slip short pieces of large steel pipe over the axle ends, but the method using wire seems to be more popular, probably due to the unavailability of large pipe of appropriate dimension in most villages. Another more sophisticated touch, which is the product of modern technology, is the application of several coats of clear wood preser-vative to the axle, though this step is often omitted in native villages because of unavailability of materials.

"The fish box is built on one side of the base, and is positioned under one end of the axle. The box is usually placed on the mid-river side of the wheel, to facilitate removal of fish into a river boat. Built of rough boards, the box is usually about six feet long, three and one half feet wide and about three feet deep. The logs in the wheel base form the floor. Some operators put a hinged, lockable cover on the box to discourage river travelers from "borrowing" fish. The hole in the side of the basket, through which fish pass, allows the caught fish to enter the box through its side.

"All parts of the fish wheel are hand fitted with axe and chisel. Pegs and joints soon swell after being placed in the water, and become tight. No drawn plans are followed, and it is a gratifying experience for the man who appreciates hand work in this day of mechanical gadgetry to see a fish wheel being built, no plans, no measuring tape or ruler.

"Once the wheel is finished, the task of placing it in the water and making it operable follows. Long spars are placed over pegs in the wheel base and a coil of heavy cable is attached to the upstream end. The wheel is then set in the stanchions and tied so it cannot turn. Fish wheels are floated with the aid of river boats to the place they are to be used. The selection of a good fishing site is critical and requires a thorough knowledge of the habits of migrating salmon and the river. The position of a wheel is so critical that a matter of a few feet one way or another can mean the difference between success and failure.

"In placing a wheel, the cable is attached to a tree or rock on the upstream riverbank, and the spars are spread and anchored in the bank to hold the wheel away from the bank. By employing large holes in the ends of the spars and slipping them over pins in the wheel base, the wheel is not rigidly attached to the bank, and can ride with minor changes in water level and current velocity. The wheel is now lubricated at its axle ends, its lashings untied so that it can turn. Minor adjustments will be made during first few days, and then the wheel begins to repay its owners.

"Salmon caught in this way often end up on native drying racks to be used for dog food. Others are dried or smoked for human consumption. Some fish wheel operators, near larger villages, sell their catch."

There are many versions of the fish wheel. This description is one.

In some drainages a permit is required to operate a fish wheel for subsistence purposes. If fish caught by a fish wheel are sold, any boat used in the operation must be licensed, and the fisherman must hold a commercial fishing license.

## Pie In The Sky?

The gear discussed in this segment has been gear that has not changed much basically over recent decades or even a century or two. It has been merely dressed up somewhat with added components and better materials. This reworking of presently used gear is a continuous process, one carried on every day by every working fisherman as he makes minute changes in the hanging of a net, the wobble of a spoon, the makeup and lie of a longline.

At the same time, more sophisticated examination of present gear and its function is made by technicians of government and private fisheries organizations the world over. The fisherman's own contributions and the contributions of trained researchers to the growing store of gear knowledge have resulted in valuable advances in gear efficiency and economy. But none of this painstaking, inch by inch revamping of long-standing designs offers much progress toward entirely new and mostly unexplored methods of catching fish inexpensively in large numbers.

What of the future of new kinds of catching gear with the stress on the word "new" and all it implies? With the exceptions of the Puretic block for net hauling, the use of pumps to handle certain small fishes, the turn toward synthetic net materials and the use of echo sounding devices, there has not been a really major advance in gear in all the world since World War II and for most of the century preceding that upheaval. Truly, these have been important additions to the gear catalog but they have been only the most important among myriad refinements and they are still just refinements of the basic gear. Troll gear, seines, gillnets, trawls and traps look pretty much as they have looked for two generations. The mechanical skipjack poles first fished by the Japanese in 1970 only substitute a machine for a man and men still must tend the machines. A hook and line are still a hook and line.

In a word, the post-war "breakthroughs" in the fisheries of the world's "developed" nations have taken place mostly above the water line. The great race by both sides to undo the enemy during the war resulted in highly effective radar, more efficient sonar, efficient loran, better radio, better RDF, refined diesel engines, reliable hydraulic systems, advanced refrigeration and it hastened the coming of age of the synthetic fibers. There is not a fisherman of the fishing nations who has not benefited in one way or another by the adoption of any or all of these devices.

But all of them have not been of much special effect on catching gear itself. They get the gear there faster, locate fish for it more easily and keep its catch in better condition but none of them catch fish.

A West Coast combination vessel in the Alaska-limit size of 58 feet overall, working only in the salmon seine fishery, enjoys radar, high-quality radio-telephone, reliable depth finders, a long-enduring, high-speed diesel engine and the strength and durability of a seine fashioned, except for its metal and plastic components, entirely of nylon. If this same vessel were to be put to work in the coastal drag fishery, loran, sonar and a minimal refrigeration system would be added to her equipment, along with an otter trawl built of nylon and polypropylene with added metal and hard rubber components. And if this same vessel were to eschew either of these fisheries and turn to the offshore albacore fishery it would have all this gear, except the nets, as well as high-test stainless steel lines and hydraulic line haulers. This last item alone saves men's backs and hands. If one figures a excellent day's jig boat haul for a man fishing alone at about 400 albacore weighing 15 pounds and up, there are three tons of albacore to handle twice. Hauling them by hand makes old men out of young ones...

But none of these navigation, propulsion, communication and other systems has much to do per se with the taking of fish. They make work and living somewhat easier, somewhat safer, and the finding of fish somewhat more precise. But essentially, modern marine fishing still is done by the chase method—the fisherman hunts the fish just as the hunter seeks the deer.

In this respect, fishing, one of the most ancient of human endeavors, has not changed perceptibly over who knows how many thousands of years. There is not yet any practicable manner by which to make the fish come to the fisherman or, at the least, a method to make the fish stand still until the fisherman comes after them, even though the fish may be far in the open sea.

It is to realize this aim that a part of modern fisheries gear research is directed. From this research by whatever nation, hopefully there will come the first truly "new" catching gear, however it may function.

Unless someone comes up with a totally new concept in nets themselves, a happening that appears improbable since nothing similar has occurred in the past century or so, fisheries gear researchers must look elsewhere for the "brave new world" of fishing gear. The most advanced nets now, all of them research gear, carry almost as much telemetry equipment as some of the primitive satellites fired into earth orbit. These nets are built of the most advanced and long-wearing materials. But beneath the frosting still lies the plain old cake of familiar design. Major advances in troll and longline gear appear just as unlikely although continuing refinement will go ahead on these gears just as it does with net gear. (One suggestion for otter trawls, one not yet seriously pursued except on paper, is the design of remotely controlled, self-powered towing units that might be placed in the position now occupied by the otter boards. The development of such a system has obvious advantages for bottom and pelagic trawls and the development of a shipboard control system could be modeled precisely on that used for the control of unmanned space vehicles such as those landed on the moon.)

It appears then that gear researchers must turn to a form of the trap, one of man's first ways to take fish, and to a comparatively untried method but one with some promise—the use of electrical current to entice fish to the fisherman. The trap antedates all mobile gear; practicable wide-scale use of an electric "come-along" lies somewhere in the future although essentially crude prototypes are in use in a rather small way in at least one marine fishery. Electricity has been used in some fresh-water fisheries for much of this century, chiefly because its deadly effect on the quarry eliminates all but the most desultory pursuit of the fish. It makes fishing (or harvesting rather) in such fisheries as those in Europe for cultivated carp mostly a matter of scooping them up.

The "fish trap" of, perhaps, the late 20th Century more properly should be called "fish control." Research in this field has been directed at finding some method of herding fish into a selected area for catching or for holding them in some spot until they are needed.

One method of moderate promise is the bubble curtain, an affair that has been used experimentally in New England to keep herring schools locked in coves until the canneries are ready to pack them as "sardines." The bubble curtain requires an air pipe or hose running in the desired configuration along the bottom, emitting an ever-rising curtain of bubbles that holds the fish within the specified confines for a time period still to be determined for location and species. The bubble curtain looks as if it might be workable for certain fish in selected locations. Holding a massed school of herring in a narrow inlet is one thing; holding a school of bluefin tuna where someone wants them to stay is an entirely different matter. Use of the bubble curtain in a big way with possible offshore application will be determined only by long experimentation and the spending of considerable amounts of money. For no return, maybe...but for a world that may be even hungrier by the turn of the century and after than it presently is, such effort and expenditure cannot be decried.

It has been suggested by various researchers that the basic bubble curtain can be strengthened by the addition to its air output of chemicals meant to attract or repel fish. In the case of the Maine herring, the added chemical would be intended to keep them away from the bubble curtain, sealed safely in their cove, shunning any approach to the bubble line because of distasteful (to herring) substances.

It has been learned that fish—or some fish, anyway—are highly sensitive to the taste or odor of substances encountered in their normal environment. It has been theorized, for example, that the salmon's homing instinct is based on its built-in reaction to the composition of the waters in which it was hatched and through which it travels on its migrations. This theory has been disputed in favor of other hypotheses, equally tenuous, but the ability of fish to detect and react to chemicals in the water around them cannot be doubted.

From this point, it has been proposed that it might be possible, at least experimentally but perhaps not commercially, to force fish along a desired sea route or into a desired location for catching by the introduction of pleasant or unpleasant substances into the water. In the case of tuna, for example, such an endeavor appears to offer, on the basis of present knowledge, almost impossible odds to work against. It might be a practicable procedure only, initially anyway, to attempt to apply the technique to densely schooling fishes such as menhaden, anchovy, herring and hake.

Coupled with the use of chemicals has been the observation that desirable fish might be moved one way or another by the use of the characteristic sounds of sea life they prey upon or that preys on them. Much has been learned during the study of anti-submarine warfare of the noises emitted by sea creatures and there appears to be a fruitful research field here in connection with fisheries. This technique does work in part. . .the reproduced sounds of killer whales on the hunt have been successfully transmitted from underwater stations at the mouth of the Kvichak River at the head of Alaska's Bristol Bay to send packing the pods of beluga whales that follow upstream migrants into this salmon-rich stream, mother lode to the Bristol Bay red run.

Another field of possible return would combine underwater lights, electric attraction and pumps to harvest fish. These might be complemented by sound and odor factors to add to their efficiency—if any. To these might be added the "floating log" element.

## Hot Seat

Central to the success of the only American work along these lines is the use of electricity as a compelling lure. Much has been made of electricity in literature dealing less than objectively with its application to fisheries. Here is the use of electric power put into perspective:[1]

"Although direct application of electricity to fishing dates back to about 1895, commercial utilization was initiated sometime in the 1920's. Currently, electrofishing in fresh water is well understood and is widely employed in many areas of the world. Its use in sea water has not, however, been as successful as some had forecast...and only a few instances can be cited where it has had practical application. It appears that considerably more investigation will have to be made in electrophysiological research and further development will be required before it can be successfully applied in ocean fisheries.

"The behavior of fish in an electric field depends upon the type of current (alternating current, continuous

---

1. Alverson, Dr. Dayton L., and Larry D. Lusz, "Electronics Role in the Fishing Industry—Past and Present," 1967.

direct current, pulsed direct current, etc.), on the voltage intercepted by the fish's body (while) its reaction to some extent is dependent upon size and species. Engineers and biologists investigating the potential of electro-fishing have developed certain jargon to describe behavioral activities: oscillotaxis, first reaction, electrotaxis, electro-narcosis and electrocution. Oscillotaxis refers to a behavioral response to the influence of normal AC in which the fish take a transverse position within the voltage gradient to tap off as little voltage as possible. Apparently, once the position is taken the fish cannot move away and will be stunned and eventually electrocuted if the position of the electrode is brought closer to the fish.

"The strong attracting characteristic of DC current on fish (toward the anode) provides a behavioral response which makes effective electro-fishing theoretically possible. Fish may thus be attracted out of shelters, burrows or even from the mud. Under a gradually rising DC current density, a fish in a position parallel to the current direction will first show a vibrating movement or if transverse it will turn toward the anode. A second reaction follows in which the fish will swim toward the anode (electrotaxis). This 'physiological must reaction' to DC power makes it much more useful for electrical fishing than AC. After a certain period, the electrotaxis response will be followed by electro-narcosis during which the fish loses consciousness...if the power is increased, the fish finally is electrocuted.

"Perhaps the most appropriate example of successful use of electro-fishing is that of the East Coast menhaden fisheries where a pulsed electric field is used to attract fish which have been caught in a purse seine to the mouth intake of a pump. The process is, then, an auxiliary operation which speeds up unloading of fish taken in a purse seine. Other examples include its use to shock fish taken on trolling lines and the electrocution of seals.

"In the Gulf of Mexico, researchers have used an electrical field to stimulate shrimp from their burrows. The system uses AC-generated primary power transmitted by electrical cables to an electronic pulse generator on the trawl door. There it is converted to DC and released at a fixed pulse rate through an electrical array in front of the trawl net. At the present time, shrimp fishing in the Gulf of Mexico is primarily a nighttime operation because the shrimp seek shelters in the mud during the day. Using the electrical array in front of the trawl, experimenters were able to make daytime catches up to 100 per cent of nighttime catches. However, efficiency of the catch varied according to the nature of the substrate and depth of shrimp burrowns.

"In spite of the wide-scale and sporadic investigation of marine electro-fishing, it has not, for the most part, found wide use in the world's fisheries. One might caution about being overly optimistic regarding the potential of electro-fishing techniques."

Nevertheless, it appears that electricity may have a useful place in techniques still to be perfected for the taking of fish in large numbers...

For years it has been known that some species of fish are attracted by submerged and floating objects presumably as places of shelter for some species and as hunting grounds for others. Skindivers have turned this reaction to their benefit by artificial bottom environments, usually by sinking old automobile hulks in suitable locations. An increase in fish populations seems invariably to follow and is maintained, according to species and migratory habits, unless over-hunting takes place. Tuna are known to congregate around surface floaters, giving rise to the "floating log" term to help describe such behavior.

This tendency of fish to orient themselves around an object is intensified when underwater lights are added. Squid, anchovy, saury, herring and others are drawn to such lights and the present Japanese Pacific saury fishery is partly based on a nighttime fishery abetted by lights. Lights are so effective with these species of fish that, for instance, their use had to be sharply restricted in the British Columbia herring fishery, prior to its downward fluctuation in the 1960's, because they attracted too many feeder salmon to them.

Fish pumps have been used for years to move small fish from vessel to holding tank and in the early 1960's pumps had been adapted to fishing vessel use. With them, as previously noted, such fish as the menhaden could be moved from the net to the fish hold in fairly short order with a minimum of manpower. In the squid and saury fisheries pumps have been used to take fish from schools swarming around the vessel in response to the lights, again a comparatively easy way to cut down on manpower.

And electricity has been built into the menhaden pumps to bring the fish to the intake...

These three elements have been put together at the National Fisheries Service Research Center at Pascagoula, Miss. To them has been added the "floating log" factor, appearing to offer shelter to some fish, food to their predators.

The entire device, its schematics bearing more than a faint resemblance to some of the simpler designs of the late Rube Goldberg, has been named the "automatic fishing platform," a drab enough way to describe a complex structure that will, if it ever becomes practicable, be tagged with a name somewhat more suitable to its capabilities.

It begins below the surface with small, stationary, suspended platforms shaped something like the old Army pup tent—the floating log element. On the submerged platforms and along the stays that secure them to the mother structure banks of mercury vapor lights are hung. Experiments have been run through combinations and light values to determine the formulae most acceptable to the desired kinds of fish. In the Gulf of Mexico waters where research has taken place, a group called the jacks—amberjack, blue runner and rainbow runner—has been found most susceptible to the lure of the sunken platforms and the lights. A second type of smaller fish, such as that known as the Spanish sardine, has been found to be somewhat less attracted to the platforms and lights but concentrations of these of more than 100,000 individuals have been observed at each of the submerged structures at one time.

The platforms and the lights are the bait, so to speak. Above them is a fish pump that might well be likened to the hook, suspended well below the surface platform. Linked with the pump is a pulsed direct current system designed to bring fish to the pump as described in the matter just quoted. Fish caught in the surge of the pump have come to the end of their watery path—up the pipes they would go to storage areas on the main platform where they would be converted by automatic means to whatever product might be wanted.

The fish come to the fisherman at last!

This description is a bare outline of a vastly complicated design. On paper it combines three things of great attraction to some fish. On paper it looks good. On paper it should work. But NMFS testing of prototype designs has been hampered by lack of men and money. In any event, years of field testing, of improvisation, of adding and taking away, lie ahead of any commercially applicable self-contained operation of this type. In all probability, the final design, if ever there is a final design, may not even resemble, in form, this pioneer forerunner.

If one of these complexes, one of these catching and processing "machines," ever does become operational, it will represent one of the few "new" catcher gears since the first net was cast from some unknown ashore. If it works, it will do what the National Oceanic and At-

mospheric Administration calls changing fishing from "a hunt to a harvest."

Much has been written, some of it drivel, on the potential of the seas and means of harvesting them of their fish, shellfish and plankton. Some of this material reads now like the Buck Rogers comic strips of the 1930's read then.

But in a world where technology and the science that creates technology are multiplying themselves so rapidly that only the advanced computer can keep up with it, it would be unwise to discount the possibilities inherent in this vast accumulation of knowledge.

How many persons, other than readers of science fiction, considered seriously, until the Soviet Union launched Sputnik I, the chance that men might some day walk unafraid on the moon?

There is no reason to believe that science and engineering are not capable enough, given proper incentive and sufficient money, to develop fish catching methods beyond the most advanced techniques in the works or contemplated in this neglected area in this last third of the century.

Certainly, the capture of some additional billions of pounds of food and industrial fish on top of the world's still-growing annual catch can not be beyond the reach of the likes of the people who created the moon flight and deep space exploration technology almost from scratch.

There is one danger, however, in any such quantum jump in the art of catching fish. Man is greedy and he is greedy enough and smart enough (and smart does not mean intelligent) to fish himself right out of business...

Remember the passenger pigeon and the Pacific sardine and keep in mind the whale.

Seattle Post-Intelligencer

*End of the fishing week . . . seiner crewmen lash down their heavy, dangerous skiff and head for town to wait out the weekly closure.*

NMFS, National Marine Fisheries Service

*Pacific Ocean pink shrimp flow from cod end of trawl on vessel fishing off Oregon Coast. This catch was made with shrimp separator trawl designed by the old Bureau of Commercial Fisheries and is more than 99 percent free of scrap fish.*

# 5/ Fish Handling and Preservation at Sea

The arrival of good fresh fish in the market or the cannery is the final step in a rather remarkable and unpredictable series of happenings where blind luck may be as important as any calculation among the people involved. Many things, most of them bad, can happen to a fish in the days or weeks before it shows up in a fish stall or goes into a can. Fish and shellfish are among the most perishable of foods, as tricky as bread pudding nine days old. Few others are as susceptible to spoilage and few others can be hurt as easily by ungentle handling.

Fish dead in a net may begin to spoil before they come from the water. Some species, because of size, body content or handling, are more easily damaged than others. The thousands of fish in the cod end of a trawl after a good tow are subject to pressures as they come aboard that do not befall the troll-caught salmon gaffed, killed, cleaned and iced with respect and dispatch. Good care of fish must begin on the fishing vessel. If it does not begin there and is not maintained through all the watches, shoreside treatment of the most advanced kind may be almost useless.

A live and healthy fish, like a man in the same condition, can fight off undesirable bacterial action while, at the same time, its life processes keep its internal balance on an even keel. But when death drops natural defenses against bacteria and upsets the internal processes that maintain life, deterioration begins at once and continues rapidly until it is slowed by icing or refrigeration or halted by freezing or heat processing.

The fisherman and the fish processor must deal with three types of quality damage—enzymatic, bacterial and oxidative—and it is up to the fisherman to get the action underway. These spoilage processes are interrelated and to a great extent they are dependent upon each other. The most important, at least over the short run, and usually the first to be noticed is bacterial action because of resultant spoilage odors. But it always must be preceded by enzymatic action in fresh fish because enzymatic action clears a path for bacterial invasion of the tissues. Oxidative action commonly is a long-term effect, generated by exposure to factors inherent in storage. The order of effect of these three upon quality depends upon the physiological condition of the fish at the time of death and upon the conditions of storage in ice, under freezing or in chilled sea water.

Enzymes are complex organic substances produced by living cells and used by all living creatures to produce desired chemical changes in such body processes as digestion. In life, enzymes are held in check. In death, enzymes begin an unbridled attack against any body substance they touch. The commonest type of enzymatic action seen by fishermen (although few can give it that name) is the quick-starting belly burn so soon apparent in herring and sardines after death. It is caused by pepsin and related digestive enzymes turning against the tissues they could not touch in life. This reversal of the life process results initially in softening of the flesh around the belly cavity. This softening is further complicated, unless controlled, by bacterial action made easier by the enzymatic breakdown of the body structure.

Oxidative action is a chemical process taking place when fatty tissues are exposed to air as in ice storage or under freezing. Without ice or freezing (or salting) bacterial action would destroy such flesh before oxidation could become apparent. But with those preservative measures oxidation supersedes bacteria and enzymes and announces its presence with a bitter taste, a "salt fish" odor, rancidity and a fish-oil smell. The mackerels and the herrings are quickly susceptible to oxidation damage because of their high fat content. Fishes like the cod or the haddock with much less fat in their tissues are not so prone to quick oxidation although no fish and no meat are immune from it.

Bacterial action, if not controlled, may make enzymatic and oxidative damage just academic. Bacterial damage may ruin a fish before the others can begin to take hold. Bacteria cannot be avoided. Without certain internal bacteria, no living thing would be able to survive. These bacteria are as necessary to life as oxygen. But these same bacteria, once removed from the body, fish or human, become a menace. And there are other bacteria occurring everywhere that are no threat to health and life but go instantly to work when death comes along. The waters in which fish swim are a source of bacteria from the outside. Bacteria in the alimentary tract of a fish turn on it when they are expelled during catching and handling. In net and on line, fish come into contact with infecting agents. In the hold they are further exposed and this exposure continues until heat processing or deep freezing put an end to the bacterial carnival.

The spoilage rate is not constant. A handful of factors plays a part—initial bacterial contamination, the fish's own body temperature, the time between catching and icing, and the efficiency of icing or other preservative measures.

## The Sooner—The Better

The preservation of fish aboard a vessel presents difficulties not associated with some other foods. That a fish comes from water of low temperature, for example, does not mean that its spoilage will be correspondingly retarded. Some fish are taken from water where temperatures are not far above freezing. The Pacific halibut likes water in the 32-39 degree F. range. Despite this cold, enough to inhibit some types of bacterial action, the halibut carries with it its full complement of internal and external natural bacteria. These creatures, once the halibut is dead, continue to work away at temperatures below those too often found among iced fish in a hold. By comparison, the slowing of bacterial action in meats such as beef is simple. The bacterial population of a steer headed for the slaughter house is accustomed to an internal temperature of about 98 degrees. Chilling a beef carcass to 36 to 38 degrees allows it to hold well for a considerable time while fish at those temperatures go bad quickly.

The interval between catching and icing is the time when major damage may get started. The quicker into the ice, the better the fish. Salmon may be kept in a seiner hold un-iced and in the round for too long simply because the tender or the cannery are only a few hours away and the skipper wants to make two or three more hauls. Trawler crewmen may be slow to get at sorting and icing because their craft is under-manned and there is other vital work to do. Tired troller men, suddenly lazy under a seldom-felt sun, may let their salmon lie in the checkers too long. That sun is not as kind to the salmon as it is to the fisherman. Even an hour of neglect may be dangerous. The fisherman might twist the old Golden Rule to read something like "Treat those fish as you would like others to treat 'em if you are going to eat 'em." That is, catch and kill them quickly, sort them and wash them, gut those that must be gutted and get them into the ice.

Icing, of course, is not a cure-all for the mistakes that have been made before icing. But proper icing does slow bacterial and enzymatic damage. Proper icing means icing down to 32 degrees or as close to it as possible. Iced fish at 37 degrees F. spoil twice as fast as fish at 32 degrees. Fish held in ice at 32 degrees for two weeks are still edible. But they lose all their appeal to most palates in a week or less if kept at temperatures much above that. It seems unlikely at first look that fish in ice can get above 32 degrees. But thin layers of ice and thick layers of fish can mean fish temperatures many degrees above the nominal temperature of the ice itself.

Here, the interval between coming from the water and icing bears directly on keeping qualities. Fish allowed to remain on deck for any considerable period will begin to approach air temperature. In the Gulf of Alaska in the wintertime, this might mean only that the fish will be half-frozen by the time they go into the hold. But in the Santa Barbara Channel, summer or winter, it may mean that a fish is heading toward a temperature of 70 degrees. Even if the best of icing techniques are used after several hours under such conditions, damage that cannot be altered already has set in.

Freezing of fish at sea is the surest way to protect them against bacterial action but it exposes them to oxidation damage. Some species preferably should not be frozen. And freezing equipment is not found on the common run of fishing vessels in the Northeastern Pacific. It is expensive, it requires a degree of skill in handling and the average vessel in most fisheries of that part of the Pacific cannot afford it despite its virtues.

Where the preservation of fish is concerned, no fishing vessel is any better than her hold. Her galley may be spotless and her decks clean enough for a picnic. These two are the vessel areas easiest to keep clean and it takes only a conscientious cook and a concerned man or two in the deck crew to keep them clean. Fish holds are another matter and often enough even the hardest-working crewman finds it easy to slack off when he has to clean and dry up a hold that recently has been filled to the hatch with fish and ice. The drainage water from iced fish is anything but enticing. The substances in it begin to decay quickly and as they decay, they begin to give off odors. Fish holds by and large are damp and smelly after even brief use. But as long as ice is used to hold fish, dampness in a hold must be counted as a fact of life for fishing vessels and their crewmen. Undue and continuing dampness is the cause for many of the things bad that take place in a fish hold. It is a source of odors apart from those directly caused by decaying fish materials. It contributes to certain types of fish spoilage and it can be responsible for damage to the structure of the vessel itself. There is an unfortunate tendency among some vessel operators to worry out loud about hold dampness in lieu of tackling the problem and solving it. There are no good excuses for a mouldering fish hold when there are available several good methods of drying holds and keeping them dry regardless of the presence of ice.

There is a danger factor for men as well as for fish in the decaying substances to be found in a dirty fish hold or in dirty shoreside holding tanks. Under certain conditions, these waste substances will produce hydrogen sulfide, a gas heavier than air, deadlier than carbon monoxide, one that works even faster than that odorless, tasteless byproduct of the petroleum fuels. High school chemistry students know hydrogen sulfide as the source of a "rotten egg" odor and for generations it has been a subject of horseplay in chemistry classes. But it also has been the source of death for many fishermen, particularly those working in warm climates. But the generation of hydrogen sulfide is not necessarily confined to the tropics or sub-tropics and a hold with decaying matter in it can produce hydrogen sulfide just as handily in Kodiak harbor as it can in Galveston Bay.

"The gas is easily detected by the human nose in extremely low concentrations. The odor...may be recognized in concentrations as low as two parts per billion. One of the most dangerous and deceptive characteristics of hydrogen sulfide is that it quickly fatigues the sense of smell, thereby stripping a person of his only source of warning. Concentrations as low as 10 parts per million are toxic, even though 600 parts per million may be survived for as long as 30 minutes. At high concentrations, collapse, coma and death from respiratory failure may come within a few seconds after one or two breaths. Low concentrations produce irritation of the eyes, nose, mouth and throat. Headache, dizziness, nausea and lassitude may also appear.

"Another dangerous characteristic of the gas is that it has a flash point of 500 degrees F. Thus if hydrogen sulfide is present in high-enough concentrations and comes in contact with a surface heated to this degree, an explosion and fire will result. The gas is potentially explosive in concentrations from 4.3 percent to 46 percent."[1]

A smelly fish hold can be regarded only as a neglected fish hold. Neglect, however, is not always intentional because the very shape of a hold may make it literally impossible to get at all the areas where contaminants can lodge and begin promptly to stink. Few fish holds in the Northeastern Pacific fleet were designed with keeping qualities and cleaning ease foremost in the architect's thoughts. Too many holds are merely the space left over after the hull was put together. Most holds on older vessels have angular construction that looks like home, sweet home to bacteria. This is as true for some modern steel and fiberglass holds as it is for the run-of-fleet wood vessels. Steel and fiberglass, however, are sealed tightly and are leakproof or should be. But many poorly-designed or hard-worked holds in wood hulls have seams sprung open enough under one pressure or another to allow passage of all kinds of junk, even fish, into hull areas where it would necessitate a major demolition job to get at the mess. What fisherman has not spent his spare time for days tracking down a stinking fish only to find the thing in some godforsaken cranny of the vessel where, it would seem, only malice well-planned could have placed it?

It is singularly easy to direct a vessel crew to scrub down and dry up a fish hold. It is singularly difficult many times to achieve that most desirable end. Simple washing is not the way to begin except perhaps for the smallest of boats where a man on his knees can get directly at every surface that must be cleansed. Most vessels, Alaska-limit seiners, for example, have holds too deep to allow even the tallest of men to reach their every surface.

---

1. Hamilton, Richard W., "Hydrogen Sulfide Kills," *Commercial Fisheries Review,* Vol. 32, No. 12, 1971.

## Clean Sweepdown

Almost every vessel of any size has a high pressure hose system and any wash-down should start with pressure hosing of every inch of the fish hold. This means clean water whether fresh or salt. Almost always clean water must exclude harbor water. Even the best-kept harbors have their waters fouled by all manner of unpleasant substances and the use of such water merely compounds the problems of cleaning a hold. Getting clean water usually means getting connected to a dockside system. But whatever the water source is, plenty of it should be used. Pressure should be high enough to pry loose and wash into the bilge every accumulation of fouling matter. This application of water should be continued until the bilge pump begins to deliver clear water. Clear water does not mean a bug-free hold, however. The visible sources of contamination may have been washed away but unseen on every surface are uncountable numbers of bacteria busily reproducing themselves. It takes more than water to do something about them. This is where the real work of cleaning starts and it is, unfortunately, the point where too many men consider that their holds are clean and walk away from the job. None of them would consider dishes to be clean if they had been washed only in cold water and left in a rack to dry. All would insist on a soap or detergent bath for their eating tools. And this is precisely what the fish hold should get next—detergent treatment and plenty of it. There are some dozens of available detergents to use and most of them can be found in the nearest supermarket. Every marine supply company stocks them too, mostly under different names and fancier prices than those in the supermarket. The ordinary household types will do the job just about as well as anything if they are used in heavy concentrations, something like 10 times the amount normally used by the housewife for a tubful of children's play clothes.

The method of application of the detergent solution makes a difference in its effect too. It can be sprayed on, the easiest and fastest way but not necessarily the best way. Detergent spraying may give the hold a uniform coating of the material but that is about the total effect, a covering only, with not much damage to the bacteria it is designed to hurt a little. A bucket of detergent and hot water, a long-handled stiff-bristled brush and strong and willing arms are the best way to get at the invisible army. The brushes, stoutly used, will scrub away any material the high-pressure water treatment didn't faze and they will make bacteria more vulnerable to the bactericidal treatment to follow. As for the brushes, they may be a prime source of contamination themselves if they are not treated as diligently as the hold. They should be cleaned frequently and kept dry when not in use. Scrubbing down a hold like this can be counted nothing but hard work and men may be strongly tempted to shirk it. The world's navies, in such situations, enjoy the services of bosuns to see that such does not happen. Vessel skippers must exercise the same prerogative.

After a hold is healthily worked over by detergent and brushes, the hoses should be brought back again to rinse away the detergent film until the bilge water again runs clear. Here again, men are ready to quit. But with all the work done so far, matters might as well be carried on to a logical ending. This is a direct and crippling attack on bacteria left on hold surfaces. Everything done so far has done no more than shake them up a bit. Application of a bactericide will not eliminate all bacteria, but it will put a dent in the population and slow the growth of the remainder. Several agents are available for this stroke and again the supermarket can supply them if necessary. A favorite and a most effective material is the common household bleach compound with sodium hypochlorite as its active ingredient. This may be sprayed on heavily in a solution delivering from 500 to 1,000 parts of free chlorine per million. The stuff smells strongly but not unpleasantly and at this concentration will not bother crewmen using it in the hold. It disinfects well and markedly improves the hold odor. There are, also, the proliferation of quaternary ammonium compounds sold under many names and widely used as disinfectants and antiseptics. These too control bacteria and fungi and deodorize in almost any flavor wanted. For a modest fee, the moldiest fish hold can smell briefly like a pine forest. These cannot be used in conjunction with any chlorine compound, however, since the interaction of the two quickly destroys the quaternary ammoniums no matter how pleasant they may smell. For lack of all else, the oldest of bacterial controls may be used. This is salt, the fisherman's friend for ages beyond count. Salt may be spread generously on all surfaces where it will stick. It does slow fungus growth and it is much better than nothing at all.

These cleaning measures are, or should be, old standards, well known to all men who crew fishing boats. They are efficient and inexpensive. They are time-consuming and their proper application calls for muscle. But technology in these matters does not stand still. Cleaning and disinfectant procedures devised for more sophisticated jobs ashore have been adapted to boats and ships of all sizes and the only limit for men who want to do the job the best way is the amount of money they want to spend to keep their fish fresher.

Once the hold is clean and dry, the question comes around to ways of keeping it dry. Any unprotected wood object exposed long enough to water begins to suffer water damage. Water is absorbed and the wood begins to swell and soften. Soon bits of wood may be scraped away with a fingernail and, in time, the entire member is rotted through. The wood surfaces of fish holds are no exception. The holds of working fishing vessels often are exposed to months on end of dampness. Rot in some degree must be expected unless protection is provided. This can be done by liquid preservatives or by moisture-proof lining. Liquid wood preservatives are of several kinds and of varying degrees of effectiveness. Substances of this nature appear to be preferable to paint or varnishes since they control the organisms that cause wood rot. For generations, paints were the only preservative applied to fish holds. But painted surfaces are subject to various forces that pierce the thin shield and allow moisture and fouling material to penetrate the wood. The consequent bacterial accumulation may spread widely under paint, raising blisters that break and peel away to leave the wood more vulnerable than ever. Every man who has tried to paint a house with poorly-vented siding will understand this condition. But the blisters on the house don't stink to high heaven as they do in the fish hold because they do not have the same bacterial origin. Paint now is considered absolutely ineffective as a wood preservative in the holds of fishing vessels. Canadian research has shown not only paints but all other coatings and varnishes to be similarly useless since only a tiny perforation is enough to allow entry to moisture.

Preservatives to be used on surfaces that later will have fish lying against them, if the fish are not properly iced and put down, must pass certain tests. They cannot transmit to the fish toxic qualities that might be passed on to the ultimate consumer, nor can they transmit disagreeable odors that might taint the flesh of the fish. And they must be effective for comparatively long spells—a salmon season or a halibut season. For these reasons, the creosote compounds, among the best of wood preservatives on all counts, must be passed up for fish holds because of odor and inimical effect on any who ingest the stuff.

The Seattle Technological Laboratory of the former Bureau of Commercial Fisheries has done much work on the problems implicit in wooden holds. These experiments show a particular compound, copper-8-quinolinolate to be particularly effective in insuring a clean hold without passing on any unwanted effect or taste. The report on the substance says:[1]

"Copper-8-quinolinolate is a potent fungicide and its mammalian toxicity is low. Suitably used, it has been approved by the Food and Drug Administration for treating materials such as cloth, paint and wood that may come into contact with food. Because it is chemically stable and relatively insoluble in water, acids and alkalis, it was applied to wood in the hold in a formulation incorporating water repellents. (Trade name 'Cumilate' in two volumes of industrial white spirits with a flash point of 103 degrees F.) These features made it an effective preservative for long periods even after continued leaching by melting ice and repeated washing of the hold. Fishermen report that it makes the wood easier to clean."

But the smoking lamp is out.

Preferably, all fish holds would be built of stainless steel or glass with rounded corners and smooth surfaces. These are easy to clean, an inhospitable ground for bacteria. They also are expensive, more costly in many cases than the vessel that would take them and the fish-

ery industry consequently falls far short of this happy state. Materials to protect wooden holds necessarily then must be considered. Among them is plastic sheeting for all surfaces that may be touched by ice or fish. The above BCF report remarks that polyethylene sheeting "used between the fish and the hold is effective in preventing undesirable changes that occur in fish when they are stored in contact with wood—in short, that the use of polyethylene sheeting to line the hold or the pens in the hold is a practical method of keeping fish and ice from contact with the hold and for keeping the hold clean and dry."

## Plastic Film Works

In test setups, iced fish kept by the polyethylene from contact with wood surfaces were in good shape at the end of 12 days while those left deliberately in contact with wood had begun to soften and acquire all the undesirable aspects of "bilgy" fish. The wood itself remained dry and free of slime or other fish substances. The researchers induced several vessel owners to use the sheeting on a season basis where it was greeted with general approval. The chief difficulty was the first installation by crews unaccustomed to handling the plastic and, perhaps, skeptical of its benefits. These initial installations on a trial and error basis found crews using sheets of the wrong sizes, encountering objects in holds and bins that ripped up the sheets as well as difficulty in fixing the sheets to hold surfaces. But when these problems were overcome, crewmen and skippers liked the plastic because of its contribution to fish freshness and clean, dry holds. The plastic is not expensive. One salmon troller hold was lined for a dollar while the cost for the hold of a halibut boat with room for 80,000 pounds of fish and ice was only about $10. A single trip with the material returns many times that amount in more first-quality fish.

A fish is a delicate creature, one of fragile flesh easily subject to bruising or other injury and consequently quick deterioration despite subsequent proper handling. Meats such as beef are hard-fibered and long-lasting by comparison. Meats and fowl become tender with careful aging and there are those who eat them in a state of tenderness so advanced that the uninitiated can't stand to be at the same table. Yet this gourmet consumer of aged meat would refuse to eat a fish in the same stage of decay. Quite properly too because fish that old shouldn't be placed in front of a hungry alley cat even. Nevertheless, fish approaching that condition are offered every day to unsophisticated consumers by retail merchants whose eagerness for a buck outweighs any sense of duty to their trade. Unjustly, the fisherman too often is blamed for this happening although the

---

1. Tretsven, Wayne I., "Care of Fish Holds," *Fishery Industrial Research*, Vol. 4, No. 6, February, 1969.

majority of fishermen are conscientious men who try hard, if sometimes misguidedly, to keep their catch in a decent state of preservation. This task involves luck, a knowledge of species and their physical reaction to the exigencies of the catch, as well as a thorough knowledge of preservation procedures. Most fishermen, either by experience or specific instruction, learn to recognize and deal with these variables from the moment the fish comes aboard. For these men, good treatment of the catch is a matter of rote. For others it is a matter of surprise.

As among humans, not all fish are equal in looks and qualities of endurance. Their spoilage rate varies among species and it may vary in the same species according to the time of year. A seine-caught salmon on Puget Sound's Salmon Bank, intercepted on its way to spawn, will spoil faster under the same conditions than an immature feeder taken somewhere outside Cape Flattery. Small fish go bad faster than big fish. Soft-fleshed fish spoil faster than firm-fleshed fish, the hake and the lordly halibut respectively, for example. Trawl fish, hauled and handled in great numbers, may not be as enduring as those like the troll-caught salmon that get individual treatment. Those like herring or sardines with food in their bellies may start to burn shortly after death. Fish killed quickly and cleanly have a head start toward good storage quality. Fish such as tuna, dead in the gear in warm water, may start to spoil before they come to the brailer while dead salmon in Bristol Bay gillnets are good for hours in that chilly water. Disease or parasites may ruin a fish for market long before it comes close to any gear.

## Head Start

To the landsman, it might seem that the proper care of fish begins only when they go to the hold. But if care has not begun even before the gear is set, irreversible damage may already have begun. Proper care of fish begins literally when they are still free in the water. It starts with good housekeeping on deck. Almost all fish, whatever the fishery, have at least one contact with the deck or deck checkers before going into the hold. Dirty working decks and checkers are a prime source of bacterial infection for fish moved around carelessly. Not only should a bactericide be used regularly on sorting bins and the deck itself, but deck gear that has no business there when fish are coming aboard should be gotten out of the way. It is a factor in bruise damage in fish and it is something of a hazard on smaller vessels to the men who must work around it. Much gear admittedly cannot be moved from the working deck with a haul going, but that part movable should be stowed out of the way.

As for clean decks and deck bins, most vessels in the Northeastern Pacific might benefit by the lessons learned during an inquiry into the uses of a chlorine solution on New England trawlers. Chlorinating gear delivering 50 to 60 parts per million of free chlorine was installed on two test vessels. The resulting sea water and chlorine solution was used to wash fish after gutting and to scrub down decks and deck bins. Holds were treated with it before icing and after off-loading. The chlorinated sea water removed blood and slime from fish more effectively than plain sea water while it seemed also to cut down the blood staining often noticed on fillets. Crews reported the regular use of the chlorine solution kept the decks slime free better than did plain sea water. Deck slipperiness was sharply reduced, a safety factor of importance in any ocean fishery. No objective tests were run during this study on the keeping qualities of the fish washed in the chlorine solution, but crewmen reported a lower rate of bilgy and edge-of-spoiling fish than on trips when the chlorine was not used. This work and that done elsewhere by the Bureau of Commercial Fisheries seems to point up quite well the value of chlorine use aboard such vessels as trawlers and halibut boats where all or part of the catch is cleaned to some extent.

Fish come aboard in a variety of ways. The trawl codend dumps a mass of them, many already dead by crushing or suffocation, onto the deck. The salmon seine delivers them lively and wiggling onto the deck or straight into the hold by way of the fish bag or the brailer. Most longline and troll fish come aboard alive too, big halibut packing a grudge and a powerful tail with them. From the gillnet come dead fish, suffocated quickly with rigor mortis already apparent. These fish do not all have the same chance for good keeping. Of all, the troll-caught salmon, all things being equal, will be in good shape when many of the others have begun to turn.

Preferably, all fish should be bled before going into the ice. This is particularly true of salmon, halibut, sablefish and cod. Of these four species, however, only troll-caught salmon and halibut customarily are gilled and gutted before icing. Sablefish and cod may be so handled if there is not too great a weight of them, but if the poundage is excessive, they are iced in the round. In the Northeastern Pacific, no other fish caught en masse are dressed before icing although most fish taken in the Atlantic trawl fisheries are gilled and gutted at sea.

The New England trawlerman is rated tops at his job of dressing fish if he can handle 16 a minute. In the main, the New Englander's fish are bigger than most of the fish taken by the West Coast dragger while, at the same time, the North Atlantic vessels are bigger than those of the West and they carry more men. The impossibility of the crew of a three-man West Coast vessel gutting 130,400 pounds of diminutive Pacific Ocean perch, 5,300 pounds of lingcod, 31,300 pounds of rockfish, and 300 pounds of Dover and petrale sole is manifest. This was the fare of a single vessel, with three men, landed at Seattle in September, 1971. Of a total landing from that vessel of 167,500 pounds of bottom fish, only 200 pounds of large sablefish were gilled and gutted.

A way out of a dilemma for the West Coast dragger when the day comes that the market may demand that his fish be dressed at sea may lie in use of one of several types of gutting machines that have been developed to the point of practicality. The work of dressing fish at sea quite rightly has been described as "back-breaking" but the use of gutting machines on anything other than large factoryships has been ruled out in the past by size and expense of the machines. But two, at least, have been brought to the point where their use on vessels in the 90-foot range or thereabouts appears economically possible. The day of the old wooden, 65-foot West Coast inshore dragger cannot last forever.

One of these is the vacuum eviscerator nursed along for several years by the BCF laboratory at Gloucester, Mass. This device is designed to handle 60 fish a minute (that beats 16 by hand), a rate that is about the maximum fish can be fed to the machine manually. The operator holds the fish against a nozzle and uses a foot control to activate the eviscerator. The belly viscera is removed by vacuum and the fish is washed and rinsed by the machine. A fully automatic system has been devised.

Another cleaning machine comes from overseas. This is the "Shetland gutter," developed by a Scots farmer and adopted by the British White Fish Authority for use on small draggers. The Shetland can handle 45 fish a minute. Early models could take fish only to 17 inches but later designs can accommodate larger fish. Either of these will accomplish its purpose, that of bleeding fish at sea as soon as they can be handled after coming aboard.

Proper bleeding adds many hours to the shelf life of all fish and months to the shelf life of frozen fish. In addition, bleeding before icing gives the flesh of all white fishes—the haddock, for example—a cleaner, whiter color. Bleeding of all fish is compulsory in some European countries such as Iceland, Norway and Denmark where the commercial fisheries play a much greater part in the national economies than they do in the economy of the United States. The buying and selling of unbled fish in these nations is punishable by heavy fines for both parties to the transaction.

The salmon troller, often a man alone far off any coast, is a cameo image of the whole world of fisheries. Most trollermen treat their catch in the best possible manner—killing, dressing, washing and icing within a few minutes. The troller's example may be hard for other fishermen to follow because of the very size of their catches, but the ideal is one to be worked at. The trollerman has a special and compelling reason for his virtue—the fresh salmon market is supplied almost exclusively by the troll fishery and salmon going into that market must meet some rather stiff standards. At least, most of the time they must. It may be a fairly simple matter to fool the consumer about the quality of a fish—until he tastes it and swears off fish forever—but no one easily fools salmon buyers. They may be loving husbands and understanding fathers and pillars of the community, but they become cold-eyed entrepreneurs the moment a troller lays alongside their places of business.

It takes only a couple of trips for a new man in the troll fleet to be labeled and remembered by the buyers for the quality of fish he delivers and a man's reputation has a way of following him along a coast. The troll fishery thus is a peculiarly individual fishery in which it is every man for himself with a man's product dependent largely on his willingness to work daylight to dark for as many days a week as he can take it. The quality of his fish depends entirely upon him. There is no one else to pass the buck to. The crewman on vessels of other fisheries of the Northeastern Pacific may be an important member of a team but his contribution to the overall state of a large catch usually is not recognizable.

A salmon troller's general efficiency and hence the quality of his fish may be measured to some extent by the frequency with which he runs his lines. Many trollers are content to loll in the trolling pit or sit in the house door drinking coffee while they listen for the tinkle of a bell to tell them they have a fish on. This may be the easy way to do it, but it is not the best way, either by the day or over the season. For one thing, slackness in running lines results in the boating of many sub-legal dead fish, those not heavy enough to register audibly or visually on stiff, heavy gear. These "shakers" drown quickly once they are on the hook. Many could be saved to grow another season with sufficient alertness in the trolling pit and minimal care in getting them off the hook. For another, there is a theory believed and practiced by some trollers that frequent or near-continual line pulling tends to give better catches. Its proponents contend that feeding salmon, conditioned to react to movement in the water, are attracted by and will strike on troll gear moving up and down in the water more frequently than they will on gear merely being dragged along before their noses. This belief seems logical enough to quite a few knowledgeable people in the business and, if nothing else, its pursuit gives some needed exercise to boat-bound trollermen and saves for next year a lot of small fish. Running four or six lines 16 or 18 hours a day becomes hard work in short order, but there are those men who do it well, whatever their reasons.

## Gear Works Blindly

When the gear goes into the water, no man knows what it will bring out with it. The fisherman knows what he wants, but wishes and their fulfillment are oceans apart and what the haul will yield no man may safely predict, be he troller, dragger, longliner, driftnetter or kid with a cane pole. This is one of the fascinations of fishing, a sublimation of man's instinct for the chase, no matter the quarry. For many men, the built-in desire for money is subordinate to the need to see what is coming in with the haul. The troll fishery combines with its utilitarian approach to fishing something of the qualities of sport fishing. In this fishery, in many cases, the fish, especially if it is a big fish, has a chance and a fair one of getting away. Is there a troller who has never seen a fish roll himself off the hook or who has not miscued his stroke with the gaff and brought up nothing but air?

Those who would be trollers quickly must learn the art of the gaff because the gaff is the only practical way of bringing a fish into the boat. The net might be desirable, but it is unwieldy and the web continually fouls on everything it touches around the trolling pit. The gaff expertly wielded, often kills as it slams into the head and the fish gets the quick death it deserves. If not, a sharp rap with the gaff handle will do the job and the fish dies with a minimum of struggle. Fish that go through a prolonged death agony experience quick rigor mortis and their keeping qualities drop off sharply.

If all affairs of fishermen were arranged by a computer, every troll-caught salmon would be cleaned the moment it comes aboard. But such matters are random and troll fishing, like other fishing, is often a thing of catch as catch can. On a one-man boat, first things must be tended to first and this means getting fish aboard as long as they are being hooked. The fish then must lie in the checkers until the knife gets around to them and this may be a considerable time if the catch is good or the weather goes haywire and the boat takes priority. Whatever happens, the fish must be kept cool and away from the sun. Much of the time, sunshine is only something to be remembered fondly in the latitudes where the troll salmon fishery is carried on. The air itself is bad enough for fish on deck, with or without sunshine, for that matter, over any extended period. Every man who ever has taken a salmon by any means has seen the quick draining away of the bright colors of the living fish. The gleaming skin fast becomes dull and with exposure to light and air, it appears to develop a glaze. This is the first sign that the fish has gone too long without cleaning and icing, that spoilage now is underway and must be checked soon lest it be too late.

The trollerman must learn the use of the knife as quickly and as well as he learns the use of the gaff. The knife must be as sharp as a good smooth stone can keep it. The stones must be handled as gently as the fish because one dropped on the deck fractures easily. The experienced troller can clean the biggest of salmon in the time it takes the novice to find his knife and gloves. It is hopeless to try to clean salmon in any number on the deck. The fish are slippery and the deck is slippery. A vee-trough of the troller's choice as to size is the only practical way to clean in quantity because the fish is held fairly rigidly in the trough and cannot slide away from the blade easily.

The cut begins at the vent and runs forward between the pelvic fins at a depth sufficient to cut through the flesh into the belly cavity, but no deeper so that the viscera will not be punctured. The cut ends just behind the pectoral fins with two or three inches of the nape left to be used as a lug for handling the fish. This cut is not easy in a sizeable fish because the belly flesh is thick and the skin is tough. A good man with the knife makes the cut in one firm, steady stroke; the novice saws away as if he were bucking wood. The gills are next to feel the knife; they are cut free of the jaws and twisted out. The knife is run in a circular cut around the gullet, hopefully without piercing the gut membrane, so that most of the viscera may be pulled through the opening. If the knife has been worked properly, all that remains in the belly cavity is the blood clotted in the kidney running from the head to the vent along the spine. Every trace of this must be removed, although most troll fish on the market will be seen to have a few small spots remaining deep along the spine at the vent end of the body cut. The troller's knife has, opposite the blade, a spoon-like scraper to clear away the blood after the membrane has been slit the length of the artery. The scraper will not get all the blood matter out and conscientious washing must be practiced to clear the remainder.

Proper washing of salmon is a demanding job and often the most difficult on the vessel. The ideal system would have two tanks, each with a continual flow of clean water, each with chlorine solution mixing into it. One tank would be used for the primary wash, the other for a final cleansing rinse. It is doubtful, however, that many fish washing installations of that type exist on vessels of the salmon troll fleets, either United States or Canadian. Most boats are too small and too poor for such a setup and their skippers may settle for something as simple as a plastic wastebasket for a wash tub. Washing is not or should not be just a quick dip and slosh in the water. The head and belly cavities should be inspected for membrane, blood clots, visceral matter and gill fragments. Once washed, excess water must drain from the body cavities before the fish are iced. Some trollers use hoses rather than tanks for their fish washing.

## Icing Styles

Nor is icing a simple procedure in its entirety. Icing should be planned before the first fish comes to the troll lure. Some salmon troll boats are day vessels delivering their fish within a few hours after the catch. Most of the time, wet gunny sacks are regarded as sufficient protection for the short time the fish are aboard. Trip boats, the majority of the troll fleet, may work a week or more before delivering. Many trollers follow the rule that the catch must be unloaded no more than 10 days after the time the first salmon was brought aboard. Icing is of utmost importance, obviously, because bacterial action, quietly and unseen, is underway. Enzyme damage began first too and only ice can check it and bacteria. And then only if the icing is done without skimping.

Men whose handling of their salmon may be excellent in every other respect may be guilty of undue economy in connection with their ice. Some of this may be charged to laziness or to fatigue on a long trip alone. Some can only be blamed on extreme thriftiness. Ice costs money certainly, but it is not unduly expensive and for most salmon trollers, it is the only available catch protection. For every troller big enough and new enough to carry even rudimentary refrigeration, there are 50 without it and many or most of those do not have any hold insulation. No matter where troll salmon are taken during the April-October season, water temperature is never below 50 degrees F. except perhaps in the vicinity of a few Alaskan glaciers. Fifty-degree water soon begins to melt ice through these unprotected hulls and the first fish down may soon find themselves with nothing but bare planking beneath them. This is not much protection.

There are a number of schemes for loading ice aboard a fishboat and distributing it through the hold area. But there is only one way to use it — plenty. When a man thinks he has put down enough ice for a layer of fish, he should then put down some more. This maxim is true for box icing or pen icing. (Box icing cuts down loss caused by crushing of the lower layers of fish, but it is not commonly used in any West Coast fishery. It is mostly not practical for the salmon troller because of small holds and the extra work it causes the lone man.) The salmon troller of 40 feet or so generally has six or even eight pens installed laterally and fore and aft of a small work area below the hatch. The trollerman funnels ice into all pens but one in weights as equal as he can make them to get maximum stability for his boat. The empty pen, which should have a foot or so of ice on its decking, is the one to take the first fish. The last ice to come aboard should go up to a couple of feet deep onto the work area deck to give first cooling to the fish as they come into the hold and to prevent heat radiation from the bilge into the fish hold.

Icing procedure for salmon on a trip boat may depend partly on the estimated length of the trip. It is best to over-estimate the time the vessel will be out because weather quite easily may keep the boat away from a buyer for quite a bit more time than the skipper had in mind when he left port. Icing is intended to reduce the body temperature of fish to 32 degrees F. as soon as possible after they leave the water. This takes six hours or more at best. To achieve this end, the first fish should be put down on a bed of ice from 10 to 12 inches deep. Fish should be iced before rigor mortis takes effect because fish in its grip may be twisted into odd and unhandy shapes that make for poor icing. To get maximum fast cooling, the belly and head cavity must be packed with ice and the fish placed on the ice so the body cavities can drain freely. Ice layers themselves should be shaped with enough crown that water and blood will drain quickly from them. Fish should be placed on ice so they are not close to partitions, pen boards or hull liner. Nor should fish be placed against

each other if at all possible. This is possible in the salmon troll fishery, but it is not practical in such as the trawl fishery where great numbers of fish must be handled with all speed. When a layer of fish is down, it should be covered with ice at least six inches deep; this dependent again on the estimated length of the trip. The procedure is repeated until the bin is full or as full as the skipper wants it. Weight on the bottom layers of fish becomes a factor to be reckoned with when fish and ice are about four feet deep. No bin should be overloaded. It is a rare troll trip when all bins must be loaded to the top of the boards.

Ordinarily, troll salmon do not face the pew until they are off-loaded at the buying station. It would be a good thing if the pew were never used on them although there are days when the volume of deliveries makes it necessary for the pew to be brought out. The pew broadcasts bacterial infection like the wind scatters ragweed pollen among hayfever sufferers. Even the most careful of fish handlers working fast is going to miss the head of his target off and on while the careless or inexperienced hand seems to miss more than he hits.

Pewing does nothing to brighten a salmon's looks either and an ugly wound on the head is enough to drive the squeamish buyer right out of the fish market. Body piercing is not as noticeable but it's another door swung wide open for bacteria. The issue of pewing has become a matter of concern (or a matter of lip service, anyway) to salmon packers over the entire range of these fish. Several have tried during recent years to convince their fishermen of the desirability of handling salmon without the pew at all times. This means taking each salmon by hand to move it about. Unloading five or 10 thousand cannery salmon is a good bit of work no matter how it is done and the prospect of shaking hands with every salmon in a hold does nothing at all to convince the average crewman of the virtue of the procedure. Many older crewmen are simply unable to endure the long periods of bending from the waist the job requires in a seiner hold. And many fishermen refuse to touch a salmon or any other fish barehanded in any circumstances because of the infection called "fish poisoning." Cuts and scratches do become quickly infected aboard fishing vessels and in many instances adequate treatment may be many miles away. Gloves are a problem. Cotton gloves quickly become soaked with blood and slime and can be smelled well to windward despite all attempts at washing. The rough-palm rubber gloves used by trollers work well but they are awkward to the man not accustomed to them and a slippery salmon can squirt right out of his grasp the first few times around. A joint industry refusal to accept pewed salmon would discourage the use of the germ-laden instrument but it undoubtedly would be construed "in restraint of trade" by some agency of the federal government.

The "no pew" rule obviously would be difficult to enforce industry-wide, even under federal order, despite its equally obvious virtues, not only in the salmon fishery but in all other fisheries where large quantities of fair-sized fish must be moved around. But a handful of salmon cannery companies began to require it among their company-owned vessels in the early 1970's, and enforced it as well among independents selling to them. Seiner crewmen, initially resentful, began to change opinions, however, when they learned that four men, working by hand in the hold, could load off much faster than two with pews. An additional benefit was soon noted—there was no danger of accidental and hazardous pewing of fishermen rather than fish. Bacteria on the pew are just as unhealthy for fishermen as they are for fish.

The preceding material has been a rather painstaking recital of "ideal" techniques and practices of the Pacific Coast troll salmon fishery. It was written with complete knowledge that, in many ways, they are not applicable to all fisheries. But they are valuable techniques requiring a discipline that might be studied with benefit by men in all fisheries of the Northeastern Pacific, especially whose who may not be duly impressed with the necessity and the inherent difficulty of getting fish from water to market in first-class condition. The troll fishery may be regarded with a bit of disdain by men of the long-distance crew vessels but that disdain is unwarranted. The troll salmon fishery delivers the freshest fish being sold anywhere. The price per pound for these fish demonstrates this clearly.

## Ice?

Every man believes he knows the nature of ice, that is, water reduced to a solid state by low temperature. But not all ice is the same and not all ice is good for fish preservation. The Bureau of Commercial Fisheries defines proper ice for fish as ice produced from water of good drinking quality for humans, filtered and chlorinated and held at sub-zero temperature in clean storage until it is ready for use. It should be in particles of a designated size.

Fish boat ice may be crushed from blocks and stored until needed or it may be produced in flake form by a quick-freezing process newer than the old block operation. In either case, it should be cooled to zero degrees F. Such ice, if properly loaded and stored, will flow well after a week or longer at sea. Under normal conditions, it melts enough to form a crust on the exposed surfaces. When the crust is broken, it should expose loose, easy-to-handle ice, easy to pack in and around fish. If the ice has not been sub-cooled, it becomes hard to handle and tends to congeal throughout its mass. There are pleasanter and more profitable ways to pass the time at sea than by chopping out ice in a hold. Ice that lumps up because of insufficient pre-cooling is called "green" or raw ice.

Particle size is important, particularly for vessels such as salmon trollers where every fish has the belly and head cavities packed. Ice that has become chunky bruises or cuts when they are being packed. Ice in particles too small holds melt water and retards proper drainage. Ice in proper particles is easy to load into pens

and minimizes the chance of a freeze-up in the hose as it comes aboard. Such blockages may take valuable time to clear. Ice not confined right in the hold may become a danger to the boat itself because of its tendency, especially that of flake ice, to shift with vessel motion.

Almost all fish boat ice is fresh-water ice. However, work with salt-water ice, with its freezing point of 28 degrees F. and its salt content, shows that it holds fish better than fresh-water ice under the same conditions. Salt-water ice in these experiments brought fish temperatures below that of fresh-water ice while the fish in the salt-water ice were in better shape at the end of the test than were fish in the fresh-water ice. For most fishermen, salt-water ice is mainly an academic matter since few ice plants are to be found on clean salt water. Harbor water and much inshore salt water are no better for ice than they are for cleaning fish holds. Even large vessels at sea with their own ice plants might find their salt-water ice fouled by their own discharges.

In the years after World War II, there was much experimentation with some chemicals and antibiotics as fish preservatives. That a handful of these substances does have a beneficial effect in retarding spoilage is clear. However, the extra cost and the extra work necessary to use them, such as incorporating them in ice, is worth neither the money nor the trouble to most fishermen. The Bureau of Commercial Fisheries decries the "casual" use of antibiotics or other chemicals and emphasizes that careful handling and icing of fish is just as productive. In addition, there has developed a strong national concern over indiscriminant use of antibiotics and a whole spectrum of chemicals as food preservatives that has caused a re-examination of the value of such use in other than selected situations. Nothing in the future appears to justify any further widespread use although any appreciable effect on the business and manufacturing community is questionable.

## Ice Cost Modest

The cost of ice is not out of proportion to its benefit to the fisherman. But ice will melt in the course of a trip and whatever can be done to save ice on the outbounder is money spent in a good cause. Refrigeration and insulation are the obvious answers to undue melt and over the life of a vessel their cost will be regained over and over again. The average fishing vessel of the Northeastern Pacific does not need an elaborate refrigeration system such as that necessary on the far-ranging tuna ship. This vessel must deal with time and with tropical temperatures. The West Coast dragger or longliner must worry about comparatively short-range trips and time on the order of a couple of weeks or so. Hence adequate insulation and a simple refrigeration system (simple compared to those of the tuna ship or the factory vessel) is enough for most needs. This installation is not intended to freeze the fish themselves but only to check the premature melt of ice. Six inches or so insulation all around would appear to be enough for most boats. Refrigeration then is required only to offset what heat may still creep into the hold through the hull, the engine room bulkhead and the overhead. When the catch is iced, the reefer system is turned off or left in the low 30's so the ice may melt enough to chill and bathe the fish.

Fishermen have been using ice to preserve fish for about as long, doubtedly, that there have been fishermen. This oldest of methods, properly done, is excellent and it will continue to be the only method for men who do not need or want anything more sophisticated. There are indicated situations, though when other methods are preferable, such as when large numbers of fish must be held for a considerable (for fish) period of time. Freezing is one method, of course, but it is not applicable in some cases. Here, chilled sea water or brine is the cooling medium. Its advantages are many and its several disadvantages are apparent. Chilled sea water or brine achieves what the theory of icing teaches—enclosing every fish individually with the refrigerating medium so it derives maximum benefit from the contact. Icing fish to meet that standard is hard enough work anyway and sometimes contrary or lazy human nature makes it even more difficult. The ideal usually can be met only by the fisherman such as the salmon troller who has usually the time and the desire to do right by his catch. The process is almost automatic with chilled sea water or brine. Every fish is bathed in its entirety by the cold water and its body heat quickly is dissipated in it. The water keeps air at a distance and slows bacterial and enzymatic action as well as oxidation. Water temperature can be regulated right to a degree.

There is nothing particularly new about chilled sea water systems. They have been used shore-side for years in various fisheries to hold certain types of fish until the processing line needs them. The major use of chilled water systems on the Pacific Coast in recent years has been to hold cannery salmon aboard tenders until a load big enough to move to the cannery has been taken aboard from fishing vessels. When the California sardine fishery was still a big affair, most of the plants used refrigerated sea water to hold these fish. Little use has been made of it in that area since the fishery collapsed. Every major salmon cannery company in British Columbia and Alaska has one or more big tenders using chilled sea water while a few have holding tanks using it at the canneries. Some of the newer tenders were designed and built as chilled sea water vessels and they are remarkably efficient at keeping their fish in good condition. The salmon are sorted by species as they come from seiner holds and are channeled into the proper tanks as they come aboard.

Refrigeration aboard a vessel using chilled sea water must be a bit more sophisticated than most such systems since it must be capable of handling problems peculiar to these systems. Each must be designed to the needs of a single vessel and a design feature that may work quite well on one boat may be found unfitted for another. Pumps to circulate re-chilled water through the fish tanks must be built of corrosion-proof materials. Piping of plastic has been found most satisfactory. The usual refrigerant is ammonia. Tanks are built of stainless steel, aluminum alloys, plastics and wood. All surfaces of these latter two must be well-coated with sealer that carries no toxic qualities. Re-circulated water quickly becomes dirty and rather elaborate filter systems are needed to keep it clean. Tanks must be as completely cleaned after each load as any other hold compartment that has carried fish. Special discharge gear must be installed to move fish from the vessel.

Introducing water into the hull of any vessel poses several problems immediately as king crab fishermen with their live tanks learned early in the fishery. The hold must be absolutely water-tight, both for the good of the vessel and for the good of the fish load. Leaks, even quite small ones, can be two-way passages with bacteria moving into fish tanks from accumulations flourishing in the leak area. Bacterial contamination will spread quite well in 30-degree water, the preferred temperature, and spoilage is as certain as it is with fish under ice.

The major threat to a vessel using chilled water or brine is impaired stability. Water in the hold maintains its own horizon around the sea movement of a vessel. Sudden and deep movement of the vessel to port and starboard throws into the roll the weight of the water in the hold and the vessel's center of gravity shifts in that direction with it. If the shift is too far in the direction of the roll, stability may be destroyed to the point where the vessel rolls over under the weight. It takes a complex gathering of forces to work this destruction on a fishing boat. But any man who has felt a heavy roll has no trouble in visualizing the consequences whether or not he is conversant with the theory of the forces themselves.

Thus all holds to be used for chilled sea water refrigeration must be broken down into a series of tanks or must have a series of baffles installed to halt the lateral and fore and aft movements of masses of water and fish. The uncontrolled movement of the water is a danger in itself and it is made worse by fish. The fish themselves can be hazardous in an unpartitioned dry hold if they shift position in a seaway and the silent disappearance of fishing vessels and tenders sometimes can be blamed on that happening.

Properly installed chilled sea water systems are not inexpensive. Nor is proper maintenance and operation to be ignored. For the fisherman concerned with such an installation on his own craft, the cost is enough to make most men think about it for awhile. Chilled sea water probably is not worth the cost for existing single-purpose vessels such as the salmon seiner whose fish almost always are off-loaded the day they are caught. It may not be practical either for any new multi-purpose vessels to be built under the "Alaska limit" restrictions although it reportedly has been tried experimentally on one vessel. But for new, over-limit vessels to be used in two or more fisheries, installation of chilled sea water equipment appears to be indicated. The added cost to a vessel specifically designed for the system is a small-enough part of the total cost of construction to make it welcome aboard. Cannery companies with their great investment in vessels and plant cannot afford to be without chilled sea water facilities on tenders that must hold fish for some time, and ashore at canneries where peak fishing often results in a great backlog of fish and a resultant catch limit on fishermen. With it, the canners can oftener attain the excellence of product their labels claim.

An alternate method of keeping fish in good condition at sea is suggested by the Department of Fisheries and Marine Technology of the University of Rhode Island, a school that has done more to disseminate new and good commercial fishing practices to a sometimes unreceptive audience than any other in the United States.

This process is called "super chilling" although it has no relationship to the methods used to induce flash freezing at temperatures far below zero. Super chill utilizes the knowledge that most bacterial growth is sharply cut when surrounding temperatures are cut from 32 degrees to 29 degrees F. and held there. Super chill was first used to protect meat en route from Argentina to Great Britain at about the time of World War I. Since then it has been used for fruit shipments and late in the 1960's it was tried successfully for fish. The Portugese have been able to extend the holding time of this non-frozen fish from 17 to 30 days. Some vessels in the Canadian East Coast fleet use it and their products enter the United States as fresh fish.

Super chill uses no ice although it does require a well-insulated hold. A large volume of chilled air is circulated through the hold where fish may be boxed or shelved, dressed or in the round. A fan system directs the air over a cooling coil carrying a secondary refrigerant, usually a brine or glycol solution. Through careful control of the secondary refrigerant temperature, over-cooling with the danger of partial freezing is avoided.

This system is less complicated and certainly less expensive than chilled sea water systems. Additionally, it removes the hazard to the vessel inherent in such systems. It does away with the hand work needed for successful icing of fish and it appears to produce high quality fish when used under a controlled monitoring system.

To the experienced and conscientious fisherman, much of the preceding comments on fish handling at sea may seem to be an unnecessary insistence on facts they have known longer than they can remember. But not all fishermen are experienced and conscientious. Every skipper knows this and so does every fish buyer—particularly in regard to skippers. Fishermen are as the run of all other men, neither better nor worse (although probably more courageous than most men because fishing is a high-hazard manner of making a living) and among them there are those who do not hesitate to cut corners, particularly in the hold, when fish are coming aboard fast. Handling fish to keep them in top shape is tedious, time-consuming and it can be hard work. Fatigued men on a stormy or busy trip may be tempted to shirk just a little. Here, a little is too much and, with an undue incidence of Number Two fish, it may be rather heavily reflected on their settlement checks. But lazy men are not the sole threat to fish quality on any boat. The inexperienced man unwittingly and unwillingly may be responsible for fish damage because of his greenhorn status. Inexperience means many things, among them that a man who may be first-rate in one fishery may be almost useless for a time in a new fishery—the tropical tuna man turning up on a Gulf of Alaska longliner, for example. And the best of men may try hard and fruitlessly to offset damage caused by the vessel herself, old, her hold unhandy to clean, the home of permanent bacterial infection. The causes of fish spoilage are many. There is just one weapon against it.

Of the freedom of the United States, someone has written:

"Eternal vigilance is the price of liberty."

It is also the price of fresh fish.

Linda Rogers

*Norwegian-made heavy-steel purse line connector, simpler and faster to use than the spring-loaded figure-eight clip.*

Jim Rearden

*Crewman of small Alaska beach seiner handhauls bag of net, seeking clue to size of catch as vessel drifts just off rock wall.*

*Brailer load of salmon swings aboard fish buyer's boat. Scale is at bottom, center. The white cap is the West Coast's badge of the old-line fisherman as well as the longshoreman.*

# 6/ Fish Handling and Preservation Ashore

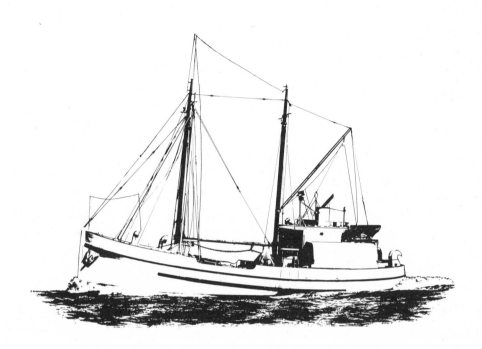

*I*f eternal vigilance may be accounted the price at sea for proper preservation of fish, it is, no less, the price of fresh fish ashore. Ashore, however, that desired standard of wholesome fish usually is somewhat easier to meet than it is at sea because of the permanence and stability of the shoreside plant. Not all fish receiving and fish processing plants are of equal merit, however, sometimes because of extremes of climate or geography or the naivete of management in such matters as fish spoilage and plant sanitation.

Ordinarily, though, a shore plant is under continuous inspection by its own management and by one or even several sets of public officials. To these may be added inspection, sometimes casual, sometimes thorough, by representatives of industry groups. Some such groups, like the National Canners Association, set standards of operation that often exceed the levels called for in federal, state or municipal regulations. And finally, wise plant management knows that carelessness in even minute details can be disastrous. Processing procedures as well as sanitation standards are or should be in continual process of examination. Whatever overconfidence there may have been in the fish processing industry in the early 1960's was quickly and painfully dispelled by a series of deaths in the United States resulting from botulism in canned tuna and vacuum-packaged smoked fish. These were commercial operations, not home processing with its usually-greater possibility of error. Carelessness in a California tuna cannery let some few cans of tuna slip out of the cannery after the fish had been precooked and packed but not retorted. The entire American tuna canning industry was in something of a state of shock for weeks after that accident because canned tuna sales plummeted as soon as the deaths were publicized in a nation-wide attempt to run down every can marked with the code of the bad batch. Similarly, smoked fish from the Great Lakes was vacuum-packed in plastic bags and marketed without freezing. The anaerobic bacterium, Clostridium botulinum, delights in just those conditions and several persons died because someone forgot or ignored one of the elementary rules of food processing.

Undoubtedly, the future holds other deaths from poor fish processing and, as regrettable as they are, they may be just the driving force periodically needed to keep processors fully alert to the dangers of poor quality control.

Nevertheless, the shoreside plant manager cannot be blamed for everything that may be wrong with his fish. If the fisherman has not used all his skill to keep his catch in the best possible condition until it is loaded off the vessel, there is not much the man ashore can do about it other than pay lower prices for usable fish below the top grade and reject all other fish of the proffered fare. This, of course, is done routinely and too often a substantial part of a load will go into meal or animal food processing, a waste of potentially good food fish.

Standards for fish buyers are largely a matter of experience and common sense. The old Bureau of Commercial Fisheries set down a set of rules as good as any and the first of these is to "observe the housekeeping practiced aboard the vessel...a clean-appearing, ship-shape vessel usually indicates that care has been taken to insure the delivery of good-quality fish and shellfish."

As for the fish themselves, the BCF suggests:

1. The fish should have bright shiny scales and characteristic colors and markings. As the quality of the fish deteriorates, the colors and markings fade.

2. The eyes should be bright, transparent and protruding. With deterioration, the eyes become sunken and cloudy and sometimes become covered with a pink slime.

3. The gills should be bright red and clean appearing. With deterioration, the gills usually fade to pink, then to gray and finally to brown or dark green.

4. The fish should have a characteristic mild, fresh odor and no off-odors. With deterioration, the odor changes from the characteristic mild, fresh odor to a disagreeable off-odor. (Like humans, fish smell differently, one from another. Fresh rockfish, for example, have a heavier and less agreeable "characteristic" odor than do fresh salmon. Some of the "soles" have perhaps the most delicate and agreeable odor of all fishes of the Northeastern Pacific—when they are fresh.)

5. The flesh should be firm and elastic. Fish that have been poorly cared for or held too long aboard the vessel exhibit varying degrees of softness. Fish frozen at sea and thawed at delivery should show essentially the same characteristics as fresh fish.

6. The fish should display no sign of body damage. Those that have been roughly handled may show damage from such handling or from pewing or from rough ice.

Absolute freshness is imperative for shellfish. Organisms and enzymes that may cause severe illness or death quickly appear in shellfish after they die and their presence is largely undetectable by taste or smell. This spoilage occurs much faster in crustaceans and mollusks than in the fishes. The slightest doubt should be cause for prompt rejection of shellfish. All shellfish, with the exception of scallop meats and shrimp, should be alive at delivery. Live clams and oysters will have their shells tightly closed if it is physically possible for them to be closed (some cannot close their shells en-

tirely) or will close them when disturbed. Crabs and lobsters should be active and ready to take a bite at anything that disturbs their peace. Iced or frozen shrimp should have the characteristic fresh shrimp odor. Scallop meats should have their own odor too. It is the responsibility of the professional to deliver fresh shellfish to the market because the consuming public generally is ill-equipped to determine for itself the freshness of fish and especially the freshness of the crustaceans and mollusks.

(Bad handling practices have been evident in certain sectors of the West Coast Dungeness crab fishery over the years. They have been practices that negate the care given their catches by the men of the crab vessels. Almost all effort to insure perfect condition of crab on shore delivery from the live tanks of the crabbers has been wasted. This bad handling is due partly to the geography of the crab coast where small outports serve as major landing points. One of the top ports of this rich fishery is a village with a wintertime, crab season population of less than 500 persons. It lies in the center of the richest crab grounds of the Pacific Coast (Alaska excepted) and millions of pounds of live crab pass over its piers each season. The hitch in the entire procedure is that only a small part of this abundance of crab can be processed there and must be transported to the major population centers to be prepared for market.

(Everything goes well for the crabs until they are dumped from wire mesh holding tanks alongside the piers into the trucks that haul them inland. Some of this hauling is done by unrefrigerated vehicles without any attempt at all at icing and it is at this point that good crab begin to turn into bad crab. These trucks are big, of the type known as semi-trailers with a net capacity of more than 20,000 pounds of live crab. On some trips, every pound of crab that can be crammed aboard is loaded into them with the driver standing on top of the growing load using feet and a fork to spread the crab evenly. When the load rises over the sideboards, it is crowned and covered with a tarpaulin. Loading takes up to three or four hours while the trip to the processor takes five or six more. From late winter to the end of the crab season, much of this procedure takes place, in this mild coastal area, under temperatures well into the 70's and often under bright sunlight. The crab on the bottom of the load suffer much crushing. The driver's boots (he wears them to escape the embrace of unhappy crabs) damage the shells of hundreds of others as he tramps around on the load. The Dungeness crab, under the best conditions, normally cannot survive more than 10 or 12 hours out of water. These crabs certainly are not being handled under the best of conditions and their mortality rate must be astronomical. It does not appear to be economically practical to haul crab almost 200 miles if a substantial portion of every load must be discarded—if it is discarded.)

## The Game's The Same

The principles applicable to preservation of fish at sea are equally applicable to the shore plant where the fisherman delivers. This means good design and sharp management—sharp management above all. Efficient management means that all other needed practices will be carried out ashore just as the good skipper sees that they are carried out at sea.

The function of the plant itself determines its design and much of its method of operation. The San Diego tuna cannery has operating methods and problems that distinguish it from the plant concerned chiefly with receiving and preparing fresh fish for market. The Bristol Bay salmon cannery with a single job to do and little time in which to do it operates like no other shore plant. The freezer plant has mechanical and storage factors to consider that need not concern management of other plants. The species of fish and their intended uses also dictate design and operation. But no matter the plant and no matter the fish, the place must be kept clean and the fish must be gently handled.

Fish handling ashore usually poses more chance for body damage than fish face at sea since fish must be handled several times from off-loading to final disposition. Halibut landed at Seattle, frozen there and shipped to the East Coast get worked over from two to four times at each end of the trip. Obviously, this calls for utmost caution in handling by all persons involved in this procedure. Halibut landed in Kodiak, and frozen there for van shipment by sea to Seattle, thence destined for the New Orleans market, for example, stand even more chance for mishandling somewhere along the line. Customarily, the salmon cannery taking fish directly from fishing vessels or its own tenders handles its fish with a minimum chance of damage.

Besides bruising or other body damage, fish that have been well-iced and held that way at sea may be exposed to undesirable temperature hikes during unloading and for some time thereafter. It takes only a little exposure to warmth and air for spoilage once checked to resume at an even faster pace. Even in some otherwise well-managed plants, fish may be allowed to go too long without re-icing, especially in times of heavy deliveries. Ice is cheap and there is no reason why it should not be used as liberally ashore as at sea even if the plant hands are having trouble keeping up with deliveries.

Fish bring ashore with them all the bacteria they carry with them naturally and all the bacteria they have picked up aboard the fishing vessel. Once ashore, they face continual further contamination right to the time they go into the freezer, the frying pan or the can. Everything that touches fish—hand, gloves, conveyor belts, carts, pews, forks, tanks, sorting tables—is loaded with bacteria. It can't all be eliminated but thorough scrubdown fore and aft several times a day with a chlorine solution of acceptable strength can do a lot toward holding bacterial population to tolerable minimums.

The Bureau of Commercial Fisheries recommends these general procedures for shore plant fish handling:

1. The pewing of fish should be discouraged; if pewing must be done, single-tine instruments should be used and fish should be hooked only in the head.

2. When fish are held briefly before processing, fish layers should be not more than 2-1/2 feet deep to prevent crushing. Shallow boxes or bins should be used. Good quality ice such as that recommended for use at sea should be packed around and into the fish. The same drainage required at sea should be applied ashore. There should be no exposure to the sun.

3. All equipment coming into contact with the fish should be scrubbed down periodically with stiff brushes, detergents and running chlorinated water. Steam should be used if available. No surfaces should be allowed to dry without scrubbing because of the difficulty of removing dried scales and slime.

4. All processing equipment—filleting machines, Iron Chinks, can fillers, steaking saws and other tools should be cleaned often with detergents and chlorinated water.

5. Before fish are cut for any type of processing, the entire fish should be washed in chlorinated water to remove blood and slime, major sources of bacterial spoilage. This is particularly important for products such as fillets that go into the market with no further processing.

6. Remove the nape from fillets. Spoilage shows there first. Packaged fresh fish products should be chilled to 30-32 degrees F. before leaving the plant.

Fish or shellfish have a market life of a comparatively few days after catching unless they are preserved by any one of the methods that have grown around particular species or groups of species through all the years they have been used by man. Drying, salting and smoking, perhaps in that order, are the oldest of these means of preservation, dating from so far back in prehistory that not even the oldest of old folk tales recounts their origin. Fish may be brine- or dry-salted and stored for considerable time under favorable conditions. Fish and shellfish may be smoked after light salting although the resulting product does not have the keeping qualities of the lengthy salting. Some of it actually is as perishable as the fresh fish or more so because hot smoking's partial cooking tends to break down tissues that would otherwise withstand bacterial assault somewhat more stoutly. Canning of fish and other foods dates from 1809, the happy result of a competition spurred by Napoleon who was not interested in the national nutrition but was seeking some way to preserve food so that his armies might ravage Europe with less impedimenta to hamper their movements. Canning was just getting a good start when Napoleon was mulling over his mistakes in the solitude of St. Helena and canning still is the major method of long-term preservation of food stuffs. Almost 80 percent of all American and Canadian salmon is canned every year while at least 99 percent of American and Canadian tuna goes into the can too although tuna is a popular fresh fish in the Asian nations where it is taken.

## The Freezer Takes Over

Freezing takes a larger percentage of the world's food fish each year, however, and this trend appears to be a continuing one as the world's fish eaters seek a product more nearly like fresh fish than they can get from a tin can. Winter-caught fish have been naturally frozen for preservation since men began to live in climates where such freezing was practical. Artificial freezing of fish in large quantities and in a uniform state is a development stemming only from the later decades of the last century. Freezing has made possible the storage of great quantities of fish that otherwise would not be harvested (and fishermen would not be working) because there would be no conceivable use for them. Practicable freezing has given sustained life to such fisheries as that for the Pacific halibut. The demands of the fresh market are filled early in the season and the surplus is frozen and stored until the market calls for it. Even in two nations as populous as the United States and Canada, 50 or 60 million pounds of freshly-taken halibut cannot be consumed as rapidly as the fish comes ashore. Other fisheries peak during a season of a few weeks or months and here again freezing allows a decent holding of the surplus until someone wants it somewhere.

Somewhere around 400 million pounds of fish and shellfish are frozen in the United States and about 150 million pounds in Canada each year. The United States plants are on all coasts and through the mid-riff of the country. There are more than 300 companies operating from one to a half-dozen plants each freezing fish in the country. These are the major companies only, those reporting their freezings and their holdings to the Bureau of Commercial Fisheries. Nor does the list take in American vessels freezing fish at sea and that includes the great tonnage of tuna taken each year by those vessels.

Although this figure includes all reported commercial fisheries, salt-water fish and shellfish account for about 98 percent of the total with fresh-water fish, chiefly from the Great Lakes, making up the small remainder. Shrimp from all coasts, fresh, fresh-frozen and "heat and eat," is by far the most important single species frozen. Others, in descending order of importance, are whiting, Pacific salmon, Pacific halibut, ocean perch (all coasts), crab (chiefly king and tanner crab) haddock, flatfishes (all coasts and all species except halibut), squid and all fresh-water fish. Although Pacific sablefish does not amount to much in total national freezing poundages, it is the third most important species, after salmon and halibut, in the Northeastern Pacific.

Raw fish are frozen in several forms—in the round (whole); drawn (gutted); dressed (gutted and trimmed),

and packaged as steaks or fillets. In addition, fish once frozen may be thawed or partially so for steaking or filleting and then refrozen. Other frozen fish may be processed into fish sticks or fish portions, then frozen again for sale to the consumer.

(The practice of partially or wholly thawing some species of fish, such as halibut, for steaking or other processing, is sharply condemned by fisheries marketing specialists. A typical reaction is this:[1]

("No fish or shellfish should be thawed even a little bit and then refrozen for any reason although it is a common practice. But it is a practice that, to a large extent, is responsible for the poor quality 'fresh' fish you find too often in (name of city) retail stores.")

No matter the way in which fish are to be frozen, they should be cleaned of all external matter—blood, ice, salt, slime—preferably by a thorough dousing in the oft-mentioned and invaluable chlorine solution. They then should be thoroughly drained and arranged for freezing and this may depend to some extent on the type of freezer to be used. There are several basic methods and each of these is susceptible to as many variations as engineers wish to contrive for them. Refrigerants include ammonia, three types of freon, nitrogen and other gases, each with its own faults and virtues. There are tunnel freezers and cylinder freezers and chamber freezers; freezers that must be loaded and emptied manually and freezers that do almost everything for freezing a fish except catching it; freezers that can take a fillet down to -100 degrees F. in a couple of minutes and freezers that strain to freeze big halibut solid in eight to 40 hours. The basic types, the freezers that do the world's freezing, are:

1. The sharp freezer in which fish are placed, in containers or not, on shelves composed of the coils through which the freezing medium is pumped.

2. The multiplate freezer in which the fish are placed between two freezing plates. This type is most used for uniform packages of such products as fillets and steaks.

3. The blast freezer in which cold air is forced around fish or packaged fish at high velocity.

4. The immersion freezer, utilizing brine chilled by ammonia or one of the freons, to freeze fish in the round, the freezer used on tuna ships.

Refrigeration and freezing are vastly complicated subjects, matters to be studied for a lifetime but not matters to be pursued here. The BCF's Fishery Leaflet 427 has enough about refrigeration and freezing to satisfy any layman. It is enough to say that the aim of all freezing systems is the same although they achieve it by different paths, at varying speeds and with varying degrees of handiness.

In our area of concern, about 27 million pounds of Pacific salmon is frozen each year with some 30 to 40

---

1. Larssen, A.K., associate editor, *Fishermen's News*, Seattle, Washington.

million pounds of halibut going through the freezers. The season for each species runs roughly from April through October, although the salmon catch peaks sharply in July and August through all areas where salmon is taken. Although 80 percent of this catch is canned, a substantial portion must be frozen because it cannot be handled in any other manner. The catch of both salmon and halibut is scattered over a vast area of ocean but holding and processing plants in this age of consolidation are usually centrally-located (comparatively, anyway) and far distant from the point of origin of the fish.

Halibut should be gilled and gutted within a short time after coming aboard the fishing vessel, carefully iced and put down in the pens. Upon landing, from 10 to 20 days or even more after the catch, the fish are headed, graded and washed down. The grades are "chicken," weighing from five to 10 pounds; medium, from 10 to 60 pounds, and large, from 60 pounds upward. (A discussion of proper on-board care of halibut will be found in Appendix III, Part I.)

But halibut—and other fish too—do not always get the care they deserve. Fish sometimes may lie on deck for hours before they are dressed and put into the hold, especially if the catch rate is high and strings of gear are being pulled continually.

An observer complains and explains:[1]

"Too many boats don't follow good practice all the time and this is because of the economics of the fishery. Small crews mean bigger man shares. Halibut are not being treated as carefully these days as they should be or as they used to be treated a generation ago. Almost all United States halibut boats employ a 'short' crew nowadays and have been doing so for quite a few years.

"The result is that fish may be on deck for several hours before they can be cleaned and iced, especially if fishing is good. Before World War II, when a 'western' halibut boat carried 11 men, including the skipper, the fish were generally dressed and iced within an hour after hitting the deck. Today, some of those vessels are manned with five or even four men. It is pretty sure that four or five men can't possibly do the same amount of work in the same time 11 men can do it..."

Whatever the freezing method, usually either the sharp freezer or the blast freezer, the fish must be positioned carefully, either on freezer shelves or in containers placed on the shelves so there is no overlap between fish. Large fish are usually placed toward the bottom of the freezer stand and the smaller fish toward the top, partly because the bottom coils in the stand usually are colder than the upper coils. Freezing in the sharp freezer may take up to 48 hours but the fish usually are left in the freezer longer so their internal temperature may be brought to zero degrees or lower. The blast freezer, with its fast-moving air currents, may cut freezing time to half that of the sharp freezer but the blast freezer induces dehydration more rapidly because of the air blasts.

Salmon usually are frozen as dressed headless fish in either the sharp or the blast freezer. The greatest single poundage is of silver or coho although all species are frozen. Large red kings usually escape the freezer because they are prized for mild-curing and subsequent processing into smoked salmon. Freezing practices for salmon are similar to those for halibut.

Sablefish, also known as black cod (which it is not) is frozen in considerable quantity in the North Pacific also. This choice fish is one of the few demersal species with a high fat content. The flesh is richly flavored, flaky and firm but little of it goes into the fresh fish stall because of the fat content. Most of the catch is smoked after freezing and thawing. The freezing procedure is essentially the same as for salmon and halibut.

A considerable poundage of fillets and steaks is frozen each year from fish of the Northeastern Pacific. This amount is divided among home and institutional type packages and large blocks for further processing into such retail items as fish sticks and fish portions. Multiplate freezers are used usually for smaller packages while the blocks may be blast frozen.

## Problems Of Freezing

The actual act of freezing is simple enough, a matter mostly of putting fish in various conditions in the right place in the right freezer and standing back to wait for it to freeze solidly with an eventual internal temperature below zero degrees F. This obviously halts (but does not reverse) bacterial damage until such time as the fish is thawed. But freezing immediately leaves fish of any kind open to other damage, and chief among the other things that can happen to fish is oxidation and oxidation-related changes. There may be dehydration too and consequent unwelcome changes of texture but these latter ordinarily occur over a longer period than does oxidation, a quick threat to frozen fish. Oxidation or undue oxidation, at any rate, causes quite undesirable flavor and color changes in frozen fish although these and other factors affecting frozen fish can be minimized under properly controlled conditions.

(It might be repeated here that fish that come out of the freezer are no better than fish that go into the freezer. Freezing is no more a cure-all ashore than heavy icing is at sea. But the best of handling to the time fish come out of the freezer cannot mitigate damage caused during cold storage.)

Of freezing and cold storage, the Bureau of Commercial Fisheries wrote in 1967:[2]

---

1. Larssen, A.K., associate editor, *Fishermen's News*, Seattle, Washington

2. Stansby, M.E., "Factors to be Considered in the Freezing and Cold Storage of Fish," Section I, Part 3, Refrigeration of Fish, Fishery Leaflet 429, Bureau of Commercial Fisheries, 1963.

"A number of changes are found...the flesh of the thawed fish will consist of two phases, the solid flesh plus a fluid known as 'drip' which was not absorbed by the flesh after thawing. The texture of the thawed fish will be soft and additional drip can be expressed (extracted) by applying only a small amount of pressure. The surface may have become desiccated...the color of the surface may have altered. After cooking, the fish may be found to have acquired an off-flavor or the normal flavor may be lacking. The cooked fish may be tough or fibrous.

"These changes...are sometimes said to be due to the freezing process. Actually, almost all such changes are caused not by the freezing of the fish but by the *cold storage of the frozen product. If fish are rapidly frozen and immediately thawed and examined, very few changes are noted and these are of a minor nature...*

"(These)...fish cannot be distinguished by the average person from fish that never were frozen. Even an expert might be unable to distinguish such fish after they have been cooked..."

Dehydration of fish can be held to a minimum by proper packaging and by maintaining cold storage room temperatures at a constant level so the temperature of the fish will not vary. No packaging except vacuum-sealed metal containers prevents the loss of moisture in cold storage fish. Balance of temperature between frozen fish and the cold storage units is theoretically possible but difficult to achieve in practice. Moisture loss may be seen first on the skin of the fish and results in the bleaching effect called "freezer burn." Fish that have been allowed to lose too much moisture suffer changes in the flesh texture that may be described as "woody." It becomes fibrous and tough when cooked and the familiar flaky texture of fish is lost.

Dehydration and oxidation could be prevented completely if frozen fish could be held in a vacuum. Since this currently is technically feasible but not economically practical, the next best thing is to keep air away from the fish to the extent this can be done by some kind of protective material. In the search for protective devices, literally hundreds of substances have been used and most of them discarded as research has groped its way through the technological jungle created by the explosive growth of the frozen food industry. These materials include chemicals, waxes, edible oils, metal foils, the myriad plastic films and all manner of treated papers. Some are good but too expensive or too unhandy; some are inexpensive, easy to use and not worth much. But all packaged fishery products, fresh or frozen, must use one or a combination of them. These are the packagings that will greet the customer so they must be attractive and receptive to all kinds of decoration and labeling and this again restricts the processor's choice.

Fish hidden away in cold storage need not be so pretty but their need for preservation from oxidation and moisture loss is desperate. (Plant management may be desperate too because the simple loss of weight through dehydration alone conceivably can turn a potential modest profit into an equal loss.) To date, there is only one good method of treating fish in bulk for long-time cold storage and that is by massive and repeated ice glazing. Such glazing "is effective in preventing loss of moisture...and in preventing ready access of air, thus retarding the onset of rancidity..." and this is just what the cold storage man wants. It took about 75 years or so after the British first used mechanical freezing in the 1870's to develop the techniques of glazing to their present utility. In fact, it was quite a way past World War I before all freezer people were convinced of the desirability of glazing.

Glazing is a simple affair in theory but like many other matters of apparent simplicity, it is a good deal harder to do on a large scale than it might first appear. The sheer labor involved in moving thousands of pounds of fish around several times during the primary glazing has set in motion a search for labor saving machinery or complete automation of an expensive procedure. The closest approach so far to automation during the actual glazing is use of a conveyor belt to convey fish through a series of sprays. But the fish still must be moved from the freezer to the belt and from there into cold storage and this involves much human muscle along the way.

Fish to be glazed must be deeply frozen with a temperature as far below zero as can be reached in a practical manner. The glaze water must be at about 32 or 33 degrees and the glazing room itself must be at a freezing temperature to minimize any warming of the fish during the glazing operation.

Glazing means simply getting a film of ice onto the fish by one means or another and this usually is accomplished by dipping the fish into a tank of chilled water and lifting it out. This can be done by hand, one fish at a time, or it may be done with a wire mesh basket or cage with several thousand pounds or more of fish in it. The essential thing is that the water must freeze on all exposed surfaces of the fish and that it must be thick enough to resist air and its corrosive and drying effect. The dipping must be repeated until the glaze gets thick enough to satisfy the man in charge or meets some standard set by management.

Not only round and dressed fish may be so treated but fish steaks and fish fillets may be glazed as well. King crab sections are commonly frozen and glazed and this product, cut to consumer size, has been one of the attractions of king crab in the retail market. Crab meat too is extensively frozen in blocks and glazed to be held until it can be processed into consumer sizes or canned or otherwise made ready for market. Shrimp are frozen, at sea or ashore, in blocks or water-filled cartons and this manner of presenting them, in proper sizes, to the retail market has been one of the reasons for the continuing growth of the North American shrimp fishery. Shrimp so processed are infinitely more attractive to most consumers than the traditional canned shrimp.

One glazing may not last the storage life time of a fish or fish block. The first glaze is good only for two weeks or so because of evaporation, and reglazing must be done. The fish may be moved through the original process or water may be sprayed on all reachable surfaces as the fish or blocks rest in cold storage. Ice glaze cracks rather easily and a number of patents exist for processes using additives to control cracking. Other types of glazing using gelatin and pectin bases are under patent but are little used commercially. Ice is still king.

Despite all practicable methods of holding fish and fish products in cold storage, there has to come an end sometime and it comes sooner, perhaps, than most persons believe. A fish, say a halibut, can stay in cold storage for a long time, five, 10, 15 years or more, with no apparent deterioration if it is kept glazed and at quite low temperatures. Of course, only a person in the last extremities of hunger would consider eating that halibut after more than a few months but the halibut still retains its outward appearance. Theoretically, that halibut could stay that way through eternity as long as the glaze and temperature are maintained. We know that mammoths and other animals of prehistory have been preserved in ice since the last glacial epoch some 15,000 years ago. This outward preservation through extreme cold is the basis for hope of new life sometime in the future for those persons today who are depending on the technology of cryogenics to keep their disease-racked bodies through the centuries until some now-unknown science can revive them and cure them. This semblance of preservation is chiefly on the surface, however, and whatever the fate of those humans who refuse to accept death and have themselves frozen for posterity, longtime freezing and storage still is not good for fish. This despite the best freezing and preservation arts.

British fisheries scientists took fresh fish, quickly iced after capture, landed and frozen with 24 hours, and ran a series of tests on their keeping qualities early in the 1960's. Their findings support what most researchers working with cold storage have pretty well known for some years, that fish held more than a few months quickly becomes unpalatable if not inedible. The British work showed too the importance of comparatively low temperature storage for any period of more than a few days. Gutted white fish species stored at 15 degrees F. showed no appreciable quality change in 30 days and was rated edible but unpalatable at the end of only four months. At -5 degrees F., the comparative periods were four months and 15 months. At -20 degrees F., they were eight months and four years. These fish, remember, were top-quality fish frozen and stored by the best methods. Remember too that the researchers were men willing, for the sake of their work, to sample long-held fish they might otherwise have hesitated to feed to a sled dog. The periods the fish might be held for market, for the man of average taste, can only be slightly longer than the period of optimum quality, that is 30 days,

four months and 15 months as noted above. The report concluded that storage at any temperatures lower than those used in the study might have "a positively disadvantageous effect" on quality because of texture changes.

## The Marketplace

Most of the fish the consumer sees in the marketplace is presented as steaks or fillets. In the fish specialty shop, these items usually are fresh, prepared and displayed without freezing. In the modern American supermarket with its thousands of items and its emphasis on none other than weekly "specials," the fish-seeking customer usually sees his fish frozen, stacked with scores of other bright packages in a freezer case where, knowledgably or not, he must select for himself. Often enough, if he does find steaks and fillets in what appears to be the fresh form, he is likely to learn, should he inquire, that they have been block-frozen and thawed by the retailer behind the scenes. (This is not meant to fault that freezing. If it were not for freezing, most people would have to content themselves with salt or canned fish and that can be precious little contentment for lovers of fresh fish. It is freezing that permits the casual bounty in the freezer cases of the supermarkets with their fish and shellfish from half the world side by side at prices most people can afford.)

Most people eat most of their fish as steaks and fillets and most people believe they know what a steak or a fillet may be, pieces of fish cut like so. . .The federal government of the United States prescribes certain standards for many foods. In 1968, it began to be apparent that the government seemed to be heading toward mandatory inspection of all fish and fish products going into interstate commerce. Such inspection now is voluntary under the Department of Commerce and this body, parent of the former Bureau of Commercial Fisheries, has definitive views of the makeup of steaks and fillets as developed by the then-BCF.

A fillet, as defined by these agencies, is a piece of fish "substantially boneless, cut away from either side of the fish along the backbone from (a point—Ed.) behind the pectoral fin down to the tail section of the fish. If that portion of flesh that lines the visceral cavity remains with the fillet, the fillet is called a full-nape fillet. Most high-quality fillets do not contain the nape."

For specific fillet, in this case frozen fillets of Pacific Ocean perch, instructions to the inspectors read:

"The product...consists of clean, whole, wholesome fillets cut away from either side of the Pacific Ocean perch, *Sebastodes alutus,* packaged and frozen in accordance with good commercial practice and maintained at temperatures necessary for the preservation of the product."

(It might be added that violation of "good commercial practice" by some commercial operators in both fresh and frozen fish is behind the growing sentiment in favor of imposition of mandatory inspection of interstate fishery products. This was the same type of thing that resulted in mandatory inspection of interstate traffic in meat and meat products two generations ago and later mandatory inspection of most meat in intra-state traffic in 1968.)

In December, 1971, the Department of Commerce, through the National Marine Fisheries Service, instituted what was called "customized inspection" of fishery products. The department described the program as one designed "to provide impartial inspection and certification of processing plants and all types of processed fishery products. . .plus specified sanitary operating requirements for plants."

(The program apparently was designed to make the best possible use of limited manpower and funds for a king-size job. The analysis of the program added that "inspection needs of a plant depend on many things—size, complexity of operation, the product or products being manufactured. In many cases, a small plant or a simple operation does not require a full-time inspector. In other cases, the one plant-one inspector rule may result in inadequate inspection for a large or complex operation."

(Thus, each plant would be inspected according to its needs—the plant that required only a half-day of an inspector's time would receive a half-day; the plant requiring two or three inspectors would get two or three inspectors. In addition, the program aimed at more closely aligning plant sanitation with regulations of the Food and Drug Administration, a practice that seemingly could have been implemented a generation or so earlier at the very least. Until the mid-1960's, however, the FDA had not covered itself with any great luster in connection with its duties as defined by Congress. In those years and following, a flood tide of consumer concern began at last to stir the FDA out of the lethargy and undue involvement with industry interests that had been its chief distinction through its history.

(Nevertheless, the fisheries products and plant inspection program (at time of writing) was still a voluntary procedure. Earnest efforts by the finally-stirring Food and Drug Administration and the inadequately staffed and funded NMFS to push much more stringent rules through Congress were sidetracked in 1971 by a coalition of fisheries states senators, particularly those of Washington, Oregon and California.)

As for fish steaks (and frozen salmon steaks specifically), the rules say:

"Frozen salmon steaks are clean, wholesome units of frozen raw fish with normally-associated skin and bone and are 2.5 ounces or more in weight. Each steak has two parallel surfaces and is derived from whole or subdivided salmon perpendicularly to the axial length or backbone. The steaks are prepared from either frozen or unfrozen salmon (*Oncorhynchus* species) and are processed and frozen in accordance with good commercial practice. . .the steaks in an individual package are prepared from only one species of salmon."

It is to produce fillets and steaks of these qualities that efforts of fisherman and processor have been directed since a point in time before the fish in question ever came out of the water.

Frozen halibut supplies more steaks than any other fish of the Northeastern Pacific. Fish are steaked as the market demands from the backlog accumulated by the processor during the season. Steaks themselves are generally distributed regionally while frozen dressed fish are shipped over long ranges such as Seattle to New York because freight is less for the fish than for cartoned steaks.

For steaking, the dressed heads-off fish move along a processing line that begins with two or three men with bandsaws and ends on a conveyor or cart for return to cold storage until shipping time. At the steaking table, the first sawyer removes gristle, belly and nape in one piece and saws the rest of the fish into steaks, working toward the tail as he does so. When the steaks become too small for a single serving, the tail piece is discarded. A second sawyer takes the belly piece, cuts away the fins and the belly wall and steaks the remainder. If three men are working together at the steaking table the third man cuts the steaks, sometimes quite large steaks, into serving sizes. A small steak may make two portions while a large one will make six servings. About 30 percent of the fish arriving at the head of this line goes into the waste bin and all of this material, in turn, is processed into animal food of one kind or another. (Seattle, indeed, lays proud claim to being the center of the world's production of halibut-flavored cat food. Cats the world over testify to the validity of this contention.) After cutting, the steaks commonly are ice-glazed before packing into cartons of various sizes. Salmon is steaked in the same manner although salmon steaks ordinarily need not be cut into smaller portions. When the collar is sawn from the salmon, the tips in it may be set aside for later processing into kippered salmon tips, the least expensive form of this tasty food.

Halibut and salmon steaks may be prepared from partly thawed or "slacked" fish. Thawing is carried out in water or in the air. The fish, still half-frozen or more, are easier to work with than completely thawed fish. As the fish move along the steaking table, they are scrubbed and trimmed and the belly flaps cut away before steaking takes place. The steaks are washed and brined and by the time they have been packaged they are completely thawed. This allows re-freezing under pressure which eliminates the air pockets found in the cartons of steaks cut from slacked fish too and similarly packaged and frozen although they are not brined because the salmon flesh with its high oil content absorbs salt more quickly than the halibut.

Filleting is a rather more complex process than steaking whether the work is done by machine or by hand. Hand filleting is an art that escapes many of those who try it. Its practitioners are among the most skilled of all workers in the shoreside fish plant and on their eyes and on their hands, if it is a hand operation,

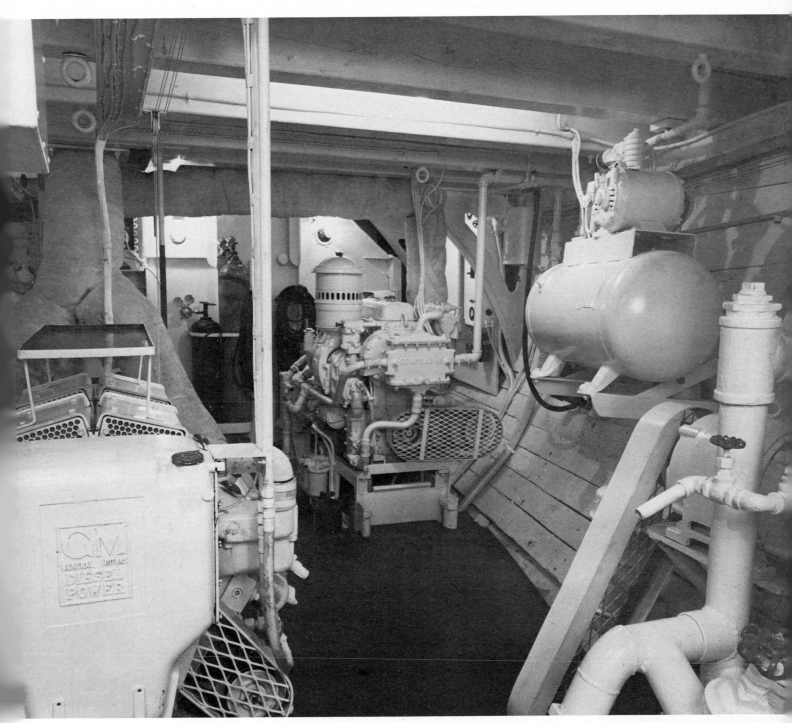

Marine Construction and Design Company

*Engine room of a modern multi-purpose vessel, the M/V Shelikof, one of the all-steel vessels designed primarily for the king crab fishery. Main engine is at left while one of two auxiliaries stands in center, rear.*

may depend the profit figure for the fillet section. Filleting is one of those things that people either can or cannot do—immediately. Men skilled with their hands at other trades may find themselves helpless when confronted with a flounder, a table and a knife. Those well-trained in the art turn out hundreds of fillets a shift, nice, trimmed, evenly-cut fillets that leave almost no flesh on the skeleton of the fish.

Written instructions concerning filleting are confusing. The BCF's Fishery Leaflet 428 directs the would-be filleter thus, warning that a very sharp knife with a flexible blade is a must:

"The filleter. . .makes his first cut behind the pectoral fin across the body and down to the backbone. When the backbone is reached, the knife is held flat and a second cut is made along the backbone parallel to the dorsal fins from the first cut to the tail. The cut then is opened by the filleter's free hand at the junction of the first two cuts. The filleter next cuts along the backbone around (along—Ed.) the visceral cavity and down to the tail in one motion. (This final cut presumably frees the fillet so the procedure can be repeated along the opposite side. That is, the fillet may be free for the professional; for most persons there still remains a lot of sawing around to get the thing done—Ed.) If the fillet is to be skinned, it is placed on the table skin side down. A cut is made across the fillet down to but not through the skin about one-half inch from the tail end of the fillet. The filleter then grasps the tail end with his free hand and pulls while cutting horizontally just above the skin."

## Machine Bests Man

That is the art of hand filleting. There are machines to do the job and several types, specifically one West German design, do it better, faster and cheaper. They produce from three to five percent more meat. The various machines associated with filleting began to appear in American and Canadian plants soon after World War II. Most had been designed years before by the Germans and as late as 1970 most of them were still being supplied by West Germany. It was the pirating of these German designs by the Soviet Union for massive use aboard its world-ranging factory ships that gave that nation what appears to be an insuperable lead in large-scale production of fishery products at sea.

The devices associated with fish filleting include filleters themselves, headers, scalers and skinners, operating separately or as one integrated machine. The principle of all is the same although actual procedure varies according to the builder. Essentially, these machines take fish of one species and of uniform size and feed them onto an endless belt or chain where they are held in position by any one of several means while they are headed. Here, some machines scale the fish after heading while others take them directly to the filleting knives. This step is carried out by high-speed rotary

and parallel ribbing knives with the fish entering the machine tail first. If the fillets are to be skinned, they next move along a belt past a horizontal band knife with a belt above it. The upper belt allows application of even, firm pressure to the fillet as the knife slices away the skin. Although all these machines can be adjusted to the size and shape of fish, uniformity of size is vital to their economic use. The plant must be assured of a steady supply of raw material to justify installation and operation. Even with the best of these machines, there still is considerable human labor involved in their most efficient use.

In good practice, all fish to be filleted are washed thoroughly in chlorinated water to remove waste matter and cut down the bacteria count. After the cutting, the fillets are run through a brine solution, sometimes with chlorine added to it. This firms up the flesh and adds to its good looks. Fillets of some species usually are candled to detect parasites in the flesh. If the defect can be trimmed out, this is done. If not, the whole fillet goes to the waste tank for conversion to animal food or meal. The parasite problem is particularly acute with Pacific Ocean perch, the major source of fish fillets in the Northeastern Pacific. If the fillets are to go into the fresh fish market, they are or should be chilled down to 32 degree F. after packaging. If they are to be frozen, this may be done in blocks, under pressure in cartons or individually. One fillet freezer being used in the Pacific Northwest uses nitrogen to take fillets down to $-100$ degrees F. in three minutes.

The Japanese through the years have many times shown themselves to be ahead of most other fishing nations in fishery techniques and utilization of the catch. The single most obvious modern exception is the American development of tuna seining vessels, gear and methods that for the decade of the 1960's, at least, placed the Yanks at the head of the pack.

But the industrious Japanese have always led and still continue to lead in the use of what they catch. The Japanese people eat species of fish and parts of fish that Americans and Canadians disdain. Too, the Japanese, soon after World War II, developed and put to use fish processing devices that almost literally utilized every bit of the fish except the wiggle.

Among these mechanical developments were several machines designed to extract fish flesh that ordinarily is lost through standard hand or machine filleting procedures. Americans eventually began to catch on to the virtues of these separator machines and by 1970 or somewhat earlier had begun to design and experiment with their own versions.

National Marine Fisheries Service laboratories in Seattle, Washington and Gloucester, Massachusetts, did the primary federal work with the separators. One NMFS report on the subject described the results of the experimentation as "a multi-million dollar bonus to industry, a complete utilization of harvested fishery resources and a substantial increase in seafood production—all without increasing the catch by a single fish!"

This bit of hyperbole may be attributed to an enthusiastic publicist rather than to the sober report of a trained researcher. But the devices do work; the problem that remains is to use the fish flesh thus obtained in ways satisfactory to American consumers or to disguise it—as in fish sticks—so that it cannot be distinguished from the sticks or fish "squares" customarily obtained from frozen blocks.

The NMFS report continued:

"Repeated tests by NMFS have demonstrated that use of the separator increases the yield of edible fish flesh by 12 to 30 percent over the yield obtained from standard filleting techniques. Ordinary processing methods for market fish recover 25 to 30 percent edible flesh from whole fish. When the separator is used, recovery rates range from 37 to 60 percent, depending upon species. At present this type of separator is used widely in this country by meat and poultry processing plants, but acceptance has been limited among fish processors.

"The device works by separating flesh from the skin and bone carcass which remains after fillets have been removed. Adhered flesh is squeezed through the perforations. The process involves feeding this waste material from filleted fish (or dressed small fish) into a hopper and thence into the machine where it is pressed between a belt and a perforated drum. The belt and the drum move in the same direction, but at different speeds, thus creating a tearing action that removes flesh from bones and skin. The soft flesh is forced through the holes and into the open-ended drum, which rotates the material into a waiting container. Skin and bones are scraped off the outside of the drum and into a waste chute.

"The highly nutritious end product consists of minced fish, unadulterated by waste products, which can be used in a variety of ways. For years the Japanese have been using their "leftovers" in fishblocks, sausages, fishcakes, pastes, and spreads. NMFS researchers have used the material produced from domestic species experimentally in similar products, including a 'pepperoni' type of fish sausage. They have also tried adding a moderate amount, 15 percent, to meat compounds such as frankfurters and meat loaves. Panels of taste testers have given wholehearted approval to almost all such experimental preparations.

"Numerous fish species ordinarily processed only for fillets have been used in the experiments, including cod, flounder, herring, rockfish, salmon, shad, and sole. Varieties of so-called trash fish and species too small to fillet—anchovy, hake, and porgy, for example—also have been put through the separator. Yield of minced flesh from cleaned, headed, and gutted fish was as high as 41 percent. Shrimp too, were subjected to testing; results indicated that a considerable quantity of shredded meat adhering to shells can be saved. As a consequence of the shrimp experiments, investigators suggested a new method for producing high-grade shrimp meal from shrimp wastes."

## On The Road

There is no point in freezing fish if there is no dependable way to move them (or fresh fish) from the place they were frozen. Until the coming of such methods, the fisheries of North America (with the exception of the cod fishery of the Grand Banks) were chiefly of local interest. Ice was used when it could be had for regional movement of some fish and shellfish—the oysters of Chesapeake Bay, the shrimp of the Gulf Coast, the salmon of Puget Sound—but the distance was little by today's standards and the keeping qualities were less. The fresh salmon of the Pacific was known only as canned or salt fish much past tidewater; the Pacific halibut was not known even by reputation east of the Cascades and most of the other fish of the West Coast, went no further from salt water than a team of horses could haul them in a day. Fisheries of great potential were ignored or were moribund because there was no safe way to transport their products widely in their fresh form.

Population increases in the United States and Canada and the dispersal across both nations of seaboard people who knew and wanted fish that was neither salt, canned nor smoked coincided with the technological ground-swell that characterized North America above the Rio Grande from the 1870's onward. This emphasis on cheaper and more efficient ways to get things done was especially evident in the United States. By that time, the frontier was gone except for Alaska and two generations of peace lay ahead for Americans. The energy that had been burned up in four wars and along the trail to the West was turned to other pursuits. The restless Americans took over the major part of the world's engineering (as distinct from pure science) and most of the significant (and insignificant) developments in that activity took place in the United States over the following years.

It was during these years that the food sciences set back the Malthusian time table by at least a century. Food production in North America north of the Rio Grande so far outstripped population that food surpluses worried politicians for most of a century far more than did food shortages. For the first time in the long pre-history and history of mankind, he sat down and studied food in all its aspects—growing it or catching it, distributing it and cooking it. More people began to eat better than people ever had eaten before and they began to eat more widely. Food sellers began to enterprise food. That's how Pacific halibut got to the East Coast.

On September 20, 1888, Benjamin and West, brokers, shipped a car load of halibut landed in Seattle by the schooner *Oscar and Hattie,* to a New York consignee. The carrier, the Northern Pacific Railway, knew no more about holding fish for the transcontinental run than did Benjamin and West and the halibut was good only for cat food by the time it reached New York. Benjamin and West grossed less than their express costs on that load but the NP prospered and has prospered since by what its people learned from that first shipment of halibut. They learned how to handle fish for the long haul. Because of the NP's knowledge, the Pacific halibut began to disappear from the grounds it had inhabited for millennia; eventually the International Pacific Halibut Commission was born to do something about fewer and smaller halibut; because of the commission, the all-time record for halibut landings was set with 75 million pounds in 1962. All because someone in 1888 tried to make a buck on halibut in a new market.

Pacific halibut was not the only fish to benefit, if that be the word, from this new technology. Fresh fish from all coasts began to cross trails in the mid-continent. Keeping time was still short enough but the trade was worth the economic risk and everywhere men began to seek new fisheries or to put life in old ones. Fishery expansion began in a fits and starts manner and by World War I, the present pattern of American and Canadian fisheries was pretty well set.

American railroads pioneered refrigerated shipping for every manner of transportation and for the whole world. The Pennsylvania Railroad put together 30 insulated reefer cars in 1859 and cooled them with block ice. They weren't very efficient but they got butter from the country into New York in tolerable condition. The first reefer ship did not appear until the British used one in 1879 to take a cargo of meat (mutton?) from Australia to the homeland. The first reefer trucks hit the road in 1905 but it was 1925 before the first mechanical refrigeration system took to the highways.

Refrigeration systems for ships are essentially the same as those for shore plants because of the size of the vehicle and convenient power. Rail car and truck systems must deal with limited size and power but their reefer technology has developed on parallel lines.

Rail, highway, ship and plane are used today to move fresh or frozen fish across the county or across 180 degrees of longitude. With decent attention along the line, receipt of these products in wholesome market condition can be guaranteed. Only extreme carelessness or major accident can thwart these well-laid plans.

The airplane, the glamor carrier of the expressways, carries the smallest amount of high-quality fish or shellfish a long distance at a pretty stiff price. Most fish products are not valuable enough to warrant the cost of air express or air freight under most conditions although substantial weights of salmon and halibut are air-shipped commonly from the West Coast to inland and eastern points. But most of the time, air travel is reserved for products of little bulk and high price wanted alive and kicking a continent or an ocean away from their native grounds. Thus, the tiny and costly Olympia oyster, at 2,000 to 3,000 to the gallon, may be found in the posh restaurants of New York while restaurants along the little oyster's Puget Sound home advertise live Maine lobsters with the customer allowed to point out his own in the holding tank. These treats do not come inexpensively.

Ships are the great bulk carriers of the world's produce. In contrast to such commodities as grain or coal, frozen fish and fish products are small potatoes on the ocean trade routes. Nevertheless, ships do carry tremendous poundages of such things as salmon and halibut from Alaska, tuna from Japan, spiny lobster from South Africa, cod fillets from Norway and shrimp from every sea where it is taken. As far as the modern American fisherman is concerned, much of these fish imports are products he would like to supply but cannot since foreign fishermen and processors work more cheaply than he does. Little that the American fisherman takes from the sea is frozen and exported.

American and Canadian fishermen are more concerned with rail and highway movements of their products, not directly concerned perhaps but beneficiaries nevertheless of the efficiency of these means of transportation. It was the reefer car that freed the fisherman from his age-old dependence on local markets and it is the refrigerated truck that carries his product along every highway and byway and into towns and villages that never have seen and never will see a train. There is little resemblance other than four wheels between the Pennsy's 30 little ice cars in 1859 and the mechanical refrigerator car or truck of the 1970's. But the job is still the same. It is merely done better.

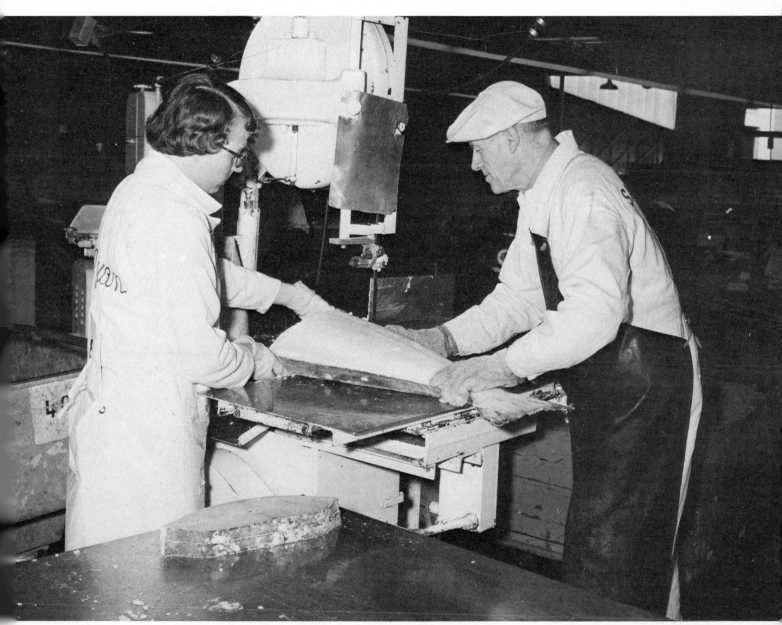

Pacific Fisherman

*Halibut steaking team works on fully-frozen fish. Note steaks on table in foreground. Some 40 million pounds or more of halibut are frozen during years of normal catches.*

*Splitters, the most skillful workmen along the mild cure process line, display their deft touch with the knife in a small plant in remote British Columbia. In foreground lie sides of bright king salmon, ready to go to the "slimer" for the next step. Note the thin line of bone left along the spinal column, the ultimate test of the splitter's ability.*

# 7/ Salting and Smoking of Fish

Drying fish in the sun or over a slow fire probably was man's first attempt to preserve fish past the few hours or few days it stayed good after coming from the water. Drying still persists in many areas where it can be done practically. It costs nothing other than the labor of cleaning fish and building a rack to hang them on. Sun drying, with or without salting, turns out a product that can be described only as pungent, a food of appeal to particular ethnic tastes but not with widespread demand outside its areas of origin.

Salting does a better job of keeping fish although some of its results too can be described as exotic in odor and taste. As early men began to explore the world around them, they must have stumbled across the near-magic properties of salt quite soon after they began to acquire something of an understanding of the arts of food preservation. Salting gives drying quite a boost as a food preservative and in all the lands where this was independently discovered, men had for the first time something of a long-term method of preserving fish.

Sometime in this dawn era and probably as soon as fire could be controlled, fishermen and hunters began to be aware of the preservative powers of smoke. With salting and smoking, they now had comparatively foolproof means to keep protein foods for weeks and months in a fair state of preservation. Actually, salting and smoking were the only ways (other than natural freezing) of keeping foodstuffs for any time at all until early in the 19th Century. All through his history down to 1809 when the French developed practical methods of canning food, man had to depend on salting and smoking if he wished to keep any meat or fish past the few days it stayed fresh enough for him to stomach it.

Primitive smoked and/or dried foods can be a shattering experience to those not accustomed to their continuing use. The friendly Indians of the Snake River, Nez Perce who had not yet become acquainted with the forked tongue of the white man, fed salmon, smoked and dried, to the near-starving Lewis and Clark party when it came out of the Rockies down Lolo Pass above the Snake in spring, 1805. The white men soon would have died had it not been for the Nez Perce and their salmon but the cure was almost as bad as the disease, a fact noted with precision in the journals of the two captains. The Indians had endured such salmon for who knows how many thousands of years and their bodies shrugged off a bacilli count that quickly felled the johnnies-come-lately of the American party. The intestinal tracts of the hungry and unsuspecting whites were not equal to the salmon and its gut-wrenching impact and it was most of a week before the company felt able to cinch up its pants with any degree of confidence.

Salt, sodium chloride, is one of the commonest of minerals and men must have begun to use it as an accidental by-product of the natural evaporation of salt-water pools along the shoreline. Such salt is not pure. It contains traces of all the other substances found in sea water. Some, such as iodine, are beneficial in minute quantities and some are potentially harmful in big-enough amounts. But there is not much chance of over-indulgence in any of them since tolerance to salt is limited and severe illness or death from salt poisoning would occur long before any of the trace elements could be felt.

Salt preserves fish and meat by driving moisture from the tissue structure. Lack of moisture retards bacterial action and the preservative effect of salting is related directly to the concentration of salt in the tissue. Some salting lasts for years. Who has not read of the men of the sailing ships and their complaints, in fact and fiction, about the aged salt meats that were their perennial fare? Quite literally, some of the more elderly beef and pork could have served as building blocks if a suitable mortar had been found. This animal flesh is tougher than the flesh of fish and fish does not salt so hard or keep so long.

279

## Shadow of Itself

The salting of fish is mostly a dying industry in North America and Western Europe although over much of the rest of the world it still is alive and well. There are better ways now to put up fish and people prefer those ways. Around 1900, more than 200 million pounds of fish was salted in the United States every year. Now salt fish production en masse is below 40 million pounds and still going down. There are certain specialty saltings that show an annual rise in weights. Among them are the many variations on salt and pickled herring.

One of the great salt fish foods of North America was salt cod from the Grand Banks and Georges Banks of the Northwest Atlantic. (It seems probable that Breton and Portuguese fishermen were familiar with Newfoundland and its fishing grounds long before Columbus and his "discovery" of the Americas. As their vessels became bigger and more reliable and their knowledge of navigation grew with them, these fishermen felt their way to the west along the northern Great Circle. Having found Newfoundland, they knew a good thing when they had it in hand and saw no reason for sharing their secret with strangers. They were publicly landing North American salt cod within a few years after Columbus' first voyage.) For much of the history of the colonies and the United States later, creamed cod on boiled potatoes was a standing Saturday night supper or Sunday morning breakfast for a good number of the more affluent people of the country. For many, it was the only salt-water fish they ever would see.

Salt cod now is of minor interest to most North Americans although the Portuguese dorymen still work the banks to salt cod as they have done for centuries. The faltering salt cod industry did not mean the end of the cod fishery itself. There is more demand now for fresh and frozen cod than ever there was for salt cod. Similarly, there was a healthy salt cod industry in the North Pacific from 1865 until the last schooner, Capt. John Shields' *C. A. Thayer* out of Seattle, was retired in 1950. The dory fishery was centered in San Francisco and Seattle and its men fished from the Gulf of Alaska as far west as the Sea of Okhotsk. World War II ended the fishery except for a few diehard skippers like Shields (who had once radioed for guns and ammunition to drive off Japanese encroachers in the Eastern Aleutians in the mid-1930's. Shields' message rang through to Washington, D. C., and Secretary of State Cordell Hull told the Japanese to get out and stay out. They did—until early in 1942).

What appears to be a catch record for any cod vessel anywhere in the world was set by another Seattle doryman, the three-master *Wawona*, with 6,830,400 cod taken from 1914 through 1940. The seasons averaged about 75 days of actual fishing time with 70,000 pounds a day looked on as a good haul. The *Wawona* still lives as does the *Thayer*. The latter ship has been restored as a vessel of the West Coast lumber fleet for which trade she was built and is berthed in San Francisco.

Men do not risk their lives any more in bobbing dories on the North Pacific or the Bering Sea to catch cod but the Pacific cod fishery still thrives. Puget Sound ports get most of the total landings of about 15 million pounds taken by the drag and longline fisheries. Cod is an excellent fish (its long history proves this) and it is a favorite fish of the fish and chip business and the ever-growing "heat and eat" stick and portion industry. It is easy enough for most people to understand why salt cod went out of wide favor when fresh or frozen cod becomes available almost everywhere. Most salt cod smells and it does not smell like roses. The dory schooners with any part of a load must have made their whereabout rather widely known down the wind.

Smoked fish was once a necessity if one wished to eat certain species many miles away from their point of origin. For most North Americans now, most but not all smoked fish is a gourmet food, a delicacy to be savored on special occasions. Some ethnic groups do eat smoked fish as a common part of their diet, of course, but most persons see it at most a handful of times each year. Salting of fish is a pre-requisite to smoking fish. Salting firms the flesh and imparts a flavor of its own as well as adding a preservative effect to the final product. In the Northeastern Pacific, the only large scale salting industry centers around salmon. This is the production of "mild-cured" salmon for smoking. The fish is lightly salted and must be held in refrigeration until it goes to the smokery. There is also a small trade, stemming mostly from Western Alaska, in hard-salted salmon, the kind of salting that once made salt salmon from the North Pacific a staple food everywhere around the world that a sailing ship could deliver it. That trade amounts at most to less than 100,000 pounds a year any more. About 5 million pounds of mild-cured salmon is produced each year in Alaska, British Columbia, Washington, Oregon and California although the smoking is done in cities as far apart as Ketchikan and New York.

## The Art of Mild-Curing

Mild-curing first was tried on the West Coast in 1889, but did not get established commercially until 1898 when two plants went to work on the lower Columbia. From there, the process spread to Puget Sound in 1901 and to Alaska in 1906. Large red king salmon are preferred for mild-curing although there is some use of large silvers for smoked salmon less expensive than the top grade. Chums and pinks are used for the production of kippered salmon, an item especially popular on the Pacific Coast. It has been written that "mild-cured salmon must be handled more carefully than any other salmon product. In few other food products is handling so important in determining the quality of the finished article."

Red kings of from 18 to 20 pounds minimum size are the salmon commonly selected for mild-curing. They must be fresh, reasonably fat, plump and bright, with no bruises or pew marks in the body. Most of the supply of kings comes from the troll catch and troll-caught red kings graded "large" bring the fisherman the highest per pound price of any fish of the entire Northeastern Pacific. Fish of such price obviously merit the gentlest of handling and salmon to go through the mild-cure are so handled from the moment the fisherman sets the gaff carefully in the head until the finished side is cooled and packed at the smokery.

The principle of fish salting is simple enough; its application is more complex. Mild-curing is a subjective affair in which a man's instinct and experience are more important than written rules. It begins with a careful heading of the salmon along the curve of the gills. As much as possible of the bony structure around the collar is left in place so that it may be used as a lug when the sides are handled during the curing process as they must be at several stages of the procedure. The headless fish then is lightly cut or scored in several places parallel to the lateral line and over most of the length of the body in order to allow rapid penetration of the salt. The cuts should be only skin deep, however, and not penetrate at all into the flesh itself.

The single most vital step in the mild-cure process, the splitting of the fish into its two sides, comes next. The splitter has been described as the most important man in the mild-cure plant and his position, economically at least, is somewhat analogous to that of the skilled filleter. His judicious use of the knife may mean the difference between profit and loss in a business trickier than most that handle food stuffs. The splitter's aim is to remove in one piece the entire side of the salmon with all usable flesh in it and little or none left on the skeleton. This may be easy enough to do with a small fish, but small fish don't get mild-cured and the exercise of this skill on a king of 30 to 40 pounds calls for meticulous care abetted by long experience. The splitter's touch must be fine enough to leave most of the backbone on the skeleton, but each side should show a pencil-line trace of bone from the spine of the fish. No machine yet devised does this job as well as a skilled splitter.

The split sides are now at their most vulnerable and they will continue to be so until the first-stage brining has toughened them up a bit. The care that has been practiced with the fish before must be intensified so they will not break up or split, a total loss as far as mild-curing is concerned. Thus, the next man in the line, the slimer, must work as gently as if he were tip-toeing blindfolded through a quarter-acre of eggs. The slimer gives the sides their last cleaning before going into the brine. Every shred of membrane and every clot of blood is removed and any final trimming is done at the sliming table. It is especially important that all traces of blood be gotten rid of because any remaining will be hardened by the salting and will discolor the flesh.

Warm or room-temperature salmon flesh will absorb salt too rapidly so the sides must next be hardened off in ice water or chilled brine. The chilling also leaches out any remaining blood and tends to seal the oil in the flesh. The chilling may last from 30 minutes to two hours, a matter dependent on the preference of individual management. The chilling time may vary widely among localities and even among plants in the same area. The temperature of the chilling tanks can range from 30 to 40 degrees, salinity from 60 to 70 percent. The chilling tank sometimes, incidentally, is mislabeled the sliming tank.

After removal from the chilling tank, the sides must be thoroughly drained before they go to the salter. Here as all along the work line, careful handling is the necessary name of the business. The salter works from a bin of salt of small, even grain with no trace substances like those the first seashore salters had to use for lack of any other. This salt is the "dairy" type used to mix with cattle feed and around fish plants is commonly called "mild-cure" salt. The salter places the sides one at a time, skin side down, and scoops salt over them. But he is careful not to rub it into the flesh because it would damage the fragile tissues and undo all the work so far. Enough salt sticks to the sides just by scooping it onto them and all the excess falls back into the bin when the side is lifted by the lug.

Mild-cured salmon is packed into a barrel called a tierce, a name stemming from the Latin word for a third of something. In this case, that something is a third of a "pipe," the great barrel used in medieval England and later as a container for wine or oil. The pipe held 126 gallons. In the fish business, a tierce should hold 825 pounds net of salt salmon although the gross weight, brine, barrel and fish, runs from 1,100 to 1,200 pounds, no light object to be handled by muscles alone.

Packing the sides into the tierce is a skill of its own and its practitioners are as deft in their work as the best brick mason. A few handfuls of salt first are scattered over the bottom of the tierce and a layer of fish placed on it skin side down, collar to tail. More salt is scattered over this layer and the next layer is placed over the first at right angles and flesh side up. In packing, the thick edge of the outside pieces in the final layer is placed skin side up and more salt is scattered over it than on the other layers. The total weight of salt runs from 85 to 120 pounds although 90 pounds would seem to be the average.

The final layer of sides comes only to the croze or ring in which the head fits. The usual practice is to head the tierce as soon as the fish have been layered and fill it through the bung with 90 to 95 percent brine. In some plants, however, the headless tierce, with its layers of fish in salt is held for periods up to 48 hours before it is headed and filled with 100 percent brine. This usage seems to depend mostly on outside temperature and rush of fish into the plant.

This salting is not enough to preserve the sides from spoilage and the tierces must be held in cool storage with temperatures around 22 degrees. Temperature fluctuation should be avoided because a rise of a few degrees will cause oil to ooze from the flesh. During the time the tierces are stored, anywhere from 20 to 90 days, they must be checked periodically for leakage and replacement of brine.

Fish held in brine such as those undergoing mild cure lose weight as moisture is extracted from the flesh by the salt and it is necessary at some time after the initial packing for the sides to be repacked. Shrinkage in the sides may run to 30 or 35 percent during the first three or four weeks after the first packing. At the same time, any imperfections are exposed by the salting and as the fish are repacked they are graded for size and quality. Before going into the new tierces, each side, still under gentle handling, is lightly scrubbed to remove excess salt and other undesirable material. Chilled brine or ice water is used for this work. During the actual repacking, the sides are placed in positions as nearly as possible like those in the first tierce with curved pieces against the side of the tierce, straight pieces in the center of it. This packing is done without dry salting but ice-cold brine of 90 to 95 percent strength is poured through the bung after the tierce is headed. Again, the tierces must be checked often for leakage and new brine may be added to them. The tierces must be held in chill storage right to the day the sides go into the smoker.

## The Oldest Profession

The historic hard-salt process for salmon in the Northeastern Pacific dates back to the Russians and their rape of the Aleutians and the Alaska Panhandle. It was used mostly for feeding their own people in the posts on Kodiak and down in Southeastern. There was little need to ship any back home because the Russians of Pacific Siberia had all the salmon there they needed and they knew as well as Baranof's looters the art of hard salting. It was the British and Americans who, between quarrels over who owned what between the Columbia and Portland Inlet, got the salt salmon business on a paying basis. It paid well, too, and after the fur trade dribbled away, salmon salting was the chief cash industry of all the land from the Columbia to Kodiak. It stayed that way until the Humes down on the Sacramento enterprised salmon canning into the dominant position it now holds over most of that same area.

Hard-salting follows much the same pattern as the process used in mild-cure. But the salting is total and lasts for as long as the saltery management wishes for initial processing. The final stage of salting, this time in barrels, lasts indefinitely, dependent to some extent, of course, on outside temperature. Hard-salted salmon obviously will keep longer in Kodiak than in Panama City. Despite the keeping qualities, there does come a point where the product loses its attractiveness, just as the salt meat of the sailing ships lost its appeal, and it must be thrown away. Because of the salt in it, nothing that walks, swims or flies can eat it.

Only first-quality fish may be used for hard-salting. All five species of Pacific salmon are used, however, or were in the vanished trade of the old days. The red salmon of Alaska's Bristol Bay are the preferred species and these fine fish, freshly caught and quickly processed, turn out a salt salmon of the first order. Butchering, washing, splitting and sliming are essentially the same as in mild-curing. The change in the approach comes with the actual salting. This is done in open tanks of varying capacity. Salt is scattered generously over the bed of the tank and fish are placed in it skin side down. The sides are layered as evenly as possible with substantial quantities of salt between them. The final salting rate is about 30 pounds to each 100 pounds of fish, a much heavier salting than that for mild-cured salmon. To allow for the inevitable shrinkage, the tank is loaded two or three layers above the rim. The final layer is heavily salted. Various methods are used to weigh down the fish to keep them covered by brine as it forms from moisture leached from the fish. It is estimated that from 15 to 20 percent or more of the moisture content of the fish escapes during the first salting. It is important that the fish be well covered at all times by the brine because oxidation causes a discoloration called "rusting." Sides so marked usually are rejected in repacking.

The first salting lasts a minimum of two weeks, but the sides can be left in the brine as long as they are well covered for many weeks or months with no damage. Temperatures should not be excessive, however. After the desired period of salting has elapsed, the sides must be repacked. This is done in barrels of 200 pounds net weight exclusive of brine. As the fish comes from the salting tanks, it is cleaned of all foreign matter and graded according to species, color and quality. The sides are packed flesh side up except for the top layer. About 10 pounds of salt is scattered on the bottom of the barrel, between the layers and over the last layer although more may be used if the curer is not satisfied with the initial salting or if the pack is to be shipped into warmer climates. After heading, the barrels are topped off with 100 percent brine. As with the newly-laden tierces, these barrels must be checked for leaks and consequent addition of brine.

Hard-salted salmon, properly cured and properly kept, may be considered the superior of most other salt fish, especially certain types of salt cod. Fat Bristol Bay reds come out the best, but all species of salmon hard-salt well. Freshness and lack of bruising contribute greatly to the looks of the fish. Good salt salmon, soaked out, steamed or poached, teamed with a white sauce and served with boiled potatoes, makes a better country Sunday breakfast than any salt cod. Hard-salted salmon lends itself to a variety of vinegar pickling processes although commercial packs are hard to find because salmon, as a basic material for such pickling, costs too much to compete with inexpensive imported herring products. Most vinegar-pickled salmon is put up by home canners who salt and pickle small quantities for themselves.

## To Smoke A Fish

A corollary of salt fish is smoked fish. Smoked fish is to salt fish as Mutt is to Jeff, as yin is to yang, as politicians are to corruption. Fish cannot be properly smoked if there has not been at least minimal salting. Salt and smoke must be harmoniously blended to produce a decent food item. The world, as a whole, eats more smoked (salted first) herring than it does any other smoked fish. Comparatively little herring goes to the smokery in the United States and Canada because the Europeans have been at it longer and do it better and more cheaply than Yanks and Canadians. Scottish kippers, dotted with butter and quick-broiled, make a noble meal despite the myriad tiny bones.

More salmon is smoked in the United States and Canada than any other fish with an annual production of 15 million pounds or more. Lake chub from the Great Lakes is next. From the northeastern Pacific, sablefish is the second species in volume of smoking although there are small lots of halibut, herring, mackerel and others put up commercially. Too often, smoked sablefish is masqueraded through the fish markets as finnan haddie. This is a libel on both these fishes since each is a fine fish on its own and it is a disservice to consumers who may find themselves paying import prices for a hometown fish, so to speak. Finnan haddie is smoked haddock, its name derived from haddock first prepared or at least first made popular in the Scottish fishing port, Findhorn. Much finnan haddie comes from New England, Canada's Maritime provinces and Newfoundland, but the Scots variety is a better product than the North American. Much of the time, unfortunately, it is not available on the West Coast.

Mild-cured salmon makes up only about a third of the salmon smoked in the United States and Canada each year, mostly because the supply of red king salmon going into mild-cure cannot be stretched to make up the 15 million pounds or so of smoked turned out annually. Thus much smoked salmon consists of quickly-salted fish, held in brine for a few days or a few hours before going through the smoke process. Such salmon is believed by some students of this esoterica to be superior in quality to that derived from mild-cured salmon held in its brine for several months. The price for first-quality smoked salmon, mild-cured or quick-cured, comes out the same in the retail market despite any infinitesimal differences in flavor or texture.

## Fine Art

The smoking of fish is a flexible process, one that can be done in a ventilated barrel in the backyard or in a fully-automatic operation in a three- or four-story building on a busy waterfront. Like mild-curing, fish smoking is a process open to as many interpretations as there are people doing the smoking. Only on one matter, that

of the woods to use for smoke, is there general agreement. That accord holds that only hardwoods can be used for smoking salmon or any other fish. The resinous woods such as the evergreens transmit their pitchy taste and greasy smoke to the fish. Alder is the wood most used in the Pacific Northwest and British Columbia although some smokers like maple and apple wood. The latter is the wood most used for smoking oysters. No matter the method, smoked salmon that turns out right is one of the finest of sea foods.

Home smoking of fish often is a catch-as-catch-can affair with luck determining the outcome much of the time. Smoking demands attention around the clock and it cannot be done successfully on an off-again on-again basis. Fish must be prepared in the right way for smoking and fires must be tended by day and night. Generally, only the commercial operator has the ability and the personnel to deliver most of the time (not always) a consistently fair product. The commercial, however, works with large quantities of fish and some of the time the fish is not of even quality or too much salt remains or something else goes wrong and once in awhile the most experienced professional finds himself with a batch of smoked fish that is not as good as it might be. Consistent high quality comes most often from small outfits, usually with management and staff encompassed in the same family or among the in-laws, persons intimately involved in an operation big enough to afford first-line equipment but small enough to inspect every fish with a cold eye.

Smokeries must allow for a number of near-intangibles in their design and method of working. By the time the fish go into the smoke chamber it is too late to do anything much about the original quality of the fish and the smoker must concentrate on working with this stock to turn out the type of smoked fish the market wants. His tools are smoke itself, temperature and humidity, factors vitally important in determining what kind of fish the batch working turns out to be. Market tastes differ, mostly according to geography. The West Coast seems to prefer a heavy smoke on its salmon with a bit of a salt flavor. The people of New York who eat smoked salmon seem to like a lighter smoke with less of the salt tang. Smoked salmon and other cured fish are not as popular in mid-continent as they are on the coasts except for such population centers as Chicago. These tastes are as variable, however, as all other matters involving fish and people.

Fish to be smoked must be salted either on a long-term basis for holding until it can be smoke-cured or it may be salted lightly and quickly for flavor and firmness. Fish rather heavily salted such as mild-cure salmon (and this discussion will be confined to salmon) needs lengthy soaking to rid it of the excess salt. The sides may be soaked in several changes of cold water for as many hours as needed until it tastes sufficiently salt free. The degree of retention of salt is a matter of taste and what may be the right amount of salt for one taster may be entirely too much for another. Most smokers go about it conservatively and soak out rather more salt than they should in order to please the most tastes. The de-salting process may be speeded by holding the fish in running water. In any event, mild-cure salmon needs from eight to up to 16 hours to free itself of salt, and hard-salted salmon, when it is used, needs even more time. Fish that have been lightly brined for four to six hours need only to be washed to free them of clinging salt.

After the necessary soaking, the sides must be lightly scrubbed to remove all traces of salt or any other material remaining after all the handling and inspection it has experienced. Smoked fish will keep under optimum conditions in direct proportion to the removal of moisture from the flesh. The home smoker does his drying usually under a tree in the backyard where the breeze may or may not do the job properly. The commercial operator cannot afford to gamble on the vagaries of the weather. He dries his fish by fans as they hang in the smokehouse before the smoke is turned in or he may leave his vents open and dry the fish over a clear fire for a day or two before getting down to smoking itself. Drying under the right conditions aids also in formation of the "pellicle," the bright, glossy appearance of the flesh that is one of the attractions of good smoked salmon. The pellicle does more than make the fish look good though because it aids in absorption of the smoke and acts as something of a protection to the delicate flesh beneath when the smoking is finished.

Salmon may be "hot" smoked or "cold" smoked. Salmon intended for distribution over a considerable area usually is cold-smoked since it keeps better than the hot-smoke product. The keeping qualities of any smoked fish are poor except under refrigeration or freezing. Hot smoking cooks as well as smokes and most commonly is used on the Pacific Coast for kippering salmon, halibut or sablefish. Any hot-smoked fish product is highly perishable unless refrigerated or frozen.

Fish smoking may last anywhere from three or four hours to three or four weeks, in extreme instances. The three- or four-hour treatment is mostly for salmon to be canned by conventional methods after the light smoke. This process is extremely popular in Pacific areas where salmon are marketed in abundance in one form or another the year around. It results in a flavorful, appealing product preferred for party treats rather than for such mundane affairs as salmon loaf.

Some salmon is hot-smoked for 10 to 12 or 14 hours with two-thirds of the time spent at 90 degrees and the remainder at around 150 degrees to achieve the cooking effect. But most salmon is cold smoked, the process that takes time and skill. It ties up a smokehouse or a part of one for days or weeks and it is another of the reasons that good smoked salmon is expensive. Cold smoking at about 75 degrees, carried on long enough, turns out a dry, dark, firm side of fish that can be held for 10 days or two weeks without refrigeration if air temperature is not excessive. It is long-time cold-smoking that brings out the true skill of the smoker.

Salmon sides, dried and hanging in the smokehouse, need a dense smoke for proper curing. The home smoker most probably will use wood up to fireplace size to get his smoke. Such pieces of wood do make excellent fireplace fires but they are not especially good for fish smoking. They turn out more heat than smoke if they are dry and they are difficult to keep low and steady if they are green wet wood. Coarse sawdust is preferable for smoking since it can be counted on to smoulder steadily and dependably, throwing off smoke as thick as the clouds navies use to hide their ships. The presence of a commercial smokery anywhere is hard to hide. The fragrant wood smoke ranges far down the wind, unmistakably the smoke from a clean fire, not industrial smoke heavy with pollution. Everything in and around a smokery attests to a location close to the smoke. The cartons in which fish leave the premises betray their origin long after the last of the fish has been consumed and forgotten.

After the long smoke, the fish ordinarily is allowed to hang in the smokehouse until it cools enough to be handled. This cooling period hardens the sides a bit more if the smoke has not been lengthy. Most establishments oil the sides lightly after cooling to give the flesh side a shiny effect.

## Poor Man's Smoking

Kippered salmon has been called the "poor man's" smoked salmon. Semantics aside, good kippered salmon does not cost as much as smoked salmon but there are many persons who hold with some justification that kippered salmon is as good and perhaps better than smoked salmon. Kippered salmon is a hot-smoked, quick-to-spoil product that must be kept well refrigerated or frozen. It should go into the cooler as soon as it is taken out of the smoke chamber and cooled.

Any kind of salmon may be kippered but white king salmon and fall-run chum salmon are the species most commonly kipper processed. First-quality kippered salmon consists of choice, evenly-shaped and sized pieces of the side. But odd pieces are extensively used too, turning into human use perfectly good salmon that otherwise might be discarded entirely or made into animal food or meal. These are such portions as the collar tips or tail chunks with excellent flesh on them but not marketable otherwise except, perhaps, in some low-income areas.

Salmon kippering utilizes short-time salting. The fish, after washing, is brined up to three hours in a 90-95 percent solution, the time depending on the size and thickness of the pieces. A characteristic of kippered salmon is its surface color of dark orange or crimson. This is obtained by dipping the pieces into a vat of dye for up to 30 seconds after they have been drained of excess brine. Several food colorings are used but the commonest is 150 orange 1, an approved food coloring.

Wire racks or large-mesh screens are used to hold the fish pieces over the smoke. The fish is smoked for three to 12 hours at an original temperature of 100 degrees. Toward the end of the cure, the temperature is raised to about 180 degrees for cooking.

Hot-smoked or kippered fish products are intended to be moist when processing is completed while cold- or hard-smoked fish intentionally is comparatively dry. Even under the best of procedures, kippered salmon, halibut and sablefish experience a significant degree of dehydration during smoking. The product suffers qualitatively and the processor suffers economically because dehydration means loss of weight. The Seattle Technological Laboratory of the Bureau of Commercial Fisheries, located in a city that does, perhaps, more smoking and kippering than any other, has done extensive work on this problem and in 1969 announced an apparently acceptable method of cutting this moisture loss by about six percent. The method essentially involves addition of sodium tripolyphosphate to the regular brining solution. The best results after considerable experimentation were achieved with a two percent solution of sodium tripolyphosphate added to the brine.

Most shellfish smoking on the West Coast is done by home processors who have the patience to toy with all manner of clams and other mollusks to get a good pack. There is a modest amount of oyster smoking, principally in Washington State, and the product is of high quality although perhaps not as attractive to the eye as the Japanese pack. The Americans use Pacific oysters that often must be cut into two or more pieces for the four-ounce containers used in the trade. The Japanese export glass and flat metal containers of the same weight but use tiny whole clams and oysters for their pack. Americans cannot compete with the Japanese effort because of the great amount of hand labor needed for such processing. There is a growing United States market for other exotics like smoked squid, octopus and whale meat as the American taste becomes more cosmopolitan but American processors offer no competition whatsoever to Japanese dominance in this field. The Japanese have been doing it longer and they do it better at prices neither Americans, Canadians nor Europeans can meet even if they were inclined to try it.

For smoking, fresh oysters in the shell are steamed for 15 minutes at atmospheric pressure or for five minutes in a retort at one or two pounds. This method is preferred over pre-shucking and steaming because the oysters open by themselves and the heat plumps and firms the meat to a point where they are better suited for handling and for absorbing smoke. When the meats have been removed from the shell and cut to size, if this is needed, they are washed and brined for five to eight minutes in a 60- to 70-percent brine. After brining, they are rinsed in cold water, drained and arranged on oiled screens for smoking. The smoke chamber is preheated to 110 degrees and smoking takes place at this temperature for about two hours. Chamber temperature then is raised to 140 degrees and smoking is continued for 30 to 60 minutes in accordance with the degree of smoke desired. After the freshly-smoked oysters are cooled, they are canned by standard canning techniques.

Robert Browning

*Modern king crab vessel loads her pots at pier at Fishermen's Terminal, Seattle, Washington, prior to departure for opening of early crab season in the Bering Sea.*

National Fisherman

*Full tuna cans move along line to get tops. Station in center foreground is quality control checkpoint. Helmeted man at right moves tops into position to go into topping machine.*

# 8 / The Canning of Seafoods

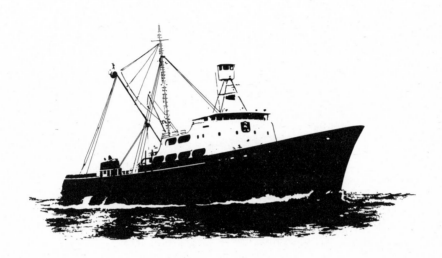

Within a period of 20 years almost two centuries ago, a restless France, shaking off the heritage of the Middle Ages, gave a stirring freedom slogan, an excellent method of beheading men, and the basis of the canning industry to the rest of the world. In 1789, the poor people of Paris cried "Liberty, equality, fraternity!" and worked the guillotine overtime to rid themselves of The Establishment of that day. In 1809, Napoleon, who had no use whatsoever for freedom slogans, unwittingly gave the world the secret of canning foods. Napoleon just wanted to carry more food and less bulk in his supply trains so his armies could move faster, and the imperial dictate of 1800 took almost a decade to produce results. It is uncertain just how much Napoleon himself benefited by the canning of foods because his empire collapsed as quickly as it had been built. Napoleon was soon gone and so was his Imperial France but "Liberty, equality, fraternity!" are still real; the guillotine works as smoothly as it ever did, and the world's canning industry cans more and more of everything every year.

Whatever the merit of Napoleon's own interest in food canning, its real benefits are more enduring than all the empires of the world. The development of a technologically and economically-practicable way to preserve food began to free a growing number of the world's people from their uneasy dependence on fresh foods or foods laboriously salted or smoked or pickled in relatively small batches. At best, foods so preserved lasted only a short time, weeks or months only for most of them. But canning allowed preservation of fish, meat, vegetables and fruit for an indefinite time, at worst a period much longer than that possible with the older ways.

Canned food is easy to distribute and this ease of moving it from here to there took it wherever there was someone to want it and some kind of transportation to get it there. (This, of course, was what Napoleon had been looking for all along.) Canning saved numberless lives and canning gave new foods, different foods, even exotic foods, to people everywhere, even if it were not a question of saving them from hunger. The canned salmon put up by the Humes and Hapgood on the Sacramento River was a gold star example of this trend in economics. Certainly, about six generations of children across the world have enjoyed canned food treats from elsewhere that they might never have seen otherwise in all their lives. It is hard to claim that canned foods in all cases are as good as the fresh material but a lot of people have worked over a lot of years to make them safe, palatable and nourishing. No one in his right mind really liked World War II C rations but they kept men strong, fit and able to fight. So maybe Napoleon was not too wrong after all.

## No Blueprint

Food canning is not an exact science and it never will be despite more than 160 years experience so far. The principle of canning had been worked out by 1809 and it has no more changed than has the validity of Newton's opinion on gravity. But canning methods and canning machinery have changed and Hume and Hapgood might have considerable trouble adjusting themselves to a modern salmon cannery. This inexactness of the art shows in the finished product and several things may be blamed for this fluctuation of quality from batch to batch. One obviously is unevenness in raw material, not only in fish but in all other foods that are canned. Sweet corn packed a week past its prime quite clearly can not be as good as corn packed on its best day. Salmon 48 hours un-iced out of the water versus salmon six hours un-iced out of the water varies as much as the corn. Management failure to update its machinery and its methods is another cause. A third is management failure to keep on its supervisory people. The skill and experience of the people who oversee actual operations may be the single most important factor in product quality. Slack supervision means slack workmanship and any man who ever has been a drill sergeant can so testify. Alert, detail-sharp supervisors in every key spot insure alert, detail-sharp workers. Skilled labor in important jobs below foreman level is a necessity too. The retort engineer in a salmon cannery may have a book full of instructions on operating his gear. But the book is worth nothing if the retort man cannot understand what the book says and follow what it says in excruciating detail. A breakdown in retorting renders absolutely useless all that has gone before. Of this particular contribution to canning, the Fish and Wildlife Service's invaluable Research Report No. 7 comments:[1]

"A few years ago, little skill was required of the man responsible for processing (retorting) and great latitude was allowed in the operation of the equipment. The processing equipment was comparatively simple and errors due to faulty equipment and the human factor were more frequent than they are now. Fish products were too often regarded as 'fool-proof.' Accurate and detailed processing records were not kept in many instances. As a result, the quality of the pack was not uniform and spoilage occurred directly traceable to preventable faulty processing."

These remarks refer only to one step in the canning procedure. But they are applicable equally to every other.

The series of events between the time a fish is caught and the time its shiny can goes into storage may be long in time and spread over a lot of geography. The potential for error is frightening. The canner of fishery products faces more hazards, perhaps, than the canner of any other food. But only twice since 1920 have fish foods canned in the United States killed people. In all truth, it must be said that the industry has been lucky as well as efficient because fishery products are more subject to unfavorable changes before canning than any other food. Fish are more susceptible to quick bacterial and chemical spoilage than vegetables and meat. Normally, most foods headed for the cannery are on the line within a very few hours after they have been picked or slaughtered (except for meats that must be hung or otherwise processed). In the pea-growing country of Washington State, canning peas are reaped and threshed often by the same machine, transported to the cannery, graded, processed and put into the can in from six to 12 hours, start to finish. But the first tuna taken by a distant-water seiner may lie in the wells of the catcher vessel or a transport for months before it nears the cannery. Salmon, iced or refrigerated, may be on the tender or in shore tanks for a week or more before they go to the Iron Chink. The lightning spoilage of crustaceans and mollusks can be nightmarish.

But all the things bad that can happen to canned seafoods do not happen before they come to the cannery. There are booby traps all along the cannery line—receiving, grading, dressing and washing, preparing for the can, filling the can, sealing, retorting, cooling, storing.

It has been said here a couple of times before that eternal vigilance is the price of fresh fish. It is also the price of good canned fish.

## Species Few

Of all the fishes and shellfish off the coasts of North America, relatively few are canned in any kind of pack. The tunas, salmon, herring, sardines, mackerel, bonito, yellowtail, anchovies, swordfish, shad, crab, lobster, shrimp, clams, oysters, squid and a handful of others—that is about the total.

About 65 percent of all the fish and shellfish canned each year in the United States comes from the four American states along the Pacific Coast. There are some 175 canneries scattered through California, Oregon, Washington and Alaska. These plants are not only the national leaders in volume but they lead also in the number of species packed, all across the list of fish and shellfish from albacore to yellowtail.

Many fish do not take kindly to canning. Mackerel is a delicately-flavored fish as tasty as any when broiled fresh or when it is lightly smoked. But all the skills of canning have not been able to pack mackerel that any more than remotely resembles the original. None of the scores of dry-fleshed rockfishes and flatfishes is canned except for a small specialty pack of halibut.

Canning does best by rather oily fish like the salmons, the herrings, the anchovies and the fishes called sardines. Immense quantities of tuna are canned but

---

1. Jarvis, Norman D., "Principles and Methods in the Canning of Fishery Products," United States Fish and Wildlife Service Report No. 7, 1943.

American canners replace its lost natural oil with vegetable oil. Some tuna actually has little natural oil.

For most of three generations after 1864, salmon was the Number One canning fish in the United States and Canada. Tuna was a fish known only vaguely in those years and until about 1890, the dainty albacore was regarded as a scrap fish and thrown back into the sea. But by the early 1920's, it had become apparent that tuna was a very popular fish indeed and the swing of tuna fishermen away from the unpredictable albacore to the other tunas, already underway, picked up fast. By World War II, tuna had well outstripped salmon on a volume basis and the imbalance still grows every year. In 1970, for example, the United States packed 22.1 million standard cases of tuna, (ranging from 18 to 21 pounds) much of it import material from Japan. In the same year, the entire American salmon pack was only 3.9 million cases at 48 pounds net per case. Salmon has been exploited for many years or so at about the maximum and the total annual pack will never exceed its previous highs unless someone figures out how to raise salmon to order in great numbers.

The North American supply of tuna comes from most of the warm seas. Much albacore is taken along the Pacific Coast from Baja California to Vancouver Island by a mosquito fleet of trollers and small bait boats. Long-range seiners, American and Canadian, fish yellowfin and skipjack on both sides of the Equator from Hawaii eastward to West Africa. The Australians have developed a lively baitboat fishery under American tutelage. The Japanese seine and longline for tuna wherever the fish is to be found, then sell much of their catch to the United States, both canned and raw-frozen.

When the Puretic Power Block and the nylon seine gave new life to the American distant water tuna fishery after the mid-1950's, the larger vessels and longer trips made necessary a refinement of the existing tuna freezer systems. This evolution was not without error and a high rate of spoilage among some fares. The Canadians, seeking a share of the tuna fishery, experienced the same growing pains. Here is an account of the tuna freezer operation and a glimpse of the difficulties experienced by the first Canadian vessels entering the far-from-home tuna fishery:[1]

"In the present American system of brine freezing tuna, vessels are fully tanked; that is the entire cargo space is taken up by many tanks or 'wells'. The procedure of freezing the fish is this. . . long before any fish are received, some wells are filled with clean sea water and some with fully saturated salt brine. The sea water is refrigerated to just about its freezing point and the brine to the same extent—to about 29 degrees and minus five degrees F. respectively. The first fish caught are fed into a tank of chilled sea water until the tank can take no more. The sea water is circulated until the latent heat is removed from the fish, then it is pumped to another tank and cold brine is pumped into the first tank and circulated and refrigerated until the tuna is frozen. Then it is again pumped into a reserve tank for

re-use while the now-dry tank with its load of frozen tuna is kept cold until unloaded by refrigerant circulated through pipe coils on the tank walls.

"A main factor in this concept is that with the tanks full of previously chilled sea water, the ship can continue fishing longer when fish are abundant. A well-designed tanked vessel can accept two or three hundred tons of fish or more in a 24-hour period. This enables her to take advantage of those periods when fishing is good.

"Of prime consideration in the entire fishing industry are the peaks and valleys of production. If a ship or plant, for whatever reason, cannot take advantage of temporary heavy production, it is not likely to be successful. Generally speaking, the fish business is such that an an operator may break even on an average production, lose money in the valleys, and only show a profit by being able to handle the peak periods.

"The Canadians, in designing their new ships, appeared to calculate an average daily production and provided brine tanks to accept only this amount. They figured about 100 tons in 24 hours, and could accept that much in a given day. In these systems, tuna was delivered directly from the seine into a tank of saturated cold brine. Each tank could accept about 50 tons but when they were loaded the fishing was over until the tanks could be pumped out and the frozen tuna transferred to a dry-refrigerated hold below. The main weakness of this idea lies in the ship being out of business while vessels with the traditional systems could accept the same 100 tons and keep fishing while the going was good.

"The fishing industry the world over is full of these failures to take into account the "peaks and valleys" concept. Many of these examples are otherwise representative of the finest in engineering and design technique. The unfortunate part is usually lack of experience in a particular fishery by otherwise capable persons.

"On the other hand, there should certainly be recognition of the Canadian contribution to the betterment of the fishing industry in many ways. They are perhaps willing to gamble a bit and depart from the traditional more than some people and many times this boldness has resulted in improved techniques.

"The U.S. has enjoyed almost unchallenged superiority in production of tuna with the seine boat. Even the Japanese with their knowledge of fishing have been frustrated thus far in their efforts to seine tuna. However, many nations are now entering the industry, and the going is bound to get tougher and tougher for the Americans. As mentioned before, the present seawater chill, then freeze system, is not ideal, and sooner or later will be improved or replaced by someone with the imagination and daring to try new concepts. There is

---

1. Eldon Grimes, manager, fisheries development, Marine Construction and Design Co., Seattle, Washington.

only one absolutely hard and fast rule for the newcomer:

"Watch those peaks and valleys."

## Tuna Packing

Americans can four species of tuna. The greatest volume is yellowfin, *Neothunnus macropterus,* taken in the tropics and not a fish of the Northeastern Pacific; the bluefin, *Thunnus saliens,* the "horse mackerel," the biggest of the tunas; the skipjack, *Katsuwonus pelamis,* the tuna of next resort when Eastern Pacific yellowfin quotas are filled (The Central Pacific potential of skipjack is put at hundreds of thousands of tons annually), and the albacore, *Thunnus alalunga.* The Pacific bonito is canned like tuna but must not be labeled tuna.

The albacore, with its white flesh, is the only tuna that may legally be labeled "white meat" tuna. The others must be called "light meat" although if the shade does not meet federal color standards, it must be called "dark meat" tuna. This dark meat tuna has nothing to do with the dark meat removed from the lateral line of the fish during the canning process. The bluefin tuna and the larger specimens of other tuna tend to have flesh of a darker shade than smaller fish. Tuna is packed as "solid pack," "chunk" and "grated." Nutritive value and flavor are equal among the three, the only difference being in the size of the chunks and flakes.

All species of tuna are canned in various ports along the Pacific coast although the industry is centered in San Pedro and San Diego where most American-caught tuna is landed. San Juan, Puerto Rico, also packs much American tuna, including much from the Pacific and most of that from the Eastern Atlantic. In years when the West Coast albacore run swings well to the north, Astoria, Oregon, receives the greatest tonnage of that species although some is shipped to California plants. Most tuna canneries in Washington do not accept albacore and confine themselves to Japanese tuna. The Washington canneries are small and usually are an adjunct to salmon canning, a reflection of management desire to keep plant and experienced personnel together the year around. The albacore run conflicts with the salmon season, the major reason for being of the companies and their cannery operations.

Most albacore from the northern runs is iced in the round at sea and delivered in that condition to Astoria where it must be frozen until the cannery can handle it. This normally is a few weeks at the very most. There is an entirely different situation with the distant-water seiners, however, ships that reckon their trips in thousands of miles and months at sea within a few degrees of the equator. Obviously were there no miracle of refrigeration, there would be no tuna fishery other than one that could deliver its catch within a week or 10 days after taking the first fish aboard. Dependable freezing allows the catching and keeping of tuna for months in heat that quite literally can drive men mad. All modern tuna vessels must have the just-described highly-sophisticated freezer systems manned by men who have been well-trained in their operation and maintenance. Outside help is a long way off.

Spoilage reportedly runs less then five percent when this freezing procedure is properly pursued. So well-developed has it become that tuna caught by American vessels off West Africa is held until the ship has a load, then is trans-shipped to packer vessels for the long haul back to San Juan or, even, through the Canal to San Diego. Trans-shipment, incidentally, allows a vessel to stay on station for many months at a time, an obvious economic advantage. Supplies come by sea, tuna leaves by sea and crewmen are rotated as their turn comes up.

Customarily, tuna is partially thawed before it is unloaded. Frozen tuna, like any other frozen fish, is glassy and brittle, easily subject to damage from rough handling. Frozen tuna stick together and it would be practically impossible to separate them without this partial thawing over a period of hours or even several days. But the thawing process must be carefully timed so that it is not carried too far too soon. Any number of intangibles can delay unloading and a vessel with its load too far into thawing might well find itself with its wells full of unmarketable fish. (The ease with which frozen tuna may be damaged is quite apparent in much import fish, tuna that have been handled five or six or even more times from the moment they were taken aboard some fishing vessel to the time they come to a cannery thawing area. These fish, or many of them at any rate, little resemble the sleek, agile creatures they once were.)

Tuna, once thawed, must be processed as soon as possible after unloading. There is more margin with tuna that have been only partly thawed. The time for thawing depends on size of the fish, of course, and on the means used to thaw it. This may be done in water tanks, in the air or by water sprays directed at tuna lying in the air. A small fish can be thawed in water in two or three hours; a big bluefin being air-thawed in cool weather will require 24 hours or more. Thawing must be carefully judged so there will be no chance of spoilage at this critical time.

Tuna are frozen in the round and must be "butchered" before the canning procedure itself gets underway. Since quality loss first becomes apparent around the gills and in the belly cavity of fish, each tuna now gets a personal inspection, one of several it or portions of it will get along the way. The entire load, of course, has been generally appraised by cannery personnel and, in California, by state inspectors before unloading and well by well as unloading takes place. The most critical test comes on the butcher line, however, as the butchers gut each fish and nose-wise inspectors test each fish for any slight hint of off-odor after the belly cavity has been cleaned and hosed out. Reject fish are deducted from the net fare. Such fish are not wasted, however, but go for industrial use. California's inspection of tuna canneries has become a high art. This rig-

National Marine Fisheries Service

*Fishery of the future? Haul of Pacific saury lies in checker of research vessel* John N. Cobb, *taken in first offshore set of experimental 150-foot seine designed specifically for saury. Saury is a welcome food fish in the Orient but is unkown to most Americans.*

orous enforcement of purity standards, coupled with the canneries' own meticulous quality control, results in a product that probably is more uniform than any other fish product put up in the United States.

The canning of salmon is an affair of great simplicity in comparison to the canning of tuna. Salmon, so to speak, simply is washed, cut to size, popped into a can and retorted. Just like that! Not so with tuna. . .tuna follows an almost-labyrinthine path on its way to the can and the retort. Tuna must be cooked twice, once before it is canned, the other during retorting. The tuna pre-cook, so called, is needed in order to put the flesh into condition for the close attention it will get from a line of woman workers who undertake the exacting handwork the fish must receive to prepare it for canning.

After the tuna leave the butcher line, they are loaded belly down (for drainage) into paper-lined wire baskets set on racks some six feet high. The pre-cook takes place in steam rooms of varying size at a temperature of 220 degrees with one or two pounds of pressure. The pre-cook time varies with the size of the fish but, discounting the lag time needed to bring the cookers up to temperature, the pre-cook runs from two to eight hours. This procedure leaches much of the oil from the flesh as well as considerable moisture. Weight loss may run up to 25 percent altogether. After the racks leave the steam chambers, the fish must be allowed to cool else the flesh would crumble away under the busy fingers of

the women. Again, the cooling period varies with the size of the fish and it may run up to 12 hours or more without aid from artificial cooling devices.

## Too Much Handwork

A single tuna plant requires more handwork than do any half-dozen salmon lines and no one yet has designed a machine even remotely first-cousin to the fabled Iron Chink of the salmon industry. The inventor of any such device, able to work tuna with the perception and sensitivity of the human eye and hand, will find himself enshrined alongside such other ingenious souls as Cyrus McCormick, Eli Whitney, Mario Puretic and that benefactor of all men who first brewed beer. Large tuna plants may require upward of 600 women in the step after pre-cooking, the one called cleaning.

The tuna go to the industrious women of the cleaning tables whose job it is to strip the fish of about a quarter or more of their remaining weight. The butchers left head and fins on. These must go. The skin is carefully scraped away and the fish is split along the back, easy enough after the pre-cook but a job that must be done gently. The spine and its adhering rib structure are lifted away and the sides are split the long way to make four "loins" of each fish. The cleaners next remove every slightest fragment of the dark meat found in a vee-shaped layer along the lateral line. (This dark flesh may be seen in salmon and similar fishes but in a shallow layer rather than the deeper vee. Even in salmon, this dark flesh has a taste stronger than that of the light flesh and many persons find it objectionable. The taste is less pronounced in canned salmon than in the fresh fish.)

American technology has never passed up a chance to improve performance and cut down on hand labor and the tuna processing industry has been studied long and hard by engineers from its inception. But it was not until the summer of 1948 that there was a machine introduced that showed quickly it could do away with a substantial percentage of the hand labor.

This was the introduction to the industry of an ingenious device called the Pak-Shaper, designed to fill automatically the tuna cans that had been so laboriously stuffed by hand for years. The Pak-Shaper was followed the next year by the Pak-Former, built by the E. H. Carruthers Co. of Warrenton, Oregon.

The value of the fillers quickly became apparent to tuna industry executives, and almost all American and many foreign tuna canners employ the fillers. The Carruthers firm builds and leases the fillers to the domestic industry and to their foreign subsidiaries,

much as computers are leased by their manufacturers to most users. In certain instances, however, the fillers are sold outright to foreign canners.

The Pak-Shaper first went into operation at the Astoria, Oregon, plant of the Columbia River Packers Association, a conglomerate since absorbed in another of those corporate mergers so common to the fishing industry. This first Pak-Shaper, equipped to handle either four-ounce or seven-ounce cans, was designed to fill solid pack cans by molding the loins under light pressure into a cylinder of uniform density. It cuts off pieces to correct weight and places them in the cans. This single machine does in one minute the work that some 50 handpackers can do in one minute while, at the same time, it cuts chunks to within one-eighth ounce of the proper weight, an accuracy that cannot be achieved by most handpackers.

The Pak-Shaper was designed mainly for filling solid-pack cans but with minor feeding adjustments it can also fill chunk and grated tuna in proper weights into the cans.

The Pak-Former can fill only chunk, grated and pet food cans but it, too, replaces much expensive hand labor. Much refinement of the original designs has

E. H. Carruthers Co.

*Tuna can filler, called Pak-Former, spews out cans filled to weight, thereby doing away with some of the hand work that is so much a part of the tuna industry.*

taken place over the years since the first production fillers went to work. In addition to their saving the costs of hand labor for filling, both machines cut down drastically on the waste of tuna meat. The average return from a tuna is about 45 percent of the round weight. The machine operator can adjust the weight of the cuts as desired.

After the cans move from the filling machines—the solid pack cut to weight, the chunk pack machine meat diced into a proper size, also with weight control—the cans move along the line where one-eighth ounce of fine salt is deposited into each open can. The oil lost during the pre-cook must be replaced for most packs and this is done in the form of vegetable oil injected into each can. Most packs also specify injecton of "vegetable broth" to each can. Some fancy albacore is packed in "spring water," actually a rather tasteless method of canning tuna, while some other tunas are packed in brine, again not an especially flavorful procedure.

Most American packs use the oil additive although the Japanese can vast amounts of tuna in brine. Something like 70 million pounds of this pack is exported annually to the United States and is sold at lower prices than the higher-quality American packs of tuna in oil.

In addition, there are small specialty packs of tuna aimed mostly at a consumer market composed of persons from the Mediterranean rim. These include tuna with a garlic flavor and tuna in olive oil, packs seldom seen except in food shops catering to those of South European and North African origin because they are quite highly flavored, too much so for Americans accustomed to the remarkable blandness of most United States canned foods. It must be suggested that this blandness is forced on processors because of the diverse national market and the inability of consumers to reach any consensus on flavor, strong or mild.

Similarly, other fish specialties, such as the many packs of herring, are sometimes hard to find except in those cities or parts of cities with concentrations of persons of German or Scandinavian ancestry.

One man's fish may be only another man's bellyache.

## Salmon Packing

A salmon cannery, a living, working salmon cannery, must be many things to many persons. To the investor, it means money earned or money lost; to the time/labor man, it is a monumental example of waste because it uses more people than machines; to the sociologist, it must represent a free-booting exploitation of a major marine resource; to the unserious student of history, it can only recall the brothers Hume, their friend, Hapgood, and the little cannery in mid-California from which all others spring.

Salmon canneries come in all sizes and ages and states of repair. There are still-operating canneries scattered from the Columbia to the Yukon that have been turning out their cases of fish for almost four generations. Some of the men of the generation of canners just after the bulk of the industry scattered north from the Columbia are still alive and knowledgeable about the old days although none of them has been close to actual operations for most of a young man's lifetime. Their own old companies, in most cases, have vanished down a corporate maw and the names survive mostly in their own memories. The salmon industry and the rest of the fish packing industry of the Northeastern Pacific have seen more takeovers than the common run of business of any kind.

There are salmon canneries like the giants of Astoria and the Naknek River and there are those like the little, half-pound-flat one-liners of villages scattered from the Kuskokwim to Ketchikan. None of these canneries, large or small, is exactly the same in design or accomplishment although their concept is precisely that of the Humes and Hapgood. Nor are there as many canneries as there were when the Depression of the 1930's crunched its way through the world's economy. That melancholy time saw many of the canneries all through the salmon country shut down for the duration or closed up forever. The Depression didn't do all the chipping away, however—some of those the Depression missed fell victim to mergers or to improved transportation or better techniques and they were boarded up too. As nature abhors a vacuum, business abhors duplicating facilities.

In season, salmon canneries do not work eight-hour shifts five days a week and call it good. When salmon are coming in, they must be packed and people who get into the business know that and accept it and do not quarrel with it. If they must, the canneries work around the clock and if they just cannot handle all the fish, the fishermen are put on limits. This gives them a chance to rest and patch up their gear and grumble about the limits while they are patching. Limits become crucial in such places as Bristol Bay during the big years when limit flags fly day after day and too many fish escape to the spawning ground because there is not sufficient cannery capacity to take all the gill-netters can catch. There is not much point in expanding plant capacity because most of the time there are two or three lean years for every good year and even the existing capacity is too much then. In 1965 and 1966 when limits were on most of the time after July 1, the canneries had so much salmon in their bins that it seemed for awhile that all of it would never get canned. In 1967 and 1968, there were hardly enough reds along the whole Bay to get the machinery properly warmed up before the season was over.

During the bonanza periods the hours and days get so long that the most willing shoreside workers begin to slow down and their efficiency dips along with the buildup of their fatigue. It is then that more things than usual go wrong along the canning lines and the foremen shut down the lines oftener and oftener for coffee and a smoke. The foremen are as bushed as their people and they show it in the tight faces and shortened tempers. But the work stumbles on and eventually

Marine Construction and Design Company

*Modern trawl winches . . . these are built for the shrimp fishery and its use of twin trawls.*
*One man controls both trawls from controls at center.*

Pacific Fisherman

*Tuna seine winch. Its size may be judged by man at left.*

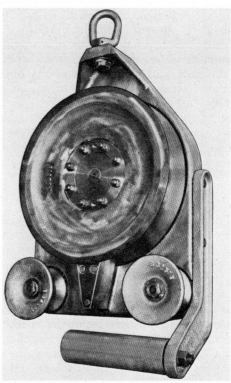

Marine Construction and Design Company
*Block used in crab fisheries.*

the bins are empty and all hands get a chance for a few hours sleep before it is time to do the whole bit all over again.

Supplies may begin to give out in the big years too, something that might have been a disaster in the old days despite all the trading round and lending back and forth among the cannery men. Now everything from cans to booze comes up by jet from Seattle in five hours or less. It may take a couple of hours longer to round up a light plane to fly a needed part from the big-plane strip at King Salmon to a place like Egegik 40 straight-line miles southwest with no roads across the tundra. Jets or no jets, a handful of ships, a fleet of power scows and a growing barge fleet handle 98 percent of cannery supplies for all of Alaska. Climate makes supply a critical problem in places like the Bay and to the north where the ice comes early in the fall and leaves late in the spring. Cannery superintendents and their bookkeepers, their Men Friday, must project a list of requirements—food, fuel, lumber, machinery—for the next season and have it in the head office early enough in August so that most can be ordered and loaded out on the last trips from Seattle in late September. As for Bristol Bay, nature was doubly unkind to the country when the bay was gouged out by the glacial ice. The ice didn't dig deeply enough and all Bristol Bay ports and ports to the north, if they can be called ports, have freight-handling problems peculiar to themselves. There are no harbors, there are no piers, the waters are shallow, the sands are shifty, the wind is shiftier and large vessels must lie miles offshore, six at least off the Naknek, and lighter their cargoes ashore, some of the time onto the beach itself.

## Convenience Vs. Comfort

Most Pacific salmon is canned in British Columbia and Alaska although the Columbia River and Puget Sound can lay claim to a modest share of the total pack. Except for these two areas and the mouth of British Columbia's Fraser River, most canneries are isolated, hours by water, minutes by air, from the next village, enclaves sufficient to themselves for most of the year. When the canneries were built, a usable site close to the salmon was regarded as of more importance than the amenities of the bright-light areas. As often as not, the people came after the cannery was built. Sometimes there are no people at all after the season and through the winter but the watchman and his wife. Much of the time, labor must be brought into the cannery because there are not enough people around it to work it properly. In the early days, Chinese were the good right arm of the labor force, manning the lines to butcher and clean the fish efficiently, patiently and silently at a small wage per hour. Not for nothing did the new-fangled gutting and cleaning machine of the early 1900's get called the Iron Chink.

Nowadays, the political emphasis on cannery labor in Alaska is on employment of the locals, both native and Caucasian. Almost inevitably, the skilled workmen in most canneries are Caucasians, many of them brought in from Seattle. These are the men that make the aging and creaky machinery of the canneries keep going through the season. Most members of the native ethnic groups have no such skills although there is a variety of programs trying with some success to teach these and other skills. But mostly they are the men and women who work the canning lines as silently and patiently as their Asian predecessors although at decent rates of pay and with working conditions as good as possible in sometimes-primitive surroundings.

All five species of Pacific salmon are canned. According to cycle years, reds and pinks take turns for the top volume. These are the smallest and most prolific of the salmons. The value of reds, can for can, outweighs that of the pinks. Nevertheless, in good pink years, the value of the pink pack is higher than that of the reds because of the volume of the pink catch. In Alaska in 1967, for example, 53 million pounds of red salmon were given a wholesale value of $29 million. That same year, 28 million pounds of pinks were worth only $10 million. But the next year, 160 million pounds of pinks were valued at $59 million while 51 million pounds of reds were worth $28 million. Chum salmon usually come next in poundage with silvers and kings trailing. Comparatively little of the king salmon catch is canned because it is worth more for mild-curing and smoking or for the fresh market. Most of the canning of kings is done in Western Alaska from Bristol Bay to the Yukon.

Although the Pacific salmons, as a genus, spread widely over the North Pacific as immature, feeding fish, their migration routes to their spawning grounds tend to follow rather narrow salt-water lanes after they come inshore and before they come to their native rivers. Through all the years salmon have been fished in the Northeastern Pacific, these lines of travel have been learned well by generations of fishermen and packers. This knowledge is exact enough that fishermen can judge pretty well to the week when salmon are likely to appear in a given area. Fisheries scientists have narrowed this even further. They are able to subdivide runs into individual races and predict with some degree of certainty almost the day when the progeny of a cycle-spawning on some creek away in the mountains will themselves appear on their own migration back to the natal creek. It is this knowledge that the International Pacific Salmon Fisheries Commission, among such agencies, uses with finesse to regulate the Fraser River sockeye and pink salmon catch in United States and Canadian waters.

The packers, too, used this knowledge of the habits of migrating salmon to build their plants as closely as possible, other requirements being met, to the point of origin of their major supply of fish. It was pretty hard, of course, to miss on the Columbia or the Yukon or the Kuskokwim (the only major rivers in all the salmon area where river fishing has not been prohibited for

70 years or more) although the runs did not always appear in the same vast numbers every year. But it took only a handful of seasons elsewhere to learn where the salmon ran, this learned easily with the help of the native peoples of British Columbia and Alaska who had been fishing those runs for quite a few years by the time the first white commercials came along.

Thus, in the first decades of  the salmon canning era, most of the fish came from the water only a few wiggles away from the cannery piers. This cut operating costs for the packers and added to the freshness of the fish in a day when absolute freshness was not as highly regarded as it is now. This does not mean that anyone intentionally packed demonstrably spoiled fish but quality control was not the fetish it is today and no one questions that a lot of salmon with a degree of deterioration that would not be acceptable today did get onto the canning lines.

(No one fooled John N. Cobb and of the practices of some salmon packers in a long-ago era, he wrote along about the middle of World War I:[1]

("If the salmon have been in the scows for from 20 to 24 hours, they are used as soon as possible after being delivered at the cannery; otherwise that length of time is usually allowed to lapse (sic), the cannerymen claiming that if not allowed to shrink, the fish will be in such condition that when packed much juice will be formed and lightweight cans will be produced. *The danger of canning fish that are too fresh, however, is of minimum importance as compared with the tendency in the other direction.")*

The freshness of cannery salmon in those early days was due in substantial degree to a device that today is politically unacceptable. This is or was the salmon trap or pound net, its name in the East. Many or even most of the fish, especially in Southeastern Alaska and on the Columbia and Puget Sound were taken in traps where they could be held alive for two or three weeks or even more. When a cannery needed fish, tenders were ordered out to the company traps (or some other company's traps if the night were exceptionally dark and the cannery superintendent exceptionally eager) and the needed salmon were brailed out with a minimum of body damage and were canned and cased before anyone could get around to questioning their origin. Traps, whatever their political faults, did catch fish more economically with less bruising than today's seines, gillnets and troll gear. Traps were condemned most loudly by those who owned no traps and had little expectation ever of owning traps. Except for two or three Indian traps in Washington and Alaska, the traps are gone everywhere now and all salmon are taken by troll gear, seines and drift and set gillnets. This arrangement gives Everyman a chance to fish and there are almost more fishermen now than there are salmon to accommodate them.

Cannery salmon obviously are doing no one, fisherman or packer, any good whatsoever when they are lying aboard a fishboat somewhere. They must be gotten to the cannery. Fishing vessels may carry their catch to the cannery under special circumstances but most commonly, fish are moved from the fishing ground to the cannery by a company-owned or chartered tender. The name "tender" covers a variety of craft because anything that can haul fish can be used as a tender. The most common tender is a fishing vessel too old or too big for salmon fishing. The "Alaska limit" on purse seiners of 58 feet overall eliminates many good vessels from that fishery because of their size but some find employment every year as tenders. Power scows are popular too as tenders, especially on Bristol Bay where they are the vessels of all work. On the Bay, the scows have one extra virtue—those that are to winter there easily can be hauled ashore above winter ice and water. If every cannery salmon were to be treated as it deserves, however, it would move around in nothing but the best of accommodations—the tanks of a refrigerated tender such as those using chilled sea water or brine as their cooling agent. But for every salmon that goes first class aboard one of these, there are a thousand that find nothing better but the hold of an over-limit seine boat earning its keep as a tender.

Most movement of cannery salmon necessarily involves too much handling. The ordinary course of events finds a salmon pewed and brailed from seiner to tender and pewed and brailed from tender to cannery pier where it may be moved to a bin by belt, bucket or even by cart. Modern tenders with semi-automated offloading minimize some of this handling by eliminating at least, the pewing from tender to cannery pier.

One of the great deficiencies of the entire salmon industry is its lack of a simple method of handling salmon under all circumstances. Just as the tuna cannery needs a machine to replace the scores of women cleaning tuna, so does the salmon industry need some way to move salmon all the way from the seine or the gillnet right to the Iron Chink. The handling of such fish as menhaden, herring and sardines is a comparatively simple operation because of their size or lack of it. Pumping systems work well with them. But with one exception, no cannery firm yet possesses, or at least operates, a practical method of handling salmon en masse. This sole, apparent exception is a Canadian-designed pumping system in use at the British Columbia Packers' cannery at Steveston, B.C., either the first or second largest salmon plant in the world, depending upon which Chamber of Commerce is consulted. This Canadian system reportedly is expensive to build and, to operate economically, it must have vast numbers of salmon to work with. The B.C. Packers installation is a one-of-a-kind system, one that can be justified from a financial

---

1. Cobb, John N., "Pacific Salmon Fisheries," United States Bureau of Fisheries Document No. 1092, 1930.

Port of Seattle

*This cluster of buildings makes up a self-contained cannery complex on the Naknek River,
near where this rich red salmon stream flows into Bristol Bay. Low tide has stranded
gillnetters on beach at right center.*

standpoint solely by the size of the plant it supports. Most canneries cannot afford such a system. In too much movement of salmon, it still comes down to pew and brailer.

Two Bristol Bay canneries exemplify the traditional methods of handling salmon in great numbers. Both canneries are owned by the same company although the approach to fish handling at each is entirely different but equally awkward. Yet it is the best these canneries can do, given the present state of development of the "art" of fish handling.

The smaller of these two canneries uses the "huff and puff" method of moving its fish from tender or dock or from the set net boats that swarm around the dock ahead of the low tide. In this traditional method of handling salmon, the fish are hand-loaded into a bucket which then is hoisted to the pier head where they are dumped into chutes to take them to the bins. The sister cannery uses a system somewhat more sophisticated but not much. From the catcher boats, the salmon are hand-loaded, not pewed, aboard scows. At the cannery, a gate is lifted from the side of the tender lying to the dock. Water is streamed onto the vessel's deck from the off side and the crewmen use "pushers," resembling a janitor's broom, to urge the fish along with the water into a hopper feeding a conveyor belt. The fish move along the belt against a stream of water and at the head of the belt, the direction of travel is diverted to individual bins. Both of these systems work but there is the inevitable over-handling of the tender-fleshed fish that, at the moment, seems inescapable.

No cannery superintendent anywhere likes to leave his fish in the holding bins for a moment longer than he must. But there are those occasions when a glut of fish, a breakdown somewhere along the canning line or other circumstance beyond the super's control (although most cannery bosses manage to control every circumstance that confronts them) give him no choice. (It was on one of these occasions, caused by a failure in the Iron Chink, of all things, that chanced to put one Bristol Bay superintendent prowling unhappily along the line of fish bins. At the Number One bin, he decided he didn't like the odor coming from it. It took two hours and the removal of two or three thousand fish before the super decided the bin and its contents were clean enough to suit him. Fish dead three or four days had lodged at the entrance to the discharge chute and had been overlooked by the crew that supposedly had cleaned the bin when it last had been empty.) But when matters are going smoothly enough, the fish are held for only a few hours at most before it is their turn to meet the Iron Chink.

Smith-Berger Corporation

*The Iron Chink, fabled labor saving tool of the salmon industry. This design of the 1970's has little resemblance to the original but it works on the same principle. Fish lose their heads in header at right and are conveyed to the bullring at left where they are finned, eviscerated and washed.*

## Labor Saver

The Iron Chink is one of the products of that engineering bloom in the United States in the latter decades of the 19th Century that produced hundreds of useful tools as well as a plethora of gadgets of doubtful value. Like most innovations in an established field, the Chink was not greeted warmly when it first appeared in a Bellingham, Washington, cannery in 1903. Even in an industry as flexible as salmon canning, there was already a body of sacrosanct tradition although as a fair-size business salmon canning was barely of age. But tradition or not, the men who ran the industry in those days could read the bookkeeper's work as well as could the bookkeeper himself and it did not take most operators more than a couple of seasons to see what the Iron Chink could do for them.

Like many other "labor-saving" devices, the Iron Chink ultimately created more jobs than it took over. All the Chinese in the world standing on each other's shoulders could not have butchered and cleaned salmon as quickly, as smoothly and as cheaply as a couple of dozen Iron Chinks which needed only a little oil to keep them happy. The Iron Chink brought order into an operation that had existed largely in chaos, and allowed for the first time a true assembly line technique in salmon canning. During its first years, the Chink replaced thousands of hand workers but the Chink helped the salmon industry grow. To grow, the industry needed people and, in the end, it put more people to work all along the chain that begins with the fisherman and ends at the retail counter as much as a continent and an ocean or two away.

The Chink is a remarkable machine but it cannot think and it needs advice from time to time on how to conduct itself. To function best, the Chink needs fish of a uniform size. It can be set to handle fish of different sizes but it cannot adjust itself from one fish to the next. But aside from this failing, the Chink does rather well. It takes a fish fed onto its belt and positions it precisely for what is to come. That which is to come include heading, finning, splitting, gutting and cleaning, while at the same time the fish is being cleansed by water jets. Most of this work takes place in the embrace of two big ring gears called "bull rings." These are three feet in diameter with 10-inch faces fitted out with pincers, knives and saws that do the job the Chinese used to do. The fish comes first to the header where a knife flashes down and beheads it neatly. From the header, the fish moves to the bull ring feed table where tail pincers, one in each ring, clutch it hard while it is being pulled around the rings. As the fish moves, it is grasped firmly by body pincers and held with its back against the rings. A circular saw removes the tail fin; one horizontal knife cuts away the belly fins and a second slices off the adipose and dorsal. The next saw opens the belly and a set of guides spreads the belly flaps. The gutting reel spins away the viscera and the blooder reel with its circular saw slits the membrane over the dorsal

*Text continued on page 31?*

# PLATE 51

*Yellowfin and skipjack tuna, frozen and held for months
in freezing wells of tuna seiner, comes ashore at cannery in Puerto Rico.*

# PLATE 52

National Fisherman

National Fisherman

*FIGURES 1 & 2— Two views, above and below, of first work on tuna after coming from the pre-cook room. Fish are headed, finned and skinned here.*

# PLATE 53

National Fisherman

*Camera angle emphasizes length of
the conveyor belt that moves tuna
carcasses along line of workers.*

# PLATE 54

National Fisherman

*Women workers deftly remove dark meat from lateral line of tuna sides as the raw material moves past them.*

# PLATE 55

National Fisherman

FIGURES 1 & 2— *Cleaned meat, left, heads on toward its rendezvous with cans as inspector keeps check. Overall view, below, of cannery interior shows complexity of modern canning operation.*

National Fisherman

# PLATE 56

National Fisherman

*Because the tuna industry has no Iron Chink to do most of the work necessary to prepare fish for canning, the human hand must do it. These workers are only a fraction of the total working in this plant. Some 600 are employed at peak periods.*

# PLATE 57

*Tuna loins glide smoothly toward filling machines.*

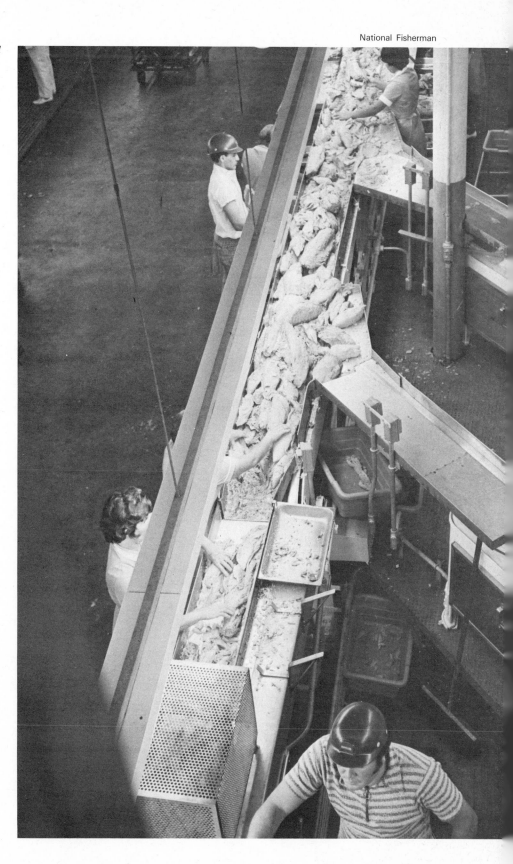

# PLATE 58

National Fisherman

*Overhead view of loins on belt as they near fillers.*

# PLATE 59

*Air view of the mouth of the Naknek River looking westerly
toward Bristol Bay. The tide, subject to great fluctuation, is at the ebb.
The bar at right is well under water at high tide. The larger vessels
are power scows used as tenders. Their shallow draft
fits them well for that job in these waters.*

# PLATE 60

*FIGURE 1— Bristol Bay gillnetters wait out closure, ordered to insure proper escapement of spawners. These boats use power rollers on the transom rather than reels because the system allows faster hauls.*

*FIGURE 2— Freshly netted red salmon, called sockeye elsewhere on the Pacific Coast, are moved from scow to cannery conveyor. Pews are not used in this fishery and sweeps and water streams nudge fish to conveyor.*

# PLATE 61

*Sweeps and water move the reds through exit slot to conveyor at rate of 40,000 an hour.*

# PLATE 62

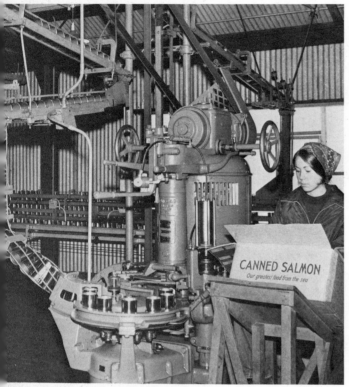

*FIGURE 1—  Can reformer turns metal sheets into familiar containers. Shipping flat saves space and money. Girl feeds bottoms to the machine.*

Port of Seattle

Port of Seattle

*FIGURE 2—  First step toward the market. . . salmon move on belt toward the Iron Chink's header.*

# PLATE 63

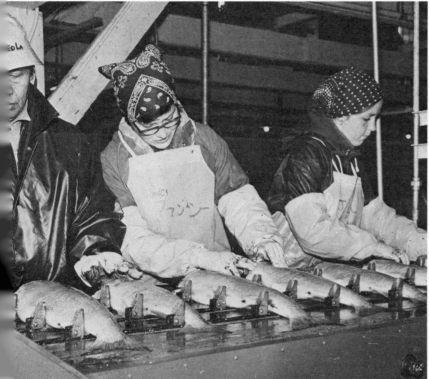

Port of Seattle

FIGURE 2— *Girls sort roe before it goes to its own curing area. This cannery, at Naknek, employs both residents and workers from "Outside." This group came from Seattle for the short season.*

*FIGURE 1— After carcasses leave the header, skeins of roe are removed for curing and export to Japan. Until the mid-1960's, the eggs, with other viscera, were discarded. The roe is as valuable as the flesh in the marketplace.*

Port of Seattle

# PLATE 64

*FIGURE 1— These cans, one-pound talls, move from filler to weigher where short-weight cans are sidetracked to separate line. Tops are placed in position here.*

Port of Seattle

*FIGURE 2— Gaping retort takes freshly packed cans for final processing, the step that makes canning a safe and practical method of preserving food.*

Port of Seattle

# PLATE 65

Port of Seattle

*FIGURE 1— After removal from the retort, the cans travel through a long trough of water for initial cooling.*

Port of Seattle

*FIGURE 2— Final cooling extends over several days in warehouse.*

# PLATE 66

*Workmen, one almost hidden, slide carton over pallet load of cans.*
*The carton, in turn, is sealed in plastic sheeting to protect it*
*against moisture on the ocean trip to the market "Outside."*

artery, cleans out that cavity and, in the final step on the Chink, a brush proportioned to the shape of the belly cavity wipes away the last of the blood and visceral membranes. But the fish is still not ready for the can.

There has been a small amount of hand labor before the fish got to the Iron Chink and there is more hand labor after the Chink. From that instrument, the fish travels along the belt where it is given another wash and a last inspection. People are here. The belt carries the fish then to a set of guillotine-like gang knives that cut it into can-size pieces, normally one-pound tall or half-pound flat. People check here. A filling machine eases the chunks of fish into the can and another machine adds the right amount of salt. The cans are weighed and those lacking are sidetracked to a secondary belt where women drop in a bit of fish to make the weight and the cans are sent back to the main line. The cans are vacuum-sealed with more people watching over and eventually they are racked and moved by fork lift to the retorts. Iron Chink, or no Iron Chink, it takes people to make the cannery line work right.

To get a one-pound can of salmon, the familiar Number One tall, the canner puts in 16.6 ounces of fish and one-quarter of an ounce of salt. Most fishery products are packed overweight to assure that the net given on the label is a minimum weight. Lesser weights than those on the label cause trouble with authorities at all levels who, peculiarly enough, insist the customer get what he is paying for.

The customer pays for a classic of simplicity among all the ways of canning fish and fish products when he buys salmon. There are no additives other than the salt. There are no added brine, oil, sauce, soup, acid, broth or preservatives. The contents of a can of salmon is salmon and salt only. Retorting releases enough oil and moisture from the flesh to amount to a couple of tablespoonsful if it is drained from the can. This secreted liquid is tasty in itself and should not be discarded if the salmon is to be used in such further cookery as salmon loaf, creamed salmon or in any of the casserole dishes using salmon. Canned salmon, particularly the bright, solid red salmon of Alaska and the Fraser River, is a meal in itself, especially when chilled and eaten with crackers or bread and butter. Salmon freshly canned is not especially appealing, however. Like many foods and beverages it improves with age, at least for a year or so. Newly-canned salmon is bland and comparatively tasteless. But in a couple of months or so, the salt permeates the flesh, the natural flavor of the fish asserts itself and a can of top-quality salmon is as good as the fresh fish and, to some tastes, even better. To repeat, the fresher the fish, the better the canned product. Everything bad that can happen to a fish shows up clearly when it is turned out of a can.

This truth is re-asserted every year, coldly and forcefully. Once each winter, the Northwest Laboratory of the National Canners Association conducts a ritual known as a "salmon cutting." It is akin, as far as the participating canners are concerned (and this includes 90 percent of the American segment of the industry) to the running of a particularly fearsome gantlet. The sole consolation is that every man is among his peers and his peers may be hurting too. The canners' chapter has its headquarters and laboratory in Seattle and the cutting is held there, usually in the ballroom of a hotel where there is plenty of room to display the merchandise and handle the crowd. The "merchandise" in this case consists of unlabeled pack samples chosen by the packers but displayed and identified by code only. Not even the packer knows which cans are his. The opened cans are judged by a panel of experts. This panel grades hard on every pertinent point and no favors are given. Men whose lifetimes have been given to the canning of salmon grow visibly uneasy as the morning wears on and the painstaking experts slash away with their point cards. As each offering is graded, the card is left with it and by the time the inspection is over and the canners are free to learn which cans are theirs, some always wish they had not bothered. This public exposition of the virtues and faults of canning practices may be accounted one of the reasons the salmon industry manages to do a bit better every season.

## Salmon Roe

Perhaps the most important development in the salmon packing industry since the acceptance of the Iron Chink has been the growing use of salmon by-products that formerly went into the discard as "gurry" or were diverted to reduction. This trend became apparent in the mid-1950's and its growth has been steady since then.

Chief among these has been the processing of salmon roe for export to Japan where it is a popular food item. The growth of the industry was described succinctly in the publication, *ALASKA*® magazine, in June, 1970:[1]

"Salmon roe, long considered a relatively worthless by-product of Alaska's salmon industry, now returns millions of dollars to the state from the Japanese market. Roe is a staple of the Japanese diet. Each year before this market was discovered, Alaska's salmon fishermen and processors dumped thousands of tons of brightly-colored little salmon eggs over the side along with other waste material,. . .as cured roe, the eggs have a wholesale value (at the writing) of about $1.20 a pound.

"Nearly $8.7 million worth of salmon eggs were produced in the state during the 1968 fishing season. . . compare that figure with the wholesale value of Alaska's halibut catch for 1968—$3.3 million.

"Prior to 1965, most of the salmon eggs processed in Alaska were used for bait. By 1968, the majority of salmon eggs throughout the state were being processed as roe and sent to the Japanese market. Production

---

1. McNicholas, Laurie, "Red Gold Harvest," *ALASKA*® magazine, June, 1970.

rose from 428,430 pounds (mostly bait) in 1962 with a value of $125,830 wholesale.

"Most of Alaska's salmon processors contract with Japanese firms to produce roe. The processor supplies the eggs, facilities and labor force while the Japanese provide technicians and handle distribution of the roe. Salmon roe processing in Alaska has created many new jobs. . .the operation involves three or four Japanese technicians and a crew of as many as 30 workers,. . .by 1972 there were some 70 such operations in the state.

"Preparation of salmon roe proceeds thus in most Alaska plants:

"Skeins of eggs, stripped from the salmon as they are being cleaned, travel along a conveyor belt (separate from the main canning lines) and accumulate in large plastic baskets. Eggs are emptied from the baskets at intervals into vats of saturated salt solution (100 percent brine) where they are stirred for 20 minutes or more. After draining, the eggs are sorted into four grades according to their size and the condition of the skein. . .chums produce the eggs most valued for roe, followed in order by silvers, reds, kings and pinks. The roe is then packed in 22-pound boxes, given a final inspection and sprinkled with a layer of salt. Boxes are stored for several days with lightly-tacked lids to allow the roe to cure, then sealed and placed in coolers to await shipment to Japan."

In 1970, the Japanese took delivery of about 6,000 metric tons of the precious roe.

There is some indication that an increasing amount of the salmon will find other, human-consumption uses, not necessarily for the United States market. Hearts and livers of salmon are a distinct possibility for such use while there has been speculation that the milt of the male fish may find such use also. It is possible that these "by-products" eventually will be of more value than the flesh of the salmon.

### Herring Roe

The Japanese also are having a hand in another growing Alaska specialty food industry, the processing of herring roe for consumption in Japan and for a limited re-export trade with other Asian nations.

About 500,000 pounds of processed roe, worth more than $650,000 wholesale, has been shipped annually to Japan from eight or 10 Alaska processors concentrated in the Seward-Homer-Kachemak Bay area.

Herring roe (not to be confused with the herring-roe-on-kelp fishery) produces a pungent product that most Americans and Canadians do not favor unless they happen also to have a taste for such strongly-flavored foods as limburger and others of the cheeses, over-aged meats and long-hung fowl.

Herring to be used in roe processing are taken in purse seines and aged in an open-top tank for five to seven days, a process that produces a certain amount of odor since they are not in brine but cure in their own blood and in the water that goes incidentally into the tank with them.

The roe is removed from the aged herring along a line employing women workers. The fish are hand-broken behind the head, then a squeezing, shaking motion is employed to extract the roe from the carcass. The ripeness of the fish when it is caught is critical (as it is with other human-use production of herring) because the stage of maturity influences to a great extent the efficiency of removal of the roe and the final quality of the pack.

The next step sees the roe dumped from the holding baskets on the processing line into a brine solution. The roe gets three soaks of from three to six hours each in three salt solutions. The first two have the salt content of sea water while the final soak is in a 100 percent brine.

After removal and draining, broken and discolored skeins are discarded and the remainder is packed in plastic-lined cartons for shipment to Japan. Each carton contains 120 pounds of roe and 18 pounds of salt.

### The Freezing And Canning Of Pacific Ocean Crabs

The freshness of seafoods, be they fish or shellfish, is of the highest importance no matter whether they are to go to the fresh market or are to be canned or frozen. Shellfish, the crustaceans and mollusks, present a special challenge to the expertise of the operator because of the speed with which they can succumb to all the ills that may afflict them in the hours or days between water and can or freezer. This susceptibility to spoilage is even more acute with the crabs than with such creatures as oysters which may survive for days with the shell tightly closed in reasonably cool surroundings. Everyone concerned with crab—from fisherman to cannery superintendent—must assure himself that all crab are alive and in a mood to use their claws when they come into the processing plant.

A crab still ready to fight when it meets the butcher is a crab in good condition, a crab that will yield the maximum amount of high-quality meat. Dead or even excessively weak crabs suffer changes in the tone of the flesh that result, at the very best, in a low-quality product, and, at worst, may cause sometimes-serious illness to the consumer. Thus, the handling of crab offers a peculiar problem to all hands and the responsibility must be divided equally. The fisherman's concern, of course, centers around the refusal of processors to knowingly accept dead crab, a refusal that makes dead crab in his tanks literally just dead weight. And it must be the processor's task to make certain that the men who do the actual work connected with unloading and acceptance of crab do not allow crab of dubious quality to get onto his lines. This applies equally to all crabs of whatever species.

No one expects fin fish to come to the cannery alive and wiggling. It is otherwise with crab and to produce live, strong crab means the use of live tanks aboard the catcher vessel and at the pier where they will be sold. Such holding tanks have been in use in the New England lobster fishery through all of this century but it was not much more than 40 years ago that the universal need for them was recognized in the Pacific crab fishery, at that time concentrated almost entirely on the Dungeness crab. There was never any question about the use of crab tanks afloat and ashore when the American king crab industry began its first growth after World War II.

The use of live tanks does not mean necessarily that all crab put into them will come out alive and strong. Excessive loading in the tanks means death or severe weakness for the crab at the bottom of the heap whether this be on the boat or at the plant. Bad weather may produce sea conditions that cause the vessel to roll or pitch heavily. Such vessel motion does crab in the tanks no particular good either while the boat pitches and rolls around the static water in her tanks. Water changes must be made in cycles of 20 to 25 minutes else the crab will die in their own wastes. All crab boats of any merit carry backup pump systems to take over the water circulation chore if the primary system breaks down.

There are three crabs of particular concern to the fishing industry of the Northeastern Pacific. These are the Dungeness crab, the familiar market crab of seacoast cities all the way from San Diego to Kodiak; the two or three species of large crabs known collectively as "king crab," and the least-known and still least-utilized species, the tanner crab, a crab smaller than the king crab but much larger than the relatively small Dungeness.

When the Americans began to get into king crab fishing in a big way, they were forced almost entirely to learn on their own how best to can and freeze this species. The Japanese had been using king crab since 1892 and by the mid-1920's, their floating processors were working as far east as the mouth of Bristol Bay. Before World War II, the United States was importing about 10 million pounds of canned king crab each year from Japan and the Soviet Union. The war put an end to that trade, of course, and by the time the Japanese were ready to try a comeback in the North American market, the Americans were producing about all the king crab the market could handle. The Japanese characteristically were remarkably coy about imparting to the Americans any of their processing secrets but in the end it didn't make any difference anyway because by about 1955, the Yanks were doing everything with king crab as well as the Japanese had ever done and in many cases, the product was superior.

Canning crab is a trickier affair than freezing it successfully and the Americans made a virtue out of necessity by freezing most of the catch during the first two decades of the fishery. King crab freezes better than other crabs (although much of the Dungeness catch of the lower Pacific coast goes into the frozen market) and

it can be held successfully under proper conditions for a year or more without deterioration. Until 1968 or 1969, there was a growing national demand for frozen king crab, a demand that had been built by a carefully planned and carefully executed promotional campaign, dating from the early 1950's, by the first successful American king crab operator. A slowdown in this demand became increasingly noticeable however, because of ever higher costs to be met by the processors. This same sales resistance was apparent too in the scallop industry, producer, like the king crab industry, of a luxury food item more and more people felt less inclined to buy.

## Color Problem

All canning and freezing of the Pacific Ocean commercial crabs necessarily is predicated, as it should be with all seafoods, on retaining as much as possible of the desirable flavor and color of the living creature. The crabs are rather more difficult than most in this respect, aside from the problems associated with control of freshness. This is particularly true with the Dungeness crab in which the orange-red pigments are highly unstable in comparison to those of the red-pigmented king and tanner crabs. It is this bright, attractive coloring of the king crab that did so much in the first years of the industry to build the demand for it in the frozen form.

A second problem, common to all the crabs but especially so with the king crab, is the occurrence of a blue or black discoloration caused by oxidation of blood pigments during cooking or during freezing and storage. The phenomenon in no way detracts from the food quality of the meat but it is troublesome to the eye of the consumer. This "blueing," so called, is the result of a complex chemical reaction that apparently is not yet completely understood and it has best been described for the lay reader by Richard H. Philips, managing editor of the now-defunct *Pacific Fisherman* in that publication's edition of June, 1965. Philips wrote:

"The blood system of the king crab (and other crabs —Ed.) contains a high percentage of copper which oxidizes and turns a blue-grey when it goes through the normal processing operations. Although it (the discoloration) is as pure and healthful as the blood found in a juicy piece of roast beef, the color is against it and it is not desirable from a consumer standpoint. . .

"Many hours and much money have been spent by the industry to combat the blueing problem. . .one of the solutions is a two-stage cook although some canners do use a single-stage cook.

". . .the theory behind the two-stage cook is that the first, rather cool cook at 150-160 degrees F. is warm enough to coagulate the blood and other undesirable liquids in the meat.

"The firm meat can be taken from the shell after the first cook without destroying its texture or shape and the blood and other fluids can be washed away before

Marine Construction and Design Company

Royal Viking, *107 feet, launched in 1972 for the king crab fishery. Note two radar antennae,
a characteristic of this fishery where twin electronics systems are standard. Crab block hangs
on starboard davit. Starboard boom is the "picking" boom.*

the second cook. The second cook "sets" the meat to prevent its leaching away in the retorting process.

"Some canners and freezers claim that a single fast cook at high temperatures (now usually 20-22 minutes in seawater at 212 degrees F.—Ed.) is a more effective way of combating blueing problems. A fast cook, they say, coagulates the blood so that it can be more easily separated from the meat during the washing process. It is claimed too that the single fast cook coagulates all the meat while the double cook allows some protein, still in liquid form, to be washed away with undesirable fluids during the washing process."

Chemists at the BCF Technological Laboratory in Seattle have studied the crab blueing problem at length and have demonstrated, at least tentatively, that the reaction can be halted or inhibited to some extent with the use of such additives as sodium sulfite and ascorbic acid. The latter is much used as a preservative for shellfish although United States laws do not allow it to appear on labels under its common name, Vitamin C.

The king crab processor buys an uncommonly large amount of crab to get out an amazingly small return in meat. Under usual conditions, the processor can expect to net only about 20 percent of the body weight of the king crab in usable meat. Only the leg, claw and shoulder portions of the king and tanner crab are utilized by the processor. The body meat of the Dungeness crab, on the other hand, is esteemed by many persons as the choicest part of this crab. With such a small return on his investment in total crab weight, the king crab processor takes unusual care to see that the crab is treated gently from the time he arrives at the plant to the time that the finished product comes off the line.

The prime meat comes from the *merus* or upper section of the leg of the crab. The *merus* produces the largest single piece of meat, one with a bright red surface coloration masking almost snow-white meat. This coloration is most attractive to the eye of the consumer and, consequently, the *merus* is used on the tops of cans and on the exteriors of frozen blocks. Meats from the large claws of the Dungeness crab are not as vividly pigmented but it has the same prestige in the processing of that species and it is similarly used in canning.

Whether king crab meat is to be canned or frozen, the processing is the same from the butchering of the crab to the end of the cook, whether that be one-stage or two-stage. Butchering is done by a worker who grabs the live crab by the legs at each side of the body and pulls it sharply agains a dull-bladed, fixed knife. This act removes the carapace and viscera from the legs, shoulders and claws. This remaining portion is known as a "section" and those of good shell color and quality may be frozen in that form after the cook to be sold as such in the retail market or to be further cut down to meet consumer needs.

The section goes immediately to the cooker where fresh or sea water may be used, according to the desire of the processor and the availability of supplies. If the two-stage cook is to be used, the first cook takes place for from eight to 12 minutes at from 150 to 160 degrees while the single-stage cook utilizes the earlier-mentioned time and temperature factors, 20-22 minutes at 212 degrees. Batch cookers were used in almost all king crab plants in the early years of the fishery but many of them have now been replaced by continuous cookers using stainless steel conveyor belts to move the product through the cook water.

Each of the three primary pieces of the crab section— shoulder, leg and claw—presents its own difficulties when the meat is to be removed after the cook. Meat is taken from the leg pieces by feeding them through a wringer resembling that of the old-fashioned washing machine. The rollers are of rubber and the meat is squeezed from the leg pieces as they go through the rollers with the meat dropping into a tray on the feeder side while the empty shell passes through the device. Shoulder meat is removed from the shell by forcing or blowing it free with a jet of water. The tough-shelled claws cannot go through the wringer so the jet must be used for it too.

After separation from the shell, the meat passes along a trough of water where impurities are washed away while clinging pieces of shell are freed to settle against baffles on the bottom of the trough. After the wash, the meat is ready for canning or for freezing if the single-stage cook has been used. If the two-stage cook is required, the meat must be steamed or boiled for about 10 minutes in a temperature range of from 210 to 220 degrees F. If the meat is boiled, the cook water may be a light brine. Some processors use ultra-violet light installations during final meat inspection to show up pieces of shell still remaining with the meat.

Processing of king crab and the other crabs requires much handwork, far too much in the view of the processors who would like to use more machines. The processors are caught in the bight of that line between increasing expense of labor and the point where total cost per pound conceivably could severely limit demand for their product. But the machine has not yet been invented that will replace the human hand and eye in much crab processing, any more than the machine has been invented to replace the human hand and eye on the tuna line. The eye still is required to detect such things as bits of shell and the hand is required to cut the big pieces of king crab leg meat into portions that will fit a 6.5-ounce can. And the cans must be filled by hand too.

After the leg meat has been cut to size, each cut portion is split and trimmed to lie smoothly in the can. The cans come to the canning line from the reform line with a parchment lining at the bottom. This parchment is intended to keep the red pigmentation of the leg piece from leaching onto the plating of the can. These attractive pieces, placed red side down on the bottom of the can, will greet the consumer cheerily when the can is opened because labeling will turn the one-time bottom of the can into the top.

The can is filled to weight with meat from the claws and shoulders and a tablet containing salt and citric or ascorbic acid is added. The tablets dissolve during retorting. The National Canners Association recommends that retorting times worked out by its technicians be followed explicitly with no variations whatsoever. The NCA says king crab must be retorted at 240 degrees with a time of 40 minutes for a quarter-pound can, 55 minutes for a half-pound can, and 75 minutes for a one-pound flat can, a pack seldom seen in retail stores. Retorting is followed by water cooling to bring the cans down to room temperature and prevent scorching of the meat. The retort times do not include temperature and pressure build-up times nor the necessary bleed-off time.

Preparation of king crab for freezing follows the canning procedure until the end of the cook, either single-stage or two-stage. When shelled meat is to be frozen, the sections are usually chilled after the cook to minimize meat fragmentation during the shelling process. After the meat is wrung or blown from the shells, it is washed, inspected and placed by hand in trays to form blocks of the desired size. Leg meat is placed on the bottom of the container, with shoulder, claw and leg tip meat making up the bulk of the block. The block is topped off with more of the colorful leg meat. These basic blocks can be sawn to desired sizes for retail or institutional use. These range in weight from six ounces to five pounds.

The growing harvest of tanner or snow crab is processed in essentially the same manner as king crab. Most tanner crab are caught in king crab pots somewhat modified to accommodate their size. During processing, the chief difficulty with this species has been that of separating the meat from the shell. The cooked tanner crab sections give up their meat much more reluctantly than does the king crab. This problem has been overcome to some extent by the use of shelling devices developed within the industry and by Bureau of Commercial Fisheries technicians. The meat of the tanner crab tastes much like that of the king crab although it sells at a lower price.

## Dungeness Crab

The canning and freezing of Dungeness crab follows much the same pattern as that of the king crab. The Dungeness is handled even more gently than the king crab because it is built on a smaller and more fragile scale than that biggest of the crabs and its meat is more prone to breaking up under necessary handling. The Dungeness weakens more rapidly than its bigger cousin under handling and holding conditions although this tendency is offset somewhat, at least along the Washington-Oregon-California coast, by the comparatively short distances the crab must be moved from fishing ground to buyer. In these fisheries, crab seldom spend more than a few hours in vessel tanks before offloading to holding tanks or directly to processing. Washington State's major crab port is Westport, a commercial and sport fishing center lying on Point Chehalis just inside the Grays Harbor bar. (Several hundred ships and smaller craft have been lost on this ill-famed bar in the years since Capt. Robert Gray first ventured across it in his ship, *Columbia*, in 1792. This was just before Gray blundered into the discovery of the Columbia River a few days after Capt. George Vancouver in *HMS Discovery* missed the river mouth. Otherwise, that great river might well be known today as the Discovery River rather than the Columbia.) The crab grounds extend no further north along the coast than Destruction Island, some 60 miles away, and scarcely that much to the south as far as Westport men are concerned. When weather allows, the crabbers pull their pots every second day and return to port with the catch that afternoon or evening. Their crab thus get minimum exposure to the vessel tanks and consequently are in better condition than crab that must be transported long distances or held on the catcher boat for several days.

The same fixed dull blade used for butchering king crab is used on the Dungeness with the crab brought smartly against it to tear away the carapace and separate the body of the creature into two sections, each with five legs on it. Gills and viscera are removed and the sections are cooked in fresh water for about 10 minutes at a temperature of 212 degrees F. The sections then are dumped into more fresh water for a cooling period, then moved to the shaking table where a bevy of women remove the meats from body and legs by cracking the shells and shaking the meats free.

The meats then are placed in a tank containing a 90 or 100 percent brine. Here the meats float freely while any clinging shell sinks to the bottom as the water is agitated. The meats move through a fresh water spray to wash away excess salt, then are placed in trays to await freezing or canning. The meats may be dipped in a dilute solution of citric acid or sodium benzoate for preservative effect if they are to be frozen.

Dungeness meat usually is canned in C-enamel cans with a net weight of 6.5 ounces. As with king crab, the more colorful leg and claw meat is packed where it will catch the consumer's eye when the can is opened. After filling, about two-thirds of an ounce of salt and citric acid solution is added to the can. This size can is retorted for 60 minutes at 240 degrees, then water cooled.

In Washington, Oregon, California and British Columbia, a great part of the Dungeness catch goes to the

National Marine Fisheries Service

*Crewmen of trawler* Washington *pick minute percentage of scrap fish from haul of pink shrimp taken off Oregon coast with experimental shrimp separator trawl. This is a good view of the West Coast drum trawler with net reel and trawl wound in it. Otter boards are secured on gallows at port and starboard while warp blocks dangle from the gallows.*

Robert Browning

*Fish bag of seine swings aboard with chum salmon in it. Catch like this is too small to break out brailer. This was a reverse or starboard set. Ring bar and other gear were moved from usual positions at port.*

fresh market. In the coastal areas of these states, fresh crab is a highly-esteemed food and is much in demand the year around. Since the commercial crab season opens in early or mid-December and all eligible males have mostly been fished out by the end of March or even earlier, it is obvious that crab sold as "fresh" crab during the balance of the year is not fresh in the sense that it has just come from the water and straight to market without intermediate holding.

During the season, the demand for fresh crab is met by use of just-caught whole or eviscerated cooked crab shipped to market in crushed ice. And during the season, a large quantity of crab is cooked, frozen and held to enter the market as "fresh" crab when the real product cannot be obtained. Similarly, a considerable amount of shelled meat is packed in hermetically sealed cans and frozen for off-season use. The useful life of this product is about six months at zero degrees storage temperature. It appears too in the market as fresh crab.

Another popular pack for institutional use is the five-pound C-enamel can held at about 32 degrees F. The life of this pack is short and it must be closely watched to see that the meat is not used past its period of best quality. Bacterial action can easily continue at this temperature. The growing demand for fresh and frozen Dungeness crab since World War II has led to a sharp decline in its canning in Washington, Oregon and California. Much Dungeness crab from Alaska also goes into the fresh and fresh/frozen market.

Live Dungeness crab may be bought on the piers in such fishing centers as Westport, Garabaldi, Oregon, and Eureka, California, and prepared by the consumer himself. This is a rather popular endeavor among visitors to the crab ports because such preparation is a simple process and it does assure the consumer that his crab is about as fresh as it is possible to get it. It is sometimes disconcerting to the novice, however, to hear his still-lively crab trying for a few seconds to scramble out of their pot of boiling water.

Preparation of Dungeness crab in the home for consumption there, as distinguished from the commercial process, is a simple matter.

The average Dungeness weighs about two pounds and it takes three or four of them to feed five or six persons adequately, assuming that the crab feast is supplemented by garlic French bread and a tossed salad.

To a large pot two-thirds full of water, add one-half to two-thirds of a cup of table salt and bring the water to a rolling boil. If four crabs are being used, cook them two at a time.

Drop them into the boiling water. . .the addition of the crabs will lower the temperature briefly. . .and when the water returns to a full boil, cover the pot and cook the crabs for from 15 to 20 minutes. When the crabs are fully cooked, cool them in running tap water, in cracked ice or by placing them in the refrigerator, making sure that they have been fully drained.

Cleaning and cracking crabs is not a difficult task although it may upset the squeamish of stomach because a crab being dismantled into his component parts is not an object of beauty.

Remove the back or carapace by wrenching it free of the body, an affair requiring little effort with a cooked crab. Remove the gills, the mouth parts and the viscera by holding the naked crab under running water and working at the reject material with the fingers. That yellowish, fatty portion, called "crab butter" may be saved if desired since many crab lovers esteem it above all else of the crab as an ingredient in crab salad.

Remove the tail flap from the underside of the crab by tearing it free, then break the remainder of the creature into halves, left and right, by using both hands in a down-pressing motion. Separate the legs from the body segments with adjacent body segments attached to each leg.

The body meat may be removed at this time from the segments attached to the leg by striking them against the edge of a bowl or by using a pointed tool (such as a nut pick) to do the job.

The remaining segments of the body, containing delicious meat in small cells somewhat like the cells of a honeycomb, may be divided and the meat picked out or this may be done by the consumer himself at the table.

The legs may be cracked with a wooden mallet or (gently) with pliers. The meat may be removed from the cracked legs or they may be served in this condition, thus saving the host a considerable amount of handwork. (The effort of picking crab meat will explain to him fully why crab meat is expensive in the retail market place.)

Crab meat lends itself to a number of easily-prepared dishes, all of them to be found in any modern cookbook.

The cardinal point to be considered in connection with home use of crab is that the meat be used as soon as possible after cooking, cooling and preparation and that any leftover meat be thoroughly refrigerated or frozen. Unprotected crab meat spoils quickly and its consumption may result in critical illness or death.

## The Canning Of Oysters And Clams

Oysters and clams hold a high rank in the fisheries of the United States both in volume and in value although the oyster and clam fisheries of the Northeastern Pacific play only a small role in the national shellfish harvest. The only exception to this limited participation in these two rich fisheries is Washington State's oyster production which usually amounts to about 10 percent of national production. (Washington also produces about 1.5 million pounds of clams a year, scarcely enough to be noticed in the total summing-up. Oyster and clam production of British Columbia, Oregon and California, as noted earlier, is negligible.)

West Coast fishermen, occupied with the exigencies of the tuna, salmon, halibut, crab and drag fisheries (including shrimp) tend to disregard the oyster and clam industries of their own areas although most have some small knowledge of the shellfish fisheries of the Gulf of Mexico and the Atlantic.

This tendency to overlook such valuable fisheries may be due, as much as anything, to the inescapable fact that oysters and clams are products of semi-cultivation, the fruit of a "farm" process to a great extent. Most of the men who seek the elusive fin-fish use the term "farmer" in a sense that is anything but complimentary.

It may come as a complete surprise to these fishermen to learn that the shellfish fisheries are richer than most fin-fish fisheries the world over. The United States clam fishery, for example, usually ranks about 11th in volume among all American fisheries. Its average harvest of 71.5 million pounds of meats places it just behind sea herring and just ahead of Atlantic Ocean perch in the tally of 54 species and two blanket classifications lumping a lot of other species together. Oyster production comes in about 15th on this list behind whiting and ahead of Atlantic cod, with 59.9 million pounds of meats.

In value, these two fisheries rank even higher, oysters running fifth place with a gross of $32.2 million on the average to fishermen and clams in seventh with a value to fishermen of $20.4 million. Neither those volumes nor those values may seem great in contrast to the same fisheries for shrimp, for instance with 307 million pounds and $103 million average but they stand for two healthy fisheries. (Volume in itself does not mean too much anyway. Menhaden consistently is the single species taken in greatest volume by United States fishermen with 1.1 billion pounds landed in 1968, for example. But its ex-vessel value was only $14.3 million. Landings of "industrial fish" other than menhaden that same year came to 211 million pounds but the total value to fishermen was only $3.1 million.)

From the average catch of clams and oysters, American packers can 507,000 cases of whole and minced clams with a net of 15 pounds to the case and 1,794,000 cases of clam chowder and clam juice, 30 pounds net. There usually are 323,000 cases of oysters canned, 14 pounds to the case. Most of this clam production takes place on the upper Atlantic coast and most of the oysters are canned on the Gulf Coast with Louisiana the center of production.

Nevertheless, the Pacific Coast does contribute in a small way to this overall figure. Eight plants in Washington and two in Oregon put up clams and clam chowder while three in Washington and a couple in California can oysters and oyster stew. Almost all these plants were (and are) small, single-purpose canneries dealing with a narrow range of products. The West's leading canner of oyster stew occupies a building of modest size and appearance in a rag-tag and bobtail industrial section of waterfront Seattle. The plant prepares oyster stew mainly and the size of the plant has no relation to the excellence of product or to its nation-wide retail sales. The canners of oysters and clams need not concern themselves with big plants such as those of the tuna packers. Canned oysters and clams in any form are semi-luxury items that retail at relatively high prices to a fairly small segment of the buying public and they rank as high in the favor of that public as do the raw products in national volume and value listings.

Pacific oysters are canned in five ways—whole oysters, fresh opened and blanched; whole oysters, steam opened; stew oysters, sliced, with milk and butter; oyster stew base containing only sliced oysters and nectar, and smoked oysters. (Preparation of smoked oysters is described in the earlier section on the smoking of fishery products.)

Fresh opening of oysters is expensive and time-consuming. Shuckers may have trouble keeping up with the capacity of the processing line but this hand-opening does preserve the oyster spat attached to the shells of the mature oysters. After opening, the meats are subjected to the regular washing procedure to remove all foreign particles from them, graded, then blanched in baskets in boiling water for 30 seconds. Oysters are canned in uniform sizes and the cans may be machine filled but the common practice is to fill them by hand. Either fresh water at 200 degrees F. and a salt tablet or 20 percent brine at 200 degrees may be added. The cans are vacuum exhausted, sealed and retorted. Round, fat oysters are sought for canning because they look better and fill the cans better. Oysters that are too thin turn up with a grayish-yellow color around the livers.

Whole oysters, steam opened for canning, come from oyster stock placed in retorts and pre-cooked for varying periods and pressures according to the equipment, experience and wish of the packer. The times run from three to 15 minutes, temperatures from 210 to 240 degrees, pressures from atmospheric to 10 pounds. The higher pressures and temperatures are used in batch processing with the oysters loaded into cars or carts for the steaming period. The lowest temperatures are found in continuous process steamers with the oysters moved along on conveyor belts. This method of opening tends to save on labor costs since the steaming frees the meats from the shells. It also preserves the natural shape of the meat.

The meats are shaken from the shells by hand or by machine, washed, graded, and packed into cans by machine or by hand. The cans usually used are the 211 by 304 size, eight ounces, and the 211 by 300, 7-3/4 ounces. After filling, the cans move on for water or brine addition and are exhausted, sealed and retorted.

Oyster stew undoubtedly is the most popular method in the United States and Canada of using canned oysters. For every person who has the courage (or the opportunity) to eat fresh, raw oysters or fresh, fried oysters or fresh oyster cocktail, there probably are a score or so who never have experienced oysters except as the major ingredient of oyster stew. Stew may be prepared from oysters canned by either of the two preceding methods. Oyster packers also make it easy for the public to prepare oyster stew by doing all or most of the work in the cannery.

Oyster stew is prepared on the Pacific Coast from sliced or diced oysters. The Pacific oyster must be cut up because of its usual large size, an imposing mouthful if it is not sectioned. The oysters are steam-opened by the previous process and the meats are chilled to firm up the flesh for the slicer. When sufficiently firm, they are sent through the machine and thence to a can filler. The usual can size is 10 ounces and the usual amount of meat added runs from 1-1/2 to 1-3/4 ounces or slightly more. Other machines along the line add milk, sometimes some nectar, and salt and butter in proportions favored by the canner. Some packers also introduce monosodium glutamate as a flavor enhancer but this practice may be curtailed by choice or federal order with the discovery in 1969 that the substance appears to have an adverse affect on human health if used excessively. Another additive universally used is disodium phosphate which is necessary to control the tendency of the milk to curdle when exposed to heat. Exhausting, sealing and retorting are standard.

An actual label found on a can of oyster stew picked from a West Coast supermarket shelf reads thus:

INGREDIENTS: *Oyster Nectar, Fresh Whole Milk, Oysters, Water, Butter, Salt, Disodium Phosphate and Monosodium Glutamate.*

This particular 10.5-ounce can contains about 1.5 ounces of sliced Pacific oysters and sold for 45 cents. Although the net weight of the solids in the can is not much, the stew is, nevertheless, a palatable and nourishing product. It is packed in the Pacific Northwest but is sold throughout the United States.

For institutional use, some packers put up oyster stew base. This consists only of sliced oysters and nectar with the consumer's kitchen free to add milk and seasonings as desired. This preparation usually is marketed on the West Coast in 46-ounce cans.

## Clam Products

The canning of oysters is a relatively simple matter in comparison to the canning of clams and clam chowder. Some body tissue must be removed by hand from most species of clams used for canning and, as with all forms of hand labor, this is a costly process.

Washington puts up a small pack of razor clams each year, both the "whole" clam, actually eviscerated, and chopped or minced clams. The whole clam may be battered and fried by the consumer while the chopped clam goes for bisque or chowder. There is no truly standard process for clam canning with each packer using methods that suit him best.

The clams, upon arrival at the cannery, go onto a conveyor belt where a water spray of from 75 to 100 pounds psi cleanses the shells of sand and other clinging material. The removal of sand is a "must" because the bane of the clam processor, whether he is operating on the East Coast or the West Coast, is the removal of sand from his product. The attempt is not always successful.

The supposedly-clean clams then are steamed open in any other of a number of devices, either a steam chest, a steam tunnel or a hot water bath. The clams are steamed or blanched for periods of up to one minute or slightly more, then are immediately cooled by a cold water spray to prevent toughening of the tissues. This again is a necessary procedure because the razor clam is a rather tough customer anyway.

Along the canning line, crushed clams are removed, shells are removed too and the meats are placed in shallow metal trays. Workers, again mostly women, use small, sharp scissors to slice off the tip of the siphon and to slit open the body from the base of the foot to the end of the siphon. The viscera is removed and the meats are shaken and rinsed in cool water.

Meats to be minced or chopped are fed through a power grinder. From there they pass to a filling machine where clam juice, heated to about 150 degrees F. is mixed with the clam meat. This juice is tasty and is a valuable nutrient. Federal regulations require a drained weight of 3.5 ounces of meat for six-ounce cans; five ounces for Number One picnics, and eight ounces for Number One talls. The six-ounce flat is the can commonly turned out by West Coast packers while the Japanese export many picnic talls of both oysters and clams to the United States. Net weights for the American product mostly are seven ounces, 10 ounces and 15 ounces. This includes the added juices.

Retorting is not standard but usual procedures for minced clams call for 70 minutes retorting at 228 degrees F. for small flats and 80 minutes at the same temperature for talls, both at five-pound pressure. The whole clams are retorted for 55 minutes at 228 degrees for Number Ones.

Large amounts of clam juice, broth or nectar are packed in the United States each year although only a small amount of it comes from the West Coast. It is agreed usually that "clam juice" is clam liquor not diluted with water; that "clam broth" is clam liquor diluted to varying degrees with water, and that "clam nectar" is that liquid obtained by heat evaporation of clam juice.

These liquids are marketed in metal and glass containers. Any of these liquids should have a "pearly, opalescent" color. Solids sometimes are found deposited on the bottom of the container but this means only that the containers have not been handled for a period of time. These liquids are retorted by standard methods with varying time ranges at 228 degrees at five pounds of pressure.

The hard shell clams canned on the Pacific Coast come mainly from Washington. These are the little

*Crewman of trawler* Washington *picks scrap fish from shrimp catch. This catch was made with standard shrimp trawl rather than experimental separator trawl. Some catches, with standard trawls, run more than 50 percent scrap.*

necks and the smaller butter clams packed usually in flat halves. These clams are too small for the use of expensive handwork to trim away the viscera and they usually are canned whole. They are small and inoffensive and a most tasty ingredient of clam chowder.

These clams are spray washed for cleaning, then steamed in the same or similar equipment used for the bigger razor clams. Some body tissue may be removed but most such removal is negligible because of the small size of the creature.

Clam chowder also is packed on the West Coast although, again, the major part of the national pack is put up in the East. Two types of chowder are common and these two do not include the clam chowder "soup" produced by two or three of the great packing companies. These "soups" have little relation to clam chowder itself.

The true chowders are divided into the "Manhattan" style, using a large percentage of tomatoes and the "New England" or "Boston" chowder. A commercial formula for Manhattan chowder lists these ingredients:

*Fine cracker crumbs, 65 pounds*
*Ground salt pork, 18 pounds*
*Ground onions, 18 pounds*
*Tomatoes, 24 No. 10 cans (a gallon each)*
*Salt, 16 pounds*
*White pepper, 5 ounces*
*Water, 55 gallons*

To prepare this mixture, the onions are simmered in a large steam-jacketed vessel and when half-cooked, the ground pork is added. The two are simmered until the contents is cooked but not browned excessively. At the same time the cracker crumbs or a portion of them are mixed with a portion of the water and mixed until smooth and then are placed in the kettle, now at the boiling stage. More cracker crumbs, the potatoes and water are added, to be followed by the tomatoes. The percentage of clams to be used varies according to the packer. If Number Three cans are used, the proportion customarily is 7.5 ounces of clams and 7.5 ounces of the remaining solid matter.

The New England-style chowder follows the above outline in general although clam juice or nectar replaces most of the water and no tomatoes are used. Various flours are substituted for the cracker crumbs. A typical label formula, giving no proportions, is this:

PREPARED FROM: *Water, Clams and Clam Juice, Potatoes, Wheat Flour, Vegetable Oil, Salt, Dehydrated Onions, Hydrolized Vegetable Protein, Monosodium Glutamate, Spice and Flavorings, Disodium Inosinate and Disodium Guanylate.*

A similar fish "chowder" recipe, a popular food seldom packed on the West Coast, calls for 50 pounds of flour, 18 pounds of diced onions, 18 pounds of diced pork, 56 gallons of fish broth, 25 gallons of water, 16 pounds of salt and five ounces of white pepper. Two ounces of fish, usually haddock in the East and cod or sablefish in the West, are added to each Number One picnic can. Citric acid may be added also, along with the usual stabilizers and preservatives.

## Canning And Freezing Of Shrimp

The insignificant shrimp makes up, in his sum, the most valuable of all American fisheries, one that accounts for about 24 percent each year of the ex-vessel prices paid to fishermen. Despite the magnitude of the catch, with 85 or 90 percent of it coming from the Atlantic, the Gulf of Mexico and the Caribbean Sea, the domestic catch cannot keep pace with demand and this catch of more than 300 million pounds must be supplemented every year by more than 100 million pounds in imports. The demand for shrimp seems to be worldwide, at least among those peoples who can afford shrimp, and the world's known stocks may not be able to accommodate that demand indefinitely. The United States, Japan and Great Britain have shown remarkable increases in shrimp consumption since World War II.

The post-war years have witnessed an equally-remarkable advance in the techniques of shrimp processing. Before the war, most shrimp eaten in the United States were canned or sold on the fresh market. Since the war, in unison with a national change in food habits, freezer technology, aided by breading technology, has joined with the somewhat-older shrimp peeler to turn the industry into a giant. Instead of shrimp canned or sold fresh in comparatively small amounts, the public demands and uses more than 200 million pounds a year of fresh frozen shrimp and shrimp breaded and marketed either peeled or unpeeled, cooked or uncooked. In addition, there is a thriving market in prepared shrimp dishes of the "heat and eat" order, a development that has brought at least something of gourmet cooking to a public that otherwise would not experience it. Most American shrimp goes to these uses.

Something like three-quarters of the shrimp taken in the Pacific Coast fishery are landed in Alaska. This fishery is built around five species of pandalid shrimp, with two species, *Pandalus borealis* and *Pandalus jordani*, dominant in the catch from mid-California to Western Alaska. These shrimp are small pinks averaging from 106 to 109 to the pound heads-on weight. (A median figure for size in the Gulf fishery comes to 30 to 40 to the pound with a fair part of the catch running below 15 to the pound, heads-off weight for both figures.) These *pandalids* are in great demand as cocktail shrimp and a premium pack comes from Petersburg and Wrangell in the Alaska Panhandle, the original site of the northern shrimp fishery. This pack is hand-peeled, cooked and put up in sealed but not-retorted cans and sent to market either fresh or frozen. This may be the best shrimp put up in the United States. Western Alaska, notably the city of Kodiak, the Pacific Coast's Number One shrimp port, freezes most of its catch although about 100,000 cases are canned each year.

The shrimp, small as it is, has a built-in problem when it is to be used in any form that demands it be shelled. This is the same problem that faces those who deal in the crabs, oysters, scallops, abalone—that of getting the meat from its protective coverings without fragmenting or otherwise injuring it to the extent that it must be discarded at worst or sold as a second-grade pack at best. Nothing has yet been devised that excels the human hand at dealing with the shells of oyster, abalones and scallops and, in some cases, those of the crab. To some extent, the human hand is still the best instrument used in freeing the meat of the shrimp from its shell. The careful handpicking practiced for the extra-high quality Southeastern Alaska fresh pack works well for this low-volume output but it is not a practical matter on a big-business basis, especially not for the tiny pinks of Alaska and the rest of the Pacific Coast. Hand labor in this era is expensive enough everywhere in the United States and Canada and it is even more costly in Alaska. The processor, then, if he wishes to compete on something like equal terms with the rest of the nation and the rest of the shrimp exporting nations, must turn to the machine peeler, a complex and expensive affair of something less than 100 percent reliability.

The peeler does do the shelling job or at least the primary shelling job pretty well and certainly it does it at less cost than it can be done by hand. If it were not for the peeler, the shrimp industry of the United States might well be stagnating at its level of the early 1930's simply because the United States today could not supply enough hand power at low cost to handle the catch. Consequently, there would not be the heavy landings being reported every year and the shrimp fishery might have trouble equalling the annual volume of the Great Lakes chub fishery.

The shrimp peeler first appeared on the north coast of the Pacific in 1957 when a new pink shrimp fishery centered around Grays Harbor on the Washington Coast led to the introduction of a single leased machine at Westport, the bay's chief fisheries town. That peeler stayed there only three or four years and, when the fishery collapsed for reasons still uncertain, was shipped to Alaska where, to this day, it may still be peeling shrimp.

The peeler with its metal fingers, oscillating rubber rollers and eccentric cams, looks, in action, like a Rube Goldberg contrivance from that artist's most imaginative period. If a figure of speech may be excused it looks —to this writer anyway—somewhat like a marimba standing with one end high and the other low. Fore and aft stands various auxiliary machinery, its multiplicity dependent on the individual installation and on the species of shrimp usually worked. Shrimp that are not deveined in processing don't need the deveining machine, for example.

Shrimp landed on the Pacific Coast customarily arrive at the cannery iced down rather than in frozen blocks like much Eastern shrimp. After washing and weighing for the vessel payoff, the catch moves by conveyor belt to a hopper at the head or high end of the peeler. During the peeling process, the shrimp are worked against the soft rubber of the rollers by the metal fingers. The rollers, with their motion, pull away the shells from the soft meat as the shrimp move, urged by machine action, water flow and gravity, to the low end of the peeler. This downward movement removes most of the waste matter but an additional peeling action takes place in a scrub trough at the foot of the peeler where other rollers take a second crack at the shrimp. But the ubiquitous human hand and eye are needed for final inspection and for most can filling so the meats, almost clean by now, pass on a conveyor belt along a line of sharp-eyed women who seek out the last tiny shell fragments before the two- to three-minute blanch in brine or steam. The filled and weighed cans are topped off with a salt and citric acid solution. The usual Pacific Coast pack has a net weight of 4.5 ounces of meat.

The mechanical peeler does most of the work in Kodiak and other "out to Westward" plants with plant capacities running from a low of about 20,000 pounds a day to as much as 125,000 pounds. The harbor waters of Kodiak city itself have suffered from the shrimp industry based there as well as from other sea-related operations such as king crab. Both shrimp and crab processing result in a comparatively small volume of meat recovered from a great weight of raw material. Shrimp meat recovery runs at about 15 or 16 percent while the crab processor counts himself lucky to get a return of 20 percent. The remainder simply goes into the water as waste and harbor bottoms around the Kodiak processing plants are increasingly covered with a growing layer of decomposing offal. No reasonable solution to this problem appears in sight.

The products obtained from Alaskan shrimp processing are varied and the range is considerably wider than offered by the original operations in Southeastern. This variety includes:

Canned meats preserved against spoilage for almost indefinite periods by having been heat-sterilized in their containers.

Meats that are usually fully cooked after having been separated from their shells and other body parts, then frozen and sealed in containers such as cans or see-through plastic bags.

Meats that are machine peeled raw, then frozen into block or log-shaped rolls which then are shipped outside for additional processing into individual serving portions such as breaded shrimp "steaks" or "cutlets."

Whole shrimp cooked "in-shell" then marketed in retail-size containers.

A relatively small quantity of larger varieties prepared heads off in the shell and frozen raw to be cooked by the consumer.

Canned products are almost exclusively packed in "half-pound" tins containing four and one-half ounces of meat although some are put up in smaller tins of two and one-half ounces net. Three grades of canned shrimp meats are produced—tiny, averaging 270 or more meats to the pound; medium and broken. Mediums bring the highest price in the wholesale market.

The large shrimp of the Atlantic and Gulf coasts lend themselves well to freezing whole in packages or individually. So prepared, they appear in the markets in most cities over the country. But the small pink shrimp of the Pacific cannot profitably be frozen singly most of the time so shrimp from the western fisheries are usually peeled, cooked and frozen in packages of varying sizes. Seattle receives large shipments of frozen shrimp meat from Alaska with shipments running up close to 100,000 pounds at a time.

Most of the breaded shrimp used in the United States originates on the South Atlantic Coast or on the Gulf or may be processed on the Pacific Coast from the larger shrimp of the western fisheries or from import material. Only breaded fish sticks and fish portions run ahead of breaded shrimp in consumer popularity. Breaded shrimp in its various forms freezes well and stores well under proper conditions. The industry asked the Bureau of Commercial Fisheries, in 1955, to set up sanitation rules for it. Compliance is not mandatory but more than 90 percent of the industry at least pays voluntary lip service to the guidelines.

## Miscellaneous Canned Fishery Products

During recent years, United States processors have canned annually more than 40 million standard cases (a standard case ranges in weight all the way from 6.75 pounds for shrimp to the maximum 48 pounds for salmon) of fisheries products worth more than a half-billion dollars a year.

Of this great mass of things from the sea (most of them come from salt water but not all of them) tuna, salmon and pet food account for more then 80 percent of the volume and value of the pack. There is no reason to believe that this ratio will change appreciably in the future. The pampered pet population grows right along with the human population and animals can be fed fish and fish scraps that would find little use otherwise. As for human consumption, tuna and salmon are the food fish caught in large tonnages and sold suprisingly at a rather modest price. Except for some quirk of nature or some rape of the fisheries, they should continue to occupy that spot.

There is, nevertheless, an appreciable pack of other fishery products of the nature of shad and shad roe, sturgeon and sturgeon paste, salmon egg caviar and similar things of luxury or semi-luxury status. And there are, also, packs of exotic items such as squid that have little or no appeal for most Caucasians but that do find favor with persons of certain ethnic groups. There is a rather utilitarian pack of the mackerels of the West Coast that sometimes goes into the pet food departments of retail stores. In one such case, a one-pound tall can of jack mackerel, packed in California, was being sold for cat food at 26 cents a can while a six-or seven-ounce can of tuna-based cat food standing a few feet away on the same shelf sold for 27 cents. This mackerel had not been intended for cat food. A distinctive label panel explained clearly how to broil it for human consumption. The market manager reported he found no

## "Anatomy of a Tuna Seiner"

| | |
|---|---|
| Main engine | One General Electric model 7-FDM-16, V-16 four-stroke cycle diesel, 3,100 hp at 1,050 rpm, 9'' x 10½'' B&S, 10,688 cu. ins. displ. |
| Reduction-reverse gear | Western Seamaster model 360-PCMR-S, 5.25:1, with Wichita clutches. |
| Shaft brake | Mathers SB8-0600, combining Mathers holder, and drum, and Fawick brake element model 32-VC-1000. |
| Shafting | Steel, intermediate 9½'', tailshaft, 10½''. |
| Propeller | Coolidge stainless steel, 5-bl., 126' x 108-112''. |
| Fuel and lube oil purifiers | Two, DeLaval 65-03, 55-03. |
| Fuel and l.o. filters | GE. |
| Air compressors | Two, Quincy, Westinghouse elec. motor driven, 5 hp ea. |
| Silencers | Maxim. |
| Main engine pyrometer | Alnor FRT. |
| Main engine tachometer, engine room | Reliance. |
| Generator sets, main | Two 350 kw ea., 450v GE generators driven by Waukesha diesels model L-1616-DSIU, supplied by Northwest Engine Service Center, Seattle. |
| Standby generator set | One, 200 kw GE generator driven by Waukesha H-1077-DSU marine diesel, also supplied by N.W. Engine Service Center. |
| Bow-thruster | One, PSI-Brunvoll model SPO-200 driven by 200 hp electro-hydraulic power system. |
| Main engine control stations | Three: Wheelhouse, port bridge wing, and engine room. |
| Automatic steering systems | Two: Sperry Rand gyro and Sperry Rand 8-T magnetic. |
| Rudder actuation system | Frydenbo model HS-40 rotary vane. |
| Shaft rpm tachometer, wheelhouse | Weston electric. |
| Engine controls | Mathers, single lever. |
| Radar | Kelvin-Hughes, Series 18-9, 65-mile range. |
| Depth recorder | Furuno 200. |
| White-line recorder | Furuno FG-11, Mark 3. |
| Transceiver, SSB | Raytheon 1275A-1275B. |
| Radiotelephone, wheelhouse | Konel 135. |
| All-wave receivers | Halicrafter, and Collins (two). |
| Ship's clock | Chelsea. |
| Water temperature recorder | Taylor. |
| ADF | Benmar. |
| Omega navigation system | Omega 700. |
| PA system | Boegen. |
| Ship's entertainment | Roberts 333-X cartridge player piped through PA system. |

**FISHING MACHINERY:**

| | |
|---|---|
| Purse winch | Northern Line model No. 314-75SH-2-3216. |
| Hydraulic power unit | N.L. model NL-3210-HPU. |
| Control console | N.L. model NL-3224-CC. |
| Corkline winch | N.L. model NL-3183-HVC. |
| Brailing winch | N.L. model NL-552-HBW. |
| Cargo & small boat winches | Three N.L. model NL-3268-HMC. |
| Vang winches | Two, Gearmatic. |
| Topping winch | Gearmatic. |
| Chocker winch | Gearmatic. |
| Boom hoist winch | Gearmatic. |
| Power block | Marco-Puretic model 42B-G-1505. |
| Anchor windlass | Northern Line hydraulic, model 3266-HAW, handling two 2,000 lb. Navy type anchors with 90 and 105 fms. of chain. |
| Purse seine | By Casa Mar, Panama, 700 fms. by 72 fms. |
| Seine skiff | One, 34' x 19'6'' x 5'9'', steel, by Antonio Mauricio & Sons, San Diego, with one Waukesha H-1077-DM, V-8 marine diesel app. 280 cont. hp at 2,200 rpm; 4.5:1 Capitol gear, HYC; 48'' x 28'' 3-bl. prop. |
| Small boats (chasers) | Four Cheetah F-glass, ea. with Mercury 65 hp model 650 outboard motors. |
| Pumps, salt brine — seawater service | Aurora GBHA, 28 units various sizes. |
| Ammonia compressors | Five Vilter 6-Cyl. units model AS-446-RCB, and one Vilter 2-cyl. model AS-446-RCB. |
| Galley range | Lang electric. |
| Galley fridge-freezer | Glenco 3-door, st. steel. |
| Air conditioning | Chillwater 45 deg. |

*Fisherman's News*

takers for the mackerel since it had moved at a glacial pace when it had been stocked on the same shelves with salmon, tuna, sardines, crab and shrimp.

Mackerel, properly fresh and treated with understanding by a good cook, has long been accepted as one of the finest of food fishes. But in the can, after retorting, mackerel loses that delicate taste that distinguishes it from the more heavily-flavored fish like salmon or tuna. California plants have been turning out about 500,000 cases (45 pounds net) annually. The record pack was 1.7 million cases in 1947.

The canning of mackerel, like the canning of salmon, is a process of comparative simplicity when one remembers the complex and tedious affair of canning tuna or the Pacific Ocean crabs. Mackerel canning lines are as mechanized as salmon lines and the process is much the same. Mackerel are of a fairly uniform size with the average commercial weight running from two to three pounds (average weights have declined over the past 40 or 50 years, apparently because of heavy fishing), allowing for feasible mechanization.

The fish are machine-gutted, headed and dressed just as salmon are. They move along the line on a belt for final inspection and for whatever cleaning must be done by hand. After dressing and washing, the fish are brined in an 80-90 percent brine for 60 minutes. They then are machine-cut into can-size pieces and are machine-packed. Larger fish are cut into three pieces. Retorting takes 75 minutes at a temperature of from 240 to 250 degrees F.

A particular delicacy to many persons is the roe of the shad, a native of the Atlantic but a happy addition to the fish of the Pacific Coast. Since the shad first was introduced into the Sacramento River in 1871, it has spread northward along the coast as far as Southeastern Alaska. The Columbia River, that mother (an ailing mother now) of so many fisheries, has become the major shad stream of the western continent. Great schools of shad come into the river to spawn in the spring and early summer.

The commercial shad fishery is conducted with drift and set gillnets and is concerned chiefly with the taking of females, the "roe shad," the source, obviously, of the eggs. The nets, of course, are not selective as to the sex of the shad they take and some hauls may consist almost entirely of buck shad. These are not wasted but go for reduction, for pet food and, to some extent, into the low-demand fresh market. Only a handful of canneries on the Columbia work shad, all of them small, hand operations. The landing of shad in 1968, for example, came to only 318,000 pounds. Largest previous catches were 1,535,000 pounds in 1962-63 and 1,245,000 pounds in one year in the 1926-30 era.

The Columbia River shad pack has averaged about 10,000 cases of 48 one-pound cans during the last several years. The shad, a member of the herring family, is well enough flavored but it is bony and soft-fleshed like all its cousins. Bones are no problem after retorting but the flesh does not hold up particularly well nor, like mackerel, does it hold its delicate flavor too well. On the Columbia, the fish are gilled, gutted and headed by hand and hand-packed into the cans. A half ounce of salt is added before exhausting and sealing.

The major interest in shad canneries is in the roe, a product of relatively high value to the packer if it is of good quality. To obtain this quality, gentle handling and a knowledge of the state of the roe is required. As the fish come into the cannery, they are washed and hand split to extract the skeins or lobes of roe. The eggs are small like the eggs of others of the herring family with as many as 150,000 or even more found in large females.

After splitting, the skeins are removed as softly as possible because a broken skein loses much or most of its value. Not all roe is acceptable for a quality pack. The eggs must be fully-developed but not unduly ripened. If too ripe, they tend to separate from the enveloping membrane and they tend at this stage to make the product tasteless and watery. If the eggs are insufficiently developed or "green," the result is much the same as with any other "green" food—toughness or at least all the toughness a small fish egg can develop. There is a definite flavor loss too with the underdeveloped eggs.

The roe is washed and all blood clots, slime, the outer membrane and other undesirable material is removed in a painstaking operation. This is the stage when the lobes are most vulnerable and one small slip can ruin them. When clean, the whole lobes are packed as evenly as possible into parchment-lined cans of six ounces net weight after retorting. The only additive is salt, a quarter-ounce of it to the half-pound can.

## The Canning Of Pacific Sardines

The canning of Pacific sardines is chiefly a matter of historic interest these days and it may never again amount to anything more. When these once-abundant fish disappeared from their vast Eastern Pacific range, their loss killed off a fisheries industry that employed more people more months of the year than any other of the West Coast. On the water, the fishery worked rather specialized boats and gear; the movement of sardines from fishing vessel to cannery utilized pumps, the first use of this tool on the Pacific Coast; the canners of sardines produced a variety of packs quite unlike that of any other fishery of the United States.

To many Americans who grew up in the 40 or 50 years before World War II, "sardines" meant sardines from California, not European imports. The California sardines are gone from the grocery shelves now, replaced by European sardines of great quality, little fish that can be found in their flat cans where no other type of fish ever has been sold. Even in those pre-war days, sardines from Maine were hard to find in the rural West where canned salmon and California sardines were just about the only canned fish most people had anything

much to do with although by the 1930's, tuna was being regarded as a real comer. As for Maine sardines, they still have a hard time competing with the Norwegian and Portuguese product because American consumers have gotten used to the imports and most Maine sardines simply do not meet that standard.

In their heyday, the California sardine packers dealt in a group of exotic packs such as sardines in brine and sardines in various oils; sardines filleted and packed in oil, sweet sauces or their natural juices; whole smoked sardines, or smoked sardines in tomato sauce as well as a handful of others. But the standard pack, the one seen most often wherever California sardines were sold, was the headed fish, pre-cooked, drained and laced with tomato or mustard sauce, in a one-pound oval can. This particular pack was inexpensive and satisfying, sufficient for a meal for a small family in a Depression era when many families found the going tough with California sardines looking good.

The sardine packers were as progressive as any people of the fisheries industries of the United States. They had to be when it came to handling their fish because it was somewhat impractical to treat each of millions of sardines with the same care accorded a salmon. Hence, their early use of fish pumps and ingenious conveyors and filleting machines together with their versions of the salmon canners' Iron Chink to head and gut their countless fish. At least three types of machines did this messy job, each developed around a slotted conveyor belt that carried the fish head first, back up, through an array of knives and eviscerators into a flume where they were washed and cleansed of blood and other waste matter.

Every canner had his own ideas on the canning of sardines and as men and methods changed over the years, such traditional steps as brining and the necessary drying gradually were shoved aside in favor of other methods. In the youth of the industry, an oil fry was almost universally used as a preliminary to all methods of packing. In the middle and last years, the oil fry was used only for a small part of the pack in deference to market demands. Instead, pre-cooking became the vogue for the bulk of the pack.

One of the earliest and simplest canning procedures was the raw pack in which 15 ounces of brined and dried fish were combined in a one-pound oval can with two ounces of mustard or tomato sauces, vacuum-sealed and retorted. The packers liked this operation because it was efficient and inexpensive. Another pack the canners liked was the broiled pack in which one-pound ovals filled with 16 ounces of raw fish traveled for 45 minutes through 320-350 degree ovens for cooking. When the broil was done, the cans were drained as they turned slowly over in any one of several machines designed for that work. The requisite sauce was added, the cans were sealed and sent to the retorts.

Pre-cooking allowed versatility in the finished product that had been lacking earlier as well as a flexibility in the canning process itself. Too, it eliminated most brining and drying which was a saving in time as well as money, matters always welcomed by management. There were three pre-cook methods used in the industry. In general, 16 ounces of clean, raw fish were packed into the one-pound oval and cooked for periods ranging from 20 to 45 minutes at temperatures running variously from 210 to 240 degrees. This was followed, of course, after addition of whatever else was going into the can, by the usual retorting.

Fillets in various combinations of oils and sauces were a popular product of the California canneries and the sardine people were using filleting machines before the rest of the American industry ever had heard of them. Whereas the fillet machines used in the American bottom-fish industry are German-designed and mostly German-built, the sardine machines were indigenous to California and were performing usefully, if not flawlessly, long before post-war Germany started shipping its gear to the United States. Filleting was a hand process in the first years of sardine canning and a laborious task it must have been with fish averaging 12 inches in length. The Californians tried the traditional pack of small sardines in oil after the style of the Europeans. But as in several other processings, the Europeans did and still do it better. In this pack, small sardines were cleaned, brined, fried in oil, cooled and packed into one-quarter-pound flat cans, in most cases, with cottonseed or olive oil added.

If the Pacific sardine fishery ever were to revive, it is doubtful that reduction use of sardines would be allowed on any scale like that of the past. It is equally doubtful that canning would ever resume in any magnitude such as that of the good old days, at least not for a long time. Most of the men and most of the plants and equipment withered away along with the fishery and the necessary skills do not exist. Any new sardine industry would be a welcome addition for the time coming when fishery resources must be increasingly used for food.

## Canned Pet Food

Federal statistics show that tuna and salmon are the leading species of fish canned in the United States. Tuna presently accounts for about one-half the pack while salmon and all other species of fish and shellfish account for a quarter. The balance consists of canned pet food, a figure that surely must make the pet animals of the United States the best-fed cats and dogs in the world. In 1970, pet food canners put up 11.2 million cases at 48 pounds each of their various products. This figure does not include another 4 million cases of such food that contained less than 10 pounds of fish to each standard case.

A substantial part of this animal food comes from the West Coast. There are about two dozen plants dealing in pet products in Washington, Oregon and California with an average annual output of more than 2 million cases.

Pet food manufacturers offer a diversified menu to their consumers. The choice includes halibut, shad, cod, haddock, Atlantic herring, Atlantic sardines, anchovies, tuna and something lumped together under the general classification of "seafood." These fish are packed singly or in combination with other fish products and cereals. They are inexpensive when compared with most fish canned for human consumption. One-pound talls of halibut-based food intended for cat food averaged 35 cents each on the West Coast in the late 1960's. Tuna in half-pound flats sold for 27 cents.

Food fishes like tuna or halibut of human use quality do not go in their entirety to pet food use. Mostly, only the waste, the trimmings, are used for these products. Tuna, for example, that does not meet the human-use standards of any of the states (and this may be a substantial portion of some fares) may be used for pet food or sent to the reduction plants. However, such fishes as anchovies, Atlantic sardines and Atlantic herring that commonly find wide industrial use are diverted to animal food in considerable quantities. Pet food canners have their own quality standards and much or most canned animal food could be eaten by humans with no strain whatsoever. In fact, social workers with experience in the poverty areas of American cities say there is a considerable use of certain canned pet food among the people of the ghettos.

## Pre-cooked Frozen Fish Products

If the men of the United States and Canada complain that their wives do not cook like their mothers did, it can be said with some justification that the wives just don't have to cook a lot of the time. In many lines, the cooking already has been done for them and it seems quite probable that a greater percentage of their cooking will continue to be done for them and over a larger field too. The use of frozen, pre-cooked foods has risen every year since their introduction and there is no indication that their use will go anywhere but up as population grows and more and more foodstuffs are turned into the pre-cooked frozen form.

Pre-cooked foods, as distinguished from canned foods, are largely a product of the years right after World War II when a kind of speed-up in living seemed to affect the greater part of the people of North America above the Rio Grande. The reasons for this acceleration of pace and a consequent divisiveness in family units are complex and are not of concern here other than in the manifestation of these happenings in a substantial change in food and cooking habits among Americans and Canadians.

The companies that deal in pre-cooked, frozen foods have shown a great versatility in putting all kinds of meats, fish and vegetables into attractively-packaged units at modest prices, all bearing the advice to "heat and eat." Such offerings, as far as fish products are concerned, probably give the average consumer, especially the one far from tidewater, a greater choice among fish and shellfish than can be found in most hinter-land places that sell fish and fish products. This choice is fairly complete among types of fish foods, that is, between fish and shellfish although the variety among the types themselves is rather limited. Fish choices for these pre-cooked frozen items usually are only among cod, haddock and Atlantic or Pacific Ocean perch. The list of mollusks includes shrimp, oysters and scallops while the crustaceans count blue crab, king crab and lobster. These latter three usually are found in such gourmet dishes as lobster or crab a la Newburg. Only shrimp and the fishes figure to any great extent in this growing industry. The crab preparations, for example, are only a tiny part of total production in the pre-cooked, frozen fish product field.

The bulk of this industry consists of the production of breaded fish sticks and fish squares or portions. Both are packed as the protein ingredient of several kinds of frozen dinners or in various family and institutional packages. Properly prepared, they are good enough on a short-term basis but not something to be favored too long by lovers of seafood. These sticks and portions, however, are almost the only fish eaten by a good many persons who have been deprived by the forces of geography of easy access to all kinds of fresh fish and shellfish. This pre-cooked frozen fish industry has been of economic benefit to fishermen of Eastern Canada but it has not done too much for American fishermen of any of the three coasts. Canada exports to the United States most of the fish used in the industry.

Fish sticks and fish portions first appeared in the late 1940's and by the middle of the 1950's, they had become an accepted part of the American cookery scene, although purists in this art like fish sticks and fish portions no more than they like other pre-fabricated foods. But to most consumers, persons with simpler tastes and complicated budgets, these rectangles and squares of fish were a welcome addition to the home menu. The United States armed forces became a major user of these products because their uniformity of size and weight simplified some of the problems inherent in cooking for and feeding large numbers of men. It might be added that these frozen products, pre-cooked or not, are distinctly preferable to some of the unrecognizable foods palmed off as "fish" on unwilling soldiers of the United States Army during World War II. This stuff called fish during those years undoubtedly bred some millions of confirmed fish haters in the Army alone.

## Breading Of Seafoods

The National Marine Fisheries Service defines a fish stick as "a uniform, compact, rectangular portion of fish flesh cut or sawed from a fish fillet block...usually about 3-3/4 inches in length, 7/8 of an inch in width and one-half inch in depth" with a weight of about one ounce. The fish portion or "square" (it seldom is square) is a "rectangle of frozen fish that typically may be four inches by three inches by 15/32 inches." Its average weight is four ounces.

The preparation of fish sticks and fish portions is an exacting procedure in all its phases, due partly to the fragility of the raw material and to the variables encountered in applying batters and breading materials and in the cooking, cooling and packaging of the finished product. Fully automatic machinery has somewhat simplified production problems by its elimination of the unpredictable human element at most points along the line.

A pre-requisite to the production of fish sticks or fish portions is the fish fillet block. Such a block is described as "a uniform, compact and cohering mass of skinless fillets frozen together under pressure." To form the blocks, the fillets are carefully hand-packed in containers and frozen in multi-plate freezers. Pressure freezing and expansion of the flesh of the fillets fills all voids in the block and provides the uniformity necessary for sticks and squares.

The sizes of the blocks vary according to the needs of stick and portion producers and these specifications are predicated on the cutting equipment in use in their plants. This equipment may be band or gangsaws or guillotine cutters. The use of any of these three must be carefully calculated by the individual manufacturer because the initial loss in the production of sticks or fillets stems from these. Either of the saws causes from seven to 12 percent loss in fish weight through the "sawdust" they necessarily create. The guillotines do away with sawdust but they have a tendency to damage blocks and turn out mis-shapen sticks when not properly adjusted. Blocks are frozen in multiples of the size of the sticks or portions to be cut from them with an allowance for sawdust waste.

Application of batter and breading material to fish sticks and fish portions is usually a fully-automated process, one that takes the raw materials from the saws through batter and breader machines and into cookers in a never-ceasing operation largely untouched by workers although it always is under strict inspection in most plants. Fish that has been run through the batter-breader process is extremely vulnerable to bacterial contamination until cooked and returned to cold storage. In addition, the materials used in these steps can easily be contaminated by many toxic substances before they get to the processing line. The United States Department of the Interior through the Bureau of Commercial Fisheries set up strict guidelines in the 1950's for such as the breaded shrimp industry with particular emphasis on the breading area.

Although the actual application of batter and the breader to fish sticks and portions is routine enough, the search for a satisfactory material or combination of materials for these purposes sometimes takes on the air of a detective thriller. There is no such thing as "standard" batter mix or breader mix because the suppliers of mixes customarily vary the composition of their merchandise to please the customer. This difference in composition shows up strongly in the final product. Most batter mixes are made of fine corn flour and corn meal and spiced, non-fat milk solids. Breaders vary widely through cracker meal, bread crumbs, wheat flours and cereals, soybean and potato flours and combinations of these. The choice is based on the manufacturer's preference in color of the finished product. Breaders with a corn base give a golden-yellow color to the stick or portion while bread crumbs or cracker meal give it a reddish-brown shade and wheat products result in a golden brown touch.

In the first years of the industry, fish stick and portion production was largely a matter of hand labor and batch breading and cooking. But engineering, as it has done so many times in so many industries, came up with better, faster and cheaper ways to do the job and modern plants work with integrated systems that turn out thousands of pounds of a particular product every hour. Essentially, such equipment consists of various types of conveyor belts moving the fish step by step in necessary directions quickly and smoothly with the least waste and the least supervision. Such critical points as the proper separation of sticks and portions as they go into the batter tank are automated, doing away entirely with the early-day chore of hand-spacing fish pieces on the belt so they would not bunch up in the batter. Now shaker or unscrambler devices sort the sticks and portions into orderly rows and keep them in that formation through batter, breader and cooker.

Batter coating is achieved in most plants by passing the fish pieces through a curtain of batter and a subsequent puddle of it that coats all surfaces evenly. The surplus is blown off by air jets and the pieces go into the breader machine where they similarly are showered with the breading material. The excess is shaken loose by a vibrator and is returned to an overhead hopper. Different designs of equipment provide variations on these themes but the principles are the same.

Cooking is a carefully-controlled process since irreparable damage can be done during the minute or minute and a half it takes the fish pieces to pass through the cooker. Undercooking opens the way for possible spoilage while overcooking causes weight loss and results in a dry and tasteless product. The color of the finished product depends to some extent on the time and temperature of cooking although the basic shade is determined by the nature of the breader ingredients. The condition of the cooking oil can react unfavorably on the taste of the cooked fish in that the flavor of old oil, oil beginning to break down from heat, is absorbed faster than that of fresh oil. Cooking oil is continuously filtered and under optimum conditions passes through the filter as often as once a minute. The cook temperature runs usually from 385 to 405 degrees. The actual cooking is done as the pieces move through the cooker on their belt with a hold-down belt securing them.

Cooling of the cooked pieces is another critical point in the processing procedure since the cooked fish must be allowed to lose water vapor in the flesh that would appear as frost in the package. Cooling must be done so that the breading will not become soft enough to drop off the cooled piece when they are handled for packaging. Modern cooling procedures take the pieces from the cooker through chilling, freezing and machine packing in about 10 minutes. The risks of older methods, still in use in some plants, are entirely eliminated.

Not all fish sticks and fish portions are cooked after breading. These pieces, still frozen, are packaged and returned to cold storage immediately. Fish portions, for example, intended for military use are packed uncooked in containers running up to 50 pounds net weight. Portions intended for inclusion in individual, aluminum-wrapped frozen dinners usually get a precook.

Of all the fish and shellfish taken by Americans and Canadians, shrimp has the highest value to the fishermen. Of all this mass of shrimp, more than 100 million pounds are breaded and sent to market both raw and pre-cooked annually. This use represents about one-fifth of total United States consumption of shrimp and it is startlingly representative of the trend of the 1950's and 1960's toward the jiffy "heat and eat" meal, as little solace as that fact may be to people who still like to cook. The industry is concentrated along the Gulf Coast, an area where more than two-thirds of all United States-caught shrimp are landed.

Shrimp to be breaded are headed, cleaned and deveined by machine. Fresh shrimp are preferred for this processing although shrimp that have been frozen can be used through necessity. Most shrimp are breaded individually even though there are specialty batches of shrimp steaks, portions or sticks prepared from two or more shrimp pieces or from two or more whole shrimp. The same factors that weigh heavily in the fish stick and portion industry are of equal importance in the processing of breaded shrimp. Taste influences must be watched even more carefully because shrimp are more susceptible to the development of off-flavors than fish.

The industry now is highly mechanized and processing is almost identical with that of fish pieces with much the same equipment in use for battering, breading, cooking and cooling. Cooking time is adjusted to the size of the shrimp on the line with undercooking and overcooking to be guarded against. Times run from 60 to 90 seconds at from 375 to 390 degrees to arrive at the color desired by the processor. The packaged product is widely distributed under "famous name" labels over the United States and Canada. Modern supermarkets stock these shrimp heavily as well as one or two forms somewhat more exotic such as shrimp Creole.

Frozen, breaded oysters appeared on the market at the same time the other forms of breaded fish and shellfish began their rise to favor through the United States and Canada. Oyster production is only a fraction of shrimp volume. Recent United States landing of oyster meats have averaged about 55 million pounds a year on all coasts compared to the enormous landings of shrimp. Both raw and pre-cooked oysters are offered on the market.

The breading of oysters is a process even more delicate and subject to error than the breading of shrimp. The unique flavor of the oyster begins to disappear shortly after the creature is taken from its shell and the loss of that flavor is compounded by the application of the necessary batter and breader materials. Flavor is compromised further when the oyster is cooked before freezing. The raw material must be as fresh as possible and all materials to be applied to the oyster must be as free of inimical tastes and odors as ingenuity can make them. But at best, the storage life of breaded, frozen oysters is no more than eight months and it is only four months for some forms of them.

Oysters to be breaded are hand-shucked, washed and batter-dipped and breaded in batches, mostly by hand in Washington State, major center of oyster production on the West Coast. They then are packed in waxed, fiber cartons and are frozen at -20 degrees F. and stored at zero degrees. Frozen oysters retain the irregular shape of the fresh oyster, allowing air pockets to appear in the container. This permits oxidation and consequent rancidity. The Pacific Coast Oyster Growers Association, chief spokesman for the western oyster industry, recommends that all persons, from processor to retailer, who deal in frozen oysters acquaint themselves with the properties of the product and insure that frozen oysters, breaded or not, be removed from circulation when their useful storage life has been attained.

Some processors freeze oysters in closely-packed poylyethylene tubes about two inches in diameter and 24 inches in length. The frozen cylinders are removed from the tubes, cut into portions a half-inch thick, breaded and packed in fiber cartons. These portions are conveniently sized and shaped but the exposed internal surfaces of the oysters are quite susceptible to rancidity and this product must be marketed in no more than four months.

Scallops from the Alaska fishery are breaded and frozen in a small way in the Pacific Northwest and offered in the market either raw or pre-cooked. This production is mostly a hand operation with the scallops distributed on a limited regional basis. Processing and storage procedures and problems are exactly the same as those of oysters.

A late comer to the ranks of processed frozen seafood products was reported in 1972 by North Carolina State University at Raleigh, N. C. There the school's Extension Service Laboratory developed crabmeat patties to be supplied to the consumer frozen but not cooked. The patties are unbreaded and are prepared from a combination of crabmeat and fish muscle with several food additives. The Extension Service described the procedure thus:

"The fish muscle portion was obtained from a combination of pre-cooked, flaked fish muscle and mechanically deboned fish muscle. . .selected from finfish of various species and sizes that are not readily usable for fillet-size portion servings. Regular crabmeat was used and the patties were prepared from a blend of fish and crab with food-grade additives and seasoning mixtures. The use of fish in combination with crabmeat results in a product relatively low in cost, conveniently portion-sized, easy to serve. Preparation for serving can be by open-flame broiling, pan-broiling, oven baking or deep-fat frying."

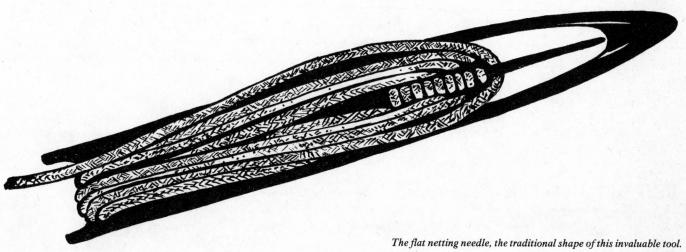

*The flat netting needle, the traditional shape of this invaluable tool.*

Linda Rogers

*History maker . . . the M/V Deep Sea, first vessel built specifically for the new king crab fishery when it began its growth in the 1950's. The Deep Sea worked both as catcher boat and processor but eventually was converted to factory work only.*

# 9/ Questions...Without Answers?

What course the future of American fisheries? Honest and concerned and thoroughly confused men of good intentions lie awake in the night trying to figure out the answer to that one. Unfortunately, most of these men are concerned directly and solely with fisheries harvest or management in one way or another. Not enough of them are highly placed and powerful enough politically to do something constructive about the problems that plague American fisheries.

It is customary among some observers of the fishing industry of the United States to write off that industry every hour on the hour (and sometimes on the half-hour) for any one of a lengthy list of reasons, many of them as specious as the observation that the chicken preceded the egg. This seems to be somewhat like the preaching of a funeral oration for a man suffering a head cold, not fatal pneumonia.

American fisheries are alive and some of them are quite well. They may not be the world's foremost fisheries in terms of vessels and men and tonnage of fish but they are fisheries that contribute to an industry that ranks sixth among the national fisheries of the world.

Some of these fisheries are supplying all the fish their present markets can absorb. Some cannot keep up with the demand for their products and processors must look elsewhere for supplies. This most notably is true in the tuna and shrimp fisheries. On the other hand, maximum exploitation of markets may be illustrated by the West Coast drag fishery which produces, with minimum effort by an underpowered and undermanned fleet, all the bottom fish that Americans of their respective markets are accustomed to eating as fresh fish, not, with slight regional exceptions, as fish processed into such products as those of the "heat and eat" persuasion. This latter fish comes almost entirely from abroad.

The bottom fishes that do get into the western American market represent only a small part of the total catch because two-thirds or more of each haul consists of fish that are perfectly edible and are eaten somewhere in the world but must be tossed back into the water dead by American fishermen because Americans will not eat them and there is no industrial demand for them. This is the state of the market and the resulting state of the fishery as reflected in a fleet of draggers composed of aging vessels manned by aging men, a remark made about the American halibut fleet of the 1950's by students of that peculiar fishery.

335

Despite their rejection of some choice species of fish taken by their own fishermen, Americans are the world's premier consumers of fishery products of all kinds over the whole range from Peruvian fish meal to Australian spiny lobster. This fondness for fishery products is true for the nation as a whole; it is not true on a per capita basis. American use of seafoods must be measured en masse. . .any small Japanese child eats more fish in a year than most Americans. Americans eat beef, pork and chicken by choice. For most Americans, seafood is a sometime, once-in-a-while treat.

Nevertheless, American consumption of food and industrial fishery products goes up each year. But most of it, too much of it, does not come from American fishermen. It comes from foreign fleets, including those of our Canadian neighbors of the Maritime Provinces whose fishing industry depends heavily on the American market. These imported fishery products come from nations whose leaders regard their marine fisheries and their ships as an economic keystone, a part of foreign policy and a ploy in international economic competition.

To illustrate the source of America's fishery products . . .in the years 1960-1970, American fishermen landed annually about 2.5 billion pounds round weight of food fish and about 2.5 billion pounds round weight of industrial fish. This was not enough to supply American demand in either category.

During those same years, the nation imported each year about 1.4 billion pounds (edible weight, not round weight) of food fish. This traffic was topped off with the average annual import of about 2.5 billion pounds of industrial fishery products. (Total imports of both types came to 13.2 billion pounds in 1968.)

These average annual figures add up to a considerable weight of fish and fish products and the figures for edible fish and shellfish merit another look. They show, if the annual American average landing of food fish is reduced from round weight to edible weight, say 1.5 billion pounds (and this is a generous estimate) that American fishermen supplied through those 11 years less than half of the nation's food fish.

Why should this be? There are two reasons.

One of them, greatly simplified, is that foreign fleets fishing off American shores have so decimated certain fish populations that their quest no longer is economically feasible for most American fishermen. Witness the attrition among every important species off New England and Eastern Canada.

Here, up to a dozen nations are competing for stocks of fish that simply cannot withstand the pressure. There comes a point in the biology of every living creature when a species reduced below a certain level cannot maintain itself, much less increase its numbers. This was the thing that faced the Pacific halibut before common sense and a strong-willed international regulatory agency took over.

New England and the Central Atlantic states have been hurt most seriously by this overfishing. Large foreign fleets have made themselves at home on the Pacific coasts of the United States and Canada but their depredations have not been as severe as on the East Coast because the foreigners concern themselves mainly with species that do not seriously concern North American fishermen.

But the potential for irreversible damage is there just the same. It appears tentatively among the stocks of Pacific Ocean perch and it appears definitely in the hake fishery where the Soviets in particular have cut deeply into the stocks. (It appears, oddly, that American salmon fishermen may have benefited, at least on a short-term basis, by the decline in the offshore hake population. It has been suggested that feeding immature salmon, competing on the same grounds for the same food with the hake, have taken advantage of the lessened pressure in the never-ending search for sustenance to maintain a higher ocean survival rate and to increase in average weight since 1965.)

## Loss By Default?

Americans have paid little attention to hake in comparison to their emphasis on such fisheries as those for halibut and salmon and several others. There have been a minor hake reduction and animal food fishery and several abortive attempts to establish going fish protein concentrate operations. It is conceivable that, if and when, Americans and Canadians do get around to worrying about hake, the species may well be on its way toward the limbo of the Pacific sardine.

Similarly, North American fishermen have barely touched the bottom fisheries of the Northeastern Pacific except for the previously mentioned limited market fishing but the Soviets and the Japanese have been doing it for them, enough so that these fisheries could be moribund by that indefinite time in the future that the North Americans do concern themselves with bigtime drag fisheries.

The second major reason for the decline of American fisheries or, more properly, for the American's failure to fill a greater proportion of consumer needs, is that it costs the American industry—NOT the American fisherman—so much more to deliver to the consumer the same tonnage of the same species of fish than it does the foreign industry.

It has been said that the American fisherman is pricing himself out of his own market. But, as far as the fisherman himself is concerned, this complaint does not appear to be justified. The entire chain of distribution, the ubiquitous "middleman" factor, apparently must be blamed. This American process of delivery of a product from the producer to the consumer may be the most efficient in the world but it also may be the most costly. It has been asked:[1]

---

1. Philips, Richard, Pacific editor, *The National Fisherman,* letter to the author May 17, 1972.

"Does it really cost us more to catch bottom fish (the food species imported in greatest volume) or is the high price (of the American industry) added during processing, transporting, wholesaling and retailing? The Soviets pay their fishermen much more per pound than we do, while the Japanese salmon fisherman gets prices equal to or better than the American salmon fisherman.

"A fish for which a West Coast trawler gets seven to nine cents a pound retails for eight to ten times that amount...our fishermen probably get less per pound for their catch than those of most fishing nations."

The individual fisherman faces another problem too. At the same time that he finds himself getting a low price for some of his species of fish, he finds himself caught in the bight between those low prices and the high prices he must pay for the supplies he must have to keep fishing for under-priced fish.

There is, also, a resistance, to the high prices the finished product meets when it finally gets to the retail outlet. Again, the chain of middlemen, efficient though they may be, have added to the cost of a product.

This may be illustrated to a degree by the varying consumer resistance to at least one pack of American salmon, and to the hard time the initial American scallop harvest from Alaska experienced during a decline in national prosperity. The salmon are the reds of Bristol Bay, the fish called sockeye elsewhere. When one-pound talls of reds reached prices above $1.15, sales slowed even during the comparative prosperity of the mid-1960's. The natural accident of a short supply of Bristol Bay reds and Canadian and Puget Sound sockeye in the years immediately following the Bristol Bay heavy-run years of 1965-66 took care of this situation in time. But, assuredly, if the years right after 1965 and 1966 had shown as many reds in the harvest as those two years, one-pound talls would not have been priced well above $1.00 or they would have languished, un-labeled, in warehouses in Seattle and Bellingham, Washington, for months or years longer than they did.

American fishermen are not doing themselves out of their fair markets by asking too much for their product. But the American fisherman feels himself just as entitled to participate in the good things of American life as his union-buoyed brother ashore. At best, his job is a hard, dangerous and thankless affair, a way of life in which loneliness and distance from home and family are compounded by discomfort and the very real and often imminent chance of injury or illness far from succor or death from a myriad of mischances. Can he be blamed for wanting a bit more of the consumer's dollar?

Just the same, that dollar stretches only minimally and the consumer has a natural and most human desire to get as much with it as he can for as little as possible. Foreign fish costs less than American fish even if that American fish is caught by United States vessels fishing side by side with East Coast Canadians. The American has been dealt out of the natural advantages conferred upon him by geography, populous market areas and professional skills of the greatest competence even though not by any design of his own.

Not all American fisheries are depressed by any definition of the term. Recovery is possible for those that do fall under any of those definitions although its prospect is not excitingly probable at a time in history when the United States government has abandoned its fisheries to the exigencies of foreign policy and the mercies of international economics. American fisheries capital got the message soon after the Korean War and began more and more to look overseas for places to go to work. It has done so successfully while, concurrently, some foreign capital has gone to work in American fisheries. The Japanese are managing in that way to get the Bristol Bay reds that escape their nets on the high seas.

Potential recovery measures are debatable; there are as many suggestions as there are men to make them. But it seems clear enough that increased productivity of the faltering sections of American fisheries cannot be helped by men alone. Nature must help too after men have been prevailed upon to give her a hand. This means, for one specific, that the crowded banks of the Northwestern Atlantic, where the fortunes of American fishermen are at their lowest, must be given a long rest, that cropping of the stocks must be reduced enough to allow nature to bring them back to a reasonable level of abundance. This relatively long-term solution will please no one.

A dimly-realizable potential of North Pacific fisheries for American fishermen—with the fisheries somehow made secure from the incursions of foreign fleets—has been startlingly outlined to the United States Congress. The proffered figures have been called "pie in the sky" statistics and are based mostly on Alaska stocks, but their proponents insist that with proper fisheries development and proper fisheries management, their realization is not impossible although it is not now a matter of great probability for several reasons. In any event, the physical dimensions of this northern state, presented along with the fisheries figures, are sufficient in themselves to awe observers more accustomed to the lesser dimensions of the "Other 49."

## Land Of Promise?

Alaska's coastline, for example, encompasses some 34,000 miles, more than the total coastline mileage of all the other states. The continental shelf adjacent to Alaska makes up 64 percent of the entire continental shelf of the United States.

Some of the tentatively projected figures for North Pacific fisheries production "sometime" are these, reduced to an annual harvest basis:[1]

Six million standard cases of canned salmon; 200 million pounds (in the round) of king crab; 250 million pounds, heads-on, of shrimp of all commercial species; 300 million pounds of Pacific Ocean perch; 140 million pounds of cod; 200 million pounds of sablefish; 1.2 billion pounds of herring; 1.5 billion pounds of flounder; 500 million pounds of other species, including presumably, the valuable Dungeness crab, plus "unknown" quantities of scallops, clams and other bivalve mollusks.

This adds up to a tremendous tonnage of marine life, some of it presently not acceptable in American and Canadian markets. Here is the pitfall, however; in the relatively near future there is little forseeable demand for these products with the exceptions of the salmon, the crabs and the shrimps. The sole factor, at least immediately, that would create a market for this great catch would be the complete cutoff of foreign fishery imports.

Only a complete, paranoid American retreat from the realities of world economics and world politics would bring about this latter happening, the only overt act that could make usable 1.2 billion pounds of herring that most Americans don't like or 1.5 billion pounds of flounder, most of it species that Americans don't like either, or 500 million pounds of "other species" they like even less. These might, of course, replace the tonnage of Peruvian fish meal shipped into the United States each year.

There is, to be sure, one other factor that could create a demand for all this North Pacific fisheries production and all the other production American fishermen could manage, a factor that some day may have to be met and braved in all its menace—a world faced with starvation.

It seems clear enough that a more liberal vessel subsidy program, one greatly expanded along the lines of the subsidy program for merchant shipping, would aid in bringing toward modernization the down-at-the-heel segment of the American fishing fleet—and that means most of it. Old age is the hallmark of the American fishing vessel. Annual new construction on all coasts does not begin to keep up with the rate of obsolescence. Such a program, obviously, would please American builders of fishing vessels. A further suggestion that Americans be encouraged to buy fishing vessels in countries where construction is less expensive than in American yards might not be as well received by American builders. But such a program could be nothing but a benefit to the men who must pay for and fish the boats.

It has been suggested (demanded, rather) by some figures in the American fishing industry that higher tariffs be imposed on imported fishery products. Superficially, this might appear enticing. But the power to impose tariffs lies on each side of every border. . .the United States does not export much in the way of fishery products in comparison to other manufactures and it is these other products that carry the weight of the

American export trade. They also carry more weight among Congressmen. In any event, higher tariffs on certain fishery imports probably would not begin to slow down their rate of flow and, anyway, even if modestly higher tariffs were to be imposed, their benefit would be little because the American fisheries industry's cost of doing business still would be going up.

It has been suggested too that a broad program of education aimed at making the American people more of a nation of fisheaters might work wonders for the industry. It would work wonders—for foreign fishermen. Any sudden and great increase in demand for fisheries products would be filled by more imports. Certainly, it would be desirable to train Americans to eat more fish but any increased consumption is unlikely to redound to the benefit of the American fisherman.

It might be pertinent to note that productivity of some highly desirable fisheries cannot probably be much increased beyond present yields. Included among these are the salmon, halibut and king crab fisheries. These appear to be producing at their maximum and no increase in that production can be expected unless some means is discovered to increase the ability of the stocks to replenish themselves. Over-fishing probably is the cause of the decline in king crab production; it is probable that the halibut, with that fishery under the strictest of controls except for the ravages of foreign trawl fleets, now is producing at or near its natural maximum. As for the salmon...man seems deliberately to have taken it upon himself to pollute his rivers and despoil his spawning grounds. It is amazing that any salmon survive in commercial quantities south of the Fraser River.

Any or all of these suggestions might do its bit toward the renewal of stagnating fisheries and the strengthening of those still in a decent state of health. But no real reconstruction of American fisheries can take place until the United States government can be made to realize, on a level somewhat higher than that of the director of the National Marine Fisheries Service, that America's fisheries are an important, if not major, part of the national economy and that their continued health is directly vital to the welfare of a million or more American citizens.

To this end, American fisheries must be regarded as and treated as an entity. They must be controlled as an entity and that control (or control of marine fisheries, at least) must be taken away from the states and vested in some agency of the federal government. Any such agency must be semi-independent of direct governmental control, perhaps one patterned after the quasi-official United States Postal Service.

As unwelcome as any such sweeping action might be to state fisheries management people or to the industry at large, it is the only way an all-embracing plan

---

1. Lauber, Richard B., Association of Pacific Fisheries, before Senate Subcommittee on Oceans and Atmosphere, July 1, 1971.

for the future of American salt-water fisheries can be worked out and effectively applied.

In this latter part of the 20th Century, the generations-old approach to fisheries management must be completely abandoned and replaced by a management scheme tailored to the era. Were it not so pathetic, the spectacle of the four Pacific Coast states attempting, fumblingly and fruitlessly, to efficiently manage their fisheries resources on a species-by-species, patch-up-here, patch-up-there basis would be humorous.

One agency control of marine fisheries might be ponderous but it cannot be any less effective than the present patchwork management so tenuously applied by the states and their politically- and industrially-dominated "management" agencies. The Canadian federal government runs Canadian marine fisheries. Not only does it regulate fisheries per se, it largely guides export market effort. The Canadians, with their central authority, seem to be handling both ends of their fisheries with considerably more competence than the Americans with their disjointed approach to the matter.

The Canadian action to reduce gear in the British Columbia salmon fishery has been cited. Similarly, the United States government or its designated independent agency must face up to the facts of limited entry and enforce limited entry in those fisheries where it is needed. The American salmon fishery should be the first beneficiary of gear limitation.

There are too many men, too many vessels and too many thousands of fathoms of webbing pursuing too few fish. Until this multiplicity of gear is reduced to a workable, professional-fisherman fraction of its present abundance, this fishery cannot long survive as a productive fishery for the men who work it. The same number of fish may be caught each season but increasingly the individual fisherman finds that he catches fewer of that total catch. The salmon fishery is not alone; there is too much gear in the American-Canadian longline halibut fishery, in the albacore fishery and in the king crab fishery.

Most importantly perhaps, the United States government must claim and exercise complete control over all fisheries along its coasts, beyond the Continental Shelf when necessary or beyond any mileage limit when necessary. This can be done by explicit and rigidly-enforced international—even global—agreement. Any such agreements perhaps could be best achieved by completely bypassing the United States Department of State.

Some nations regard their fisheries as an arm of the national government. This is especially true in the case of the Soviet Union. Others, Great Britain, Japan and Norway, for example, recognize that their fisheries are indeed major contributors to the respective national economies and the fisheries are treated with the importance they deserve.

Among those nations, fisheries do not suffer the buffets of an unfriendly economy without some degree of official concern. Where the private effort is insufficient, the government offers a shoulder to lean on. In the United States, the government looks the other way if it realizes at all that the United States does possess commercial fisheries.

To the conservative rugged individualist and to those who remember George Orwell's "1984," too much government is as bad as too little. But it would seem desirable that a government that once conned its Congress into appropriating $250 million to bail out a failing private concern on the sole ground that the company was a major "defense" contractor might be at least equally concerned with an industry scattered along all its coasts, one that employs thousands of persons and puts something like $2 billion dollars into the national economy each year.

That a sick fishery can be nursed back to health can be demonstrated in the case of the California-based distant water tuna fishery. During the years between World War II and the mid-1950's, the tuna people found themselves with both feet firmly in the bight of the line between cost and return. During those years, an old and relatively inefficient bait boat fleet tried fruitlessly to deal with rising costs, falling productivity and the rising tide of Japanese tuna imports. (The canning industry could not long survive without those imports in any event.)

By the book, the industry should have withered away just as the Pacific sardine industry was withering during those same years. But matters did not go according to the script. . .by 1970, the Japanese found themselves bound in the same cost-return squeeze with most of their fleet old, tired and inefficient while the Americans were outfishing the world for tuna.

What happened here?

In fairness, it must be said that good luck as much as anything, a happy quirk of fortune, saved the Americans in the 1950's. In 1954, Mario Puretic brought to fruition his Power Block while, at the same time, the net manufacturers began to recognize the mid-20th Century by adopting synthetic materials for their products. This is not meant to mean that the American tuna men did not quickly realize their good fortune. They took advantage of their new tools as rapidly as they could be produced.

The Puretic block gave them a smooth and fairly fast method of hauling their great seines. The synthetics gave them tough, comparatively light nets that fended off sharks better and handled tuna more gently. These developments, abetted by great fishing skill, so increased the productivity of the American seine fleet that the distant water bait boat became almost as obsolete as the dory schooner. Increased productivity led quickly to design and construction of the far-ranging, beautiful vessels that characterize the America tuna fleet. This newly-profitable fishery attracted new money just as any other enterprise with a promise of profit attracts investment. By the late 1960's the Japanese were coming to the Americans to inquire, with unaccustomed humility, how the thing was being done.

The tuna men had something else going for them too. They had a fishery completely free of "legislated inefficiency." For a change, these American fishermen found themselves harassed more by the governments of the West Coast of South America than by their own government or by the State of California, home base to most of them. Unfortunately, freedom from all governmental interference will not work in all American fisheries. It makes one shudder to realize how long unrestricted Pacific salmon and halibut fisheries would last. But this one time it worked. . .but the time seems to be here when someone is going to have to start regulating all the world's tuna fisheries because that fishery is no more inexhaustible than any other fishery.

It is too much to hope that this kind of dynamism might strike every ailing American fishery. It seems to have become the policy of the federal government and the states to hinder rather than to help technological advances in domestic marine fisheries. There must have been great soul searching among fisheries administrators concerning the wisdom of admitting the Puretic block to the salmon fishery.

## The Marine Jungle

The continued survival and well-being of American fisheries does not depend upon the Americans alone. That survival, in essence, depends upon every other marine fisherman in the world.

There is a degree of rivalry among some fishermen and their fisheries in the Northeastern Pacific. Salmon seiners and gillnetters, especially those of Washington State, probably would be at each other's throats if it were not for criminal and civil law. The trawler tries to get the International Pacific Halibut Commission to permit him to drag for halibut; the longliner lobbies against him with every weapon he can command. There is a rivalry for a limited salmon resource between the fishermen of Alaska and their cousins from "Outside" that northern state.

But these internecine rivalries are as nothing in comparison to the international rivalries among the Pacific Rim nations, most especially among those nations that fish the North Pacific. International rivalry pits American against Canadian and it pits American and Canadian, shoulder to shoulder, against the fisheries nations of East Asia. In turn these Asians (and the Russians in this context must be considered Asians) contend bitterly and sometimes bloodily for position in the western portion of this rich sea. This means the Soviet Union, Japan and South Korea headlong against each other in a race to ravish the fisheries of the whole Pacific Ocean, not solely those of the Northeastern Pacific.

The very meaning of the conservation of fisheries resources seems to escape the Japanese entirely. The Russians, thoroughly and cynically aware of the meaning of conservation, practice it only at home. The South Koreans, Johnnies-come-lately in distant water fishing, have never heard of conservation either but it is doubtful that they would observe its tenets at home or elsewhere.

There are other nations too with a developing interest in the fisheries east of the International Date Line and there is no reason to believe that these late starters will not move into them. Eventually, the Red Chinese will break free of the psychological ties that bind them to the mainland and move into the far-offshore fisheries just as the Philippines and the Taiwan Chinese have begun to do. This latter two nations, under the friendly persuasion of the United States pocketbook, may be somewhat more amenable to observance of sound fishing practices than others among their rivals. Somewhat and perhaps. . .

Some half-dozen Latin American nations claim territorial jurisdiction (not just fisheries jurisdiction) up to 200 miles off their coasts and it has been urged by some American and Canadian fishing interests that those two nations make such a claim in order to block out the trawl fleets of the Asians. The 200-mile conceit is tolerated by the big maritime nations, including the United States, because there is no effective international agreement to prevent it nor any effective international judicial body before which to argue it.

Neither the United States nor any other maritime power is likely to challenge this 200-mile claim by force which would appear presently to be the only effective, if not acceptable, manner of dealing with it. Thus, the United States, with its freedom of the seas tradition, would face overwhelming problems of its own in voicing any similar claim despite the hysteria of the 200-mile partisans and consequently will do nothing more than it has done by placing its disapproval on the record.

The nation's leaders believe or seem to believe that the United States has international interests surpassing those of fisheries. This undoubtedly is true but it does not seem fair that fisheries constantly should be denigrated in favor of such ephemeral fancies as the friendship of Peru or Ecuador, the 20th Century pirates of the Eastern Pacific Ocean.

The present United States concept of exclusive fisheries jurisdiction out to 12 miles was hard-won over the protests of the tuna and shrimp fisheries whose people have an understandable fondness for three-mile limits and, like the United States Navy, no liking whatsoever for anything beyond three miles.

This rush toward chaos can be halted only by a genuine (and doubtful of realization) international understanding of and agreement about the Pacific Ocean fisheries and the fisheries of all other oceans too. Every nation with a stake in those fisheries, even those who have not yet put in their chips, must be party to a wide-ranging agreement about fishing rights, fishing areas and conservation and of these, the latter is most important. They must be parties as whole-hearted in their observance of treaties, conventions, agreements, or

Pacific Fisherman

*The invader . . . a 252-foot Japanese stern trawler photographed off the Washington coast.*
*This vessel characterizes the big and efficient Asian fleets that have invaded North*
*American waters.*

whatever they may be called, in the same spirit in which the United States and Canada regard the International Pacific Halibut Treaty or the Fraser River salmon treaty, this latter an affair that increasingly rubs the Canadians the wrong way but one, nevertheless, that they observe scrupulously.

The absence of even one fishing nation from the company of those endorsing any fisheries agreement could be fatal. The likelihood of Red China agreeing to any such restriction on its movements appears remote. The specter of Red China roaming the fisheries of the Pacific Ocean, taking what it can where it can, is frightening. But even more frightening is the thought of all other fishing nations going the same route because the failure of one nation to conform to some rather elementary rules. Even if the Red Chinese cannot be brought to heel as far as ocean fisheries are concerned, it is no reason for other nations to turn into mad dogs too...but, other than the United States and Canada, they will. And it is too much to ask of these latter reasonable nations to run on the slow bell when the rest of the Pacific nations are sweeping clean the seas.

The fisheries of the Northeastern Pacific and the fisheries of the entire mother ocean are not inexhaustible; there is a limit to what they can produce. For evidence, see the whaling industry and the Pacific

sardine fishery. These examples show beyond dispute what man and nature working together, although not with purpose, can do in any fishery at almost any time.

The Soviets, thorough-going scientists fully aware of man's impact on the oceans and their crops, have published some intriguing figures concerning the potential of the oceans, all the while their working fishermen are raping them. Their assumptions only confirm what Americans and the British decided long ago, but the Soviet marine scientific community's espousal of them fits poorly with their national fishery effort.

The Soviet calculations, more optimistic actually than some Western calculations, say that only 33 percent of the total ocean area is hospitable to fish and shellfish life, and most of this welcoming water is over the continental shelf and the near-slope where some 70 percent of the world's phytoplankton grows. The Soviet assumption is that these same relatively-shallow waters produce enough zooplankton to support 300 million metric tons of fish and large invertebrates. They estimate further a 90-million-ton maximum annual sustainable yield of all marine fishery products.

The irony of the Soviet writing is this, paraphrased in translation from "Vodnyi Transport," of June, 1970, facts that seem to be apparent enough to Soviet scientists but incomprehensible to Soviet fishermen:

"The biological equilibrium...can be disrupted by large-scale commercial fisheries which cause decline in abundance of stocks...in one hour of trawling, a large stern trawler takes an annual 'crop' of 10 square kilometers of the Continental Shelf; one purse seine haul takes the 'crop' of 100-500 square kilometers."

The paper adds that "intensive combined fishing efforts of several nations in relatively small areas have depleted resources." The author cites halibut off the North American Pacific coast; flounder off Australia and in European waters; Pacific and Atlantic salmon, and the Atlantic Ocean perch.

## Death Of A Fishery?

The publication fails utterly to remind its readers that Soviet fleets have been as guilty as any in the world in the matter of large-scale fisheries and their disruption of the orderly cycle of marine life. The Soviets (and the Japanese too) should think long and carefully about their respective whale fisheries.[1]

"In early July, 1971, the United States Senate unanimously passed a joint resolution instructing the State Department to negotiate a 10-year international moratorium on the killing of whales. A similar resolution, though in a weakened concurrent form, was reported from a House committee.

"This action follows centuries of unbridled carnage. Eighteenth and 19th Century whalers, harpooning from small boats, ravaged the initially enormous populations of right and bowhead whales, leaving only scattered survivors. With the development of steam-driven "catcher boats" and cannon-fired harpoons tipped with explosive shells, the whalers turned on the faster rorquals, such as the blue and finback. Antarctic waters, populated seasonally by a vast host of previously unmolested whales, were invaded in 1904, first from shore stations, then by pelagic fleets with floating factory ships. For over 35 seasons, fleets from an increasing number of nations steamed south for the Antarctic summers.

"World War II afforded the whales respite, as their persecutors turned their attention to one another. By this time the blue whale had been reduced to a fraction of its former Antarctic population, and whale numbers had everywhere declined before the brutal mechanized assault.

"In 1946, a whaling convention was signed by 17 nations, establishing an International Whaling Commission charged with 'conservation and rational utilization' of the world's whale resources. What followed the advent of 'conservation management' late in 1948 was not a reduction of the kill to provide for a sustained yield but a dramatic acceleration of the massacre. The past two decades have been the most sanguinary in the slaughter-glutted history of whaling. During the late 1950's and early 1960's, new killing records were set. In 1962, the worldwide kill reached 67,000—well above the maximum annual kill of laissez-faire whaling.

"The International Whaling Commission, since its establishment, has been dominated totally by the commercial whaling interests, who have simply ignored any suggestion of restraint. Blue and humpback whales were still common in 1948. They now exist in pathetic remnants, near the abyss of biologic extinction. Finbacks were previously the most abundant of the baleen whales. Having borne a decade of unprecedented technological havoc following the "commercial extinction" of the blue whale, finbacks are now scarce. The industry, using spotter helicopters, radar and sonar, presently focuses its attack on the smaller sei and minke whales, and is engaging in a relentless, oceanwide pursuit of the sperm whale. In 1970, Japanese whalers took 2,000,000 dolphins and porpoises.

"Since 1960, nation after nation has been forced to cease whaling because of a dearth of whales. Today, only the Japanese and Russians continue large-scale pelagic operations. The end, long predicted, is now clearly in sight; an end of whaling, and to the travesty of 'conservation management.'

"This 'final solution' cannot have been, and is not being, unknowingly imposed. The Japanese and Russian whaling industries are now quite deliberately 'whaling themselves out of business.' An entire order of unique and magnificent mammals is being destroyed, fed into the pitiless jaws of the factory ships. The reason why resides simply in the decision of men involved in the whaling business that it is much more profitable to whale on a very large scale for a few years until the whales are exterminated than to whale indefinitely on a limited sustained yield basis.

"For the overwhelming majority of humans, the loss of the whales is a total loss, unrelieved by the slightest gain. We must declare a moratorium on whaling to permit the slow restoration of a previously enormous marine resource, mindlessly decimated. We must permit the survival of living beings with immense and convoluted brains, with a greater cortexual mass than our own.

"Yet the Whaling Commission coldly parcels out this once-vast marine population. The whaling nations override the objections of the U.S. and United Kingdom, and again set quotas far above those considered sustainable. The State Department opposed the joint resolution for a moratorium, on the grounds that it might 'jeopardize the vigorous role of U.S. leadership' within the commission. During the House hearings, Government officials complained that an attempt to preserve the whales might make the U.S. 'look foolish.'

"While human quibbling over propriety and profits continues, the last chance to halt the killing in time to save the great whales may be fading. One hundred and twenty-five times each day, on an average, a whale is located, and killed with an explosive harpoon. All that the whales are, all that we might learn from them, is

---

1. Garret, Tom, untitled article on unrestricted whaling, *New York Times*, September 1, 1971.

being lost. Their profound living songs, at the moment we have at last heard them, and might hope to understand them, are being effaced, irrevocably, from the earth's oceans."

The previously cited Soviet paper concludes with the suggestion that countries with major marine and distant-water fisheries get together "to organize a scientifically supported commercial fishery." Since this group would include most of the nations of the world, it is difficult to see that it would be any more fruitful in sweeping efforts toward internationalization of fisheries than is the hamstrung, ineffective and fretful United Nations in affairs of world import.

The Soviet author also supports an international study of the biological resources of the ocean and the conditions, methods and techniques for "rational utilization and multiplication of marine fauna." A study of this nature, carried out with the same whole-hearted fervor among major nations that characterized the International Geophysical Year of 1957-58 could not help being of the greatest importance. For the first time, man would know pretty well what he can depend upon in the way of food and other articles from the sea.

Attempts at international cooperation in regard to fisheries have had a fair amount of success in the Northeastern Pacific during much of the past half-century. Agreements as workable as possible have been arrived at by the Americans and Canadians in the International Pacific Halibut Treaty and the International Pacific Salmon Fisheries Commission. Workable too, although friction simmers beneath the facade of diplomatic smoothness, is the International North Pacific Fisheries Compact, that three-nation agreement among the United States, Canada and Japan. Russian and Japanese attempts to regulate fisheries on their side of the Pacific use gunfire and confiscation as a means of enforcement on too many occasions. The North and South Koreans carry on a guerilla warfare at sea.

Suggestions as to international cooperation and regulation of fisheries range across the whole spectrum of possibilities. Some envision naval escort and consequent gunfire for such as the United States tuna ships fishing the West Coast of South America. Most, like this one, are self-seeking, the pleadings of special fisheries interests.

Most of these suggestions have a common thread running through their patterns—they are concerned mostly with divvying up the take. They concern themselves not at all with perpetuation of a favored species, with putting back into the water as many or more fish than they take out. This replacement of the catch, in the present state of knowledge of pelagic fisheries, may not be the most practicable of endeavors but it does deserve an increasing amount of study. Attempts to rear tuna being made by Japanese and American researchers may come to be a going concern.

## One Answer?

It is apparent that all the world's exploited fisheries must some day be exploited and propagated just as artfully. A possibly-workable suggestion to this effect came in July 1970, in that month's issue of *ALASKA*® magazine. The publication is influential in the northern state from which stems its name; it is widely circulated through the other 49 states and is read in more than 70 countries abroad.

The magazine does not concern itself with policies and it is guided by its masthead slogan reading:

**"We labor in faith that man will accept controlled use of his wilderness resources and in so doing he will not only preserve but he will also prosper."**

But conservation of fisheries cannot be divorced from domestic politics or from international politics, latterly especially in the case of those fishes that roam across a half or more of the world's greatest ocean.

The *ALASKA*® editorial, written by its editor and publisher, Robert A. Henning, recognized this simple fact of life and proposed that the world do something about it. He did not call for naval escort, punitive action or any other of those senseless half-measures that appeal to the thoughtless who are concerned with revenge as much as they are with fish.

Henning suggested instead that the world begin to put back into the sea a measure of fish equal to that taken from it. His specific subject was the salmon, both of the Pacific and the Atlantic, but implicit in his argument was the thought that such good works need not apply just to salmon.

He explored the general problems of the North Pacific fisheries and suggested what may become a reality —that Russia and other North Pacific fishing nations be brought into the present International North Pacific Fisheries Commission although he observed, as mentioned a few paragraphs back, that "regulations and agreements for and between nations may only serve to delay depletions."

Then, with no bitterness and with complete logic, he wrote:

"One of the fish species in greatest jeopardy is the Pacific salmon, a preponderance of which are hatched in North American rivers and lakes and which go far to sea to grow to maturity before returning to parent streams where they spawn and die.

"It used to be that nobody had the slightest idea where the salmon go and, until the fish returned, fully grown, to continue the business of renewing their kind, fishermen got no crack at them. That is different now.

"In the Pacific, it has been determined there are vast nursery grounds for salmon in the mid-Aleutians area and that nets and longlines can catch large numbers of immature salmon, a practice condemned by many biologists who argue that to catch a fish weighing only a few pounds in its youth instead of waiting until it is several pounds heavier at maturity is the worst kind of fish exploitation.

"The same situation, incidentally, has also resulted in the case of the Atlantic salmon. Hundreds of early-day Atlantic salmon spawning grounds have been wiped out by pollution and civilization, and now the long-hidden mid-Atlantic nursery 'growing-up' salmon grounds have been discovered off the Greenland coast just as the mid-Aleutians feeding grounds of the young Pacific salmon have been discovered. In the Atlantic, Danish and Faroe Island fishermen apparently have refused to go along with fishing restrictions designed to protect the young Atlantic salmon.

"In the Pacific, Japanese are exploiting high seas salmon. Russia claims it is doing little high seas taking of salmon but cannot help but take some—and perhaps the mere dragging of the bottom for crabs and flatfish and taking of hake and saury in midwater trawls may in itself be as destructive as actual salmon taking, subtracting from the marine environment various supporting components of the complex biological food supporting chain necessary for growing salmon...

"So what will we do? Pressure of gear is mounting. Numbers of fish are declining. The chance of political solution to the many problems involved is a poor bet. It would seem then that perhaps a new tack is indicated. Perhaps we should concentrate on putting fish back instead of just reducing the number of fish we take.

"We say the Japanese are catching our salmon. So? Maybe if they would put enough fish back...? The Danes are great farmers of trout. They ought to be good salmon farmers too.

"Perhaps we should all combine our technological abilities and perhaps governments and fishermen could pool money as well as brains toward creation of an unprecedented salmon rearing program. If we could all submit our waters to an international group committed to a joint effort toward rearing far more salmon than we ever dreamed we might have had—farm the great reaches of the Yukon, the Kuskokwim, the Anadyr, the Amur and dozens of other major rivers; develop for spawning (and protect them from other use or misuse) the myriad little streams of Alaska, of Sakhalin, of Hokkaido, of British Columbia, Finland, Scotland, Iceland, Labrador, Nova Scotia, Maine—wherever salmon run or lived in years before. . .to create giant hatcheries in a variety of locations to supplement improved natural spawning potentials. . .in short to internationalize the development of all existing and potential parent salmon streams, to further create vast supplementary hatchery systems, and to finally conserve and protect the international salmon on his international nursery grounds with international fisheries police.

"We could learn much more together about the business of building salmon numbers. We could create salmon runs beyond any in the memory of man.

"We might even learn to get along with each other."

# Notes on the Literature

The world's literature concerning its commercial fisheries antedates that of its sport fisheries by about 3,500 years. But we can take a few liberties and extend the story of commercial fishing even farther into the past if we date its first "literature" from the Upper Paleolithic era when later Cro-Magnon man rather skillfully depicted fishes and scattered the likenesses among great animals of the hunt on the walls of some of his caves in the south of France and in Spain.

In times of surplus, fish must have served as well as any medium when there came a chance for barter among members of the band or with members of a neighboring band and the exchange of a fish for a bone of meat or a spear head would seem to qualify the procedure, from the taking of the fish to the consummation of the trade, as commercial fishing. For the exchange of the catch for something of at least equal value is commercial fishing.

Sport fishing and its mass of literature had to wait until mankind or parts of it had something of an assured food supply with its consequent freedom from the never-ending drive to fill empty stomachs. It took comtemplative leisure to produce Izaak Walton's 17th Century writings and the earlier work of Dame Juliana Berners, her 1496 "Treatyse of Fysshinge With an Angle." These are almost spot news when measured against the record.

In the Western World, the true history of commercial fishing begins with the Egyptians of the Middle Kingdom, its rise set at about 2,000 B.C., when chiseled inscriptions and the written word record the catch and sale of fish from the Nile and from the Northern Sea. Even the Assyrians, mostly a land-bound people, told of such fishing and the story runs fairly continually through the histories of the later civilizations of the Mediterranean rim.

By 100 B. C., during the final years of the Roman Republic, fish and shellfish farming was being practiced in favorable sites around the boot of Italy. The first oyster farmer of record in the West was a Neopolitan. But in that time, the peoples of Eastern Asia were old hands in the art of aquaculture and it is only lately that the West has begun to catch up with them.

The works on which the historical matter in this book is based are relatively few because this is not a history. Those used date from 1900 although they, in turn, have their bases far back in maritime history. Note the 1376 complaint about trawls, carried to Parliament by English fishermen, related by W. L. Scofield in his "Trawl Gear in California." Among the most valuable single volumes were John N. Cobb's "Pacific Salmon Fisheries," United States Bureau of Fisheries Document No. 1092, 1930; A. T. Pruter's "Commercial Fisheries of the Columbia River and Adjacent Waters," Bureau of Commercial Fisheries, 1966, and the above-mentioned Scofield monograph and its companion works, "Trolling Gear in California," 1955, and "Purse Seines and Other Round-Haul Nets In California," 1951, publications of the California Department of Fish and Game.

The most rewarding source of the month-by-month history of the American and Canadian fisheries of the West Coast lies in the file of the magazine *Pacific Fisherman* from its founding by Miller Freeman in 1903 to its merger with the *National Fisherman* on January 1, 1967. No one seriously interested in Pacific fisheries can overlook this massive accumulation of facts, written contemporaneously with events—or as near thereto as the mechanics of publishing allowed—by men well-trained in their field. The coming of the Iron Chink or the floating salmon trap or the birth of tuna packing are history now; they were events of the day to the early staff of *Pacific Fisherman*. Much of the magazine's advertising material, especially that prior to 1920, was as illuminating as its text.

The comparatively small amount of biological material came from varied sources, chiefly from publications of federal and state agencies. Francesca La Monte's "Marine Game Fishes of the World," Doubleday, 1952, with its global survey of species, proved rewarding because of its many illustrations in color. Most species referred to in *Fisheries of the North Pacific* are found in the La Monte book.

No written work served materially in the writing of the segments on fishing vessels and on fishing gear construction and use with the exception of the Scofield monographs on gear. These have become somewhat dated, however, because of the continuing evolution of gear and the introduction of new materials. Nowhere else, at the time of this writing, except in a handful of research papers, could be found other such matter specifically applicable to the small-vessel, fiercely individualistic coastal fisheries with which this book concerns itself.

The Bureau of Commercial Fisheries' 1956 five-part series, "Refrigeration of Fish," furnished background for the sections relating to that many-faceted endeavor. This series was supplemented by a number of research papers from the BCF and its successor, the National Marine Fisheries Service, while from across the Atlantic for consultation came G. H. O. Burgess' "Developments in Handling and Processing Fish," a result of the author's work at the famed Torry Research Station, Aberdeen, Scotland.

Every person involved in the processing of sea foods, and I as well, must acknowledge a debt to Norman D. Jarvis for his two definitive works on that subject. The first is "Principles and Methods in the Canning of Fishery Products," United States Fish and Wildlife Service Research Report No. 7, 1943, and "The Curing of Fishery Products," Research Report No. 18, 1950. In the years since their appearance there has been no basic change in the canning of fish and shellfish or in their salt, smoke and pickle preservation. There has been, however, the appearance of a multiplicity of cooked and uncooked fish and shellfish preparations based on the popularity of the home freezer that were barely on the horizon when Jarvis did his writing. A single-volume survey of the procedures commonly used or in development in this field would appear in order.

Fisheries research is a multi-tiered structure based on the work of many individuals. Its results appear every year in thousands of research papers of greater or lesser import on every phase, no matter how specialized, of biology, vessels, gear and gear use, and fisheries-related techniques of every nature.

To grasp some of this effort, I have read, at least in part, more than 700 publications in addition to the sources already mentioned. To the authors of these laboriously researched and written efforts, I offer my thanks for their contributions to my work and my apologies for their necessary anonymity.   ■■■

*Robert Browning*
*Edmonds, Washington,*
*1974*

# *Appendices*

## APPENDIX I
### Part 1

### Characteristics of Fishes

*By Dr. Dayton L. Alverson*
*Director, Northwest Fisheries Center,*
*National Marine Fisheries Service*

In the systematic arrangement of animals, fish are grouped in the phylum *Chordata* which encompasses all vertebrates, including lampreys, sharks, bony fishes, amphibians, reptiles, birds, and mammals. The fishes are generally subdivided by taxonomists into two major classes: *Selachii* and *Pisces*. The *Selachii*, which include the sharks, rays, and chimaeras, differ considerably in external shape, but have a fundamental similarity in that all have cartilaginous skeletons with primitive ribs and a complete brain case (condocranium).

In *Pisces*, or the true fishes, the skeleton is normally bony and comprised of a vertebral column, ribs, and a skull having many distinct bones. Through general use, both shellfish (*mollusks* and *crustaceans*) and the whales and porpoises (*cetaceans*) are at times included with the fishes. However, these animals, structurally, are far removed from "fin fishes."

### Form

Fishes possess a wide variety of external shapes or forms. The most common fish shape is that normally referred to as fusiform, that is, roundish and somewhat compressed, and tapering towards the head and tail. The fusiform is considered to offer minimum resistance in water. The Pacific salmon is a good example of this shape.

In some fishes the bodies are greatly compressed such as in the *Bramids* and sunfishes. Those fishes that are thin and flat are referred to as depressed, and include such fishes as the skates and rays. Flatfishes or flounders are often thought of as being depressed because they appear flat and thin. Actually, these fish are compressed, and the adults orient themselves with one side down during swimming and feeding. In the eels, which are also true fishes, the body is roundish and snake-like.

### Fins

The appendages of fishes may be considered as analogous to the limbs of mammals. There are normally two pairs of lateral fins; the pectorals which lie just behind or below the gill opening and the pelvic or ventral fins which lie near the median ventral line. The pelvics may be positioned well back on the fish as in the sharks, herring, and salmon, or forward near the pectoral fins, such as in the tunas. Fins on the median line of the back are termed dorsals, and depending on the species there may be one, two, or three such fins. Along the median ventral line there may be one, two, or three such fins. Along the median ventral line there may be one or two fins which are termed anals. The tail of the fish is referred to as the caudal fin. Various fins are, at times, absent, depending on the species or may be (in the case of the dorsal and anal fins) continuous along the back or mid-ventral. In some fishes such as smelts, trout, and salmon the second dorsal is characterized by a complete lack of supporting rays, the fin being a fleshy protuberance. A fin of this type is referred to as an adipose.

Fins of fishes are characterized by the nature of the supporting rays. Those fishes whose fins lack spiny supporting rays such as salmon and trout are called soft-rayed fishes, while those having fins supported partly or totally by spinous rays are referred to as spiny-rayed fishes.

The greatest number of median fins generally occurs in the cod family which has three dorsals and two anals. The *cluepoids* (herring, sardine, menhaden, etc.) possess only one dorsal fin.

There are a number of adaptations of fins for specialized uses. In the remoras, the first dorsal has been modified into a sucker-like disc which the fish utilizes in attaching itself to other marine animals. By contrast, lump suckers have a modification of the pelvic fins which form an adhesive disc. In the trigger fishes (*Balistidae*) the first dorsal has a strong, stout spine which may be locked into an open position when assuming a defensive posture. In some of the more rapidly-swimming fishes (tuna and marlin) grooves occur along the median and ventral line into which the dorsal and anal fins may be recessed during swimming. The elongation and enlargement of the pelvic and pectoral fins in flying fishes is an excellent example of specialized use of the paired fins. Although these fins are not used for actual flying in the sense of propulsion, they are employed for gliding or soaring.

A fish generates most of its propulsive power through undulations of the body and rapid lateral motion of the caudal fin in the water. The remaining fins generally serve in balance and turning. However, in some fishes the lateral fins are also utilized for propulsion.

### Protective Covering

In the majority of modern fishes the body is covered with scales, although a few are naked. Fish scales vary greatly in shape and size and are usually grouped into two categories: *cycloid* and *ctenoid*. Cycloid scales are characterized by smooth posterior margins and are frequently found in fish having soft dorsal fins, such as salmon and trout. *Ctenoid* scales have comb-like (spious) posterior margins. They are commonly found in fishes having spinous rays in the dorsal fin. Both types of scales occur in some fishes. In the more primitive, bony fishes such as sturgeon, scales consist of bony plates covered by ganoin *(ganoid scales)*. The protective covering in sharks is in the form of scale plates, each containing a small spine which stands upward or curved slightly back. The scale-like features of the sharks have a dentine base and a spine normally capped with hard enamel. They are formed in a similar manner to teeth and may be considered as minute teeth set into the skin. The scales of sharks are known as *placoid* scales and the minute teeth-like structures are referred to as denticles.

## Eyes

The eyes of most *elasmobranch* and *teleost* fishes follow the normal vertebrate structure. They are provided with a lens capable of throwing an image upon the photosensitive retina, and according to Nicol, show the following layers: a felt-work of optic nerve fibers, two layers of ganglion cells separated by a synaptic layer, a basal layer of rods and cones, and the photoreceptors proper.

Fishes generally have their eyes positioned on the sides of the head. However, in some they may be placed in a superior position (on top of the head). Most of the bony fishes are myoptic (adjusted for near vision), and the eyes at rest are set for near vision. It has generally been considered that fishes do not have acute vision. However, there is increasing evidence that the sharks have distant vision (hypermetropia) and may be able to distinguish between the shapes of small objects at a considerable distance.

Of the two types of photosensory cells, the rods and the cones, the rods are most effective at low intensities and the cones, being less sensitive to light, are able to function over a range of higher intensities.

Bony fishes which inhabit areas where wide ranges of light intensities occur have both rods and cones in the eyes. Forms which inhabit dimly lighted areas at great depths may contain pure rod retinas. The eyes of fishes inhabiting dimly lighted regions are considered to be the most sensitive in existence; however, in some of the very deep-water fishes the eyes may be degenerate and the fishes blind.

Considerable debate has occurred regarding the capabilities of fish to detect difference in color. There seems sufficient evidence now to indicate many species of fish do detect color differences although considerable variation between species may occur in the ranges of colors that can be delineated.

## Smell

The chemoreceptors in bony fishes are localized in the nasal pits forward of the eyes on the dorsal surface of the snout, and in sharks and rays the olfactory pits are located on the ventral surface of the snout. Olfaction in fishes is important in determining certain behavior patterns such as providing clues for orientation, searching for food, and avoidance reactions. Chemoreception in the *selachians* and bony fishes is highly developed, and experiments have indicated that the fish can detect extremely small differences in concentrations of dissolved substances. Brett and Mackinnon demonstrated that adult salmon were repelled by water in which human hands had been rinsed, and Hasler relates an experiment in which fingerling salmon *(Oncorhynchus)* were capable of detecting solutions of morpholene in concentrations as low as one part in a billion.

## Other Sense Organs

Although fishes do not have external ears, an elaborate inner ear (labyrinth) system exists. The labyrinth consists of two vertical and one horizontal semi-circular canals which connect to a membrane sac, the sacculus, containing the earbone organs *otoliths)*. In the bony fishes there is no direct connection with the exterior. However, in sharks and rays there may be an external opening through a small duct.

Lowenstein states the labyrinth has four functions: (1) to maintain and regulate muscular tone, (2) to act as a receptor for angular accelerations, (3) to act as a gravity receptor and (4) to act as a sound receptor.

The lateral line system in fishes is normally well developed and may be seen along the sides of most fishes. Most often the lateral line can be seen starting back of the gill opening on the flanks of the fish and extending back toward the caudal fin. In some fishes several lateral lines may exist. Although not always detected, branches of the lateral line extend onto the head. The lateral line system acts in a sensory capacity and is considered by most physiologists to be concerned with orientation or "distant touch." Lowenstein states: "In aquatic environment where optic orientation is of reduced accuracy, these organs are capable of supplementing vision by helping to localize objects at a distance. These objects may be moving (prey animals) and thus constitute the focal point of a mechanical disturbance, or their presence and localization may be perceived and computed from the time relations of reflected water waves set up by the swimming movements of the fish itself."

In the sharks and rays there is a system of jelly-filled canals in the rostral region called the *Ampulla of Lorezino*. The exact function of this canal system is not well understood. It has been suggested that it functions in detecting changes in hydrostatic pressure. There is more evidence, however, that it may be a thermoreceptor.

## Reproduction

Fishes may give birth to living young which develop as embryos in the uterus. They may incubate the eggs internally, or they may spawn the eggs into the water where they are subsequently fertilized. In some sharks the embryo is completely developed internally and the female gives birth to fully-developed young. The gestation period in sharks is relatively long, and in the dogfish shark (*Squalus*) it may be as much as 20 months. In some of the sharks and rays the embryos are partially developed internally and subsequently the female deposits the developing embryo in a capsulated "egg case" from which further development and hatching takes place. In the bony fishes the viviparous perch (*Embiotocidae*) are an example of a group which gives birth to living young. In the majority of fish, however, the eggs are spawned and fertilization takes place in water. The number of eggs spawned varies considerably between species and may exceed a million eggs (cods and halibut). The eggs in some species are deposited directly onto the sea floor *(demersal);* in others, they may be laid in large sticky masses where they adhere to rocks or plant growth. In many of the pelagic (offshore) fishes the eggs drift freely with currents.

Some fishes such as the Pacific salmon (*Oncorhynchus*) spawn only once and then die; however, the majority of fishes spawn several times, and fishes such as halibut which have a long life span may spawn many times. Generally a female will spawn only once each year, but the peak spawning period of a species may extend over several months.

## Air Bladder

Any fisherman or sportsman who has cleaned marine fishes has probably noticed an elongate, double lobed sac lying close to the back bone in the abdominal cavity. This thin-walled, semi-transparent sac is referred to as an air or swim bladder and it presumably functions as a hydrostatic organ. Fishes living at mid-depths or in surface waters are required to expend energy to maintain their position whenever their body density differs from the surrounding water. When a fish moves vertically, it passes through pressure gradients and thus the gasses in the air bladder are expanded or contracted, and the fish's density altered, depending on the direction of movement (up or down) Because of the change in the volume of gasses in the air bladder, the fish then must bring itself back into equilibrium with the ambient water. This is accomplished through exchange of gasses across the body wall of the bladder or by means of a duct which connects directly into the gastrointestinal tract. Gasses are added when the fish moves into depths of higher pressure, and gasses are removed when the fish travels to regions of less pressure. Through the function of the air bladder the fish is capable of maintaining energy expenditure. The air bladder also functions as a sound receptor, and in some fishes as a sound producer, as a lung, and in sensory functions.

## Respiration

The gills of fishes are analogous to the lungs of higher vertebrates and serve as an organ of respiration. Free oxygen enters into the circulatory system and carbon dioxide is expelled through the slender gill filaments.

## Migrations

The migration of fishes is a topic which is of considerable interest to the general public and the homing of salmon to parental streams is well known even to the school boy. In recent years it has been found that the Pacific salmon *(Oncorhynchus)* inhabit the ocean waters at great distances from shore through most of their marine life, and on approaching maturity migrate vast distances to the rivers of their birth. Although homing is well developed in the salmon, some straying to other streams does occur. The king salmon is known to migrate even to headwaters of large rivers such as the Columbia, Yukon, and Sacramento. Fish such as salmon, and shad *(Alosa sapidissima)*, which migrate up rivers to carry out spawning are referred to as *anadromous* forms, and fish such as the eel *(Anguilla)*, which migrate from fresh water to salt water to spawn are referred to as *catadromous*.

Although the homing of salmon to their parental stream is one of the most often discussed migrations of fishes, many marine forms show similar spectacular migratory patterns. Atlantic cods undertake extensive migrations. Such migrations include movements to and from nursery or feeding areas and distance migrations such as from West Greenland to Iceland and from the Norwegian coast to the Barents Sea. From some of the tunas the migrations may be transoceanic. Albacore *(Thunnus alalunga)* tagged off the Pacific Coast of the United States have been recovered off the coast of Japan. Not all tunas, however, seem to display extensive migrations. The yellowfin *(Neothunnus macropterus)* of the eastern Pacific do not appear to be involved in transoceanic movement. At least one species of Pacific flounder, *Eopsetta jordani*, has been noted to undertake rather extensive migrations along the coasts of Washington and British Columbia, the movement being north in the summer and south during the winter months. There is some evidence that the adults of this species may return each year to the same spawning area. Sharks may also undertake long migrations, and there is at least one record of a dogfish shark *(Squalus)* tagged off the Washington coast, subsequently being recovered off Japan.

Although geographic migrations of the type noted above are the best documented movements of fishes, a number of species are known to undertake seasonal vertical movements, and in some, large-scale diurnal vertical movements occur. The Dover sole *(Microstomus pacificus)* may move from relatively deep water along the continental slope (200-400 fathoms) in the winter, to the shallower water (60-150 fathoms) of the continental shelf and upper slope during the summer months.

The patterns of migration of fishes vary considerably between species, and may vary within species, depending upon age. Among bottom fish there is a general "rule of the thumb" that the larger, older fishes will inhabit deeper waters than the young (Heinckle's Law). There are many causative factors responsible for fish movements. Local movements may be related to changing physical or chemical conditions in the ocean or varying food supplies. The homing of fish to certain areas for spawning is probably most closely related to the need to provide a requisite spawning medium for the developing egg.

The patterns of migration and behavior of fishes may have considerable influence on fisheries which seek to harvest them. Some of the more pronounced migratory patterns such as the movement of adult salmon, shad, and other species to fresh water make them easily accessible to man. Similarly, in ocean fishes the seasonal movement of fishes along coastal areas, and the movement of fishes into inshore areas for spawning, are patterns which have made these species susceptible to harvest. In addition to migrations, patterns of distribution play an important role in fish harvest. Those species which aggregate into large schools such as herring, cods, tunas, etc., can be taken in large quantities. However, those species which are dispersed through much of their adult life are more difficult to harvest and must generally have a high market value if they are to come under exploitation.

■ ■ ■

*(Note: Dr. Alverson's detailed bibliography of literature quoted has been omitted as not essential to his premises.)*

## APPENDIX I
### Part 2

### Names of Fishes

*By Daniel M. Cohen*
*Research Zoologist,*
*(then) Bureau of Commercial Fisheries*

Commercial fishermen, the food processing industry, anglers, scientists, writers, federal and state agencies, students and teachers and many others use names of fishes. Communication about these animals is impaired because some kinds of fishes have no names, others have more than one name, and some names are used for more than one kind of fish. The obvious solution would be for every species of fish to have one name that was universally recognized as referring to it alone.

Because they are essentially less complex, let us first consider scientific (Latin) names. The rules for the formation and use of scientific names are governed by the voluntary adherence of zoologists to the International Code of Zoological Nomenclature, most recently revised and published in 1964. In essence, the Code tells us that a zoologist who finds a species that lacks a scientific name may describe the species and give it a Latinized name, subject to certain rules and recommendations.

The name is composed of two parts. Let us take as an example the goldfish, *Carassius auratus*. *Carassius* is the generic name; one or more species may be included in the genus and will have *Carassius* as the first part of its scientific name and refers to only one species of *Carassius*. Both names together, *Carassius auratus*, make up the scientific name for the species that we recognize as the goldfish.

The starting point for scientific names is a book by the Swedish biologist Linnaeus, published in 1758. No scientific names published before that date are admitted to the system. If for any reason a zoologist gives a scientific name to a species that already has one, the name with the earliest date after 1758 takes precedence. If for any reason the same scientific name is given to two species, the last-named one must be given a new name. This system offers a relatively stable method of communication. *Poisson rouge* in French, *chin-yu* in Chinese, *chrusoparon* in Greek, *aranyhal* in Hungarian, *kingyo* in Japanese, *zolotoi ribki* in Russian, and *dorado* in Spanish, are all different names for what we call the goldfish. Communication about goldfish is difficult without the universally recognized Latin name, *Carassius auratus*. It is a worldwide code word.

International currency notwithstanding, scientific names cannot replace common names for several reasons. Latin has no meaning for the average person; having two words in a name is cumbersome; and scientific names are subject to change, for as well as being a way of communicating they serve as a working tool of the scientist who classifies animals, and as classifications change scientific names may do so likewise.

Common names serve a variety of purposes and arise in many ways. In fact, the only characteristic they share is that they are not Latin. To understand common names properly, we should consider the different kinds.

Local or folk names are the largest class of common names. They are deeply entrenched in the language of a region, and are often obviously descriptive, but sometimes their origins are lost in the past. They may present as much variation within a single language as do goldfish names between languages. An example is *Micropterus salmoides*, widely known as the largemouth black bass. In a study of the common names applied to the fishes of the bass and sunfish family, Smith in 1903 listed 53 different common names for this species. A few of them are: Big-mouthed trout in Kentucky; chub and welshman in North Carolina and Virginia; cow bass and moss bass in Indiana; grass bass in Minnesota; gray bass in Michigan; green trout in Louisiana; marsh bass, bride perch and pointed tail in Ohio; and perch trout and jumper throughout the South. Of course, many of these names have died out, but the fact that they once existed and were useful in communicating within a region illustrates what one writer described as ". . . colloquial names that have grown up spontaneously among ordinary people."

Another category of common names might be called coined or invented names. Many kinds of fishes are known to scientists alone and have only Latin names. If, in writing of one of these animals a common name is required, one is invented. The American Fisheries Society (1960) has listed all known kinds of fishes living in the United States and Canada to a depth of 100 fathoms. Some of the fishes on this list previously lacked any common name while others shared a common name with one or more species. In order to insure a single common name for every species on the list, a number of names were invented. Another reason for inventing names is the importation into the United States of species from non-English speaking regions. The aquarium trade is the best example: A brief perusal of any authoritative book on aquarium fishes will show many fishes from South America and Africa for which English language names have been invented. A recent popular booklet on California deep sea fishes invented common names for species that previously lacked them. In some situations, scientists who describe a previously unknown species and give it a Latin name also invent a common name. This practice is very common in Japan.

The chief problem, however, lies with fishes that have too many names rather than with those that require invented ones. The commercial fishing industry, state and federal agencies, and writers communicate about fishes chiefly by using common names. When a species has more than one common name, and there is a clear need for only one, it may be a major undertaking to decide which should be used. In some instances one of many local names is selected, in others an invented name is chosen. The basic reason for the choice of any name should be that it is understood by the widest audience.

In the Bureau of Commercial Fisheries publication 'Fishery Statistics of the United States' (Lyles, 1966) a glossary is presented, which lists scientific and common names, including for many species alternative common names. The names used are those with which the bureau is best able to communicate with the various segments of the fishing industry.

The Food and Drug Administration is concerned with names of food fishes and deals with a set of names that might be termed semi-legal. This agency is charged with maintaining standards of identity and its regulations require that labeling must not be false or misleading. In deciding what common names may be used by the food processing and distributing industries, they select (when such exists) a name that is common or usual from the viewpoint of the general public who use and purchase fish products. Allowable names are decided on a case-by-case basis.

Because they often write for a wide audience, sportswriters are another group requiring common names that do not vary regionally. The Outdoor Writers Association of America has attempted to promote stability by publishing a list of scientific and common names of principal American sportfishes. Although they hope their common names are widely accepted, they have annotated their list and presented many widely-used alternative names.

The scientific community depends chiefly on the American Fisheries Society list of United States and Canadian fishes, a comprehensive and authoritative guide to scientific names. However, its common name section is of limited value because of inadequate coverage of alternative common names.

Users of common names have strong attachments to the familiar. Names of objects are so important to us that we tend to merge the name with the idea of the object. The idea of a piece of leather tied around the foot, and the name of the piece of leather as a shoe, are virtually inseparable. Therefore, in addition to serving as a shorthand way of communicating, names become part of the total concept of an object. Consider, for example, an angler who associates the fish that scientists know as *Micropterus salmoides* with the name "green trout." If he is served in thinking about *M. salmoides* or in communicating with others about it by the name "green trout," and if the name "largemouth bass" has no meaning, then to him "green trout" is that kind of fish, official pronouncements notwithstanding.

If communication problems increase, the number of official lists of names may do likewise. When common names are required for legal reasons or other special purposes, a single name for each species is clearly desirable, and special lists will fill a real need in designating names that offer the best communication value for a particular purpose. A general list of fish names should serve a very different purpose. It may recommend a preferred name, but its chief function should be to report on and cross-index names that actually are used. The worth of any general list of names as an aid to communication and understanding is only as great as the scope of its coverage of alternative names and the basic documentation it presents. A general list should first of all tell its users whether names are invented or folk names. The source of invented names should be described and also the degree to which they are used—that is, whether they are found only in books or have entered the spoken language as well. Folk names should be presented by region and their degree of usage should also be indicated. A properly compiled and documented general list will present the basic information of useful special lists.

In summary, names of fishes are basically of two kinds—invented and folk names. Scientific names are invented and are usually, but not always, stable; however, they are not suitable for everyday use. Some common names are also invented and may be important, as for fishes imported from foreign language regions. Folk names may vary regionally. They originate in many ways and their usage is often deeply rooted. Various segments of the common-name-using public often use different names for the same species or the same name for different species. Because many common names have a high communication value and have also become part of the idea of the animal, it will probably be impossible for each species to have one common name that refers to that species alone. Users of common names for special purposes have attempted to list the names that serve them best. A well-documented general list, including alternative names, is needed.

(See "Mackerels"—Page 28.)                                    ■ ■ ■

## APPENDIX II
### Part 1

## Worldwide Trends in Fishing Vessel Design and Equipment

*By Edward A. Shaefers*
*and*
*Dr. Dayton L. Alverson*
*National Marine Fisheries Service*

In recent years many countries have become aware that the sea is the most logical source for increasing the production of animal protein. Marine fisheries have had a phenomenal growth. Such a growth would not have been possible had not the fishing industry, which is noted for its conservatism, contained more than a few farsighted individuals. These people realized that if the fishing industry was to thrive or even survive, it must undergo the sociological, economic, and technological changes that have been so successful for land-based industries. It is only in recent years, however, that throughout the world the fishing industry has begun to accept the fact that the "good old days" are a thing of the past. No longer can fish resources be profitably harvested by inexpensive vessels that are designed by rule of thumb and that utilize human power instead of mechanical power. Now, most fisheries people agree that if a nation is to assure itself of an adequate harvest of food from the oceans, it must devise or adopt more efficient and more effective fishing methods, equipment, and vessels.

Present progress in fishing has been achieved through combining the talents of naval architects, engineers, fishermen, economists, equipment manufacturers, and fishery scientists. Today's modern fishing vessel, whether it is a 50-foot Pacific Coast combination vessel or a 286-foot factory stern trawler, would not exist had a unilateral course been pursued. Additional factors that helped bring about much of the progress in recent years are:

1. Technical meetings that resulted in the dissemination of technical information on all aspects of fishing. These included the three World Technical Meetings on Fishing Vessels and two on Fishing Gear and Methods that were sponsored by the Food and Agriculture Organization of the United Nations, the 1966 Canadian Atlantic Offshore Fishing Vessel Conference, the 1963 Whitefish Authority Meeting on Stern-trawling and the annual meetings of the Gulf and Caribbean Fisheries Institute.

2. The financial assistance programs sponsored by the governments of many nations.

3. The extension of fishing effort to new grounds and the harvest of new and under-utilized species on existing grounds.

In addition, various international organizations have played a rather subtle role in influencing fishing vessel design to date. We are quite certain, however, that their influence will be more pronounced in the future. These groups are:

1. The International Labor Organization whose conferences led to the 1966 Convention on Accommodations Aboard Fishing Vessels.

2. The Intergovernmental Maritime Consultative Organization, which supports the studies of the recommendations on fishing vessels adopted at the 1969 Safety of Life at Sea Conference.

### Overall Trends

Let us now review general and specific trends in fishing vessel design and equipment.

Based on available information, the general trend can be summed up as MORE:

1. More automation.
2. More stern trawlers, both large and small.
3. More size.
4. More conversions and more combinations.
5. More horsepower.
6. More use of mechanical refrigeration.
7. More electronic aids.
8. More attention to accommodations, comfort, and safety.
9. More capital.

We will now consider these factors individually, beginning with automation.

### Automation

The main purpose of automation aboard fishing vessels is to increase the productive capacity of the individual fisherman, so that catches can be maintained or increased with fewer fishermen. In most nations the most serious problem in operating fishing vessels is the lack of trained fishermen. The pros and cons of automation have been discussed at great length with regard to its economic and social impacts in nations having highly developed fisheries, as well as in those with developing fisheries. It appears, however, that it is not a case of the vessel eliminating the crew, but rather the crew eliminating the vessel. Does this seem unreasonable when only recently in this country has there been a tendency to consider fishermen as being entitled to at least some of the comforts found at home. Why, it has been asked, should men spend from 4,800 to 7,200 hours per year on a vessel, away from their families and subject to all the hardships attendant with life at sea, when their counterparts ashore are spending about 2,000 hours per year on the job and often at higher pay? Many predict that if living accommodations and working conditions are not improved and pay increased, fishermen will seek jobs ashore.

One of the first actions to effect increases in productivity per fisherman on trawlers is to mechanize and automate the engine room and trawl handling equipment. A good example of this concept is the 90-foot English stern trawler *Ross Daring*, launched in 1963. After several years of operations it was found that this vessel could be handled by a skipper, a mate, three deckhands, and a cook, with bridge control of both the engine and the winch. The cost of the skipper and the crew is 27 percent of the total operating cost; on conventional trawlers such costs are 35 percent of the total. The automation of the engine room, however, was too complex to be successful and furthermore, the owners state that they are unable to find an engine that is simple enough to run trouble free and without attendants. Consequently, the owners have trained skippers to look after the engine and have suggested that perhaps the success of such a fishing operation depends more on skippers' skill in maintaining an engine than on their fishing abilities.

The application of mechanical equipment has increased the efficiency of menhaden vessel operations so that the crew now averages 14 to 16 compared with 27 to 29 a few years ago. This remaining complement includes that member many individuals consider the most important man aboard, the cook.

At least one naval architect, however, does not agree. He has pointed out how strange it is that the fishing industry is quite advanced in the production of frozen foods but completely ignores the advantages of pre-cooked frozen meals on its own ships. He also made the point that a vessel is out to catch fish and not to feed the men—an observation that is sometimes not always readily apparent.

To our knowledge, no fishing vessel has been designed with a galley containing a micro-wave oven instead of a cooking stove. Thousands of pre-cooked meals are served each day aboard airlines. From a purely economical standpoint, this concept is in its infancy. For example, the food cost per man per day for Boston trawlers was $3.20 exclusive of the cook's salary. At the present time, serving pre-cooked meals would help solve the problem of obtaining trained cooks. Some of the large tankers operated by international oil companies have adopted this method of feeding their crews, which may have mixed nationalities. The preferences of each, therefore, ranging from finnan haddie through arroz con pollo to meatless Irish stew can be easily satisfied. Please do not get the idea that we are recommending that the ship's cook should be replaced by a stewardess; that vessels should be built without a galley range or that we feel this concept of meal preparation will be adopted by the fishing industry overnight. We are merely trying to point out a possible trend.

## Stern Trawlers

Let us now examine the trends in stern trawlers. One of the striking features is that some of the smaller ones (under 100 feet) are turning to drum trawling. This trend has extended from our Pacific coast to the east coast of Canada and to several European countries. This system, coupled with midwater trawl and accessory equipment developed at the Bureau's Gear Research Base, at Seattle, Washington, has enabled combination-type purse seiners only 53 feet long to effectively harvest Pacific hake in Puget Sound, Washington. For example between January 16 and June 10, 1966, one such vessel, the *St. Janet*, made 126 drags, caught 1,750,000 pounds of hake, and averaged 19,000 pounds per hour. The similarly equipped 73-foot *St. Michael* landed 2,500,000 pounds between November 20, 1965, and April 6, 1966, and averaged 17,000 pounds per hour. The results of these two vessels are excellent examples of the trend to consider the entire fishing operation as a system. For example, the drum reel itself was not responsible for fishing success, but rather its combination with the midwater trawling system, which in turn is composed of various integral parts. Two conventional drum trawlers fishing the same area with high-opening bottom trawls caught only 4,000 to 5,000 pounds of hake in drags of one to two hours. Vessels with midwater gear caught about 10 times more in less than half the time than did vessels with the bottom gear—these results occurred while the four vessels were fishing close to each other.

## Size

Another trend, which has been documented in previous papers has been the increase in the size of vessels. Until recently, increases in size were usually associated with increases in length. The trend towards raised forecastle head type vessels with the shelter roof extending aft, such as in certain West Coast combination style vessels, as given some 80-foot vessels the spaciousness of former 100-foot types.

## Conversions

Changes in vessel design are not related specifically to new construction. A combination of factors, including the increased use of hydraulic systems that operate whole arrays of deck machinery and gear handling apparatus, along with the adaptation of existing gear to both traditional and new fisheries, has brought about conversions that are startling. For example, who would have thought that the whale catching vessels of the Norwegian fleet would be suitable for other fisheries? However, during the first nine months of 1966, eight of these vessels, from 127 to 168 feet long, were converted to herring purse seining. Far less striking, but probably more numerous, has been the conversion of former purse seiners, draggers, and other types of vessels for king crab fishing. Existing California pilchard-type seiners were converted to king crab fishing with modifications varying from the installation of circulating sea-water tanks to a complete redesign of the deckhouse and upper pilothouse structure. Equipment for these vessels consists of a hydraulic pot-hauling device and cargo gear to swing the pots aboard and stack them on deck.

This prototype king crab vessel has the enclosed forecastle deck which provides more enclosed space and has a shelter deck aft. Such a design is also typical of the newer trawl vessels being constructed in Canada. One might think that modern high-seas stern trawlers were the first to recognize the concept of more enclosed space. The standard Great Lakes gill net tug developed many years ago, however, is an earlier example of the use of an enclosed deck aboard fishing vessels. Some of these vessels have been converted to trawlers to harvest the tremendous stocks of alewives that now inhabit Lake Michigan. This type of vessel, along with a few former Gulf shrimp boats, has taken considerable quantities of alewives.

While it is rather difficult to consider conversions and combination vessels as separate topics, in general, conversion entails major alterations and structural changes. A combination-type vessel, however, requires little major alteration to shift to multipurpose fishing. These are exemplified by the Aluminum Association's design and the various types of combination type hulls described at this meeting. An example of a combination type vessel that has fished successfully for almost a decade is the 63-foot combination purse seiner-bottomfish trawler *Silver Mink* which, although originally built for shrimp trawling, has engaged in these fisheries off New England without significant structural changes.

An example of a larger combination vessel is the Bureau of Commercial Fisheries exploratory fishing and gear research vessel *Delaware II*. This 155-foot vessel is designed for efficient handling of a variety of different types of standard and experimental fishing gear in the North Atlantic. Special effort was devoted to designing and developing a system of stern trawling to achieve a simple, rapid, and efficient trawl and catch handling procedure. To achieve this, the designers provided a long trawl ramp extending from the vessel's stern through the deckhouse to the trawl winch in the bow. This ramp permits the entire trawl including the cod end and its catch, to be hauled aboard with one continuous pull. The trawl and cod end are not hitched and hoisted successively as they are aboard coventional side trawlers and most stern trawlers.

The vessel was designed to be able to operate side trawls, drag heavy scallop and sea-clam dredges, and handle purse seines. By removal of the portable (bolt-on) trawl gallows at the stern of the *Delaware II* and attachment of a power block to the central hydraulic system, the vessel is easily convertible to purse seining operations. Purse seines for herring, mackerel, tuna or other pelagic schooling fish can be readily set from the broad open stern of the vessel.

A foredeck space is available for hauling longline gear and gill nets; these units can be set out smoothly through the trawl ramp at the stern.

Power is available from a central hydraulic system and from two 150-kw electric generators for powering various gear hauling units. The main winch is powered by a 225-hp diesel engine driving winch drums through a variable speed, high-pressure hydraulic system.

## Horsepower

An example of the trend to more horsepower is indicated in the Gulf of Mexico shrimp fleet. In 1964, the typical double-rig 55-foot shrimp vessel fished two 45-foot nets and had 197 hp. The relation between vessel size and main engine horsepower varied considerably from the typical. The reason for this variation may stem from the popularity and preference for certain engine models, available only in specific horsepower. It may also indicate, however, that data needed to determine specific horsepower requirements are not available. The present trend is toward larger vessels in the 65-to 80-foot class and 275-to 300-hp main engines. Accordingly, the use of larger shrimp trawls follows.

## Mechanical Refrigeration

The trend towards replacing ice with various systems of mechanical refrigeration has been well documented and is certainly worldwide. Let us consider for a moment, however, a preservation system that will enable sea-fresh finfish and shellfish to be sold in cities not only at seacoast ports but throughout the entire Midwest. We refer to the experimental shipboard irradiator aboard our exploratory fishing vessels *Delaware* and *Oregon*.

A "portable" 17-ton shipboard irradiator was installed aboard the *Delaware* in January 1966. The purpose of this irradiation program is to determine the effect and value of irradiating fishes immediately after they are taken from the sea. Shore-based irradiation experiments on fish products have shown that over 99 percent of the bacteria responsible for spoilage can be killed without adversely affecting the quality of the fish flesh. Reduction of spoilage bacteria to a very low level immediately after catching the fish will make possible a greatly extended storage or shelf life of fresh fish. Sea-clam meats and haddock have been irradiated aboard the *Delaware*, and preliminary reports indicate excellent results in quality improvement and extension of the "fresh life" of these products.

## Electronic Aids

Naval architects and boat owners are becoming increasingly aware of the need to design vessels with enough pilothouse space and generating capacity to accommodate the wonderful array of electronic devices now in common use aboard vessels, as well as provide room for devices not yet adopted by the fleet. One such device, called a portable radio-facsimile recorder, which receives synoptic sea surface temperature data, is now used aboard the *Delaware*. The recorder is used aboard the vessel during tuna work in the western North Atlantic and ashore between tuna trips to assemble data for planning future explorations. The unit is about 5-1/2 feet high, 2 feet wide and 15 inches deep. Other than a good whip or wire antenna, and a 110-volt power supply, no special installation is needed, provided space is available. Charted information is recorded on a 11- by 19-inch display. Synoptic environmental data for our area of interest are transmitted from U.S. governmental facilities several times daily. Information on these daily charts of potential use to fishermen are:
1. Sea surface temperature analyses and forecasts.
2. Sonic layer depth analyses and forecasts.
3. Selected bathythermograms.
4. Surface weather analyses and forecasts.
5. Sea wave analyses and forecasts.
Temperature data are collected, digested, plotted, and broadcast by the Navy. Sea surface temperature charts are made up from data radioed in daily to the Naval Oceanographic Office. Five hundred participating merchant and naval vessels report four times daily from the North Atlantic area—a total of 2,000 daily temperature observations. Charts are made up from 5 days of observations and are updated with new information every 3 days.

Sea condition and weather information pertaining primarily to vessel operations; sea surface temperatures (SST), layer depth (LD), and selected bathythermograms (BT) can be useful in selecting areas that have environments favorable for fish. Such an environment, for example, is the warm convergence that shows as a characteristic gradient pattern on the synoptic SST chart, and a characteristic thermocline structure on the LD and BT charts. This information is the basis for advance selection of the fishing area. Succeeding verification of the structure by vessel transect across the area is followed by test or commercial fishing.

## Comfort, Accommodations, Safety

As indicated by the vessels described in various publications, there has been an increasing trend to design greater comfort and safety into new and converted vessels. Greater crew comfort is accomplished by designing more enclosed space and shelter decks in all size ranges and types of vessels from the smallest combination vessels to the large stern ramp factory trawlers. Even so there are some complaints. For example, enclosed working spaces are fine during winter, but some complaints have been voiced that complete enclosures on factory-type trawlers during summer may be likened to a jail where the August sunlight and refreshing breezes above deck are considered the outside. In addition to improved firefighting and lifesaving equipment more attention is being given to designing deck layouts to insure as much built-in safety as possible. Considerable attention has been devoted to developing anti-rolling devices. While the so-called "flopper stopper" or paravane type stabilizer, suspended from outrigger poles, works quite well in vessels of the 30-to 75-foot range, no practicable system has yet been devised for fishing vessels in the 75-to 150-foot range. Most attention is being centered on development of a passive system involving anti-roll or surge tanks. Some authorities feel that vessels without efficient anti-rolling systems will soon have difficulties in getting crews. Various naval architects agree there is little chance of reducing the present crews aboard, unless a practicable anti-rolling device is developed so that automatic processing equipment, which requires a fairly quiet ship, can be used. A reduction from 17 to 8 crewmen could be achieved aboard 100-foot English trawlers if automatic processing equipment were used.

## Capital

Some of the cost figures for recently constructed vessels indicate a considerable increase in the amount of capital investment required due largely to the 'mores' we have mentioned as well as present overall economic conditions. In general, the trend towards automation has resulted in a significant increase in the ratio of capital to labor. Studies of industries that have been characterized by increasing productivity and efficiency show one common factor—an increasing capital to labor ratio.

## International Organizations

Without going into a long history of the factors that led to the International Convention on Accommodations Aboard Fishing Vessels resulting from the 1966 International Labor Organization Conference in Geneva, it is apparent that the provisions contained therein will influence vessel design in many of the nations in which it is ratified. This document applies to all fishing vessels 75 gross tons or larger (except sport fishing vessels, whaling vessels, and vessels operating in fresh water).

A Bureau of Commercial Fisheries study of accomodations on U.S. fishing vessels shows that present vessels of 200 gross tons or larger would have no difficulty meeting the standards; however, many vessels in the 75- to 199-gross ton category would be affected. For example, the convention requires that vessels less than 250 gross tons or under 115 feet long should have no more than 6 crewmen in a sleeping room. Each crewman must have an individual berth, and no berths shall be placed side by side in such a way that access to one berth can be had only over another. Further, all vessels carrying a crew of more than 10 persons are required to have mess room accommodations separate from sleeping quarters. With regard to sanitary facilities the convention specifies that for each department of the crew wherever practicable the following shall be provided.
a. One tub and/or shower for every eight persons or less.
b. One water closet for every eight persons or less.
c. One wash basin for every six persons or less.
The Intergovernmental Maritime Consultative Organization (IMCO) is a United Nations specialized agency, which has a membership of 60 nations. The main objectives of IMCO are to

facilitate cooperation among governments in technical matters of all kinds affecting shipping and to encourage the general adoption of highest practicable standards of maritime safety and efficiency of navigation. IMCO organized the 1960 Safety of Life at Sea Conference and administers the resulting Convention. Although fishing vessels are not included in the Convention, a recommendation was made to study the problems of stability of fishing vessels. IMCO subsequently established a panel on the stability of fishing vessels in July, 1964, and formulated a work program in which 17 countries are involved. The program has numerous studies, all aimed at the ultimate objective of establishing criteria for judgment of stability and ensuring that the skipper is guided by adequate and understandable information. In its work to date, the Stability of Fishing Vessels Panel has

compared national criteria for stability, issued recommendations concerning freeing ports and the closure of deck and superstructure openings, dealt with the construction and arrangement of fish bin boards, considered heeling forces produced by fishing gear, and issued some suggestions to fishermen about construction and operating practices that affect stability. The formulae concepts used for determining stability criteria, being from the Greek, are, as the old saying goes, Greek to us so we will not offer an explanation.

In summary, although it is apparent that tremendous progress has been made in developing more efficient vessels and equipment, it is hardly a start compared to the adaptation of available space age technology to shore-based resource extracting and transportation industries. ■■■

## APPENDIX II
### Part 2

## Method of Modifying a Gulf of Mexico-Type Shrimp Trawl To Sort Out Fish and Bottom Debris

*By Jerry E. Jurkovich*
*Equipment Specialist, NMFS Exploratory Fishing*
*and Gear Research Base, Seattle, Washington*

The National Marine Fisheries Service Exploratory Fishing and Gear Research Base in Seattle has been undertaking the development of experimental shrimp trawls that will separate fish from shrimp catches.

Many designs were tested and have shown good to excellent results. However, until recently, catches of shrimp with the experimental separator trawls did not compare favorably with those made with conventional shrimp trawls, usually the 57-foot Gulf of Mexico trawl. In addition the separator trawls occasionally captured excessive amounts of smelt which required considerable time and labor to sort out. A series of separator trawl

experiments were, therefore, carried on during 1969 to provide answers to these problems.

One test was aimed at determining what effect overhang had on catch composition. It was found that elimination of the overhang resulted in a considerable reduction in smelt catch. Conversely shrimp catches did not appear to be influenced by overhang reduction.

Additional work was conducted with a vertical distribution sampler to determine the vertical position of shrimp in the water column during various weather conditions and degree of light penetration in the water. The sampler indicated that the original

*FIGURE 1*

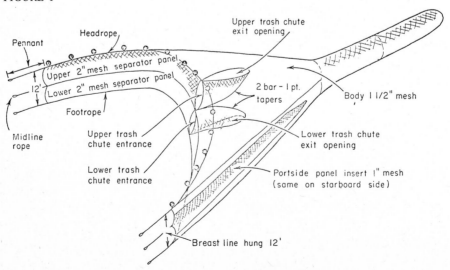

*Schematic conversion of a Gulf of Mexico 57-foot shrimp trawl into a shrimp separator trawl, according to designs worked out by the BFC-NMFS Seattle gear base.*

wing-type shrimp-separator trawls, which opened vertically only three to four feet, were inadequate because they did not open sufficiently to capture shrimp under all fishing conditions.

The 57-foot Gulf of Mexico shrimp trawl was selected for modification because adaptation of this trawl could be accomplished for a fraction of the cost of fabricating an entirely new trawl. This modification was accomplished by adding a webbing insert in the side panels to increase the vertical height, and by removing the overhang and rehanging a new headrope of 70 feet to equal the footrope in length. A fish-separator panel, made of two-inch mesh plus two trash chutes (upper and lower) was installed. The two-inch mesh separator panel was supported in the center by the addition of a midline which runs parallel to and is of equal length to the headrope and footrope.

The modified trawl was used on the final cruise of the *M/V Baron*, in which comparative tows were made with commercial shrimpers working off Westport, Washington. Following the *Baron* cruise, this same modified trawl was placed aboard the *M/V Corsair*. The trawl's catch rate was made known to other vessels working in the area and considerable interest was shown by fishermen in this new design concept.

This paper attempts to describe the detailed modifications necessary to make revisions from a 57-foot Gulf of Mexico-type trawl to a shrimp-fish separator trawl. Conversion of other types and sizes of shrimp trawls can also be easily made utilizing the

design principles described later. Although we have had initial success with this modified Gulf trawl, gear research personnel recognize that it is not the ultimate. However, it is one model that has been competitive with regard to catch as well as eliminating most unwanted bottom fishes when fished against conventional shrimp trawls on shrimp grounds off Oregon during October.

## Materials Needed To Convert a 57-Foot Gulf-Type Trawl

**Headrope**—Three-eighth-inch wire rope, Manila or polypropylene-wrapped (70 feet needed for hanging) plus any additional to make comparable with length of footrope and length of pennants. This assumes that the footrope is wrapped steel cable or a combination of steel and Manila.

**Breastlines**—Two new 12-foot breastlines.

**Riblines**—100 feet one half-inch diameter "Samson Stable Braid" dacron-covered polypropylene core or equal.

**Other rope**—Five-eighth-inch diameter "stable braid" rope. Seventy feet are needed for hangings plus any additional length for pennants and eye-splices to equal length of headrope and footrope.

**Webbing**—One piece, 1051-1/2 meshes long by 150 meshes deep, one-inch mesh No. 9 or No. 12 nylon. Two pieces, each 721

FIGURE 2

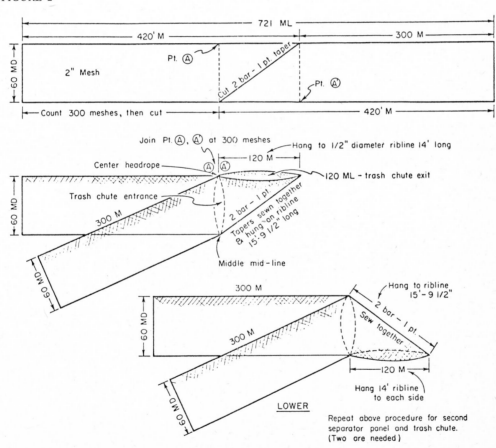

*Cutting diagram for the separator panel and the trash chute for the 57-foot Gulf of Mexico trawl conversion.*

*Figure 3*

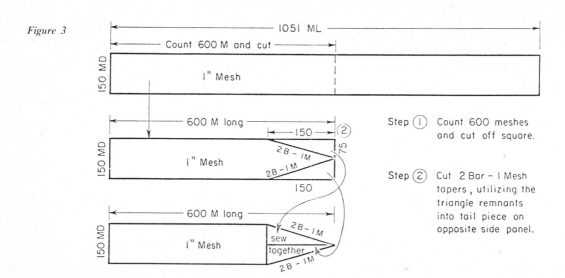

*The cutting diagram for the side panel inserts for the travel conversion. This conversion has been found to be an inexpensive way to convert a standard trawl to a separator trawl.*

meshes long by 60 meshes deep, two-inch mesh No. 12 or No. 15 nylon.

**Hanging** twine and sewing twine. Preferably white untreated so that shrink may be utilized to help tighten knots.

### Fabrication Instructions

After carefully examining the footrope of your trawl, cut out the overhang to exactly the same webbing dimensions as the footrope. Caution: Cut the top panel *exactly* the same as the bottom and carefully check the corners to see if they have cross-mesh wedges or tapers. Some 57-foot Gulf-type trawls may have them.

Rehang the new headrope in exactly the same manner as the footrope, being careful to leave slack in the loops between hanging knots. The headropes and footropes should be exactly the same dimensions and a 70-foot hanging length.

Count 400 meshes along the side panel, starting at the wingtip breastline, (starting at a point not more than 10 meshes above the bottom seam). Tie a string or marker at the 400-mesh mark. At that point (400th mesh) count the meshes between the top and bottom of the side panel and find the midpoint (approximately 12 meshes). Then cut 400 straight meshes toward the wingtip breastline. Remove the breastline because a new 12-foot breastline will be installed after the side panel has been inserted.

Refer to Figure 1 and cut out the side panels. Insert the finished side panels (600 meshes long, one-inch mesh), into the 400 mesh cuts, and carefully lace three each one-inch meshes opposite two one and a half-inch meshes in the original side panels. When you reach the two-bar, one-mesh taper, remember that two bars are equal to and counted as one (1) mesh.

Refer to Figure 2 and cut the separator panels and trash chutes.

Starting at the center of the headrope, midline, and footrope, mark off six each 70-inch sections toward each wingtip. A 57-foot Gulf of Mexico trawl with a 70-foot headrope and footrope will result in twelve 70-inch sections for a total of 70 feet. (The number of sections in other trawls may be more or less than the number in the 57-foot trawl; however, the same measurements and counts

apply.) Tack together points A and A1 per Figure 2. Starting at the center of the midline, tie a clove hitch with the one half-inch trash chute ribline precut to 32 feet, six inches, so that each piece is approximately 16 feet long. Measure 15 feet, nine and one half inches and tack to the bitter end of the upper trash chute exit. Repeat same on the lower trash chute exit. Tack the points opposite A and A1 of both upper and lower trash chutes to the clove hitch in the center of the midline. Hang the two point-one-bar taper distributed evenly along the 15 foot, nine and one half inch ribline. Starting at A, A1 tack upper trash chute to the center of the headrope and the center of the lower trash chute to the center of the footrope. Opposite each previously marked 70-inch section, tie 50 meshes of two-inch separator web (see Figure 3 for details). Following the division of the fifty two-inch meshes opposite every 70 inches, lace the two-inch separator panel selvage to the selvage of the one and one half-inch mesh hung to the headrope. Repeat the same procedure on the footrope. The upper and lower separator panels are then laced together mesh upon mesh. The midline shall be hung along this seam, measuring 70 inches on the midline and 50 meshes shall be hung to each 70-inch "mark off".

The trash chute exit is 120 meshes long of two-inch mesh on each side. The exit should be laced to the 160-mesh cut made in the top and bottom panels starting at center of the headrope and center of the footrope. The lace ratio is four one and one-half inch meshes opposite each three two-inch meshes. After lacing is completed, each double-laced web will be hung to a ribline. The ribline is cut 29 feet long. This ribline is marked in the center and is tied with a clove hitch so that two riblines, 14 feet, six inches each, will extend back from the headrope and footrope. Each side of the trash chutes will then be hung on 14 feet of ribline.

The breastlines should be rehung on 12 feet of breastline at each wingtip.

The modified 57-foot trawl is now ready to fish. Rig it in the way you would normally fish a shrimp trawl.

Note: It may be necessary to add additional looped chain or dropper chain *and* additional floats to counteract the collapsing effect of the sorting panel webbing. Each eight-inch float will support approximately seven pounds of chain. ∎∎∎

# APPENDIX II
## Part 3

### A Method for Tapering Purse Seines

*By Jerry E. Jurkovich*
*Fishery Methods and Equipment Specialist,*
*NMFS Exploratory Fishing and Gear Research Base,*
*Seattle, Washington*

For the past 10 years, vessel owners in the salmon fishery in Puget Sound and Southeastern Alaska have favored using wedged or tapered strips in one or both ends of their seines. The taper is inserted between strips of identical mesh size, and equal twine weight, and as near the corkline as possible. A tapered seine can be fished near shore without having a deep section of web dangling near the bottom. Deep gavels or breast lines have a tendency to foul the purse line, and a tapered strip reduces this tendency to a minimum. Although the need is obvious in seining for salmon, tapering of seines may be advantageous in other fisheries.

Purse seines are made up of either horizontal or vertical strips of webbing. In the United States, most purse seines, other than those used in the menhaden fishery, have horizontal strips of webbing; in northern Europe, purse seines have vertical strips.

Nearly all experienced fishermen can cut tapers or wedges when they are needed, but few can compute a taper to a predetermined length before cutting. Haphazard cutting of netting can be costly if mistakes are made. This report presents an easy method of computing and cutting a taper.

### Computing the Taper

First step: Draw a sketch of desired taper (Fig. 2).

Second step—Convert 24 fathoms SM (stretch measure) to number of meshes by dividing 1 fathom or 72 inches, by 4-inch mesh, to get 18 meshes per fathom; 24 fathoms x 18 = 432 ML (meshes long). The 432 meshes are rounded to the nearest 50, to get 450 ML.

Check a 1-fathom piece of webbing with a fathom stick, then count the meshes per fathom. This will prevent any mistakes caused by shrinkage or other variations quite often found in machine-made netting.

Third step: Set up a proportion.

The fraction (1/9) indicates a proportion of one mesh in vertical direction opposed to nine meshes in a longitudinal direction. Vertical direction can only be achieved through the use of bar cuts. Because two bars make one mesh, multiply the fraction (1/9) by 2 (see Fig. 3).

A netting fundamental is that two bars constitute one mesh. Therefore, all tapers must include two or four bars. Use of a two-

Figure 1

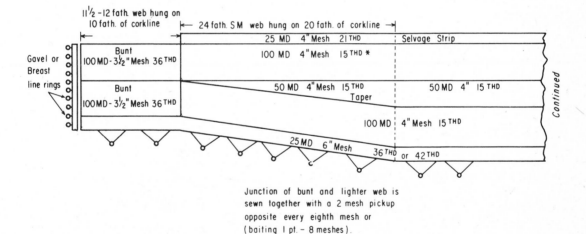

Junction of bunt and lighter web is sewn together with a 2 mesh pickup opposite every eighth mesh or (baiting 1 pt. – 8 meshes).

### Purse Seine With Uniform Taper

For the first example of a taper, I use a salmon purse seine from the Icy Strait district of Southeastern Alaska (Fig. 1). The bunt end begins with 225 MD (meshes deep) hung 11 1/2 to 12 fathoms in length to 10 fathoms of the corkline, which is the longest side because the leadline is usually 10 percent shorter than the corkline. At the completion of 10 fathoms of bunt, 25 more meshes of corkline selvage webbing are laced in. A taper starts at the junction between bunt and body between two adjacent strips of four-inch stretch mesh. (See Figure 1) This taper is to be 10 fathoms (hung measure) long, so allow 16.6 percent additional length for "hanging in" length (24 fathoms of web hung on 20 fathoms of corkline).

bar formula is preferred for two reasons: (1) two bars reduce the chances of making mistakes; (2) four bars produce larger steps, resulting in a taper that is not quite so smooth. Either a two- or four-bar formula can be used with good results.

Result: Fraction 2/18—the numerator (2) indicates a cut of two bars. Next, compute the number of meshes in this taper:

Denominator - Numerator = No. of Meshes

Substitute: 18 - 2 = 8 meshes

The taper is two bars and eight meshes. Figure 4 reveals that in a two-bar, eight-mesh taper the total number of legs cut is 18, which is the denominator. The first two legs cut form bars, or numerator of the fraction 2/18. A cut of two bars eight meshes; two bars eight meshes; repeated until the opposite selvage side is reached will result in a taper about 24 fathoms long.

At this point, it is obvious that tapering is not precise; at best it is a compromise. The aforementioned taper was computed for 432 meshes long, then rounded to 45 meshes in length. The re-

*Figure 2*

First step: Draw a diagram of taper

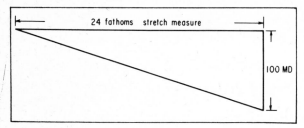

*Sketch of desired taper.*

sult is a taper that is 18 meshes, or 1 fathom, too long. Regardless of this minor variation, the system is far more accurate than random cutting. As one becomes more familar with this system, a minor compromise in length can reduce the 1-fathom variation.

Note also that two bars are equal to one mesh and are laced to the adjacent strip in this manner (Fig. 4). Note the dotted lines between the taper and adjacent web.

*Figure 3*

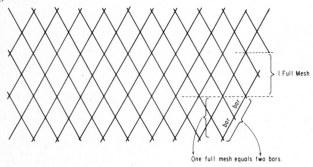

One full mesh equals two bars.

When a bar-cut (bias cut) is made on a 50 mesh deep piece of webbing we have cut into the web 50 meshes in length, the resulting number of bars equal 100, proving 2 bars equal one (1) full mesh.

*Basic netting fundamentals.*

### Procedure for Cutting a Taper

Place a strip of web 50 meshes deep on a flat surface, spread evenly, and keep the opposite double selvages almost parallel.

Face the web from the end, start at the left side, and count one full mesh to the right, then cut two bars eight meshes, and continue with two bars eight meshes until the opposite selvage is reached. The full meshes are cut parallel with the selvage. This is a crucial point. The strips are used in a longitudinal direction, and the knots on the full meshes used in this taper will clean to

form a rounded loop. The opposite is true when vertical strips are used; the knots in the full meshes will untie to form two separate legs. In vertical strip seine construction, the reference to full meshes will become points (two bars eight points).

When a taper is cut in the aforementioned manner the remains may be used at the opposite end of the seine.

An alternate method for better utilization of webbing should be used if a taper is needed only at one end of the seine.

The 50 MD webbing is spread in the same manner as above. Count 25 meshes starting from left to right and begin a two-bar, eight-mesh taper on the next half mesh. This cut is continued until the right selvage is reached. The severed piece is 12 fathoms long. It is then matched selvage to selvage, taper to taper, and sewn together to eliminate all waste (see Fig 5).

After the taper has been cut, I strongly advise selvaging the tapered edge prior to lacing it to the adjacent strip. The most experienced net maker should lace the taper because it must be sewn properly with the same number of meshes used on the tapered side as on the adjacent side. Two bar cuts equal one mesh is a simple but most important rule to remember when lacing the taper. Once started, the person lacing should never stop without making a secure tiedown so the meshes cannot slip. The best procedure is to lace the entire taper without any stops.

To make taper computation complete, one more example has to be explained. In the first example, Figure 4, the proportion of 2/18 is an even number in the denominator. In the next example, the denominator is an odd number.

Note that the net is 350 MD and the taper is 100.

### Computing the Taper

First step: Draw a diagram of taper (Fig. 2).

Second step: Convert 24 fathoms SM to number of meshes long.

72 fathoms per fathom 4-inch mesh = 18 meshes per fathom.

24 fathoms x 18 meshes per fathom = 432 meshes which, rounded to the nearest 50 meshes, will be 450 ML.

A piece of webbing should be checked for possible shrinkage after treatment by a preservative. Stretch the webbing to be used and measure along the selvage with a 1-fathom stick and count the meshes in 1 fathom. Do not be surprised to find a different count than your computation. Always use the count obtained from the measured length.

Third step: Proportion:

MD = 100 (MD) = 2

2 in the numerator = 2 bar cuts.

Denominator—Numerator = No. of meshes in taper.

Substitute: 9—2 = 3-1/2 meshes

It is impossible in net making (with tapers) to cut 1/2 mesh as each full mesh is made up of two legs. Whenever a fraction is encountered, use the nearest smaller whole number. In this example use three meshes and four meshes.

Fourth step: The taper formula is:

Two bars three meshes; two bars four meshes; two bars three meshes; two bars four meshes; repeated, until the double selvage at the right is cut. Please note that two bars three meshes and two bars four meshes can be added together to result in the formula: four bars seven meshes. The tapers are identical and the ultimate 24-fathom taper is achieved with either formula. However, I prefer to use the two-bar formula for two reasons: (1) two bars reduce the chances of making mistakes; (2) the four-bar, seven-mesh makes larger steps, and the taper is not as smooth.

### Cutting Procedure

This taper will be started in the same manner as the first. Spread the 100 MD webbing on a smooth flat surface, spread evenly, and keep opposite double selvages almost parallel. Count one full mesh then start two bars three meshes; two bars four meshes; two bars three meshes; two bars four meshes, alternating three and four meshes after each two bars. This will produce a 24-fathom taper from 1 to 100 MD.

**Conclusions**

1. This taper method will work equally well in altering old seines or constructing new nets.

2. The taper is inserted between strips of identical mesh size and equal twine weight, and as near the corkline as possible.

3. On a seine using horizontal strips, full meshes—not points—are cut because the cuts are made in a longitudinal direction. When the knots are cleaned, complete loops are formed.

4. On European nets using vertical strips, the tapers consist of bars and points because the cut, full meshes will untie at the knot.

5. Always count the single legs as you cut them as bars. All tapers with two bars may appear to have three bars, but the "third" bar is part of the next full mesh. ■ ■ ■

*Figure 4*

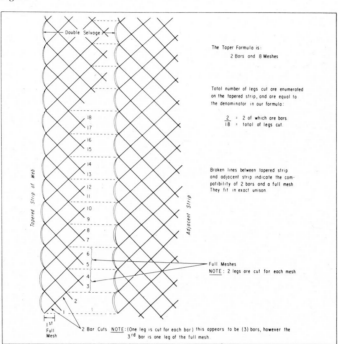

*Method of cutting a uniform taper.*

*Figure 5*

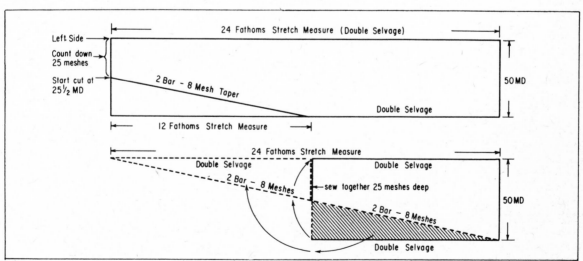

*Method of cutting a taper when only one end of the seine carries a taper.*

## APPENDIX II
### Part 4

## Universal Trawls

*By Jerry E. Jurkovich*
*Fishery Methods and Equipment Specialist,*
*NMFS Exploratory Fishing and Gear Research Base,*
*Seattle, Washington*

If a trawler is equipped to fish in midwater as well as on the bottom, why should two different trawl designs be needed? The NMFS Exploratory Fishing and Gear Research Base at Seattle, Washington, has been experimenting with a single trawl to determine if it can be used to fish effectively in midwater and on the bottom.

Differences between midwater and bottom trawls are concerned with their design and support paraphernalia. Any existing trawl could loosely be termed midwater when being lowered away or retrieved. However, the configuration of most bottom trawls would not be effective in midwater. On most bottom trawls the vertical height between headrope and footrope is narrow, thus the net requires an additional section of web, locally called an "overhang." Midwater trawls have the same top and bottom symmetry because the "overhang" is considered unnecessary owing to greater distance between headrope and footrope. Accurate depth readout equipment is, however, a must for successful midwater trawling.

The Exploratory Fishing and Gear Research Base's midwater trawl, the Cobb pelagic trawl, was based on the principle that a large net towed at relatively slow speeds (two to three knots) would capture a wide variety of midwater species. This midwater trawl caught large quantities of Pacific hake (*Merluccius productus*) while being towed off the bottom at speeds of two to three knots. However, several disadvantages were noted. When echograms showed heavy traces on the bottom, our attempts to take these near bottom dwellers often resulted in ripped webbing and owing to the resistance caused by the large size of the Cobb pelagic trawl, it was difficult for our small local trawlers to hold a course when towing across strong tides. These disadvantages led us to design several trawls incorporating the best features of the Cobb pelagic trawl and the 400-mesh Eastern bottom trawl. The new trawl was named the BCF Universal trawl MARK I.

The BCF Universal trawl MARK I is a four-seam trawl which has top and side panels of No. 21 nylon and belly panels of No. 36 nylon. All webbing is two and one half inch mesh,

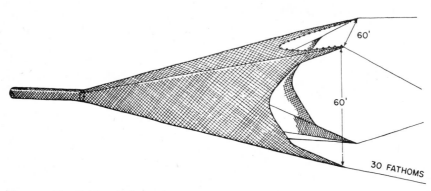

**Above:**   The Cobb pelagic trawl

**Below:**   BCF Universal (MK.I) trawl

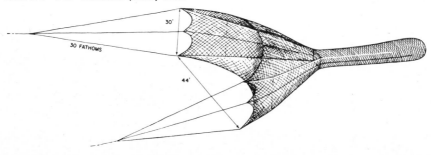

stretch measure. All mesh measurements are stretched measure inside of one knot over and including opposite knot measured parallel to the selvedge.

### Heavier Material

The side panels begin with 384 meshes, and taper to 50 meshes at the cod-end junction. The footrope and headrope are 121 ft. long. A one-half inch diameter 6 x 19 galvanised wire rope with a thimble spliced in each end, served with three-eighths-inch diameter manila, was hung to the footrope. Twenty-six feet of five-eighths-inch chain were hung to the footrope at the wing tips. The headrope has 67 eight-inch diameter aluminum trawl floats. Three bridles, 30 fathoms long, were used between the otter doors and the trawl. This trawl was constructed of heavier material than the Cobb pelagic trawl to withstand the abrasion of the bottom. However, it retained the wide mouth opening necessary for midwater trawling. This trawl was made with 41 percent less webbing than that used in the Cobb pelagic trawl.

When situations permit, scuba divers make observations on all new trawls. Measurements are taken, as well as still pictures with underwater cameras. Owing to poor visibility, most test dives are limited to 20 fathoms, therefore, observations made at diver depths must be corrected to correspond to fishing depths. Theoretical corrections were verified by empirical measures made with transducers placed on the trawl or trawl doors with the readings being made on the fathometer recorder via telemetry wires. Divers reported that the Universal trawl, MARK I exhibited good configuration when being towed in shallow water. It had a 30-foot vertical opening on the sides, 42 feet between central footrope and central headrope, and a 44-foot horizontal spread.

In February, 1967, the Universal trawl MARK I was placed aboard the *St. Michael*, a 73-foot western style seiner dragger powered by a Caterpillar D-353 diesel engine rated at 380 b.h.p. The *St. Michael*, was fishing on a large stock of Pacific hake, in Port Susan, which is in the inland waters of the State of Washington, about 25 miles north of Seattle. The first trip lasted two days, in which eight drags caught 155,000 pounds.

The following week, five drags over a one and one half day period took 168,000 pounds of Pacific hake. One of the five tows picked up a large log in the trawl which required considerable mending and resulted in only an 8,000-pound catch. These catches were the result of heavy concentrations of spawning hake rather than being entirely due to gear efficiency.

From the results of the initial trials with the BCF Universal trawl (MARK I), we theorized that this net should be effective in taking Pacific ocean perch *(Sebastodes alutus)*, a species sought by the fleet from May to November. In April, 1967, the *Westness* was chartered, even though April is normally too early for the appearance of perch on the grounds. The testing

was limited because the trawl was torn so badly on the second day that it was not repairable aboard the vessel. These tests indicated that the new trawl was effective in taking roundfishes (rockfishes and true cod); however, it was not catching as many flatfishes as other boats in the area. Comparative tows with a partner boat indicated that the BCF Universal trawl MARK I was unwieldy when dragged against or across strong tides and it did not move over the grounds as rapidly as the 400-mesh Eastern trawl.

Minor rigging changes could be made to adapt this trawl for catching flatfishes, but nothing short of major alterations could correct its drag resistance. The only exception would have been to eliminate one bridle on each side by coupling the top and centre lines together, decreasing the vertical opening to reduce resistance. Because of commitments to other research projects, further testing was halted.

At a seminar held in September, 1967, we decided to design a new combination bottom and midwater trawl to replace the 400 mesh Eastern trawl which has been the mainstay of our local trawl fleet for many years.

### New Trawl Designed

At this point, a comparison of the relative resistance of the various trawls used in the local trawl fishery was determined. The total number of meshes in each webbing panel of various trawls was calculated, then multiplied by the twine cross section diameter in inches and by the mesh size. The resistance was computed first for the 400-mesh Eastern trawl, which can easily be towed by any trawler in our local offshore fleet. This trawl was used as a standard and other nets were compared with it. From these measures and from the experience developed with the Cobb pelagic and the BCF Universal MARK I trawls, a new universal trawl was designed.

### Resistant to Abrasion

The new BCF Universal MARK II represented an integration of bottom and midwater trawl designs. The headrope and footrope of the BCF Universal MARK II are both 94 feet. The entire trawl was made with polyethylene webbing. The webbing in the body and wings was reduced by 20 percent (see BCF Universal MARK I) and the side panels 48 percent. The wing and forward body panels were made of five inch mesh. The after-body panels plus the intermediates and cod-end were made of three and one half-inch webbing. The top and sides were made of No. 21, and the belly and wings of No. 36. The intermediate panels were made up of three pieces—each 120 meshes long, 60 meshes deep, three and one-half inch mesh, No. 60. The cod-end was 110 meshes long, 60 meshes deep, three and one half-inch mesh, No. 84 twine.

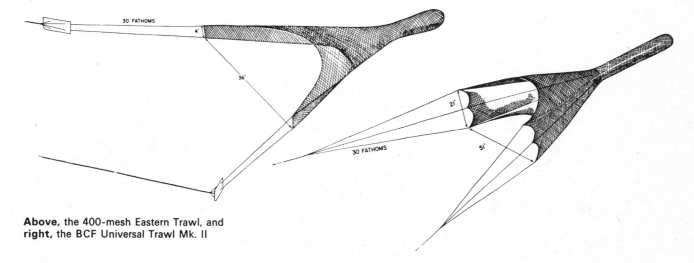

**Above,** the 400-mesh Eastern Trawl, and **right,** the BCF Universal Trawl Mk. II

The headrope and footrope were hung, using five-eighths inch diameter spliceable braided Dacron rope, which was selected for its low permanent elongation, low-working elasticity, and high abrasion resistance. This rope was soaked in water for 24 hours prior to being used to ensure maximum stability. This eliminated incremental "hang-in," and is a standard procedure when using nylon with high working elasticity. One ribline was hung along the centre of each side panel, starting at the centre wingtip and ending just short of the splitting rings in the cod-end. The "hang-in" coefficient was 0.866 on this ribline. To the footrope was hung a one-half inch diameter galvanised 6 x 19 wire rope, with a thimble spliced in each end and served with three-eighths-inch polypropolene rope: 145 pounds of chain were hung at each of the wingtips on the footrope. A total of 67 eight-inch diameter aluminum trawl floats were spaced with 25 of them at 12 inch intervals over the wingtips, and the remainder equally spaced along the headrope.

Preliminary diver reports, pictures and measurements of the BCF Universal MARK II looked extremely good. The mesh shapes achieved approached our optimum which is described as the fore and aft length (with the run) being twice the lateral width or forming 30 degree to 60 degree diamonds.

The vertical height was 21 feet at the wingtips. It measured 34 feet from the center footrope to center headrope in midwater, and 28 feet when on the bottom. The horizontal width corrected to fishing depth was 51 feet. These measurements are amazing when one considers that the calculated maximum vertical height is 24 feet and the maximum width 54 feet. The Bureau vessel *John N. Cobb* was able to make a 180 degree turn in midwater in six minutes with this trawl when using 6 ft. x 10 ft. "V" otter doors.

A short series of comparative drags was made between the *John N. Cobb* using the BCF Universal MARK II with 60 square feet "V" doors and a commercial trawler using a 400-mesh Eastern trawl with smaller "V" doors having an area of only 28 square feet.

Until the merits and shortcomings of the BCF Universal MARK II have been firmly established, it is difficult to forecast the ultimate benefits of a trawl used for both midwater and bottom trawling.    ■ ■ ■

---

## APPENDIX II
### Part 5

### Methods of Net Mending

*By Boris O. Knake*
*National Marine Fisheries Service*

Net mending has been practiced by fishermen for centuries. There are many known methods of mending nets throughout the world but all are basically similar.

When fishing, the crew must mend the nets night and day under the difficult conditions of cramped space, awkward positions, rolling seas, decks awash, rain, snow, and bitter weather. Whenever a tear is found, which usually occurs everytime the net is hauled in, it must be mended. Because of these conditions, a mending method suited to most fishing has been developed. Most nets are made by machines, but since the mending of these nets must be done by hand, this probably will remain always a part of the fisherman's work.

### FUNDAMENTAL PRINCIPLES AND SIMPLE MENDING

#### Terms Used in Net Mending

Fish netting or webbing is a sequence of loops, known also as bights or half-meshes, which are interwoven by knots. These form a series of meshes as shown in Figure 1. The knot is known as the sheet-bend-hitch, weavers' knot, fishermen's knot, or mesh knot.

A single mesh is the combined upper and lower half-meshes tied midway by a hitch. As illustrated in Figure 1, the weaving is from right to left (dark strands) to the number of meshes required, and back again from left to right (light strands) and so

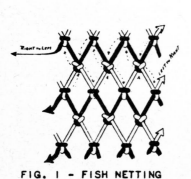

FIG. I - FISH NETTING          FIG. 2 - DAMAGED AREA

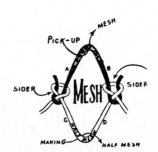

FIG. 3 - SINGLE MESH

forth until the desired depth is reached. Reversing the direction causes the knots in one row to look different from those in the next row. It is the same knot, however—front view in one case and back view in the other.

In net repairing, the knot is referred to by name, according to the position it has in the damaged area. As numbered in Figure 2, the names used are:

(1) Mesh knot
(2) Pick-up knot
(3) Sider knot
(4) Starter three-legger
(5) Finishing three-legger

## Definitions of Knots

1. **Mesh knot**—Also referred to as a half-mesh knot; that is, in making a half-mesh a mesh knot is tied. The white lower half (Figure 3) is the half-mesh; C and D are the legs of the half-mesh. After a knot is tied the legs are referred to as bars or strands.

2. **Pick-up knot**—A knot tied to a half-mesh on the base or the lower part of the damage. Tying the lower mesh forms the pick-up knot on the pick-up mesh. The pick-up mesh is shown in black in Figure 3. The legs of the pick-up mesh are A and B.

3. **Sider knot**—Refers to a knot of two separate strands as shown in Figure 3. These are only found on the sides of the damage; that is, on either side of the webbing when held or hung straight. Sider knots are of two types, called "sider on the left" or "sider on the right" depending on which side of the damage they are located.

4. **Starter three-legger**—A knot having three strands intact and only one strand cut off (4 in Figure 2).

5. **Finishing three-legger**—A similar type knot to the starter three-legger (5 in Figure 2). The importance of the three-legger knots is that the mending is started and finished on those knots. This particular point cannot be over-emphasized.

## Net Size

Netting is designated by the size of the mesh in a stretched form (Figure 4) and is measured by the number of meshes in length and in depth. The length is often expressed in feet or fathoms when a large quantity is ordered.

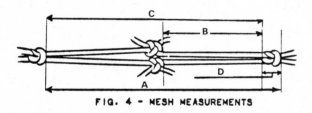

### FIG. 4 – MESH MEASUREMENTS

The stretched mesh size (A in Figure 4) is a recognized method of measuring by the manufacturers. The length of the bar, leg, or strand (B in Figure 4), the actual inside opening of the mesh (C in Figure 4), and the size of knot (D in Figure 4), are units of measure which are sometimes used in netting specifications.

## Net Damage Inspection

The first step in repairing damage to a net is to determine the type and the extent of the damage, so that the best mending procedure can be decided upon. The proper procedure allows the mender to restore the meshes by weaving in an uninterrupted sequence. The next step is to determine the trimming necessary. To do this, the damaged section of the net is stretched so that the strands line up easily. This is referred to as "straight twine" (Figure 5).

The wrong way to stretch netting is known as "cross twine" (Figure 6). Note how the strands tend to loop. To further illus-

trate trimming, Figure 7 shows a damaged section with several broken strands (dark section). In correctly trimming this type of damage all the dark shaded twine must be cut out as shown in Figure 8. Usually the resulting stumps in the half meshes and pick-ups are removed by cutting them out, when there is suffi-

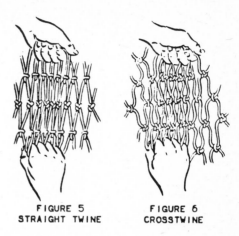

### FIGURE 5 STRAIGHT TWINE    FIGURE 6 CROSSTWINE

cient time for it. This serves two purposes. The mending will look neater because of the reduction in the number of bulky knots, and also in mending badly torn nets it helps the mender to avoid mistaking the half-mesh knot for the sider knot which would result in a disorder of the mesh sequences known as "getting into cross twine". If this mistake is made it will be necessary to cut out the repaired meshes and start all over again. Figure 9 represents a correctly trimmed hole ready for mending. Note that knots, 6, 7, and 10 are missing. A detailed description of the mending procedure is given later.

## Method for Cutting Out Stumps in the Half-Meshes and Pick-up Meshes

Figure 10 shows how the cuts are made when cleaning up the half-meshes and the pick-up meshes. When the loop AA is to be retained, cutting out the strand BB is done quite simply, by cutting at point C where the strand leg B-1 serves as a cushion, thus preventing injurious cuts on the A strand. Figure 11-A shows the method of cutting and the position of the knife.

It is more difficult to cut off the A strand stump, which must be cut at point D, Figure 10. The knife is placed between the legs B-1 and B-2, so that the blade cuts between the two legs BB (Figure 11-B). This operation requires a little practice to get the feel of the snap, when the last fibre of stump AA has been cut without the knife reaching and injuring loop strand BB.

When trimming the unwanted strands, care must be taken not to cut the strands too close to the knot, on three-leggers and siders. These knots are not always firm and may untie. Therefore, at least a 1/2 inch of strand should be left when trimming the knot. If cut too close they frequently will untie and it will then be impossible to retie them. Then the cutting of additional meshes will be required to regain the proper mending sequence.

## Filling The Needle

Only one special tool is required in net mending. This tool is known as a net needle or shuttle and is made of wood (ash, hickory, bamboo, or dogwood), metal, plastic, or even ivory. Figure 12 shows the shape of the net needle customarily used. A is the eye, B is the tongue, and C is the fork or heel. The size of the needle for mending depends upon the size of the mesh and the twine to be handled. The needle used to repair otter trawls is about 8 inches long by 1/4 inch thick. Filling "single needle" means that single twine is wound on the needle (Figure 13-A). The twine is

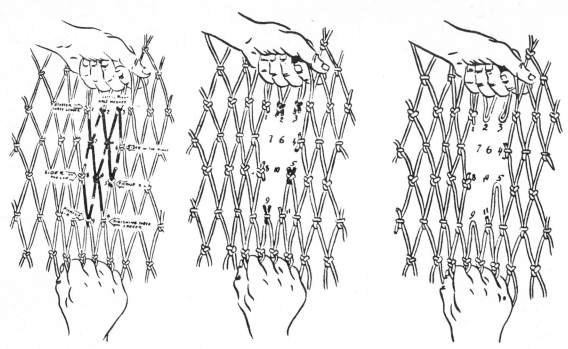

FIG. 7 - DAMAGED SECTION    FIG. 8 - PARTIALLY TRIMMED    FIG. 9 - CORRECTLY TRIMMED

wound under the heel and over the tongue on the other side of
the needle and back under the heel, turning the needle in a rock-
ing fashion each time the twine goes under the heel of the needle
until the needle is filled. Filling "double needle," means that instead of single twine, double twine is used (Figure 13-B). In
starting to fill "double needle" a length of single twine sufficiently long to fill the needle is measured off. The twine is then doubled. The middle bight is placed over the tongue of the needle
and the filling is done the same way as when single twine is used.
Some needles have a flexible tongue. This makes it possible to
push the tongue slightly outward permitting the twine to slide
under it (Figure 14-A). In other types the eye of the needle is
flexible. In filling this needle, the right thumb pushes the eye
downward, passing the twine over the tongue (Figure 14-B).

FIG. 10 - POINT OF CUTS

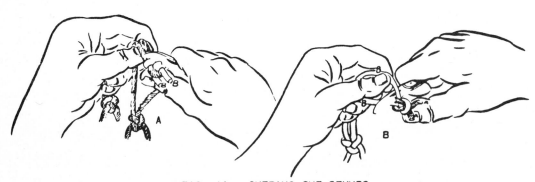

FIG. 11 - CUTTING OUT STUMPS

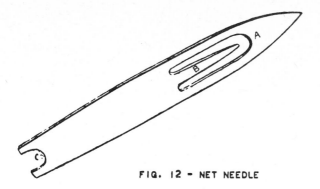

FIG. 12 — NET NEEDLE

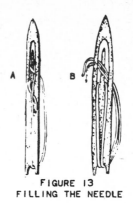

FIGURE 13
FILLING THE NEEDLE

FLEXIBLE
TONGUE
NEEDLE

FLEXIBLE EYE
NEEDLE

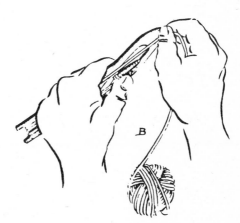

FIGURE 14

## Mesh Gauging

When mending nets the mesh size is gauged with the fingers of the left hand. The new mesh is constantly gauged for uniformity with the previous mesh. Figure 15 illustrates the use of the fingers in gauging the mesh size. Of course, this limits the size range of the mesh to the size of the hand, but this is quite satisfactory as the mesh sizes are usually more or less within that range. They can be varied a little by the tautness with which the twine is held. With some practice it becomes easy to gauge the mesh size almost precisely. When mending a wet net, it should be remembered that the twine has shrunk; therefore, the dry twine meshes mended in should be slightly larger than the wet net meshes.

FIG. 15 — GAUGING MESH SIZE

**Handling The Needle**

Shuttling the needle through meshes of large size, the needle
is always held in the hand in the progressive positions A, B, and
C, shown in Figure 16. When mending smaller size mesh, the
needle remains secure in between the meshes, without slipping
out. Sometimes the left hand forefinger may aid in holding the
needle while changing the grip for the completion of the
shuttling.

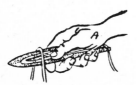  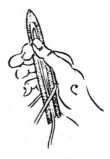

**FIG. 16 — SHUTTLING NEEDLE THROUGH LARGE MESHES**

**Supporting The Net For Mending**

In starting to mend a damaged net, it is more convenient for
the mender to tie the area of the damaged net section to something secure so that the meshes will line up easily. This is done
very simply by running a length of twine through the meshes
about one or two meshes above the starter three-legger of the
damage in the net (Figure 17). These meshes are tied in a bunch,
and fastened with a slip-double-half hitch knot (A in Figure 18).
The other end of the twine, B, is then secured to something solid
that may be within easy reach. Often at sea when many meshes
are damaged in a very large tear there will be someone who as-
sists the mender by holding the net properly.

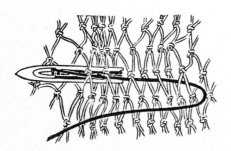

**FIG. 17 — RUNNING SUPPORTING TWINE
THROUGH MESHES**

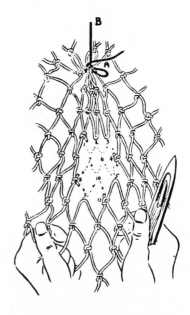

**FIG. 18 — SUPPORTING
DAMAGED NET**

## How To Mend

In repairing damage to a net, there must be a definite starting point and a finishing point, both of which must be three-leggers. Figure 19-A shows a simple tear with only two three-leggers, which become the starter three-legger (1), and the finishing three-legger (11). Figure 19-B shows the starter knot completed and Figure 19-C shows the finishing knot with the damage repaired. Figure 20-A illustrates a tear known as "tear on the siders." Here, in addition to the starter three-legger (11), is "sider on the right" (4) and "sider on the left" (8). Figure 20-B shows the damage mended. Sider knots are very important as a signal to reverse the direction of shuttling on the mending procedure. Often when mending badly damaged nets where the netting has been under severe strain, the knots become deformed and hard to recognize. It is then good policy to cut the knot open to find out whether it is a mesh knot or sider. Sider signifies the end of the row.

The tear shown in Figure 21-A is known as a "tear along the meshes." Here is the starter three-legger (1), and the finishing three-legger (11), in addition to the half-mesh (2-3) and the pick-up mesh (9). In Figure 21-A the starter knot is on the left side of the tear, therefore, mending proceeds from left to right (Figure 21-B). In this case the needle is always taken from *under* the loop when making the tie. Figure 21-C shows a tear similar to that in Figure 21-A. The only difference is the location of the starter three-legger on the right side of the tear. Therefore, mending proceeds from right to left (Figure 21-D). In this case the needle is always inserted into the loop from *above*. If the proper mending direction is not carefully noted, the twine will not line up in the proper direction when each knot is made with the result that the knot will be upside down or twisted.

The smallest possible tear requiring all of the variations in mending is illustrated in Figure 22. Beginning with the starter three-legger (1), (also see Figure 23), the needle follows the arrow of the dotted line to the half-mesh (2), (also see Figure 25), from there to another half-mesh (3), then downward to the sider on the right (4), (also see Figure 27), which is the signal to reverse and mend from right to left. First the needle is shuttled through the pick-up mesh (10), (also see Figure 25), and then downward to the finishing three-legger (11), (also see Figure 37). This completes the mending.

The sketches beginning on Page 9 (Figures 23 to 38) illustrate step by step details of tying the various types of knots. The numbers in the sketches show the order in which the movements should be made. The tear illustrated is the same as the one shown in Figure 22. The knots as completed are illustrated below.

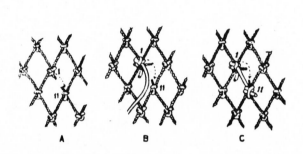

FIG. 19 - SIMPLE TEAR

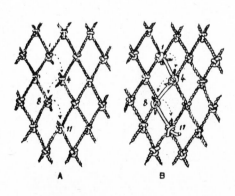

FIG. 20 - TEAR ON THE SIDERS

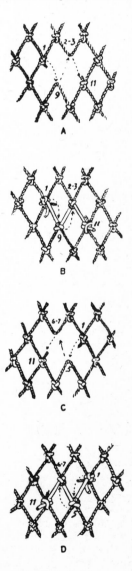

FIG. 21 - TEAR ALONG THE MESHES

**FIG. 22 – SMALLEST TEAR REQUIRING ALL MENDING VARIATIONS**

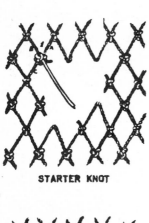

**STARTER KNOT**

**HALF-MESH KNOT – L TO R**

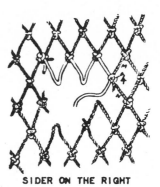

**SIDER ON THE RIGHT**

**PICK-UP KNOT – R TO L**

**HALF-MESH KNOT – R TO L**

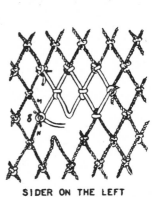

**SIDER ON THE LEFT**

**PICK-UP KNOT – L TO R**

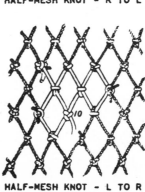

**HALF-MESH KNOT – L TO R**

**FINISHING KNOT**

### Starter Knot

**(Sketch A)** Shuttle the needle with sufficient length of twine, between A and B legs *from underneath* (1), and with the left hand fore-finger checking the twine at junction (2) allow the twine to slip by until only about one inch of the end (E) extends from the knot (D).

**(Sketch B)** Grip tightly at junction (3) with the left thumb. With sufficient slack of the mending twine form a loop to the left (4) and shuttle the needle (5) to the left under the A and B legs and over the twine loop (6) and follow through.

**(Sketch C)** Swing the needle to the right and downward tightening the hitch (7).

**(Sketch D)** Make another loop to the left (8) and shuttle the needle under one leg (B) and follow through to the left (9).

**(Sketch E)** Pull twine downward to the right and tighten the hitch (10).

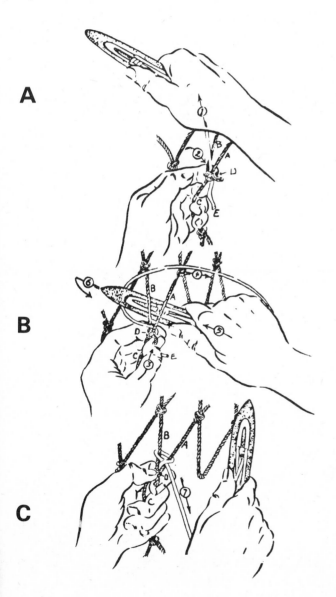

FIG. 23 - STARTER KNOT

FIG. 24 - STARTER KNOT (1) COMPLETED

**FIG. 25 - HALF-MESH KNOT
L TO R WEAVING**

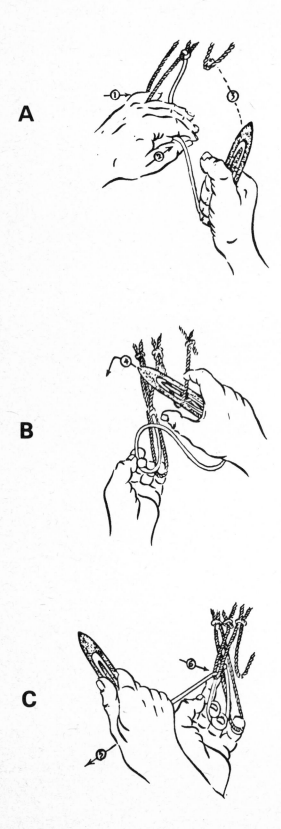

**Half-Mesh Knot**
*(Left to Right Weaving)*

(**Sketch A**) Holding the needle twine somewhat stretched, get hold of the twine with the left hand so that the little finger will be in the mesh (1) next to the one in the making; the twine is held between the forefinger and thumb (2), then shuttle the needle through the mesh loop (3) from underneath.

(**Sketch B**) Follow through (4) and then downward.

(**Sketch C**) Pull the twine (5), at the same time checking with the forefinger (6) until the loop in the making is of the same size as the one of the little finger.

(**Sketch D**) Grip tightly with the thumb at the junction (7) and with sufficient slack twine form a loop (8) and shuttle the needle from right to left under the two legs, and over the mending twine loop (9) following through (10) swinging the needle to the right.

(**Sketch E**) Pull the twine taut to the right and slightly downward (11).

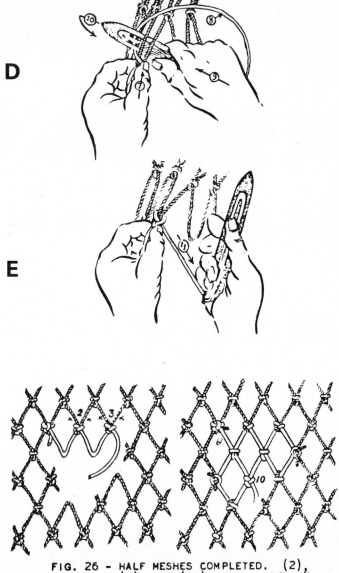

**FIG. 26 - HALF MESHES COMPLETED. (2),
(3), AND (10) ARE ALL OF THE
LEFT TO RIGHT TYPE.**

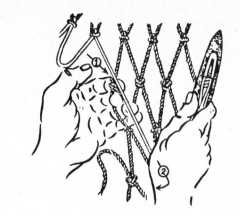

**A**

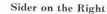

### Sider on the Right

(Sketch A) Grip the lower leg of the sider, so that the fore-finger is under the knot (1). Swing the mending twine to the right holding it quite taut in a position close to the right side of the knot (2).

(Sketch B) With the thumb, roll (3) the sider knot over the mending twine and hold tightly (4), then begin to swing the needle with sufficient twine slack (5 and 6).

(Sketch C) Continue swinging the needle to form a loop to the left of the knot (7) and shuttle the needle (8) under the upper leg of the sider knot and mending twine, and over the twine loop; follow through.

(Sketch D) Pull the twine to the left and somewhat downward to tighten the hitch (9).

**B**

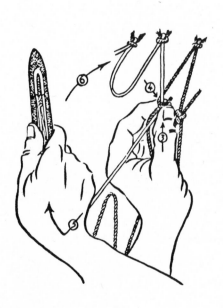

**D**

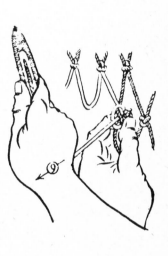

**C**

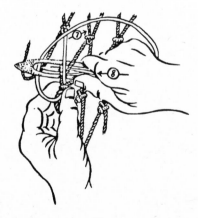

FIG. 27 - SIDER ON THE RIGHT

FIG. 28 - SIDER KNOT (4) COMPLETED

**Pick-Up Knot**
*(Right To Left Weaving)*

**(Sketch A)** Shuttle the needle through the pick-up mesh (1) following through and upward (2).

**(Sketch B)** With the left hand, grasp at the junction (3), pull the twine until all sides of the mesh are equal, then tighten the grip and pick up the mending twine with the little finger (4). Swing the needle downward (5) to the right, and with sufficient slack twine form a loop.

**(Sketch C)** Still supporting the twine with the little finger shuttle the needle under the pick-up mesh legs following through and over the twine loop (6) while holding the twine clear of the needle with the little finger.

**(Sketch D)** Pull the twine to the left and upward (7) to tighten the hitch.

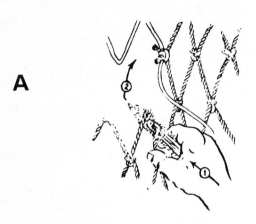

A

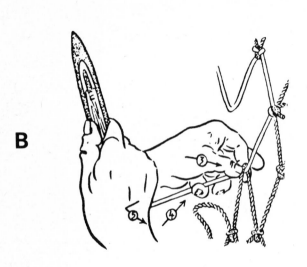

B

C

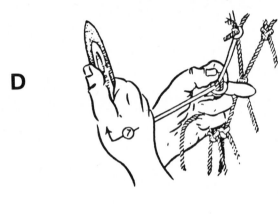

D

FIG. 29 - PICK-UP KNOT
R TO L WEAVING

FIG. 30 - PICK-UP KNOT (5) COMPLETED

## Half-Mesh Knot
*(Right To Left Weaving)*

**(Sketch A)** Place the left hand little finger into the mesh next to the one in the making (1), and throw the twine under the third finger for the size check. Hold the needle to the left and move upward (2).

**(Sketch B)** Shuttle the needle from above (3) and follow through.

**(Sketch C)** Pull the needle and mending twine slightly downward (4) checking the twine at the junction (5) with the forefinger until the bight being made is the same size as the half mesh (6) held by the little finger.

**(Sketch D)** Clamp tightly with the thumb (7), then with sufficient slack twine form a loop to the right (8) and shuttle the needle (9) under the two legs of the mesh and over the twine loop, then follow through.

**(Sketch E)** Tighten the hitch to the left and downward (10).

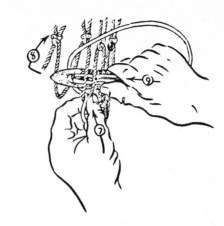

**D**

**A**

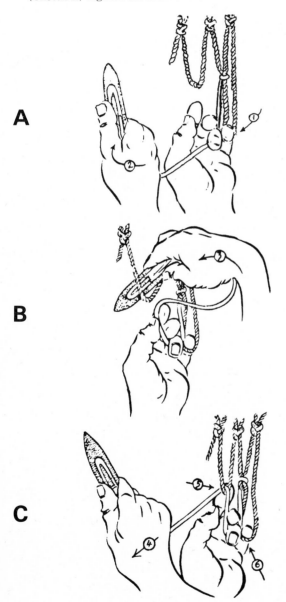

**B**

**C**

**E**

**FIG. 31 - HALF-MESH KNOT
R TO L WEAVING**

**FIG. 32 - HALF-MESH KNOT COMPLETED**

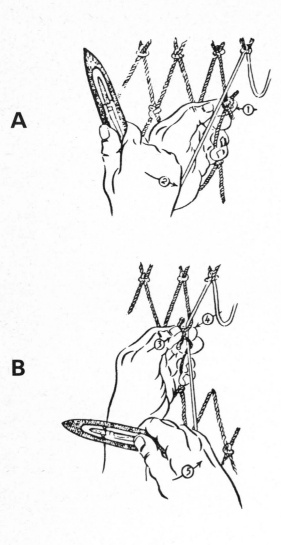

A

B

C

### Sider on the Left

(**Sketch A**) With the left hand get hold of the sider lower leg, and with the forefinger under the knot (1), swing the needle and twine to the right (2), with the mending twine somewhat taut place it on top of and to the left of the sider knot.

(**Sketch B**) With the thumb, roll the sider knot to the left over the mending twine (3) and hold tightly (4). Swing the needle to the right and upward (5).

(**Sketch C**) With sufficient slack in the mending twine form a loop to the right (6) and from above shuttle the needle (7) under the upper leg of the sider knot and mending twine, and over the twine loop, then follow through.

(**Sketch D**) Pull the needle and twine to the right and downward to tighten the hitch (8).

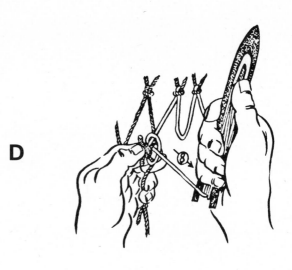

D

FIG. 33 - SIDER ON THE LEFT               FIG. 34 - SIDER ON THE LEFT COMPLETED

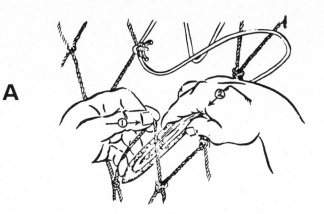

A

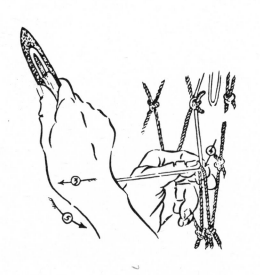

B

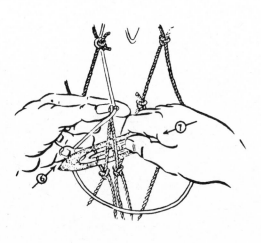

C

### Pick-Up Knot
*(Left to Right Weaving)*

(Sketch A) Hold the top of the pick-up mesh with the left hand (1) shuttling the needle (2) from underneath and follow through.

(Sketch B) Pull the twine to the left and slightly upward until all sides of the mesh are of equal length and hold tight at junction, (4), then swing the needle downward to the right.

(Sketch C) Hook the mending twine on the little finger of the left hand and with sufficient slack twine form a loop, then shuttle the needle (7) under both pick-up mesh legs, and over the twine loop. The little finger keeps the loop clear for the needle to pass over.

(Sketch D) Drop the twine off the little finger and pull the needle upward to the right, tightening the hitch (8).

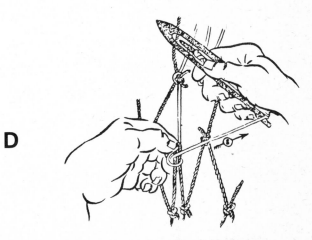

D

FIG. 35 - PICK-UP KNOT
L TO R WEAVING

FIG. 36 - PICK-UP KNOT COMPLETED

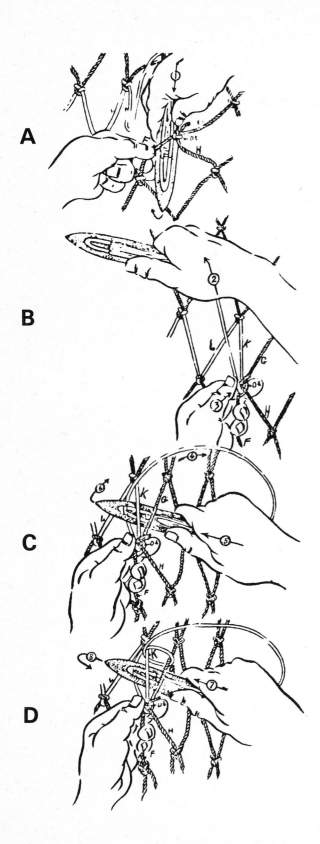

**FIG. 37 - FINISHING KNOT**

## Finishing Knot

(**Sketch A**) Hold the leg (F) of the finishing three-legger as shown (1). Shuttle the needle from *above*, between legs F and H and follow through.

(**Sketch B**) Pull the twine (2) upwards until leg K (in the making) becomes equal in length to the legs L and G. Then hold fast with the left thumb at junction (3).

(**Sketch C**) Form a loop with sufficient slack of the mending twine (4). Shuttle the needle under legs G and K and over the mending twine loop (5), and follow through (6). Do not tighten the hitch.

(**Sketch D**) Repeat for the second time, shuttling under (7) legs G and K, making another hitch (8). Be sure that the second hitch is below the first one, otherwise it will not tighten smoothly, the overlapping hitch will jam.

(**Sketch E**) Pull the twine to the right slightly downward to tighten the hitch evenly (9). When cutting off the twine leave about a half-inch stump.

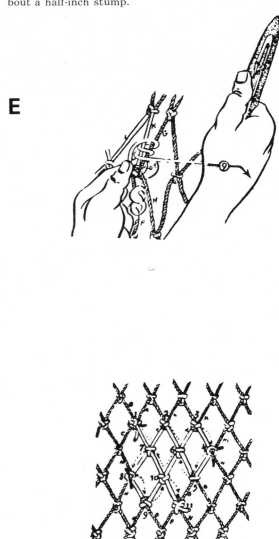

**FIG. 38 - FINISHING KNOT
COMPLETED AND THE
TEAR MENDED.**

**Double Hitch Sider Knot**

Often in a damaged net that has been under extreme strain the sider knots have become loose or defective in some other way. Then the double hitch knot is used for additional security.

**(Sketch A)** Bring the mending twine into position to the left of the knot as shown in the sketch, hold firm with the left thumb (1), shuttle the needle from underneath between the upper leg of the mesh and the twine (2), and follow through, then swing the needle to the left and slightly downward.

**(Sketch B)** Pull the twine (4) until a half hitch is formed under the knot, then hold fast (5) by pressing the mending twine with the thumb against the knot, then with sufficient slack in the mending twine form a loop to the left of the knot (6).

**(Sketch C)** Shuttle the needle (7) under the sider leg and twine then over the mending twine loop and follow through.

**(Sketch D)** Pull to the left (8) and slightly downward to tighten the hitch. ■ ■ ■

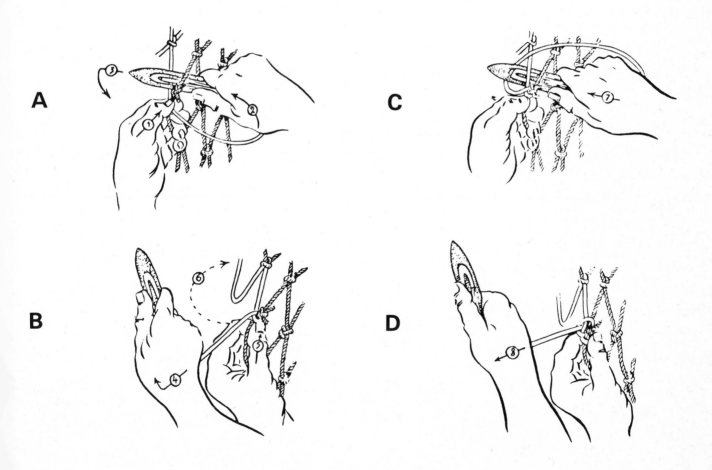

FIG. 39 - DOUBLE HITCH SIDER KNOT

# APPENDIX III
## Part 1

### Recommendations for Handling and Icing
### Fresh Pacific Halibut Aboard Vessels

*By Wayne Tresven and Harold Barnett*
*Research Chemists, National Marine Fisheries Service*

## 1. FACTORS THAT AFFECT THE HANDLING
## AND ICING OF HALIBUT

During the five years 1962 through 1966, the Bureau of Commerical Fisheries Technological Laboratory at Seattle, Washington, evaluated the quality and condition of Pacific halibut landed at ports in Alaska, British Columbia, and Washington. When the fishing vessels were unloaded, we occasionally noted that the halibut showed an excessive loss of quality and that this loss of quality was more serious among the small halibut than among large ones. We also noted that these small halibut usually had little or no ice in their pokes (the poke includes the mouth cavity and the visceral cavity), whereas the large halibut ordinarily contained at least some ice. In addition, we noted in the summer that the vessels which usually brought in halibut of good quality, occasionally landed halibut of poor quality because of inadequate icing. One other thing we noted was that the methods of handling the halibut greatly influenced their quality.

These observations prompted us to study factors involved in the handling and icing of halibut. The purposes of this article are to report our findings and to give recommendations that will increase the effectiveness of the handling and icing of halibut and will thereby help to ensure that the consumer will have halibut of uniformly high quality.

Because the quality of halibut is so closely related to the method of icing used, you can tell much about the quality of halibut simply by observing how well they have been iced. In the following discussion, we consider first the factors that affect the handling and icing of halibut and then the factors that indicate the quality history of halibut.

Halibut are a relatively large, thick-bodied fish having good storage qualities; however, when stored in ice, undesirable discolorations of the white skin and indentations due to particles of ice may occur. The discoloration is a "yellowing" or "greening" and is due to the growth of *Pseudomonas fluorescens*, a mobile bacterium that grows on the slime at low temperatures in the presence of free oxygen. The discoloration is on areas of the skin where oxygen is available, as in the proximity of pieces of ice or open spaces, and it doesn't occur where oxygen isn't available.

Special techniques have been developed to overcome these undesirable conditions. Emphasis has been directed toward filling the poke with crushed or flake ice and toward limiting the amount of ice distributed between and around the halibut to only that which will melt and provide adequate chilling without any pieces of ice remaining to cause discoloration or indentation of the skin. Instead of icing halibut on shelves where they would be exposed to oxygen, the halibut are now iced and packed tightly in pens to exclude the presence of oxygen. As in the icing of other fishes, however, a layer of ice is used to keep the halibut away from the sides of the hold, and ice is applied to the top of the load.

Except for the top layers of halibut, the iced halibut are usually placed on their side with the dark side down and the white side up. Small hemorrhages and bruises that objectionably affect the appearance of the white side tend to disappear when the halibut are iced in this position.

For economic reasons—primarily, the time and labor required —some fishermen do not attempt to ice the pokes of the small halibut as well as they do those of the more valuable large halibut. Unfortunately, however, when small halibut spoil, they contribute to an increased rate of spoilage of the entire lot.

Because we found that inadequate icing was associated primarily with the smaller halibut, we studied the icing of halibut of various sizes to find why the smaller ones were sometimes poorly iced.

We assumed that the size of the poke is proportional to the size of the halibut and, hence, that the amount of ice that can be packed in the poke is also proportional to the size of the halibut.

We found variations in both the size of the poke and the shape of the poke among halibut of similar weight in our sampling of commercial halibut. Halibut of various sizes caught in July near Goose Island, which is about 75 miles north of Vancouver Island, were quite plump and had relatively small pokes that represented about 16 percent by volume of the whole halibut. On the other hand, those caught in November in the Bering Sea were big-bellied halibut whose pokes represented 22 percent of their volume. In general, we found that the size of the poke is proportional to the size of the halibut and that the poke represents about 18 percent of the volume of the whole halibut.

When the poke is iced, the sides of the head and body walls bulge, thereby permitting more ice to be added to the poke. In spite of this bulging, the amount of ice that can be packed into the poke is, of course, limited by the size and shape of the poke. With smaller halibut, bulging of the body wall is more limited than it is with larger halibut, and proportionately less ice can therefore be packed into the pokes of small halibut. On a weight basis, the ice equalled about 17 percent of the weight of the large halibut and only about 6 percent of the weight of the small halibut.

The bulk density of the commercial crushed ice and flake ice obtained at various times and locations during this study varied considerably. It ranged from 28 pounds to 39 pounds per cubic foot. As the amount (weight) of ice in the poke is affected by the density of the ice, the use of low-density ice may contribute to inadequate icing. The amount of low-density ice that can be packed into the poke of a small halibut may not be enough to chill the halibut adequately.

### Other Important Factors

We consider first such handling factors as stunning, bleeding, cleaning, removing body heat, and draining away melt water and then consider mechanical refrigeration.

#### 1. Handling Factors

During our study of temperatures aboard commercial halibut vessels, we simultaneously observed the handling practices aboard the vessels. On some vessels, the method of handling caused the halibut to lose quality.

a. Halibut that had not been stunned often bruised themselves when struggling on deck. Dressed halibut were frequently bruised by being dropped into the hold.

b. Halibut were not always bled adequately. (Bleeding results in halibut of a lighter, more desirable, color.)

c. Halibut were not always cleaned adequately.

d. Halibut were sometimes purposely left on deck for as long as several hours, because the fishermen believed that the body heat should be dissipated before the halibut are chilled with ice in the hold.

e. Incoming iced halibut sometimes lay in melt water in the pens, because the water had accumulated faster than it had drained out.

After finding that the internal temperatures were higher in small halibut lying on deck than in larger ones, we studied temperature in greater detail.

The internal temperatures of halibut caught on adjacent hooks on the same line or on nearly adjacent hooks were determined at the following times: (a) immediately after the halibut were landed on the deck; (b) after they lay on the deck 6 hours; (c) after they were iced 24 hours; and (d) after they were iced 48 hours.

We learned that immediately after the halibut were landed on the deck, the internal temperature of the small halibut was 4 degrees F. higher than that of the larger halibut. This difference in temperature applied to halibut brought to the surface by a gurdy operating at full speed—that is, at 146 feet per minute. Lengthy delays in pulling in the line caused the temperature of the halibut to approach more nearly that of the surface water. This finding applied both to large halibut and to small ones.

Changes in temperature occur more rapidly, however, in the smaller halibut. This fact is reflected in the more rapid rate at which they deteriorate in quality, and it emphasizes the adverse effect of holding them at a relatively high temperature for even a short time.

Allowing the temperature of halibut to rise by leaving them on deck should therefore be avoided. Not only does the quality of the halibut deteriorate rapidly when their temperature rises, but both more ice and more time are then required to chill the halibut adequately. Thus, from the standpoint of both quality and economy, the temperature of halibut should be kept from rising.

Aged ice is sought by halibut fishermen because they believe that it preserves the halibut better than freshly made ice does. Aged ice is usually kept in cold storage, where it is gradually cooled considerably below its melting point. Aged ice is usually available at the start of the halibut season but is seldom available during the summer, when spoilage is more of a problem. Undoubtedly the favorable reputation of aged ice is due to its lower temperature making it easier to handle and last longer.

Use of colder ices in sufficient quantities can chill halibut to temperatures below 32 degrees F., and holding at the lower temperatures can result in retaining quality for a longer time. Most research workers advocate the use of ice at its melting point because it is at this temperature that it absorbs the most heat. In addition, water from the melting ice is probably useful in washing slime and bacteria from the fish.

Storage at temperatures lower than that of melting ice, however, permits fresh halibut to be stored for longer times with less loss of quality due to bacterial action and to loss of fluid from the flesh than when the halibut are stored in melting ice. With mechanical refrigeration, fresh halibut can be maintained at temperatures of 29 degrees F. to 32 degrees F.

If mechanical refrigeration is used, care must be taken not to lower the temperature of the halibut below 29 degrees F., its initial freezing point. Otherwise, problems may be encountered in unloading the frozen halibut.

Most fishermen, in attempting to control the temperature of the halibut, dress them and put the dressed halibut into the hold as soon as possible. As a number of halibut are accumulated in the hold before they are iced, some of the halibut undergo a slow and limited pre-chilling before being iced. Prechilling by immersing the dressed halibut in slush ice, refrigerated sea water, or refrigerated brine is recommended because the chilling is not only rapid but the pre-chilled halibut can be stowed with less ice, thereby permitting the stowage of more halibut in the same space.

Because fresh-water ice melts and refreezes at 32 degrees F., its condition within iced halibut indicated how well the halibut have been chilled and stored. Loose and dry particles of ice with little or no evidence of melting and of freezing indicate that the halibut were maintained at a temperature below 32 degrees F. and that the ice used has melted little or not at all, owing to its initial low temperature or to the prechilling of the halibut, or to both. Dry clumps of ice indicate that the ice melted partially, probably during the chilling of the halibut, followed by freezing of the ice due to the lower storage temperatures attained by mechanical refrigeration. Clumps composed of large aggregates of ice indicate that more ice has melted and resolidified than do clumps composed of many small particles of ice. Ice in melting condition indicates temperatures of 32 degrees F. or higher, and if the ice is dirty, is discolored, and has a foul odor, it indicates that the quality of the halibut has probably deteriorated. When little or no ice is present, the halibut have been inadequately iced or have been held too long a time. Under these circumstances, the halibut could have spoiled.

### Recommendations For Handling and Icing Fresh Halibut

1. Stun and bleed the halibut as they come aboard the vessel.

2. Immediately after bleeding them, dress and clean them inside and outside.

3. Keep the temperature of the halibut from increasing.

4. Immediately after cleaning each halibut, slide it down a chute into the hold; do not bruise it by dropping it.

5. Prechill the halibut.

6. Ice the pokes of all halibut—small halibut as well as large.

7. Lay iced halibut with the dark side down and the white side up.

8. Because the pokes of small halibut are often iced inadequately, use additional ice around the small halibut. (Note: To risk discoloration of the skin resulting from the use of the ice around the halibut, which permits the growth of aerobic bacteria, is better than to risk spoilage of the flesh.)

9. Use ice of high density and low temperature.

10. Use mechanical refrigeration to lower the temperature of the hold before obtaining ice; maintain the iced halibut close to 29 degrees F., but *do not allow them to freeze.*

11. Use thermometers placed throughout the hold to measure the temperature.

12. Use more ice in warm weather than in cool weather.

13. Reduce the amount of heat entering the hold by using effective insulation, by providing curtains as heat barriers about the chute leading from the hatch opening to the ice, and by flooding the deck with sea water.

14. Provide drainage facilities throughout the pens and the holds to prevent blood, slime, or melt water from accumulating and from thereby contaminating the halibut.

(These recommendations may be applied in general to the preservation of all fish at sea. Ed.) ■ ■ ■

## *APPENDIX III*
### *Part 2*

**Recommendations for the
Sanitary Operation of Plants that
Process Fresh and Frozen Fish**

*By J. Perry Lane
Supervisory Research Food Technologist,
National Marine Fisheries Service*

In recent years, the spotlight of public attention has focused on the operations of food processors. This spotlight has shown on fish processors, as is evidenced by the succession of bills on compulsory inspection and regulation of the seafood industry that have recently been introduced in the Congress of the United States. These bills are concerned primarily with protecting the consumer from health hazards and aesthetically undesirable practices. For the most part, the need for protection results from problems in sanitation.

The guidelines presented here represent an attempt to assemble the existing requirements for food plants in general and to relate them to fish plants in particular. A processor of fresh fish may find that many of the suggestions are difficult to put into practice without rebuilding his plant. These suggestions or closely related ones, however, are actual requirements for some food processors, such as packers of meat and poultry and producers of dairy products. So, similar types of regulations are likely to be forthcoming for the fishing industry. These guidelines thus can serve as a preview of some of the things that may be required.

Apart from any regulatory activities, another powerful factor now at work suggests that the fish processor should take a lively interest in sanitation. This factor is the consumer's increased awareness of quality factors and processing conditions. With the elimination of the captive Friday market for fishery products, seafoods have to compete on their own merits for the consumer's food dollar. The fishing industry, therefore, is fortunate in having a product that is both nutritious and flavorful when it is handled properly.

## I. PLANT RECOMMENDATIONS

In this part of the guidelines, we are concerned with plant premises and plant construction.

### A. Plant Premises

Both the location of the plant and the plant surroundings are important factors in a sanitary food-processing operation.

### 1. Location

a. If possible, locate the plant away from such sources of odors, dust, and air contamination as refineries, chemical plants, and dumps.

b. Locate the plant where it will be accessible to a supply of potable (here potable means that the water supply meets the criteria in the current edition of the "U.S. Public Health Service Drinking Water Standards") water and to a sewage system.

c. Locate the plant in a well-drained area.

d. Pave the entrance roadways.

e. Physically separate the plant from any plants that process nonhuman food.

### 2. Surroundings

a. Keep the surroundings free of unkempt vegetation capable of harboring insects or vermin.

b. Keep the grass and shrubs trimmed and neat.

c. Keep refuse areas separate from the processing plant.

d. Do not pile waste containers and fish boxes, for example, in the open area outside the plant.

### B. Plant Construction

In the construction of the plant, we are concerned with both the facilities for processing the fish and the facilities for the employees.

### 1. Processing Facilities

We can divide the processing facilities into what we might call the basic facilities and the equipment used in the processing operations.

**a. Basic facilities.** — What do we mean by basic facilities? We consider the basic facilities to be building construction, water supply and waste disposal, refrigeration, and lighting and ventilation.

a(1) Building construction.

Construct the buildings large enough to accommodate the operation without hampering sanitary cleanup.

In areas where food is processed or stored, use building materials that are impervious to water, easily cleanable, and resistant to wear and corrosion.

Keep all exterior openings such as doors, windows, and vents in good repair, and equip them with screens or other devices, such as air curtains, to prevent the entrance of insects, rodents, and other animals.

In this section, we consider the following subjects specifically: floors, walls, ceilings, and entrances.

*(a) Floors*—Two aspects of floor construction are of concern: the floors themselves and their drains.

(1) Floor Construction

Construct the floors of hard material such as waterproof concrete or tile. Do not make the floors extremely smooth. To prevent the workers' slipping on the floors, give the concrete a rough finish or use embedded abrasive particles.

Apply an approved latex synthetic resin base on concrete or mortar floors to increase resistance to corrosion.

Install drainage covers at the junctures of floors and walls.

(2) Floor drains

If any are where water is used, install at least one drainage outlet for each 400 square feet of floor space.

Give floors a slope of 1/4 inch per foot to drainage outlets.

Make the slope uniform with no dead spots.

Provide drains with traps.

In areas where the water seal in the traps is likely to evaporate unless replenished, provide the drains with removable metal screw plugs.

Construct drainage lines of galvanized iron or steel.

Use drainage lines with an inside diameter of at least four inches.

Vent drainage lines to outside air.

Screen the vents to prevent rodents from entering the plant.

Do not connect drainage lines from toilets to other drainage lines; be sure that the drainage lines from toilets discharge directly into a sewage system.

*(b) Walls.*

Make interior walls smooth and flat.

Maintain the walls in good repair.

Construct the walls of water-impervious materials, such as glazed brick, glazed tile, smooth-surfaced portland cement plaster, or other nontoxic, nonabsorbent material. (Poured concrete walls are satisfactory if they are trowled to a smooth finish. Marine plywood or metal walls (stainless steel, aluminum, or galvanised iron or steel) also are satisfactory if seams, nail holes, and junctions of floors, walls, and ceilings are watertight.)

Do not allow the supporting structures of walls to be exposed.

(c) *Ceilings.*

Place ceilings in work areas in such a way as to prevent foreign material from falling from overhead pipes, machinery, and beams onto exposed fishery products.

Make ceilings 10 feet high or higher in work areas.

Construct ceilings of portland cement plaster, large-size asbestos boards with joints sealed with a flexible sealing compound, or other suitable material impervious to moisture.

(d) *Entrances.* — In this section, we are concerned with the construction of doorways and doors and with pest control.

(1) Construction of doorways and doors.

Make doorways through which products are transferred on handtrucks, dollies, or forklifts at least 5 feet wide.

Make the doors and frames of rust-resistant metal or of wood sheathed completely with rust-resistant metal having tightly soldered or welded seams.

(2) Pest control—Both insects and rodents need to be controlled to prevent contamination and destruction of food products.

2(a) Insect control.

Screen and maintain in good repair all windows, doorways, and other openings that would admit flies and other insects.

Provide "fly-chaser" fans and ducts over outside doorways in the food handling area.

Limit the use of insecticides to those that are approved by the U.S. Food and Drug Administration.

In the application of insecticides, take care to prevent their contact with fish or other food products and with any working surfaces that come in contact with food products.

2(b) Rodent control.

Provide all exterior openings with screens that are rodent proof as well as insect-proof.

Except for solid masonry walls constructed or lined with such materials as glazed tile or brick, embed expanded metal or wire of 1/2-inch mesh or less in the junction of walls and floors.

Routinely inspect beams and storage areas for evidence of rat runways and nests.

If you find signs of rodents, call in a professional exterminator.

*Exercise extreme care so as not to contaminate fish or work surfaces with rodenticides.*

a(2) Water supply and waste disposal.

(a) *Water supply.*

Use only potable water for cleaning fish in any form or for cleaning any surface that could come in contact with food products or that could contribute to their contamination. Sea water may be used for the fluming of whole or dressed fish if the source meets local health requirements and if the water itself meets the microbiological requirements of the "Drinking Water Standards."

If water from a public water supply is used, test it against the "Drinking Water Standards" at least every six months to ensure that no in-plant contamination has occurred.

If water from private well is used, make sure that the source is free of contamination. Test the water for purity monthly.

If chlorinators are required to insure a continuous supply of potable water, use an automatic type equipped with warning devices to signal when it is not functioning properly.

Throughout the plant, provide both hot and cold water under adequate pressure and in quantities sufficient for all operating needs.

Install all equipment so that liquids will not be back syphoned into lines carrying potable water.

In general, except as provided above, do not use nonpotable waters in the plant. If such water is used for fire protection, steam lines, and the like, supply it in separate lines with no cross connections with potable-water lines. Clearly mark nonpotable water lines and outlets, and instruct all plant personnel that nonpotable water is a *deadly hazard* if it comes in contact with food products.

(b) *Waste disposal.*

Insure that waste-disposal systems meet the pertinent requirements given under the section on "Floor drains." page 66.

If permitted by local ordinances, discharge plant wastes into the municipal sewer system.

If you use a private septic tank or sewage disposal system, insure that it is effectively designed and operated so as not to produce objectionable conditions.

Do not discharge gurry and processing waste or plant sewage directly into harbors or other water areas without explicit written permission of Municipal or State Public Health Authorities.

Have any sewage-disposal facilities approved by the appropriate health authorities. Get the approval in writing, indicating when the facilities were inspected last.

Store gurry and other fish waste that cannot be carried by a sewage-disposal system in insect-proof and rodent-proof containers outside the plant or in physically segregated refrigerated rooms that are not used for any food products. Empty or remove unrefrigerated gurry from the plant premises at least once every 8 hours. If the containers are to be reused, wash and sanitize them before using them again. Do not store gurry in refrigerated rooms above freezing (32 degrees F.) for more than 48 hours. Dispose of frozen gurry as expeditiously as possible, and do not keep it on the premises for more than a week.

a(3) Refrigeration.

Make the refrigeration adequate to handle raw materials as will be discussed under the section on "Receiving raw materials."

Provide a temperature of 50 degrees F, or less in work areas where fresh fish are processed.

Maintain freezer rooms at −10 degrees F. or lower for the storage of the finished product.

For the initial freezing of finished products, use plate freezers or blast-freezing tunnels that provide contact temperatures of −20 degrees F. or lower.

If refrigeration wall coils are used in chilling rooms, provide, beneath the coils, a drip gutter of concrete or other moisture-impervious material properly connected with the drainage system.

Provide overhead refrigeration coils or plates in chilling rooms with insulated drip pans connected to drains placed beneath.

Use potable water in making ice used for holding fresh fish or other food products. Store, transport, and handle the ice in a sanitary manner. Do not reuse ice after it has been in contact with fish or fish products or with contaminated work surfaces or holding areas.

a(4) Lighting and ventilation. Both proper light and ventilation are important in maintaining sanitary surroundings and comfortable employee working conditions.

(a) *Lighting.*

Provide unrefrigerated workrooms with direct natural light where possible. In windows and skylights, use uncolored glass having a high transmissibility of light.

Use heat-absorbing (blue) glass to reduce glare in windows and skylights that are exposed to considerable sunshine.

In a workroom, make the glass area at least one-fourth the size of the floor area.

Provide well-distributed artificial lighting of good quality where natural lighting is not available or sufficient. In work areas, make the overall intensity of artificial illumination not less than 20 foot-candles.

Lights over processing areas should be covered by clear shields to prevent glass from falling into food products if a light bulb should break.

In candling for parasites, use lights that provide at least 50 foot-candles of illumination. Cover the light by a clean glass surface so arranged as to prevent any moisture from seeping down to the light fixture.

*(b) Ventilation.*

Provide sufficient natural or mechanical ventilation to control visible molds, objectionable odors, or excessive condensates.

Provide ventilation by means of windows, skylights, mechanical air conditioning, or a fan and duct system.

Supply mechanical ventilation in refrigerated workrooms where natural ventilation is lacking.

Locate fresh-air intakes so that the air is not contaminated with odor, dust, or smoke.

Where mechanical systems are used as the sole means of ventilating non-refrigerated workrooms and employee welfare rooms, use systems that can provide at least six complete changes of air per hour.

Install the ventilation systems so that air does not move from raw material or preparation rooms into processing or packaging rooms.

**b. Processing equipment.**—Having considered the basic processing facilities, we shall now consider processing equipment. Design all equipment and utensils of such material and construction that they are smooth, easily cleanable, and durable, and that any surfaces in contact with the product are free from pits, cracks, and scale. In addition, design and construct the equipment and utensils to prevent contamination of fish and fishery products with fuel, lubricants, metal, and other extraneous material.

b(1) New equipment—Insure that new equipment conforms to the applicable standards cited in the AFDOUS Food Code (Association of Food and Drug Officials of the United States, 1962) and with any more recent revisions of these standards.

b(2) Materials.

Use stainless steel as far as possible in all metal equipment that will come in contact with seafood.

In general, do not use galvanized metal, because it is not sufficiently resistant to the corrosive action of food products and cleaning compounds. If you must use galvanized metal for economic reasons, use it for such purposes as the construction of waste containers. If galvanized metal is used, make sure that it has the smoothness of high-quality commercial dip.

When fish are handled, make cutting boards or table tops of synthetic rubber-thermoplastic or of other hard, nonporous, moisture-resistant, synthetic material. If you use hardwood cutting boards, be sure that the surfaces are smooth and intact.

Do not use copper, cadmium, lead, painted surfaces, enamelware, or porcelain on surfaces in contact with the product. (The first three materials are toxic, and the last three may chip or flake off into the product.)

Make certain that any plastic materials and resinous coatings that you use are abrasion- and heat-resistant, shatterproof, and nontoxic, and that they do not contain any material that will contaminate the fish or fishery products.

b(3) Conveyor belts.—Make conveyor belts of moisture resistant material that is easy to clean, such as nylon, hard-finished rubber, or stainless steel.

b(4) Equipment design.—Design the equipment in such a way as to eliminate dirt-catching corners and inaccessible areas. Install equipment capable of rapid and complete breakdown for cleaning. To facilitate cleaning, use steel tubing rather than angle or channel iron.

b(5) Motors, bearings, and switches.

On food-handling equipment, locate all motors and oiled bearings in such a way as to prevent oil or grease from contacting the product.

Protect motors and switches from contact with water.

Raise motor mounts high enough to permit you to clean under them and between them.

Protect drivebelts and pulleys with guard shields that can be readily removed for cleaning.

b(6) Welding.—Make all welding within the product area continuous, smooth, even, and as nearly flush with adjacent surfaces as possible.

b(7) Stationary equipment.

Install all parts of stationary, or not readily movable, equipment at least 1 foot from walls and ceilings to provide access for cleaning.

Mount this type of equipment at least 1 foot above the floor, or else have a watertight seal with the floor.

b(8) Water-wasting equipment.—Install water-wasting equipment—such as flumes, brining tanks, and wash tanks—in such a way that waste water from the equipment is delivered through an uninterrupted connection into the drainage system without flowing over the floor.

b(9) Cutting tables.—Turn up the edges, at least 1 inch, of cutting tables or other equipment having water on the working surface.

**2. Employee Facilities.**

Having considered the processing facilities, we now consider the facilities for the employees. To get plant personnel to recognize the importance of sanitary practices and to obtain their full cooperation, make proper provisions for their personal needs. Considered here are the dressing rooms, toilet facilities, hand washing units, and eating facilities.

**a. Dressing rooms**

Provide separate dressing rooms for employees of each sex.

Separate the dressing rooms from the toilet and work areas.

Ventilate the dressing rooms, and provide receptacles for the disposal of cigarette stubs and other waste.

Provide each employee with a metal locker that is at least 15 by 18 by 60 inches. Place the lockers on legs 16 inches high to enable you to clean all areas of the floor.

**b. Toilet facilities**

Separate toilet rooms from dressing rooms by tight, full-height walls or partitions.

Do not permit toilet rooms to open directly into food-processing areas; instead, separate them by a ventilated vestibule with two sets of self-closing doors.

Provide elongated water closets with open split seats, in the following ratios:

**c. Hand-washing units**

Locate hand-washing unit (lavatory) conveniently and make sure that they meet the appropriate requirements discussed under the section on "Plant sanitation and cleaning provisions."

Make the minimum size of bowl 16 by 16 by 9 inches, and supply each lavatory with hot and cold water delivered through a mixing faucet fixed at least 12 inches above the rim of the bowl so that an employee may wash his arms.

Locate liquid soap and sanitary towels in suitable containers at each was basin.

**d. Eating facilities**

Provide clean, well-lighted, and ventilated eating facilities that are separate from work areas and toilet areas.

If eating facilities are provided in the dressing room, set the space aside separate from the immediate locker area.

Provide tables and chairs or benches, washing facilities, and drinking fountains.

Clean the area after regularly scheduled work breaks and lunch periods to prevent food particles from attracting vermin and insects.

## II. PROCESSING RECOMMENDATIONS

We now have given detailed consideration to the guidelines on the plant—both its premises and its construction, including the processing facilities and the employee facilities. We turn next to the guidelines on the processing itself. In so doing, we consider methods of guarding against microbial contamination of the plant and product and then give attention to the handling of the product.

### A. Guarding Against Microbial Contamination

In guarding against microbial contamination, we are aided by a knowledge of certain procedures for testing for the presence of microbial contamination and of sanitation principles involved in microbial control.

## 1. Bacteriological Testing Procedures

Have the microbiological tests listed below (under "Microbial tests to be performed") made periodically on samples of the finished products from all processing lines. These tests will serve as a guide in determining whether you have a sanitation problem in your plant. Do not consider the numbers as being an absolute standard of product quality, but rather as being levels that, if exceeded, indicate that a more thorough microbiological survey of raw material, processing equipment, and personnel should be made. This survey will help you decide whether you do have an area that is a source of serious contamination or that could become one.

Here we are concerned specifically with five subjects:

a. Directions for microbial tests.
b. Microbial tests to be performed.
c. Sampling.
d. Corrective action.
e. Resurvey.

**a. Directions for microbial tests.** — Have the microbiological tests carried out according to the procedures given in "Standard Methods for the Examination of Dairy Products": current edition, prepared by the American Public Health Association (1967). These procedures are suggested for fishery products but are *not* standardized for such products. There are still no generally recognized methods for testing fishery products, but the procedures given in Standard Methods for Dairy Products will prove generally satisfactory for the type of microbiological survey recommended in this section.

**b. Microbiological tests to be performed.** — The microbial tests to be performed include the following: (1) total plate count, (2) coliforms, (3) Salmonella, (4) E. coli, and (5) coagulase positive staphylococcus.

a(1) Total plate count. — Take remedial action if the total plate count exceeds 200,000 organisms per gram. Consider that the total plate count indicates the entire bacterial exposure of the product. Furthermore, consider that it also indicates the level of spoilage organisms present. Although a direct relation between total plate counts and organoleptic quality or storage life has not been established, excessive counts do indicate that the storage life of fresh fish with such counts will be reduced materially.

a(2) Coliforms. — Take immediate remedial action if the MPN (most probable number) is more than 360 per gram. The presence of coliform organisms indicates contact of the product with water and soil contaminants and warns that the product may possibly be polluted with sewage.

a(3) Salmonella. — Take immediate remedial action if the sample is not free of this organism. Salmonella is a bacteria that causes food poisoning. The presence of this organism indicates human or animal contamination.

a(4) E. coli. — Take immediate remedial action if the MPN is greater than 50 per gram. This organism is one of the coliform group and is a specific indicator of fecal contamination. Although this organism does not cause disease, it indicates that the product probably has been contaminated with organisms that are pathogenic.

a(5) Coagulase positive staphylococcus. — Take immediate remedial action if the MPN is more than 5 per gram. The test for this organism should be the confirmed test. The presence of the organism indicates human, infectious contamination. It is a toxin-producing organism that causes food-poisoning. The organism is readily killed by heat, but the toxin is quite heat stable.

**c. Sampling**

Make the initial survey of the finished fresh or frozen product at the point where it is ready to leave the plant.

Sample precooked products before they enter the cooker. Cooking will destroy many organisms but not all toxins; for this reason, microbial tests on cooked products give a misleading indication of their microbial exposure.

Take samples separately and place them in sterile containers.

Store fresh samples at 33 degrees F.

Store frozen samples at 0 degrees F. or lower.

Test all samples as soon as possible after they are taken.

If the results of any of the bacteriological tests exceed the suggested limits, make a complete microbiological survey of the plant.

Take samples of all raw materials.

Take samples of the products after each stage of processing (that is, after initial washing, filleting, skinning, brining, tempering and cutting of frozen blocks, after applying batter and breading; and after packing).

Take swabs of all equipment during processing and after cleaning up.

**d. Corrective action.** — From the results of the complete survey, take corrective action if any trouble spots were identified. Corrective action may range from a general cleanup of the entire plant to something as specific as cleaning up a single piece of equipment, discarding certain raw materials, or having one of the employees change his personal habits.

**e. Resurvey.** — To determine the effectiveness of the corrective action, repeat the product survey after the corrective sanitizing and cleanup measures have been instituted. Make periodic surveys to determine if the plant-sanitation program is continuing to be effective or if new problems in sanitation have developed.

## 2. Plant and Personnel Sanitation

Having outlined the bacteriological testing procedures, we now consider plant and personnel sanitation. In so doing, we consider provisions relating to plant sanitation and cleaning and those relating to personnel practices.

**a. Plant sanitation and cleaning provisions.** — Keep in mind, that, although proper sanitation is the direct responsibility of the plant manager, an effective sanitation program can be obtained only when every employee in the plant is instructed in proper sanitary precautions and is fully impressed with the reason for proper sanitation in terms of product quality and protection from public-health hazards.

Plant sanitation requires the services of a sanitarian and adequate cleaning methods and facilities.

a(1) Sanitarian. — Assign one person as a sanitarian for the plant. The sanitarian must be tactful and have good judgement. If the plant is large, assign one or more assistant sanitarians for specific work areas. Make the sanitarian responsible for supervision of all plant-cleaning operations. Have him thoroughly inspect all processing areas and equipment before each day's operation, and see that any deficiencies are corrected before operations are started.

a(2) Cleaning and cleaning facilities.

*(a) Cleaning.* — To clean adequately, adopt a schedule and carry out the cleaning steps in proper sequence.

(1) Cleaning schedule.

Adopt a cleaning schedule for each area in the plant, and adhere to it unless conditions warrant more frequent cleaning or sanitizing operations. Thoroughly clean continuous-use equipment such as conveyors, filleting machines, flumes, batter and breading machines, cookers, and tunnel freezers at the end of each working shift, or oftener, if conditions indicate the need.

Clean batter machines and other equipment in contact with milk or egg products more frequently, depending on the temperature at which the batter is maintained, the type of material going into the batter (fresh or frozen), the ambient temperature of the work area, and the microbiological level of the raw materials and fish. Ascertain these factors, and design the cleaning schedule accordingly.

(2) Cleaning sequence.

Mechanically or manually remove loose dirt by scraping and brushing floors and equipment.

Rinse with cold or warm water. Because fish residues and other proteins coagulate at high temperature and may become baked onto the contact surface, remove these materials at temperatures below 100 degrees F. early in the cleaning process.

Wash with an acceptable detergent.

Rinse twice with hot water at a temperature of at least 170 degrees F. Hot water is more effective than cold water in removing fats, oils, and inorganic material.

Sanitize with an acceptable bactericidal agent. Chlorine compounds are the most widely used—recommended strengths are given later. Other sanitizing agents approved for use in food-processing plants are also effective.

Rinse twice with hot water. A thorough rinse with potable water should follow any operation involving a chemical sanitizing agent.

*(b)Cleaning facilities.*—Here are considered detergents, chlorine solutions, single-service articles, and hand-sanitizing units.

(1) Detergents.—In using detergents, we are concerned about their characteristics and about which of them are approved for use with food products.

1(a) Characteristics of detergents.

Keep in mind that the desirability of a detergent usually is determined by the degree to which it exhibits the following characteristics.

High wetting or penetrating action, which causes rapid washing away of the soil.

Good rinsibility, which results in the detergent and soil being rinsed from the equipment freely and rapidly after the desired cleaning has been accomplished.

High emulsifying power for oils.

High deflocculating or dispersing power, to bring deposits of precipitates into suspension so that they can be washed away.

Water conditioning or sequestering properties in alkaline solutions, to prevent deposits on equipment of any calcium and magnesium compounds from the water.

Dissolving and neutralizing power, for the purpose of dissolving or neutralizing tenacious deposits and saponifying fats to make them soluble in water.

Low corrosiveness to the surfaces on which they are used.

1(b) Approved compounds.—Detergents and sanitizing compounds approved for use in food processing plants may be found in the current edition of "List of Chemical Compounds Authorized for Use under USDA Poultry, Meat, Rabbit, and Egg Products Inspection Programs," U.S. Department of Agriculture (1968).

(2) Chlorine solutions.

Use the following suggested concentrations of chlorine solutions in fish processing plants.

Keep in mind that it is important to rinse with clear potable water after using any sanitizing agent and that, to prevent corrosion, it is especially important to rinse metal surfaces after chlorine solutions are used.

(3) Single-service articles.

Store materials intended for one-time use, such as paper cups or towels in closed containers, and dispense them singly and in such a manner as to prevent their being contaminated.

Provide closed containers for the disposal of such articles.

(4) Hand-sanitizing units.

Locate wash sinks and sanitizing hand dips outside of lavatories and adjacent to work areas, such as filleting lines or packing tables, where fish are handled.

Supply sanitizing dips with 100 part per million available chlorine. Keep filleting knives and steels in sanitizing solutions when not in use, and have each filleter rinse his hands and change knives frequently.

**b. Personnel provisions.**—Employee health and employee practices are important in controlling microbial contamination.

b(1) Employee health.

Have all food handling employees examined physically prior to their starting work at the plant and at least annually thereafter. Comply with local health requirements regarding the physical examination and see that each employee has a current and valid health certificate showing no evidence of any communicable disease. Have the employee take a physical examination before returning to work after any contagious illness.

Do not allow employees with open sores and lesions into food-processing areas.

b(2) Employee practices.

Prohibit employees from eating, using tobacco in any form, and spitting in food-handling areas.

Do not allow employees to wear jewelry, except plain wedding bands and unadorned pierced earings (not screw-on type), in food areas.

Have all food handlers wear clean outer garments, preferably white, that cover personal clothing. Have fish filleters wear easily cleanable rubber or plastic aprons or coveralls and boots. Such garb should be worn by personnel working with fresh fish or on cleanup crews using large amounts of water. Require that all clothing worn during working hours be clean, and maintain an adequate supply of replacement garments.

Have all employees wear clean head coverings (caps or hairnets) that cover or hold the hair in place.

Require that each employee wash and sanitize his hands after each absence from a work station. When rubber gloves are worn, have them washed and sanitized in the same manner.

## B. Product Handling

In our guidelines on processing, we now have completed our suggestions for guarding against microbial contamination. We turn now to our guidelines on how to handle the product. Here, we consider the receiving of the raw material and the processing of it.

### 1. Receiving Raw Materials

By raw materials, we mean both the fish and any other raw materials used in processing.

**a. Fish.**—We consider first the fresh fish and then the frozen fishery products.

a(1) Fresh fish.

Check fresh fish for signs of spoilage, off odors, and damage upon their arrival at your plant. Discard any spoiled fish.

If the fish are to be scaled, scale them before you wash them.

Unload the fish immediately into a washing tank. Use potable, nonrecirculated water containing 20 parts per million of available chlorine and chill to 40 degrees F. or lower. Spray wash the fish with chlorinated water after taking them from the wash tank.

If incoming fresh fish cannot be processed immediately, inspect them, cull out the spoiled fish, and re-ice the acceptable fish in clean boxes; then store them preferably in a cold room at 32 degrees to 40 degrees F. or, at least, in an area protected from the sun and weather and from insects and vermin.

Wash, rinse, and steamclean carts, boxes, barrels, and trucks used to transport the fresh fish to the plant if any of these are to be used again. If disposable-type containers are used, rinse them off and store them in a screened area until you remove them from the premises.

a(2) Frozen fishery products.

Use a loading zone that provides direct access to a refrigerated room.

Check the temperature of the product at several areas in the load. When the product arrives, it should be 0 degrees F. or lower.

Place the product on pallets and assign a freezer lot number to it to ensure that the rule of "first-in, first-out" is observed.

Keep the freezer storage at −10 degrees F. or lower, and use a separate blast freezer capable of rapidly lowering to −20 degrees F. any product that arrives at a temperature higher than 0 degrees F.

**b. Other raw material (dry).**

Unload other raw material in an area separate from the fresh or frozen products.

Store the material in a dry, ventilated area on pallets or shelves that will keep the material away from the floor and the walls.

Screen the storage area adequately to prevent entrance of insects during loading or unloading operations.

## 2. Processing Raw Material

Keep in mind that fish is a highly perishable food. The primary cause of deterioration of fish flesh and the resulting loss of quality is bacterial contamination. Every step in the recommendations for sanitary plant operation and fish-handling procedures is designed to reduce this contamination and thereby protect the health of the consumer and maintain the quality of the product. The basic rules for handling a fishery product are:

Keep the product cool, as near 32 degrees F. as possible for fresh fish and below 0 degrees F. for frozen fish.

Keep it moving. It is the combination of time and temperature that permits bacteria to grow and build up. Even under optimum conditions, quality will be lost, so you should get the product into the consumer's hands as rapidly as possible.

### a. Fresh fish.

Handle incoming fresh fish as was described in the section on "Receiving raw materials."

Cool the filleting room to 50 degrees F. or lower. If the room is not cooled, then ice the fish so as to maintain their internal temperature of 40 degrees F. or lower.

During hand-filleting operations, scrub the filleting boards at least twice a day. Use water containing 2 to 5 parts per million available chlorine to flush continuously the filleting boards and conveyors used to transport whole fish.

When cutting fillets by hand, handle them so that the cut surface does not come in contact with the filleting board; then immediately place them on a fillet conveyor or in a container.

Furnish filleting machines with a continuous supply of water on the surfaces in contact with the fish.

Have the fillets discharged directly onto a conveyor or into a container.

Use a machine to prepare skinless fillets and spray water on the skinning machinery.

Complete all trimming operations before sending the fillets to the final wash.

Because certain species of fish (such as cod, ocean perch, and some Pacific rockfish) may contain parasites, candle the fillets from these fish before brining them in the final wash. Do the candling on a clean glass surface well illuminated from below. Because heat from the lights may cause bacteria to grow rapidly on the surface of the candling table, clean the surface thoroughly and sanitize it frequently. Continuously flush it with chlorinated water.

A brining tank, to be effective, must be used correctly. Use brine as a final wash in order to help reduce the loss of moisture. Chill the brine to 35 degrees F. or lower, chlorinate it, and change it at least once an hour, so that it will decrease bacterial contamination. Convey the fillets through the tank so as to regulate their time of exposure to the brine; after they leave the tank, pass them through a multijet mist spray. Keep in mind that the strength of the brine and exposure time should depend on the species of fish being handled. Only mildly brine the fillets of fatty fish, especially those fillets that are to be frozen; otherwise,

the residual salt on the fillets will accelerate the development of rancidity during storage.

See that fresh fillets that are to be packed in bulk have an internal temperature no higher than 35 degrees F. before the fillets are packed in a bulk container. If need be, use an adequately refrigerated brining tank as a prechiller. Promptly pack and ice well the prechilled fillets, or place them in a cold-storage room at 30 degrees to 35 degrees F.

Promptly pack fillets that are to be frozen, and place them in a freezer in less than 30 minutes after they are packed. If it be necessary to transport the fillets to another building for freezing, transport them under refrigerated conditions if the elapsed time from packing to entering the freezer exceeds 30 minutes. Do not expose the packaged fillets to sun and dirt.

### b. Frozen fishery products.

Handle incoming frozen fish as was described in the section on "Receiving raw materials."

Where frozen fishery products such as fish blocks must be tempered (brought up to a higher temperature) before being processed, temper them in a refrigerated room under controlled conditions. (Uncontrolled tempering in work areas causes blocks on the outside of the load to become excessively warm while the blocks at the center of the load remain too cold for efficient processing.) Once the blocks are tempered to the desired temperature (not higher than 20 degrees F.), process them as soon as possible. Slake out, in refrigerated water, blocks or bulk packs of fillets or whole fish that must be thawed and separated for further processing. Prechill the water to below 40 degrees F. Remove the fish or fillets as soon as they are thawed sufficiently to be processed. Change the water and clean and sanitize the tanks before you put more fish in them.

Good product handling practices require that breading lines be given particular attention. Maintain the temperature of the batter below 50 degrees F. Discard all unused batter at the meal break and at the end of each work shift, and clean and sanitize the batter container before reloading it. Discard unused breading at the end of each work shift. Place drip pans and dust shields around breading and batter machines. Remove any spillage from the floors at once.

When processing precooked products, pass cooking oil through a continuous filter to remove any food particles in the oil. Locate adequate exhaust fans in the working areas to remove smoke, odors, and excess heat. Pass precooked products directly from the cooker to a freezer before packing them. Maintain temperature-control charts for all cookers.

Handle all frozen products as expeditiously as possible to prevent them from thawing. Do not allow the time between bringing the frozen blocks to the processing area and placing the finished product in the freezer to exceed 1 hour. (In a well laid-out plant, this time will be less than 30 minutes.) Because of this short time interval, the work area need not be refrigerated, but prevent it from becoming warmer than 75 degrees F.

Show the date of packing on the primary code containers of all finished products. ■ ■ ■

## APPENDIX IV
### Part 1

#### Operations Involved in Canning

*By Norman D. Jarvis*

*(The following discussion of canning procedures and canning spoilage, Parts 1 and 2 of this section of the appendices is, perhaps, one of the best of its kind in English. Norman D. Jarvis, in his Research Report No. 7, for the Interior Department's Fish and Wildlife Service, established himself as a leading authority on fish packing and processing. The study was first published in 1943 but its basic premises are still as valid as they were in that year. Some of the writer's references have been made out-dated by progress in the field but this does not make his general thesis any less valuable.)*

The canning of fishery products may be divided into a number of definite steps applying equally well to all types, although details of methods and processes vary with the individual product. To discuss points of importance in certain operations common to all canning procedures in the description of individual products would cause duplication, or the information would be presented in disjointed fragments whereby most of its value would be lost. Therefore, a general discussion of the steps in canning follows.

### Securing The Raw Material

The quality of the product and the price range at which it is to be sold usually determine the choice of fishing gear and the method of fishing. For instance, tonging could hardly be used extensively in obtaining oysters for canning. It would be too slow for mass production or would require the efforts of so many men that production costs would be raised so that canned oysters could not be packed at a price the consumer could afford. Since the oyster tonger can operate only in the shallower waters, this method would also limit the catch by restricting the fishing area.

The pack of canned shrimp has been increased and the price reduced largely because it has been found impossible to procure a larger supply at less cost by adopting nets of the otter-trawl type instead of the haul seine formerly used. The trawl nets have been found to be more efficient since they can be operated at depths and in areas previously unfished, and can also be operated more cheaply.

Raw material taken from the same area by different types of gear may differ in quality when landed at the cannery. Fish taken by gill nets are "drowned" and always receive more handling than fish taken by other types of gear such as the trap, which means that decomposition is accelerated and the length of time the raw material will remain in good condition is reduced. Therefore, gill nets are suitable for use only where the fishing grounds are in close proximity to the cannery and tranportation facilities permit frequent pick-ups from the fishermen and prompt delivery at the cannery.

In canning salmon, the trap or pound net is considered the best type of gear since this apparatus is the most efficient where its use is practicable or permissible. It is also easier to control the flow of raw material which is usually of better quality when it reaches the plant because the fish may be held alive until needed. The fish are not handled so much and fewer water-marked salmon, fish closely approaching the spawning stage, are taken.

As a rule, the method which supplies the largest amount of good-quality raw material most cheaply should decide the type of gear to be used. The most efficient apparatus cannot always be adopted; sometimes it is too efficient and causes depletion of the fishing grounds and therefore may be prohibited by law or regulation. Finally, due to political considerations, the use of certain types of gear has been forbidden in some states.

In some areas the cannery owns the fishing craft, and the men who operate the gear are paid a fixed monthly or daily wage. On other fishing grounds the cannery supplies the equipment, contracting to purchase the catch on a poundage basis. In still other instances the fishermen may own boats and gear, selling the catch to the cannery operator according to a scale of prices previously arranged by contract. In some cases the catch may be sold on the open market to the highest bidder.

Open-market purchase is not practicable for the majority of packers and is only economical when the fresh market is in a depressed condition. The packer should have a steady supply of raw material and should be able to estimate his packing costs accurately, especially if the pack has been sold under contract previous to manufacture as is often the case. As a rule the packer cannot risk depending on fluctuating market prices or on securing a steady and adequate supply of raw material on the open market.

It is also to the advantage of the fisherman, equally with the packer, to have a steady market for his catch at a definite price. Costs of operation, especially of boats and gear, are high. This is one reason why many fishermen do not operate their own gear. The fisherman may "catch the top of the market," but too often he finds the price that he obtains at open sale leaves him no return for his efforts.

Canners of fishery products, especially those operating on a small scale, often find it more desirable to lease rather than to purchase fishing craft and apparatus at least for the first few seasons of operation. Where the canning season is short and it is difficult if not impossible to turn boat and outfit to other fisheries for the balance of the year the fisherman probably should not attempt to operate his own outfit.

### Transporting And Receiving

All raw material in any one lot should be equally fresh on receipt from the fisherman. Sometimes when fishing is poor the fisherman is apt to delay delivery, hoping to secure a better boat load. Unless it is possible to enforce regular deliveries there is considerable variation in the freshness of the material.

Since fishery products spoil quickly, the time elapsing between the time of catch and that of receipt at the cannery should be as short as possible. It is not possible to set a certain specified time for this period; it is affected by many variables such as temperature and humidity. For example, in Alaska, salmon will usually remain in good condition for canning for 48 hours, but during certain seasons some fish or portions thereof will show an appreciable degree of staleness after 12 hours.

If fish or crustaceans are piled to any great depth in the hold of a vessel, pressure will cause overheating which rapidly advances the rate of deterioration. Heavy pressure also contributes to softening the texture of the lowest layer, mechanically tearing and bruising the flesh, and increasing contamination due to the greater amount of slime dripping down from the excessive upper layers.

Raw material is sometimes transported in boxes holding about 500 pounds, net weight. These boxes may be piled to any depth since the pressure is divided and no individual fish is subjected to a heavier pressure than the weight of the other fish in that particular box. The holds may also be divided into pens, which are capable of further subdivision by removable floors at different levels.

The raw material should be handled as little as possible and any carelessness in handling should be penalized. "Forking" or "peughing" should be forbidden absolutely or restricted only to the head of the fish. Peugh holes in the body not only provide an easy entrance for spoilage organisms, thus hastening deterioration, but also cause dark discolored streaks in the flesh which are often visible in the canned product. Trampling upon, throwing or buising the cargo not only lowers the quality of the raw material but often makes it unusable in products of any grade.

### Grading

The raw material should be graded and sorted into different bins on arrival at the packing plant. A standardized system of grading has not been developed in the preparation of any canned fishery product but a number of methods are in use varying according to locality, experience of the packer and the product. In general, grading is dependent on freshness, size and species of fish and color of the flesh. Careful grading is essential to the preparation of a high quality canned product.

The condition of the fish on arrival may be determined by a number of factors. Odor is important in judging freshness, but the sense of smell varies with the individual. While some are hyper-sensitive, others cannot recognize the difference between fresh and slightly stale fish. The ability to judge by odor also seems to vary with the health of the individual and is affected by smoking and drinking.

A rapid determination of freshness may best be made by judging firmness, appearance of the gills, eyes, and the flesh of the belly cavity near the backbone. The flesh of fresh fish is firm and resilient, so that when pressed with a thumb in the thick portion of the back, the impression will gradually disappear when the thumb is removed. If fish are fresh the gills will be clear pink to deep red in color, firm, free from slime, and with an odor which may best be described as a "salt water odor," rather agreeable and free from any suggestion of taint. The eyes of fresh fish are bright, clear and protuberant. The flesh near the belly cavity around the backbone should be free from discoloration and may be stripped away from the backbone only with difficulty, leaving many shreds clinging to the bone.

When deterioration sets in, the flesh becomes soft and flabby in texture, and the impression will remain if a thumb or finger is pressed into the thick back flesh. The gills acquire a faded brown color, which may become gray or grayish-green in a rotten fish, and are covered with a thick lumpy slime. The odor of the gills is rank and unpleasant if the fish is stale, changing to a characteristic tainted or putrid odor as decomposition continues. The eyes are dull, opaque, sunken and often bloodshot. The flesh around the backbone shows a dull red discoloration and may be stripped away easily and cleanly.

In some packs, such as shrimp and sardines, size is an important factor in grading. Carelessness in sorting to size will result in poor appearance and may cause difficulties in filling.

A catch may include several species. A single delivery of salmon may include five, some of which are more highly regarded than others though all are about equal in food value. If carelessly graded according to variety the contents of a can may be composed of pieces from two species of noticeably different appearance.

The color of the flesh is of importance in the production of several fishery products. In the Columbia River chinook salmon there is variation of color within the species and fish of several shades may be taken in the same delivery. The deeper shades of color are considered more desirable, so separation according to shade is necessary in the production of a "fancy" pack. Variation in color also occurs in some species of clams but in this instance the lighter color is preferred for "high quality" packs, especially of minced clams. Mixing grades operates principally to the loss of the packer since such products are sold at the price of the lower grade.

### Dressing and Washing

Dressing and washing is the first step in the actual process of manufacturing the canned product and consists of removing viscera and other waste material, and of freeing the raw material from blood, slime or dirt by the use of generous quantities of water. Dressing or cleaning must be carefully and closely supervised at all times in order to prevent needless loss of edible material which is sometimes sufficient to destroy the margin of profit. Careless cleaning may also result in the inclusion of waste in the canned product. This is considered as adulteration by the U.S. Food and Drug Administration.

Washing may be of three general types, soaking or tank washing, washing by agitation, and spray washing. Tank washing may act as a source of contamination rather than a means of cleansing unless the water is changed frequently. Soaking or tank washing is effective in removing blood but softens texture if the fish are left too long in the tank, unless a salt brine is used. Oversoaking in salt may toughen the texture or make the canned product unpalatable.

Agitation increases the efficiency of washing. The earliest development of this type of washer in fisheries is the wooden flume conveyor in which the fish are carried from the cleaning and trimming section to the pre-cooking or preparation stage by a current of rapidly running water. Another simple type of agitating washer is the tank equipped with a propeller. Unless the propeller is guarded by a heavily screened cage or is geared to move slowly it may bruise the fish. Compressed air is sometimes used to agitate water as in the oyster shucking "blower" which is probably the most efficient type of agitating washer.

The drum or squirrel-cage washer is also used widely in the canning of fishery products. It consists of a drum of heavy small mesh wire screen with a central axle and equipped with longitudinal baffles of angle iron at intervals around the inner circumference. The drum revolves in a tank of water. This type is used in washing herring, Maine sardines and shrimp. Comparatively few fish are bruised, it requires little water and also acts as a scaler.

Washing fish mechanically by means of strong jets or sprays of water has increased in fish canning establishments during the past few years, and should be used even more extensively, being more efficient and economical. Spray washing methods depend on pressure of water rather than volume. A combination of a spray wash with the revolving drum has been found satisfactory in washing some types of fish which must also be scaled.

### Preparation For The Can

Some articles such as salmon and shad are simply cut into container-length pieces after washing and undergo no other preparatory treatment. Other packs, such as tuna and sardines, undergo several additional steps.

Close control of precooking, mixing, grinding and other preparatory steps is essential. For example, in steaming oysters to open the shells an excess shrinkage of several ounces in the bushel occurs if either the time or temperature exceed the normal requirements. This loss can and should be avoided.

Preparation should be continuous with the remaining steps in canning and delays should be avoided as far as possible. In some canneries material is prepared before it is required and held for several hours in order that the plant may go into full operation in the minimum length of time. In warm weather a very brief delay is sufficient for spoilage to set in and an appreciable amount of deterioration may occur which is not visible to the naked eye.

## Filling

Filling may be accomplished either by hand or by machine. Filling machines have been greatly improved in recent years but are not yet adapted for use with odd sized containers or products requiring careful handling. The principle of the filling machine is to deliver a certain volume rather than a definite weight of material. There is, however, little variation in the weight filled into individual containers. The principal objection to machine filling is that it is not always done as neatly as is desirable; for instance, salmon may occasionally be cross-packed and will not have the appearance of a smooth cylinder when removed from the container. When good workmanship in packing is desired, hand filling should be used, but it is slower and more costly.

In filling, a headspace of about 1/3 to 3/16 inch should be allowed in the top of the can in order that a proper vacuum may be obtained. The tendency is to over-fill rather than under-fill a container. Canned fishery products usually weigh somewhat more than the amount stated on the label as the packer does not wish to be penalized for packing an under-weight product. Studies indicate that while the excess weight per container given away by the packer may be only an ounce or even less, the total amount lost in a season's production is considerable. Careful control of filling will reduce this source of loss. Accurate scales or weighing machines are an economy in filling.

Filling should be carefully supervised for other reasons. Foreign objects may get into the containers at this time, through carelessness or deliberate sabotage. A nail or other fragment of metal may have been left in a can previous to filling; a portion of a cotton glove may be drawn into a can during the filling process or trays of filled cans may be left unscreened and exposed, with the result that a fly may be included in the product. These objects have all been found in containers. Even where damages are not awarded, legal costs are expensive and each instance means a loss of sales through newspaper publicity. Cans should be absolutely clean when they reach the filler while the filling area should be screened and protected against flies or similar sources of contamination.

## Exhaust Or Vacuum And Sealing

Sufficient exhaust or vacuum may be obtained by (1) filling a hot precooked product into the container and sealing immediately, (2) by heating after the product has been packed into the container and (3) by mechanical means. The choice of method will depend on the product, space available in the cannery and scale of production.

Products such as soups and chowders are precooked, filled into the container while hot and sealed immediately. Others, such as fish roe, are "exhausted" by passing the container through a steam heated exhaust box. A mechanical apparatus (vacuum sealing machine) which combines the functions of exhausting and sealing is used in most salmon canneries and is especially suitable when space is at a premium as on floating canneries. Vacuum-sealing machines have not yet proved to be economical for small scale operation as in the packing of specialty articles but they are being improved, and their use may soon extend even to small scale operations. The older system of using heat exhaust to obtain a partial vacuum, sealing the cans in a second operation, has not been entirely displaced, but is apparently fast becoming obsolete.

## "Processing," "Cooking" Or "Sterilizing"

A few years ago little skill was required of the man responsible for processing and great latitude was allowed in the operation of equipment. The processing equipment was comparatively simple and errors due both to faulty equipment and to the human factor were more frequent than they are now. Fish products were too often regarded as "fool-proof." Accurate and detailed processing records were not kept in many instances. As a result the quality of the pack was not uniform and spoilage occurred directly traceable to preventable faulty processing.

It is difficult, if not impossible, to eliminate faulty processing and insure against human error if sterilizing equipment is operated manually, with a pressure-gauge and indicating mercury

*Recommended canning procedures, Part I.*

| Product | Season | Cleaning loss | Pre-treatment | Blanch or precook | Total loss | Exhaust | Closing temperature |
|---|---|---|---|---|---|---|---|
| | | *Percent* | | | *Percent* | | *Deg. F.* |
| Salmon:<br>Chinook | Varies from year to year according to necessity for conservation of supply. See fishery regulations for Alaska, British Columbia, Washington and Oregon. In general, June through September. | 30 | Wash and cut in container length pieces. | None | 30 | Vac. seal | Room temp. (60-70) |
| Red | | 33 | do | do | 33 | do | do |
| Coho | | 33 | do | do | 33 | do | do |
| Pink | | 35 | do | do | 35 | do | do |
| Chum | | 33 | do | do | 33 | do | do |
| Sardines:<br>California | Northern dist. Aug. 1 to Feb. 15. Southern dist. Nov. 1 to Mar. 30. | No data | Brine 45 to 90 min. | 1. Raw pack<br>2. Fry—large, 7 min.; small, 4 to 6 min.<br>3. Steam exhaust, 25 to 40 min.<br>4. Steam exhaust, 20 to 25 min., then through super-heated steam, 240 to 260° F.<br>5. Oven cook, steam and gas 20 to 40 min.<br>6. Oven cook, gas, 45 min. "Ullman" Cooker. | 50 | Vac. seal<br>12 to 15 min. ex<br>Sealed hot<br>do<br><br>do<br><br>do | 70<br>150<br>150<br>150<br><br>150<br><br>150 |
| Maine | Apr. 1 to Dec. 1, most of canning during period July 1 to Aug. 1. | 15 (small fish) | Salted in boat, 280 lb. to hogshead (1200 lb.) | Steam 18 to 20 min., 212–220° F | 35 to 41 (small fish), 55 to 60 (large fish). | Cold fill, no exhaust | 65 |
| Mackerel:<br>Boston | July 15 to Oct. 1 | 30 | Brined 1 hour in 100° brine. (salt mackerel—12 hours in 100° brine). | Raw | 50 | 15 min. at 210° F | 160 |
| California | July 1 to Oct. 1 | 30 | Brine 1 hour in 80 to 90° brine. | Raw | 60 | 20 to 45 min. at 210° F. | 180 |
| Tuna:†<br>Albacore | July 1 to Aug. 1 (Calif.)<br>Aug. 1 to Oct. 1 (Oregon) | No data | Clean and wash | 10 to 14 lb. 3–3½ hrs. 216–220° F<br>18 to 40 lb. 4–4½ do | 60 to 64 | 3 min. 210° F<br><br>(some packs vac. seal). | 70 |

† The season given applies only to California. Tuna of the various species are found on some part of fishing grounds throughout the year.

*Recommended canning procedures, Part I, (continued)*

| Product | Season | Cleaning loss | Pre-treatment | Blanch or precook | Total loss | Exhaust | Closing temperature |
|---|---|---|---|---|---|---|---|
| | | *Percent* | | | *Percent* | | *Deg. F.* |
| Bluefin | June 15 to Nov. 30 | do | do | 8 to 18 lb. 2 ......do<br>18 to 50 lb. 3 ......do<br>50 to 60 lb. 4 ......do<br>60 to 200 lb. 5–9 ......do | 60 to 64 | | |
| Yellowfin | Jan. to Dec. 31 (generally more abundant in summer months). | do | do | 8 to 18 lb. 2 ......do<br>18 to 50 lb. 3 ......do<br>50 to 60 lb. 4 ......do<br>60 to 200 lb. 5–9 ......do | 60 to 64 | | |
| Striped | Aug. 15 to Nov. 30 | do | do | 5 to 12 lb. 2–2½ ......do | 60 to 64 | | |
| Alewife or river herring. | Apr. 1 to May 15 | 30 | Brined 8 to 12 hours 100° brine. | None | 30 to 35 | 3 min. 210° F. also filled with hot brine or water. | 100 (approx.) |
| Fish flakes (haddock) | Apr. 15 to May 15 or at other times depending on fresh fish market. | 40 | Brined 10 to 14 hours 100° brine. | Steam 30 min., 240° F | 70 | Hot fill | 150 |
| Finnan haddie | Nov. 1 to Feb. 15 | 40 | Brined 20 to 40 hours 100° brine then smoked 10 to 14 hours. | Fillets steam 15 min., 240° F. (whole fish 30 min.) | 70 | Hot fill | 150 |
| Shad | Calif.<br>Apr. 1 to May 15.<br>Columbia River<br>May 1 to July 1. | 25 | Wash, cut in container length pieces | Raw fill | 35 | Vacuum seal | 60 |
| Clams:<br>Razor | Apr. 1 to May 31 | 65 | Wash | Steam | 65 | Vacuum sea and hot fill. | 150 |
| Soft | Oct. 15 to Jan. 1 and Mar. 15 to Apr. 15. | 65 | Wash | Steam 20 min. 212° F., 15 min. 228° F | 75 | Hot fill | 150 |
| Oysters:<br>Atlantic or Gulf | Jan. 1 to May 1 | 75 | Wash | Steam 5 to 8 min. 245° F | 93 | Hot fill | 150 |
| Pacific | Nov. 15 to May 1 | | Wash | Steam 3.5 to 4 min. 240° F | 80 | Hot fill | 160 |
| Crab, Dungeness | Apr. 1 to Nov. 1 (Alaska—closed season during this time) Jan. 1 to Jan. 1 (Ore. Wash.) | 73 | Shelled alive, washed | Boiled 20 min | 75 | Vacuum seal | 80 |
| King (Japanese) | May 1 to Aug. 1 | No data | Shelled alive, washed | Boiled 15 to 20 min. 25° brine | No data | Heat exhaust 7 to 10 min. 210° F. | 165 |
| Shrimp | Apr. 15 to June 30. and Aug. 15 to Mar. 15. | 55 | Brine soak 30 min. 50° | Wet pack 5 to 7 min.<br>Dry pack 7 to 12 min. | 75 | Hot fill<br>Vacuum seal | 150<br>85 |
| Clam chowder | Oct. 15 to Jan. 1 and Mar. 15 to Apr. 15. | 75 | Solid ingredients diced | Potato 2 to 3 min | 75 | Hot fill | 180 |
| Fish chowder | Apr. 15 to May 15 | 40 fish, 20 potato | Fish brined 10 to 12 hours 100° brine, potato peeled, washed, diced. | Potato 2 to 3 min., fish, steam 30 min. at 240° F. | 70 | Hot fill | 160 |
| Fish cakes | Aug. 15 to Jan. 15 | 20 potato | Wash and peel potato. Soak salt cod 10 hours. | 200 lb. potato, 100 lb. cod. Boil 30 min. (212° F.) | 20 | Hot fill | 150 |
| Fish roe:<br>Alewife (river herring). | Apr. 1 to May 15 | 5 | Rinse | Raw fill | 5 | Heat exhaust 3 min. 209° F. and hot fill brine. | 80 |
| "Deep sea" (Cod & haddock). | Dec. 15 to Apr. 15 | None | Ground and mixed with brine. | None | None | 15 min. 210° F | 180 |
| Shad | California<br>Apr. 1 to May 15.<br>Columbia River<br>May 1 to July 1. | 10 | Wash, skin and slime | Raw fill | 10 | 8 min. 210° F...<br>Some use no exhaust. | 150<br>60 |

*Recommended canning procedures, Part II*

| Product | Brine or sauce | Fill* | | Process | | Yield | · Remarks |
|---|---|---|---|---|---|---|---|
| | | Weight | Can size | Time | Temperature | | |
| | | *Ounces* | | *Minutes* | *Deg. F.* | | |
| Salmon:<br>Chinook | ¼ oz. salt added to each can. No brine. | 16.6<br>16.2<br>8.0<br>3.9 | No. 1 tall<br>No. 1 flat<br>No. ½ flat<br>No. ¼ flat | 90<br>do<br>80<br>70 | 240 to 245<br>do<br>do<br>do | 1 to 5 fish per case | Col. River chinook only part of pack graded for quality and is mostly hand packed. |
| Red | do | 16.0<br>7.9<br>3.9<br>64 | No. 1 oval<br>No. ½ oval<br>No. ¼ oval<br>(602 x 403) | 90<br>80<br>70<br>195 | do<br>do<br>do<br>242 | 12 to 13 fish per case | This species in demand because of deep red color and excellent flavor. Use of lighter colored species has given rise to rumors that some fish are dyed and sold as salmon which is entirely false. It is not permitted by law and technical difficulties of such a process are practically insurmountable. |
| Coho | do | | | | | 9 to 10 fish per case | Is not very abundant. Forms only 7 percent world's pack canned salmon. More important in fresh, frozen and smoked fish trade. |
| Pink | do | | | | | 17 fish per case | Smallest and most numerous of salmon. Forms 41 percent world's pack. |

*Recommended canning procedures, Part II (continued)*

| Product | Brine or sauce | Fill* Weight (Ounces) | Fill* Can size | Process Time (Minutes) | Process Temperature (Deg. F.) | Yield | Remarks |
|---|---|---|---|---|---|---|---|
| Chum | do | | | | | 9 to 10 fish per case | Has less color in flesh and lower oil content than other species so is not as popular. Sells at lower price than other species salmon but has very high food value and can be made into palatable and nutritious dishes. |
| Sardines: California | 2 oz. tomato or mustard sauce to each 1 lb. oval — 1 oz. to each ½ lb. oval or 8 oz. rect. | 16 | No. 1 tall | 75 | 240 | Average of 13 cases of sardines per ton of fish required by Ca'if. law. Fish in good canning cond. about 20 cases to ton. 20 cases per hogshead (large fish); 30 cases per hogshead (small fish). | Packing of natura and smoked fillets becoming important feature of pack. |
| | | 16 | No. 1 oval | 65 | 240 | | |
| | | 9 | ½ rect. | 50 | 240 | | |
| | | 8 | ½ oval | | 240 | | |
| Maine | Cottonseed oil | 3½ | (quarter oil) | 45 | 240 | | Maine law requires; use of winterpressed cottonseed oil 4 lb. per case (100 quarter); minimum of 4 fish per can (keyless), 5 (keyopening). |
| | | 11 | (¾ mustard) | 60 | 240 | | |
| Mackerel: Boston | 3 percent | 14 | 1 lb. oval | 75 | 240 | 66 percent | Declared net weight on 1 lb. ovals 12 oz. but fill is always heavier. Mackerel must be firm and not over 24 hrs. old. Some mackerel packed raw pack without brining, ¼ oz. salt added to can. Not recommended by canning technologists who urge brining be employed. |
| | | 14 | 1 lb. tall | | | | |
| | | 16 | No. 2 short | | | | |
| California | 3 percent | 17⅝ | No. 1 tall (301 x 411) | 9 | 240 or 250 | No data | |
| | | | | 75 | | | |
| | | 17 | No. 1 meat (301 x 407) | 75 | 250 | | |
| | | 11½ | No. 1 standard (211 x 400) | 75 | 240 or 250 | | |
| | | | | 60 | | | |
| Tuna:† Albacore | oil and salt | 9⅛ | 8 oz (211 x 304) | 75 | 250 | | Small amounts of tuna packed in glass tumblers, aluminum cans. Number of specialty tuna packs of which "tonno" is probably most important. Other specialty packs "ventresca," creamed tuna, garlic flavored tuna. Bonito also packed tuna style but may not be labeled as tuna yellowtail packed tuna style, but may not be labeled tuna. Some canned tuna imported into U. S. from Europe packed in salt brine without oil must be labeled "Packed without oil and in salt solution." |
| | oil ¾ oz.; salt 1/14 oz. | 3½ | No ¼ tuna | 65 | 240 | 76 cases; 48/¼s per ton | |
| | oil 1½ oz.; salt 1/7–3/14 oz. | 5½–5¾ | No. ½ tuna | 75 | 240 | 47 cases; 48/½s | |
| Bluefin | oil 2; salt 9/14 | 11–11½ | No. 1 tuna | 95 | 240 | 23 cases; 48/1s | |
| Yellowfin | oil 9; salt 6/7 | 46 | 4 lb. tuna | 230 | 240 | | |
| Striped | | | | | | | |
| Alewife or river herring | 3 percent | 16 | No. 1 tall | 50 | 244 | No data | Labeleed "fresh river herring" but is of canned salt fish style. Fish wrapped individually in parchment paper and packed in No. 2 tall cans. |
| | | 26 | No. 2 tall | 60 | 250 | | |

*The fill given is actual weight filled into container and should not be considered a recommendation for declared weights to be used on the label of the container.
†The season given applies only to California. Tuna of the various species are found on some part of fishing grounds throughout the year.

| Product | Brine or sauce | Fill* Weight (Ounces) | Fill* Can size | Process Time (Minutes) | Process Temperature (Deg. F.) | Yield | Remarks |
|---|---|---|---|---|---|---|---|
| Fish flakes (haddock) | None | 7¼ | (211 x 300) | 55 | 240 | 66 cans per 100 lbs. (211 x 300) size; 40 cans (307 x 208) size per 100 lb. | Fillets sometimes used instead of whole fish. Cod may be mixed with haddock 1 to 3, but is not often packed alone because of its soft texture. |
| | | 12 | (307 x 208) | 75 | 240 | | |
| Finnan haddie | None | 4 | (211 x 109) | 55 | 240 | About same as fish flakes not definitely determined. | Also packed in "nappy" glass tumblers. Imports from England in 1 lb. oval cans. |
| | | 12 | (307 x 208) | 75 | 240 | | |
| | | 16 | (300 x 307) | 75 | 240 | | |
| Shad | ½ oz. salt | 16½ | No. 1 tall | 90 | 240 | 75 lb. round—fish per case 48/1s. | Pack resembles salmon but flesh is rather dark and soft. Some kippered shad also canned. |
| Clams: Razor | Clam juice | 5¼ | No. ½ | 45 | 220 | 1 bu. clams = 40 No. ½ flat cans; = 20 No. 1 tall. | Sold mostly as "minced sea clams." |
| | | 7½ | No. 1 picnic | 45 | 220 | | |
| | | 12½ | No. 2 | 60 | 220 | | |
| Soft | 3 percent | 6 | No. 1 picnic | 20 | 240 | 48 No. 1 picnic cans per bu. | Darkening is principal difficulty in canning. If this occurs blanch in 1.5 percent citric acid and add 0.5 percent citric acid to pack. |
| | | 10 | No. 1 tall | 20 | 240 | | |
| | | 11 | No. 300 | 20 | 240 | | |
| | | 12 | No. 2 | 25 | 240 | | |
| Oysters: Atlantic or Gulf | 4° | 3 | (211 x 300) | 8 | 250 | 20 to 25 No. 1 picnic (211 x 400) cans per bbl. | 1 bbl. = 3 bu. in Miss.; 1 bbl. = 4 bu. in La.; Miss. bu. = 2826 cu. in.; La. bu. = 2150 cu. in.; Std. U. S. bu. = 2150 cu. in. 9/0 loss = shell, mud, oyster juice (nectar). |
| | | 4 | (211 x 304) | do | do | | |
| | | 5 | (211 x 400) | 9 | 250 | | |
| | | 8 | (307 x 400) | 10 | 250 | | |
| | | 10 | (307 x 409) | do | do | | |
| Pacific | 3 percent | 7 to 9 | No. 1 picnic (211 x 400) | 29 | 240 | | Pacific oysters filled by count as well as weight. Certain number oysters must go into can for each grade size. Count per No. 1 tall can usually 6 to 7 if oysters are large, 8 to 14 if grade is medium, and 15 to 20 if grade is small. |
| | | 10 to 13 | No. 1 tall | 35 | 240 | | |
| | | 14 to 16 | No. 2 | 42 | 240 | | |
| Crab (Dungeness) | 4° | 6½ | (307 x 202½) | 70 | 228 | 20 lb. meat or 4 doz. crabs to case 48/½s. | Dry salt may be used instead of brine. Speedy operation without delay, esp. important in this pack. |
| | | 13 | (401 x 211) | 80 | 228 | | |
| | | 17 | (307 x 408) | 80 | 228 | | |
| King (Japanese) | See remarks | 3½ | (211 x 109) | 80 | 220 | No data | 2 oz. weak buffer sol. of organic acid (lactic or citric) added when discoloration may occur. |
| | | 6½ | (307 x 202½) | 90 | 220 | | |
| | | 13 | (401 x 211) | 80 | 228 | | |
| Shrimp | 4° | 5⅜ | No. 1 picnic and squat. | 10 | 250 | 190 No. 1 picnic cans per bbl. (210 lb.). | Processes for dry pack increase as liner type changes from 3 pc. to 1 pc. |
| | | 9¾ | No. 1½ | 12 | 250 | | |
| | | 5⅛ | No. 1 picnic and squat. | 70 to 85 | 240 | | |
| | | 9¼ | No. 1½ | 75 to 90 | 240 | | |

*Recommended canning procedures, Part II (continued)*

| Product | Brine or sauce | Fill* Weight | Fill* Can size | Process Time | Process Temperature | Yield | Remarks |
|---|---|---|---|---|---|---|---|
| | | *Ounces* | | *Minutes* | *Deg. F.* | | |
| Clam chowder | Hot soup added | 4 solids ⎫ 6 soup ⎪ 7.5 clam ⎬ 7.5 potato ⎪ 11 soup ⎭ | No. 1 picnic / No. 3 | 60 / 85 | 240 / 240 | 550 No. 1 picnic to 168 lb. solid ingredients. | Down East clam chowder similar to Manhattan, except D.E. has whole clams, no tomatoes and flour instead of cracker meal. |
| Fish chowder | Hot soup added | 2 fish ⎫ 2 potato ⎬ 6 soup ⎭ | No. 1 picnic | 60 | 240 | 550 No. 1 picnic to 168 lb. solids. | Darkening or discoloration principal difficulty to be guarded against. |
| Fish cakes | Solid pack | 10 | | 75 | 240 | 460 cans to batch (200 lb. potato, 100 lb. fish). | Green mountain potatoes best variety in fish cakes. Discoloration caused by over-processing. |
| Fish roe: Alewife (river herring). | 3 percent (sometimes hot water used). | 8 / 16 | (211 x 300) / (307 x 400) | 50 / 60 | 240 / 240 | 1 case 48/8 oz. cans per 20 lb. bucket "green" roe. | From 13½ to 15 oz. green roe required as fil lin weight to give drained weight 16 oz. Variation depends on condition of roe. |
| 'Deep sea" (Cod & haddock). | 3 percent | 14 | (300 x 407) | 75 | 240 | 400 cans to 300 lb. roe | Fill of cans must be watched carefully. Leave ½ in. headspace in filling. |
| Shad | ¼ oz. salt | 7¾ | ½ oval | 110 / 55 / 90 | 230 / 240 / 240 | 26; 6 lb. per case 24/½ ovals. | Roe must be fully developed but not over ripe. If roe is too ripe, is watery and lacks flavor. If too green is hard and tough. |

thermometer too often untested for accuracy as the only control devices. Retorts should be fitted with at least the following equipment: (1) An automatic control for regulating temperature (2) An indicating mercury thermometer of a range from 170 to 270 degrees F., with scale divisions not greater than 2 degrees F. (3) A recording thermometer of a range from 170 to 270 degrees F., with scale divisions not greater than 2 degrees F. (4) A pressure gauge of a range from 0 to 30 pounds with scale divisions not greater than 1 pound. (5) A blow-off vent of at least 3/4 inch inside diameter in the top of the retort. (6) Bleeders not less than 1/8 inch diameter. (7) Adequately perforated steam inlet pipe running through the length of the retort. (8) A drainage valve.

Indicating mercury and recording chart thermometers must be installed either within a fitting attached to the shell of the retort, or within the door or shell. If the thermometer is installed within a fitting, the fitting should communicate with the chamber of the retort through an opening at least 1 inch diameter and should be equipped with a bleeder at least 1/8 inch, inside diameter. If the thermometer is installed within the door or shell of the retort, the bulb must project at least two thirds of its length into the principal chamber. The pressure gauge is connected to the chamber of the retort by a short gooseneck tube. The gauge must not be more than 4 inches higher than the gooseneck. The bleeders should be spaced not over 1 foot from each end of the retort and not more than 8 feet apart.

Processing times listed in canning literature do not include the entire period from the time the product is placed in the retort until it is removed but only the length of time after the retort has reached the required temperature and until steam is shut off. The time required for bringing the retort up to processing temperature and for reducing pressure to atmospheric level are in addition to the processing period and it is important that sufficient time should be allowed for each of these steps. If properly processed the product should be neither overcooked nor should it have an "understerilized" flavor, while the texture should be reasonably firm yet not stringy or woody.

## Cooling And Washing

Since a sudden release of pressure after processing causes severe strains on tin containers, which may result in leakage and spoilage, the pressure should be reduced slowly and gradually. From 5 to 10 minutes should be allowed to bring the retort pressure to atmospheric level and before opening the retort doors. Water cooling in the retort will also cause buckling and distortion, especially of larger sized containers, unless the water is admitted from below. The water should be admitted slowly at first and under air pressure sufficient to maintain a pressure in the retort equal to the steam pressure required for processing.

Most canneries packing fishery products still cool and clean the cans after leaving the retort. The pack should be cooled as rapidly as possible. If cooling is unduly prolonged, especially by stacking cans very shortly after removal from the retort the product will be darkened in color and overcooked in flavor. The necessity for cleaning the cans after processing has been reduced but not entirely eliminated by the general use of can-washing devices previous to processing.

## Coding

It should be possible to identify any container as to species, grade, date and place of pack and in some instances the origin of the raw material, or method of catch. This may be accomplished by "coding" or marking by a system of numerals, letters or special symbols, using a simple, carefully worked out system with a minimum of characters. In canning fishery products the codes are marked on the containers by stamping the cover with a die usually operated as an attachment of the closing machine.

As far as possible code lots should be segregated in the warehouse both before and after casing. The cases should also bear the code mark, placed so that it may be readily observed in handling the product. Some packers regard coding merely as a means of avoiding seizure of condemnation of large parcels made up of several code lots when inferiority or spoilage may be confined to a minor fraction. However, the most important function of coding is to enable the packer and distributor to better determine the grade of the product, and to improve the quality by correcting faults in workmanship and packing.

## Lacquering

For a great many years nearly all cans of fishery products were coated with an asphalt-base brown lacquer as a rust preventative. Cobb (1919) stated that this practice originated through demands of English buyers who constituted the principal market for canned fishery products in the earlier days of the industry. Little attention was paid to conditions of storage at that time. Ship holds were damp and on long voyages there was much variation in temperature and humidity so that cans were often heavily rusted on arrival at destination. Only a minor portion of the pack is lacquered at present, usually for special orders, since it has been found that a well labeled can with enameled ends has sufficient protection against rusting, under proper conditions of storage and shipment.

## Warehousing

Changes in buying methods during recent years have increased the importance to the packer of proper warehousing. The tendency of the distributor and retail trader is to buy for immediate needs only, forcing the packer to warehouse a greater portion of his pack over a longer period of time. The external appearance of the containers deteriorates unless the pack is well housed. Many consumers will not buy cans with stained or rust spotted can surface or label, therefore, a bright attractive appearance of label and can is an important factor in the sale of the product.

The prime essentials for a warehouse are that construction must be of a strength calculated to withstand strains in excess of any loads it may be expected to bear; that it be dry, well lighted and reasonably cool. Canned fishery products will resist a fair degree of heat or cold for short periods, without serious injury, but continued heat or repeated alternate freezing and thawing are injurious to quality. The product becomes flabby in texture and loses its flavor. Chemical changes double in activity with each increase of 18 degrees F. in temperature. Storage in a warm moist place promotes rapid deterioration.

Canned fishery products deteriorate very slowly if well stored. Samples of canned fish ranging from 5 to 12 years old have been examined and were found to be still of satisfactory quality. Most canned fishery products require a few months storage before distribution, as salt and other ingredients are absorbed gradually and therefore the flavor may be uneven and judged unfairly if the goods are consumed shortly after packing. The "ripening" of sardines is an illustration in point.

However, the packer cannot afford to hold his pack in storage over a long period. Goods in storage represent unproductive capital on which interest must be paid, and warehouse charges must be added to the cost of the product. Packs held longer than a few months are often subject to several forms of taxation which might otherwise be avoided.

## Labeling

The product may be labeled and packed in cases for shipment immediately on cooling or it may be stacked for labeling and casing later. A portion of the pack may be cased unlabeled for the buyer using his own label. The packer should guard against labeling and casing too soon after packing because labels do not adhere well if applied while containers are still slightly warm and fiber cases insulate the cans, unduly prolonging the cooling process.

Hand labeling, except in the case of oval cans and other odd shaped containers, has been largely replaced by automatic machines of light construction and simple operation which may be easily transferred from one point to another in the warehouse as needed. A workman places the cans on a conveyor and they roll through the machine by gravity. The cans travel over small rollers which apply a small amount of adhesive, either glue, a casein preparation, dextrin mucilage or other types of label paste. They then roll across a stack of labels, one of which is picked up by the adhesive on the can and is fixed in place automatically by the machine. Finally, adhesive is applied automatically to the end of the label which is then sealed to the can.

The value of a well designed label and a well-known brand is undoubtedly very great. The design should be simple and clear, the brand name easily remembered and pronounced, for customers will not remember or call for a name of difficult pronunciation. Customers cannot read the label and depend on the picture in a number of markets in the South, or in foreign sections of cities in the United States and in certain export markets. If the picture is misconstrued the sale of the product may be adversely affected. Canned fishery products have been refused in some markets because they showed a fish with a tail not curved upward and the retail buyers therefore claimed that the fish was not fresh when packed. In one instance a good sized shipment of salmon sent to the oriental market had to be relabeled because it bore a picture which led the buyers to believe that the cans contained cat meat.

Brand names are protected by law under the "Trade Mark" Act. For a small fee the records of the U.S. Patent Office at Washington, D. C., may be searched to determine if a name or brand is registered. Attempts have been made to imitate distinctive designs in order to promote an unknown brand. In other instances the design has led the customer to believe that he was purchasing one species when the container held an entirely different variety. Such labels do the packer more harm than good since the products acquire the reputation of being second-grade imitations.

The label must comply with the regulations of the U.S. Food and Drug Administration. It must bear the net weight of the contents, conspicuously placed, the name of the articles, the packer's or distributor's name and address, and the grade. The use of labels which lead the buyer to believe that the species is superior to that contained, or are otherwise deceptive as to quality, is considered as misbranding, when grade is determined by species. The Food and Drug Administration does not approve labels but will advise packers as to labeling regulations with which they must comply. There is one exception to this rule; it must approve all labels under which inspected shrimp are packed.  ■ ■ ■

---

## APPENDIX IV
### Part 2

### Spoilage in Canned Fishery Products

*By Norman D. Jarvis*

Certain changes which occur in canned products after sterilization or processing are beneficial and improve the quality of the product. Canned salmon eaten immediately after packing may taste "flat" as if insufficiently salted or the taste may be excessively salty, depending on the portion of the contents tasted. Other cans of the same pack sampled after a short storage period will taste sufficiently salted. The salt added in filling requires time for an even and complete distribution throughout the contents. Sardines or tuna packed in oil may taste "raw" or "flat" if sampled immediately after canning. The oil in which these fish are packed is only absorbed gradually.

There are other changes which affect the product adversely. Strictly speaking, many of these changes should be called deterioration rather than spoilage. The commercial value of the food is lowered and it may be unappetizing but not absolutely inedible. However, since the product is of lower grade and frequently unmerchantable, it is considered spoiled. We have then two general types of deterioration in canned products; first, that due to physical and chemical changes, and second, spoilage brought about through micro-biological action. It is necessary to determine whether the product was properly prepared, of suitable raw material, or to locate faults in materials or methods.

## Classification Of Spoiled Cans

There are certain easily identifiable evidences of spoilage in canned foods which have been classified as follows:

*Flipper*—A can which may be normal in appearance, but if one end is struck on a box or table, the other end becomes convex, though the convexity may be pressed down again. A flipper is the initial stage of a swell, but may also be caused by overfilling or lack of vacuum.

*Springer*—A can having convex or bulging ends, which may be pressed flat again with the fingers, but will spring out again after pressure is released.

*Swell*—A can with badly bulged ends resisting pressure with the fingers or if the ends are pressed down, they spring back immediately on the release of pressure.

*Flat sour*—A can whose contents have been spoiled by microbiological action without the formation of gas and therefore gives no external indication of spoilage. The product has a sour taste and may or may not have a sour odor when the can is opened.

*Hydrogen swell*—A can with swelled ends caused by the formation in the can of hydrogen gas as a result of corrosion of tin plate. Varying quantities of metal are usually dissolved in such cans. The contents are almost always sterile and often fit for food. Such swells are externally indistinguishable from swells caused by micro-biological action, and can only be identified by analysis of the gas in the head-space, and through the heavily etched interior of the container.

*Buckles*—A type of swelled can which may be the result of improper cooling. The internal pressure during processing may be so great as to bulge or distort the can ends so that they cannot return to their normal position after cooling. The seams of such cans are usually strained so badly that they subsequently leak and the cans spoil through the entrance of micro-organisms. A buckled can may also represent the final stage of a swell.

*Leakers*—These are cans exuding a portion of the contents. Cans may become leakers through (1) faulty seaming, either by the can maker or canner (usually the latter); (2) defective tin plate; (3) internal corrosion or external rusting; (4) buckling; (5) excessive pressure within the can as a result of gas formation caused by micro-biological action in decomposition, or by hydrogen gas through corrosion; (6) external damage such as battering caused by excessively rough handling in manufacture or shipping; (7) nail holes occurring when cases are poorly or carelessly nailed or are damaged in shipment.

*Paneled cans*—These are cans ruptured or distorted through excessive external pressure; that is, they are the opposite of buckles.

## Physical And Chemical Deterioration

The most important factors in deterioration, loss, or spoilage brought about through physical or chemical means are: (1) Discoloration, (2) perforation and corrosion of tin plate; (3) foreign tastes; (4) undesirable textures; (5) freezing; (6) rusting; (7) faulty technique; and (8) unsuitable products.

## Discoloration

The problem of discoloration is most serious in packing shellfish and crustaceans, but it may also be met with in canning salmon, chowders, fish cakes, kippered herring and other fishery products. It is usually due to physical or chemical action but may be caused by micro-biological processes, for example, the "angry" or deep red color sometimes observed near the backbone in tainted canned fish. This color tends to fade on exposure to the air.

## Blackening

Blackening of the contents or inside of the can is most often encountered in packing crab, shrimp and lobster, but may also be found in other canned products. It occurs most readily where the product has an alkaline reaction. Sulfur compounds in the flesh of these species break down in processing and unite with the iron base of the tin plate to form iron sulfide. This substance is not injurious to the consumer, but the product acquires a most unappetizing appearance and unpleasant flavor.

Formerly, sulfide blackening in canned marine products could only be combated by the use of parchment paper can liners, preventing contact between food and container, and by the addition of small amounts of organic acid. Studies by the Research Laboratory of the National Canners Association indicated that zinc salts would reduce the formation of black in canned corn by combining with sulfide compounds to form zinc instead of iron sulfide. Zinc sulfide is harmless, white and is therefore unnoticed. This laboratory then developed a lacquer containing small amounts of zinc, to be used as an inside lining for cans used for products liable to blackening. As the zinc is contained in the lacquer and the sulfide formed is also trapped there, little or no zinc is found in the product. Difficulty in packing vegetables was the primary incentive for this research, but the enamel developed has done more than anything else to reduce or inhibit blackening in canned fishery products. Parchment paper linings and the use of organic acids are still necessary to some extent as blackening may otherwise occur at the side seams, where the enamel lining is occasionally fractured in can making.

## Copper Sulfide Discoloration

Some discoloration in fish and clam chowders have been traced to the use of copper lined can-filling machines. A thin film of copper oxide or copper salts gradually forms on the copper surface. The chowder coming in contact with the copper dissolves some of these copper salts which then react with sulfides formed in processing, resulting in copper sulfide and causing serious darkening throughout the product.

Some blackening of canned products has been traced to the use of rubber conveyor belts. There is a certain amount of sulfur on the surface of new rubber belts, which may be converted to sulfurous acid by water and heat, then into tin or iron sulfide by union with the metal of the can.

## Discoloration Caused By Processing Times And Temperatures

Processing temperatures and pressures may be a cause of discoloration. For instance, minced razor clams processed at 240 degrees F. (10-lb. pressure) are appreciably darker in color than those processed at 236 degrees F. (8-lb. pressure). Crab processed at 240 degrees F. will acquire an unpleasantly dark color, while the color will not be affected if a longer process at lower temperature is used. Norwegian-style fish balls lose their white color if processed over too long a period or at temperatures higher than 228 degrees F. (5-lb. pressure). Packers of these products and others of similar type must control processing times and pressure very closely to prevent serious loss through discoloration. The use of processes at 10 pounds or higher pressure, considered necessary in packing most non-acid products, must be foregone if a merchantable product is to be secured. While processes used for clams, crabs and fish balls are of the "borderline" type, loss through insufficient sterilization is generally slight.

## Stack Burning

This type of discoloration is similar to that caused by overprocessing; in fact, it is a form of over-processing. A considerable amount of heat is retained over a long period when canned products are stacked or cased before they are sufficiently cooled. Cooking goes on over a much longer period than is intended, which affects both color and flavor unfavorably. Stack burning is usually thought of in connection with the canning of fruits, or possibly such marine products as clams or lobster. Nevertheless, Clough (1937) points out instances of deterioration through stack burning in canned salmon and warns that discoloration through this cause must be guarded against in canneries with a large daily production, where it is a temptation to warehouse the pack at the earliest possible moment.

## Perforation And Corrosion

Loss in canning may occur through perforation, the product "eating" through the wall of the container, or the inside may become corroded or etched so that the product is unmerchantable. Loss is usually greatest in the canning of acid fruits, but may occur in the canning of fishery products though more rarely.

Corrosion depends to a large degree on the presence of oxygen. Cruess (1938) reported that in the presence of oxygen the can acts as a primary cell of the oxidation type. That is, the reactions which occur in the can may be explained upon the basis of an electrolysis in which oxides of tin and iron are formed and hydrogen is liberated.

Slack-filled cans are therefore more liable to corrosion than well-filled cans of the same product held under the same conditions, for the headspace is greater and the volume of air contained is larger. Other things being equal, cans with low vacuum are more liable to corrosion than containers with a high vacuum. Air should be expelled from product and headspace as completely as possible.

The rate of corrosion and perforation in tin cans, like other chemical reactions, is dependent on the temperature; that is, it increased in rapidity as the temperature of storage is raised. For this reason it is advisable to store canned goods in a cool place, thus minimizing perforation by reducing the rate of corrosion.

Increased corrosion as well as stack burning may be the result of packing the cans while still warm or stacking them in large piles while insufficiently cooled. Cans should be thoroughly cooled before packing or stacking to reduce corrosion as well as for reasons given elsewhere.

## Foreign Tastes

Canned fishery products have been ruined through such causes as lubricating oil dripping into the cans while passing through filling or sealing machines. Lack of sanitation in canning is apt to cause foreign tastes, as in poor cleansing of brine tanks or pipe lines leading to fillers in products such as soups or chowders. Foreign tastes may also be due to the presence of microorganisms. Canned salmon is believed to have acquired disagreeable flavors because the salmon had consumed certain odoriferous plankton such as pteropods. Insufficiently cleaned raw products such as clams may cause "foreign tastes" in the canned products.

## Undesirable Textures

In some areas hardness of the water used in canning may give the product an undesirable texture. Calcium salts are absorbed from such water, toughening the product. This defect is most apt to occur in vegetable and fruit canning, but is also possible, theoretically at least, in the canning of fishery products.

Important instances of undesirable texture occur where fish are brined or salted previous to canning. If the degree of salting is at all excessive, the texture becomes fibrous and stringy after processing, in addition to being much too salty in flavor. Some fish such as albacore are canned after being frozen. If the fish are not properly frozen or are thawed out carelessly, the texture of the flesh is impaired. This is evidenced by perforations which are the result of the formation of large ice crystals in the flesh. These are quite distinct from the "honeycombing" sometimes found in the canning of stale products.

## Freezing

Storage of canned foods at too low a temperature is just as injurious as at too high a temperature. Much loss is caused by the freezing of canned goods. Data on loss through freezing are not extensive, but it is believed that some canned foods, including salmon and tuna, are not affected. The amount of damage is usually greater when the products are thawed rapidly.

Condensation of moisture may occur, resulting in the formation of drops of water on the exterior of the containers, if there is too great a difference between the temperature of the frozen product and the room in which it is held. This may cause rusting of the cans or spotting of the labels, even if no greater damage is done. For this reason canned products should be thawed slowly at a temperature only slightly above 32 degrees F.

In some products there may be a separation of liquid, causing a watery appearance. Freezing progresses from the outside of the can toward the center and forces out of solution and concentrates near the center of the can, the salts and other soluble constituents. The separated materials may not again regain their uniform distribution on thawing.

## Rusting

Cans may rust both internally and externally. Rusting is a form of corrosion, but must be considered separately as the formation is influenced by factors of special importance and it is the most common yet most easily corrected type. Rust is formed through a combination of iron with oxygen in the presence of moisture, the rate of formation being influenced by temperature and also by acidity or pH. The latter is rarely, if ever, a factor in internal rusting in canned fishery products.

Internal rust may be formed during storage of the empty container, especially along the side seam. If conditions after packing are favorable, the presence of small flecks of rust may accelerate rusting. Rust may also be formed after packing. In this instance it is usually due to oxygen in the headspace, as the result of improper, that is, slack fill, with excessive headspace, and too insufficient exhaust. In some products such as oily fish like salmon, tuna or sardines, where the oxygen is absorbed by the product, there is little danger of internal rusting but in non-oily products, with considerable liquid, such a fish roe and clams, this possibility must be kept in mind.

External rusting is largely caused by poor storage and to a lesser extent by faulty packing methods. There should be no opening in a storage warehouse through which rain, snow or other atmospheric moisture may enter. Floor areas are apt to be moist in buildings of ordinary construction which may be otherwise dry. If cases are resting directly on such a floor the moisture gradually seeps through the case bottoms and may eventually rust the bottom layers of cans. The most common cause of rusting during storage is believed to be due to sweating. Sweating is most apt to occur if the temperature of the warehouse is high in contrast to the temperature of the product, or if the relative humidity of the atmosphere is high, and if there is excessive variation in temperature and lack of ventilation. Three rules should be observed to prevent rusting during storage:

1. The warehouse should be dry and well constructed.
2. The temperature should be uniform.
3. The warehouse should be well ventilated.

External rusting may also be caused by faulty packing procedures such as:

1. Processing in retorts: a. Improperly vented. b. Using a long coming-up time. c. Using low pressure steam containing considerable moisture.
2. Water cooling: a. To temperatures below 100 degrees F., when residual surface moisture will not evaporate. b. Failure to remove surface water on cans after cooling mechanically.
3. Casing: a. In wooden boxes made from green lumber. b. In wooden cases which have become damp.
4. Chemical composition of water used in processing and cooling.
5. Label pastes. Instances have occurred where rusting was traced to label pastes with high hygroscopic (moisture absorbing) properties.

## Faulty Technique

Faulty technique is due to ignorance, carelessness and dependence on outdated or obsolete methods or equipment. Most of these factors have been mentioned incidentally in the discussion of physical and chemical factors in deterioration, but poor technique must also be considered a direct as well as a contributory cause of deterioration.

Ignorance may be corrected if the packer is willing to study the principles on which canning is based, rather than place dependence on the tedious and expensive trial and error method which often leads to false conclusions. Information on canning technique is available to the industry through the efforts of the National Canners Association, the research departments of the can and canning machinery manufacturers, trade journals and governmental agencies.

Carelessness may be combated only by vigilance on the part of the packer, who must insist on exactness in methods and penalize any infraction of rules. Dependence on outdated methods is itself a type of ignorance. Certainly a man who, having once learned the trade, refuses to study, cannot be called intelligent. Dependence on obsolete equipment is poor economy. Such equipment is apt to be defective, resulting in high packing costs and lower quality.

Faulty technique includes such factors as slack and overfill, insufficient exhaust of vacuum, excessive cook, insufficient cooling, and other factors too numerous to list. Strictly speaking, poor workmanship in packing, such as cross fill in salmon canning, visceral material in canned roe, or bits of shell and antennae in canned shrimp, cannot be included as this discussion is concerned with changes occurring after processing.

## Unsuitable Products

Because of changes taking place in canning, due to the structure or composition of the flesh, certain species of fish are not suitable for canning by any known method. Grayfish (dogfish) is appetizing and edible if used fresh, although the flesh contains urea. In processing, the urea is transformed into ammonia making the flesh inedible. During World War I grayfish were canned as a food conservation measure, but the pack was a total loss.

Seafood cocktails are packed in glass and are hermetically sealed without sterilization. Attempts to lengthen the period of preservation by processing have thus far failed, since the degree of heat necessary to process the seafood portion of the ingredients causes caramelization of the sauce, with consequent darkening in color and "burnt" flavor. Products such as bismarck herring or rollmops are prepared from salt herring or are salted in preparation. Processing these articles causes the texture to become tough and fibrous. Other instances may occur to the reader in his own experience.

## Spoilage By Micro-Organisms

Spoilage through micro-biological action may occur before or after canning. Spoilage before canning is dealt with at length in discussions on raw material and sterilization. This discussion is concerned only with changes caused by micro-organisms *after* processing. Spoilage after canning may be caused by organisms of low or high heat resistance.

Spoilage by organisms of low heat resistance is due in most instances to leaky or improperly closed cans. Deterioration in these instances is primarily due to physical and chemical causes and bacterial spoilage is secondary, occurring only because these changes have already taken place. If living organisms of low heat resistance are found, the containers should be carefully examined for defects. Processes now in use are sufficient to destroy any organisms of this group. The organisms of low heat resistance in canned seafoods are predominantly of the cocci type.

The bacterial flora of canneries packing marine products have been carefully studied by Lang (1935), who concluded that micro-organisms in marine product canneries are largely of the cocci group and states that it is also apparent that they are readily destroyed by heat, and that the cleaning up of the premises after the butchering operation may be made sufficiently effective to prevent their dissemination to other departments of the packing plant. Assuming that these deductions are correct, it may be con-

cluded that the presence of cocci is a more important factor in the preservation of the raw product after catching, in transit, and prior to precooking than it is in the canning and sterilization procedures.

Bacterial spoilage in canned fish and shellfish is caused almost entirely by organisms of high heat resistance, and may be divided into two general types, gaseous and non-gaseous.

## Gaseous Spoilage

Swelled or "bulging" can ends are a common indication of gaseous decomposition. The ends of the can may be pushed out by other causes but this cannot be determined by external examination. Any container with swelled ends should be regarded as unmerchantable.

The gas-forming organisms found in canned fishery products are almost always spore-formers of the anaerobic or facultative anaerobic types. Organisms found quite commonly are *Clostridium welchii* and *Clostridium sporogenes*. Gas formation is accompanied by an extremely foul and offensive odor. Cans of clams cultured with sporogenes and insufficiently processed became swells in from 10 to 90 days. The contents were almost entirely liquefied and discolored to an inky blue hue (Jarvis and Puncochar, 1940).

Another gas-forming heat-resistant organism is *Clostridium botulinum*. Spoilage may not always be accompanied by the excessively disagreeable odor, but the product will have lost its normal odor and may have an "off" smell difficult to describe. *Clostridium botulinum* has been isolated from home canned fishery products within the last few years but it has not been found in any commercially canned fishery product since 1925. The toxin produced by this organism is a causative agent of food poisoning. In studying the bacterial flora of marine products Lang found evidence of the presence of gas-producing anaerobic thermophiles on fish carts and in slime previous to canning, but no gas producing thermophiles were isolated from among organisms found present after processing.

The resistance of these organisms varies. Some resist boiling for 4 and others for 12 hours. Certain organisms are destroyed by exposure of 110 degrees C. (230 degrees F.) for 10 minutes while others withstand 120 degrees C. (240 degrees F.) for 10 minutes. Heat processes in canning must be determined on the basis of maximum heat resistance in order to secure commercial sterilty and eliminate the possibility of spoilage by putrefactive gas-producing organisms.

## Non-Gaseous Spoilage

There is no external indication of non-gaseous spoilage. The ends of the containers are flat and the contents may be normal in appearance. An "off" odor may or may not be noticeable, but the product is sour in taste. This type of spoilage is known as "flat souring". It has been studied most extensively in relation to the canning of vegetables, but also occurs in canned fish and seafoods. Fellers (1927) found spoilage of this type in chinook, coho and pink salmon; in crab, and in shrimp. It has also been found in clams and fish roe and its presence is possible in other seafoods.

Some of the organisms causing flat souring are aerobic spore formers such as *Bacillus cereus*, *B. mesentericus* and *B. vulgatus*. While flat souring has been most common in cans with a low vacuum, it has been noted in cans with vacuums of 8 to 10 inches. It is believed that flat souring by aerobic spore formers is due to a heavily increased contamination of the raw material, insufficient process, or a combination of the raw material, insufficient process, or a combination of both. Fellers questions the theory that high vacuum is effective in preventing spoilage in certain marine products, and studies made during the past few years tend to confirm his belief.

It is also suggested that souring may occur previous to processing, especially in plants packing clams, crabs, oysters and shrimp. The rate of operation is often slow and trays of material may stand an hour or more before being filled into the cans. When operating at a slow rate, 45 minutes or more are required to fill three standard retort baskets. Where the partially filled retort baskets are also allowed to stand over the noon hour without being processed, the time may be much longer, sufficient for souring to develop, especially on hot days. Subsequent sterilization prevents decomposition but cannot remove any souring which has occurred.

While flat souring may be caused by aerobic spore formers if conditions are favorable, much of the flat souring in canned fishery products is due to the action of thermophilic micro-organisms which develop only at temperatures higher than those at which these products are normally held. Cans of salmon or other seafoods may be piled in large stacks while the cans are still warm, and left for several days. Cooling is delayed to such an extent that the contents remain at a temperature favorable to the growth of thermophilic organisms, long enough for spoilage to develop. Vegetable canners have long been warned against this practice, but its importance has not been recognized as a spoilage factor in canning marine products.

Storage temperature is also important. Thermophilic spoilage may develop in canned fishery products held in warm storerooms while it would not have appeared if the temperature had not been usually high. For example, canned fishery products may be placed in attic storerooms under a sheet iron roof, where the temperatures may reach 100 degrees F. or more in the summer. Such temperatures are favorable to the development of thermophilic organisms as well as being a causative agent of chemical and physical deterioration of texture, such as softening. High storage temperatures are encountered most often during summer weather in inland regions and among retailers or small distributors. The packer may be unable to control storage conditions, except as he is able to educate jobber and retailer on the importance of proper handling in maintaining quality.

## Examination Of Canned Fishery Products

Laboratory examination of canned fishery products is made for two purposes, for the enforcement of food and drug regulations and for control purposes to enable the manufacturer to improve the quality, or as a means of determining the quality when the manufacturer is disposing of his pack.

The U.S. Food and Drug official uses much the same methods as the trade examiner to determine whether a canned product is good and wholesome, but if spoilage or inferior quality is indicated in addition, chemical analysis with elaborate and complicated analytical techniques are sometimes required. Chemical analyses are necessary in legal proceedings as the evidence must be sufficiently strong to convince the court that the canned food is not good and wholesome. Methods of analysis which are recognized in court are found in the manual of the Association of Official Agricultural Chemists (1940). In addition, food and drug officials and manufacturer's chemists may offer evidence based on analytical methods from other sources.

In most instances the food technologist in making examinations for control purposes does not require an elaborate analytical technique. He makes use of a few simple tests and depends principally on accurate observation and keenness of the physical senses.

The advantages of an adequate system of grading have been acknowledged from time to time, but little progress has been made in the establishment of grades. With a few exceptions, at the present time, grade depends on species and canning area or is based on the reputation of a brand, rather than on established quality factors. Methods of examination may vary between laboratories and the criteria of the individual examiner may vary from time to time. Published information of value is limited. Only in the case of salmon has a determined effort been made to work out a systematic practical method for the examination of a canned fishery product. The results of the study of the Northwest Branch, National Canners Association, on the examination of canned salmon should be carefully studied by all canning technologists.

## Description Of Lot

A systematic method for the examination of any canned fishery product begins with a description of the particular lot to be examined. Descriptive data must be very complete and should include the number of cases, size of the container, code mark, variety of product, brand or label and the location of the parcel. The place where the parcel is stored should be noted for several reasons, of which the most important is that the quality of the product may be affected by conditions of storage. The packer, cannery, date packed and any other points that may aid in identifying that particular lot should also be noted. It is suggested that a label from one of the sample cans be attached to the record sheet for reference in questions arising after disposition of the sample.

Coding has been developed to the extent that it is possible to determine even the retort load in which the cans were processed. However, a great deal of improvement is needed in the coding of fishery products. Some products are only coded to give the packer and season when packed, while other products are not coded at all. An accurately coded product not only simplifies the work of the food examiner, but enables him to furnish more accurate and detailed information to the packer.

A sample of any fishery product should consist of at least 24 cans, if the lot is less than 1,000 cases. If the pack is new to the examiner or if it is questionable, a larger number of cans is required. One half of the sample should be set aside for reference, while the remainder is examined. If the results of the examination are questioned, authentic samples are then available for use in any controversy, or as evidence in case of legal action. If the cans are found to be uniformly good in quality, the unopened portion of the sample may be returned to the packer if not needed.

When the sample is taken, the cases opened should be inspected and the external condition of the cans noted, especially the condition of the enameled ends, cleanness of the cans, condition of the labels, and the presence of swelled, leaky, rusted and dented or battered cans. While the percentage of defective or damaged containers should be determined, no can which is obviously abnormal should be included in a sample taken for examinations.

A sample should not be examined immediately after it is drawn from the pack, but should be held in the examining laboratory until it has reached room temperature, preferably about 65 degrees F., or for a period of about 48 hours. Samples examined immediately after removal from the warehouse may give misleading results. If they have been stored at very warm or cold temperatures, the significance of the vacuum obtained is difficult to detect abnormal ("off") odors, or odors of decomposition.

## Bacteriological Examination

A bacteriological, or microscopic examination of a canned fishery or other food product is made to determine the number and types of any organisms present, to confirm data in heat penetration tests, to determine the quality of raw material used, the quality of the pack, or the cause of spoilage, if present. The number of samples examined bacteriologically will vary with the type of pack and character of information desired, but should not be less than 12 cans as a minimum and a sample of 24 cans is preferable. Data obtained as a result of bacteriological examination may be misinterpreted unless correlated with the results of a careful physical examination and is of the fullest value if processing data are also available.

Samples intended for bacteriological examination should be first incubated for at least a week at a temperature of 55 degrees C. (131 degrees F.), as thermophilic organisms develop only at high temperatures and physical changes are also hastened under such conditions. When possible, bacteriological examinations should be conducted in a room separated from the rest of the laboratory to reduce the possibility of contamination.

### Preparing The Sample

The tops of the cans should first be washed thoroughly with soap and water, then dried. The top of the can is usually sterilized by flaming with a Bunsen burner, though it may also be done by pouring a small amount of alcohol on the can and igniting.

When the top is sufficiently cool, the vacuum reading is taken, disinfecting the tip and rubber stopper of the vacuum gauge by dipping it in formalin. The can may then be opened with a sterile can opener flamed in a Bunsen burner each time used, or by punching a hole one-half inch in diameter using a screw driver or brand awl. The can top should be flamed again as an added precaution against contamination. If the container is abnormal, that is, a swell, a sterile towel should be wrapped around the tip of the punch when making the hole to prevent any of the liquid contents from spurting into the room.

### Taking The Sample

If the contents include considerable liquid the sample may be drawn out through a straight piece of glass tubing, sterilized before use. Tubing is preferable to pipettes as it is more easily cleaned, is not blocked as readily and a more representative sample may be obtained. A sample of approximately 1 cc. of liquid and small particles of solids is required for each inoculation. If there is not sufficient free liquid, a cut is made into the center of the contents with a sterile scalpel and a portion of about 1 gram is removed with sterile forceps for each inoculation. Aerobic and anaerobic plates and tubes should be inoculated for each container. Special media are not required in most cases. Dextrose agar and peptone broth are usually satisfactory culture media. As a rule plates and tubes are incubated 72 hours at 86 degrees F. (30 degrees C.) before being examined. If results are negative, an additional incubation of 48 hours at 131 degrees F. (55 degrees C.) may be given to confirm the negative results and test the presence of thermophilic, heat-resistant organisms, or two sets of cultures may be taken, one incubated at each temperature.

### Direct Microscopic Examination

Microscopic examinations are made by removing a drop of liquid from the can with a sterile loop and placing it in a hanging drop or by preparing and staining a smear on a side.

Clark, *et al.* (1923), describing the microscopic examination of canned salmon, a type of examination equally applicable to all canned fishery products, stated "...that direct microscopic examination of the liquor...is sometimes of value, in that often one may determine at once without waiting for results of the bacteriological tests, whether active spoilage is present. It is also of value in confirming results of the physical and chemical examinations, because in canned decomposed salmon, even when properly processed, large numbers of dead bacteria may be found by use of the microscope. Canned fresh salmon show few or no dead bacteria by this method. By the use of the microscope, certain parasites and pathological conditions may also be studied."

The results of microscopic examination are then interpreted in the light of available details of the process of manufacture and in correlation with data secured from physical and chemical examination. It should be possible to determine (1) whether further spoilage will occur, (2) whether bacteria gained entrance through defective seams or leaks, and (3) whether bacteria resisted the processing temperature. Inactive or dormant spore forms are not of great significance, as they do not develop due to unfavorable conditions. Bacteria common to the dust and soil, if present, probably come from dust in the empty can. The most common causes of non-sterility are believed to be poor seaming by the canner and under-processing.

### Physical And Organoleptic Examination

Factors included under a physical and organoleptic examination may be separated into three groups: (1) Quality of the raw material when caught; (2) quality of the raw material when canned; and (3) workmanship in packing.

### Quality When Caught

The quality of raw material when caught is not uniform for any species. Variation is best illustrated, and has been most thoroughly studied in salmon canning, but is found in other products. Salmon for canning are caught only during the spawning migration. Once started on the journey, they no longer feed, but depend on stored-up body fat for nourishment. "Therefore, the amount of this fat within their bodies gradually decreases, and fish which are caught late in their migration period are poorer in quality and lower in food value. Not only does the amount of fat decrease, but the color of the flesh fades out and the canned product has a poorer color. In general, salmon packed early in the canning season are of better quality than those packed later" (Clark, *et al.,* 1923).

In some products size is important in determining the quality when caught. Marine sardines are considered of higher quality if small, shrimp if they are large. Size may vary with the locality and time in the season caught. It is also affected by the density of population on a fishing ground and amount of food present in the water, which may vary with the season in any given area. Degree of temperature and change in salinity may affect size ranges, as with oysters.

Color of the flesh is important for canned salmon, shrimp and crab. In fact, a characteristic color has been established for these species. Color is noted in the examination of other canned fishery products, but in them the color is a factor in workmanship in packing or in determining the quality when canned.

The amount of free liquid and oil may be used as an indicator of quality in such species as salmon, shad, alewife or river herring and mackerel. The average amount of oil and liquid per can has been carefully worked out for each species of salmon. In living fish the amount of water in the flesh tends to increase as the fat content decreases. If the amount of liquid is much higher than is usual, the amount of free oil is low, indicating low quality fish. This factor is not significant in products such as "wet-pack" shrimp and oysters packed in brine.

### Quality When Canned

Fish and shellfish are more delicate in structure than other flesh foods, are readily injured and decompose rapidly. A fish may be of good quality when caught but poor when canned. The original quality may have been destroyed by the method or conditions of handling the catch, as described in the discussion of handling and transportation. Examiners for the enforcement of U.S. Food and Drug Administration regulations are especially interested in the condition of the fish when canned, *as the law stresses fitness for food, rather than standards or grades of quality.*

The factors used in determining quality when canned are odor, texture, reddening of the flesh, "honeycombing" and turbidity of liquid.

### Odor

Odor is the most important and most reliable indication of decomposition in the examination of canned fishery products, and is usually the factor which determines whether a can of fish shall be considered as fit for food. In determining the odor a great deal depends on the keenness of this sense in the examiner. Some persons have an abnormally sensitive sense of smell, particularly those with a "weak stomach." Others have a deficient sense of smell. The sense of smell also may be affected by the use of liquor, or by smoking just before an examination.

In smelling salmon or other fish a handful of the contents of the can is broken between the hands, immediately after which it is opened up and held close under the nose. Odor may be classified as good, stale, tainted and putrid. Fishery products canned when fresh have a normal "marine" odor, not at all unpleasant or disagreeable. The odor may be lacking, but unless there is a definite odor of staleness, the classification should be "good."

If stale fish are canned, there is a definitely abnormal odor, which is best described as a slight odor of decomposition. The odor disappears after the contents have been broken up and exposed to the air for a few minutes. Such a product may be considered to be of poor quality, yet not unfit for food.

Fish canned when tainted, that is, after decomposition has definitely set in, will give off an unmistakable odor of decomposition which does not disappear after the cans have been opened and allowed to stand for a few minutes. The examiner lists such an odor as "tainted" and the product is regarded as unsalable, although it may not be actually harmful if eaten.

The classification of "putrid" is reserved for an extreme degree of decomposition. In some cases the odor is persistent and disagreeable, noticeable even at a distance as soon as the can is opened and by a person unacquainted with the examination of canned foods. Such cans, of course, are unmerchantable and the contents are inedible. Shellfish canned when spoiled, have the most offensive odor of putridity and if this odor is found in a sample of canned clams, for example, it would appear in almost every can, making a detailed examination impossible and unnecessary.

Some odors may be encountered which, while perhaps abnormal, are not odors of decomposition, and should not be considered as indications of spoilage. Canned crustaceans, such as "dry pack" shrimp, crab and lobster may give off a slight ammoniacal odor, which is probably due to degree of process and does not affect the quality or fitness for food. Sometimes a "musty" or "muddy" odor is detected, usually in fish taken in fresh water, which are soft fleshed and have not been given a pre-treatment to improve the texture. Clams may have a "seaweed" odor, but are perfectly edible, though the appetite appeal may be lessened.

Canned salmon sometimes has a "weedy" or "grassy" odor, which is usually associated with strongly "water-marked" fish. Fish which are caught with hook and line in salt water usually contain partially digested food and often have a characteristic odor not in any way due to decomposition of the flesh. Sometimes a "scorched" or "caramelized" odor is noted which may be due to over-cooking. Canned smoked fish have a "caramelized" or "creosote" odor if smoked too heavily before canning.

## Texture

The texture of the flesh may be used as an indicator in some products, but there is such variation that it must be used with caution and only in conjunction with other factors. Softening of texture usually accompanies and corresponds roughly in amount to the degree of decomposition as judged by the odor. It should also be remembered that fat fish will be softer than thin fish of the same species. The portion of the body may make a difference in texture, as for instance, belly flesh may be softer because it is more oily than other sections of the body. Excessive handling of the cans before examination may also adversely affect the texture.

## Reddening of the Flesh

Reddening of the flesh is especially important as an indicator of spoilage in the examination of canned salmon but may be used with other products. According to Clark, *et al.* (1923) "The flesh of raw salmon if not promptly canned takes on an unnatural bright red appearance and this color persists through the processing; but when the can is opened this unnatural or so-called 'feverish' red color quickly fades and can usually be distinguished from the true color of the fish. Furthermore, this unnatural reddening is unevenly distributed and most likely to be observed at the tips near the gills, next in the belly walls, and least of all in the back flesh."

Reddening of this type may also be observed in white fleshed fish such as canned herring or sardines, especially along the belly cavity. Reddening in canned roe may be due to poor washing, but roe packed when sour also shows a definite "bloody" color, evenly distributed throughout the mass.

## Honeycombing

Occasionally in packs where the contents of a can consists of two or three pieces solidly packed, it will be noticed that the flakes of flesh are perforated by small holes. The accepted theory accounting for this condition is as follows: When more or less decomposed fish is canned, there is a considerable production of gas in the flesh caused by the growth activity of gas forming bacteria. When such fish are processed, the gas expands and makes little pockets in the flesh. On cooling, the pockets remain and the flesh seems to be filled with small holes or air spaces. This appearance is called "honeycombing," and was originally found in canned salmon but may occur in other products such as tuna and sardines. If a small piece of "honeycombed" flesh is placed on the tip of the tongue and held there for a short time, a sharp biting taste is sensed, a flavor like that of "sharp" or "old" cheese. This test is not recommended if the can is obviously spoiled.

## Turbidity of Liquid

The degree of turbidity or color of liquid is sometimes used as a factor in determining condition when canned. For instance, in canned oysters a grey liquid is considered a sign of old or inferior oysters. If the product is packed when fresh, the free liquid extracted in processing or the added brine is fairly clear. The degree of turbidity tends to increase with advance in decomposition. However, the liquid may also be turbid if the product has been handled roughly, is examined at a very low temperature or has been stored at a high temperature. Turbidity is useful as a corroborative index of spoilage but should not be considered apart from other evidence.

## Workmanship in Packing

Workmanship in packing is an important factor in establishing the quality of the canned product and under this head are included all those canning operations, the effects of which may be observed in the canned product.

## External Appearance

Examination for workmanship begins with a determination of the external appearance of the container and is usually made in sampling or as the first stage in laboratory examination. If there are rust spots, the sides are dull and dirty, the enameled ends are scratched or show "peeled" spots, or if the cans are dented or battered, then the external condition of the container does not show good workmanship in packing. Labels should be fresh, clean and properly placed on the cans. Defective labels should not be used, nor should they be torn or scratched. The product may be of fancy grade, but if through carelessness or inattention it reaches the grocer's shelf with the faults listed above, the appearance will unfavorably attract the consumer.

## Vacuum

Vacuum is determined just before bacteriological samples are taken or the cans are opened. A gauge equipped with a hollow piercing point surrounded by a soft rubber gasket is pressed down into the lid, giving a reading indicating the amount of vacuum in inches. A vacuum of 7 inches or more is generally regarded as satisfactory, but it should average 9 to 12 inches to be classed as good. An average vacuum meeting this standard is not satisfactory if any considerable number of cans in the sample show an extremely low or no vacuum, though there may be no signs of internal pressure. Any appreciable percentage of low vacuum cans indicates defective seaming in the cannery, poor vacuumizing, overfilling or similar mechanical defects.

## Headspace

The cans are opened by any cutter which will remove the entire end, either just within the top seam or cutting around the side just below. Headspace is measured from the top of the contents to the bottom of the lid and is recorded in fractions of an inch, usually sixteenths. The amount of headspace necessary for good canning practice varies to some extent with sizes of cans. For the great bulk of fishery products, 3/16 inch may be considered satisfactory. If the variations are too wide, the amount of headspace is an indication of over or under filling.

## Amount of Liquid

When the headspace has been determined, the drained weight may be taken, if necessary, after which the contents of the cans are emptied into dishes. Shallow white enamel pans are very satisfactory for this purpose. The amount of liquid is measured by pouring into glass cylinders, graduated in cubic centimeters. In products where turbidity determinations are made on individual containers, a 100 cc. cylinder is used, but when, as in the examination of salmon, free liquid and oil are judged on the basis of a dozen containers, a 1000 cc. graduate is necessary.

## Fill

Fill includes not only the total area of the container taken up by the contents as determined by headspace and net weight, but also *how* the contents are packed. Fill varies with the product, and to some extent, depends on whether the can has been filled in mechanically or by hand. In products such as salmon, shad or tuna, there should not be more than two or three pieces, and the cut ends should be packed facing the can ends with the sides parallel to the sides of the containers. Pieces should not be jammed or crumpled in. Ends should not be ragged and uneven but clean and smooth.

In products such as Maine sardines, the fish should be packed with tails to the center and with sides at an angle, so that only the silver skin shows. In sardines and herring the skin and flesh should not be broken in filling. Fill is judged by the drained weight, amount of liquid and headspace in products such as shrimp and oysters. In products such as chowders, fill is judged by the proportion of the ingredients, that is, has an even mixture been obtained?

As a rule, under proper supervision, hand filled packs show a more attractive appearance. If large herring or mackerel are packed in tall cans, the contents should be removable in an almost solid single cylinder. It is bad filling practice in this type of pack to use small ends to fill the center of the can or spaces caused by the size and shape of the fish. If the fish are properly selected for size, trimmed and filled in carefully, alternating heads and tails, there should be little if any need to use scraps and the fill will be more attractive. An occasional unsatisfactory fill may be disregarded, as a small percentage will be found in most carefully filled packs, but if such cans occur with any degree of frequency, the fill must be classed as poor.

Consistency of the product is distinguishable from texture in some products, where the degree of consistency is affected by and is an indicator of workmanship in packing. In chowders or soups, for instance, consistency, that is, the amount of liquid or "thickeners," is variable at will. The consistency is poor when the chowder is too thin or if too much thickener has been used. In products such as fish pastes, if the texture is rubbery and the pastes are difficult to spread, the consistency is regarded as unsatisfactory. If canned fish cakes crumble when formed into cakes, consistency is poor.

## Cleaning

Cleaning varies to some extent with the type of product. The presence of skin in any amount in fish flakes, fish chowder or fish cakes would be considered poor cleaning, but not in other products. The presence of fins is normal in sardines, but indicates poor cleaning in salmon. As a rule, intestines or other offal should not be found and all clotted blood which is removable should have been washed out. Blood may settle behind the backbone and sometimes fish are bruised before the blood has congealed, leaving a discolored bloody area in the flesh. Such blood clots are not removable and are not considered as evidences of poor cleaning. The presence of foreign objects in the container or extraneous dirt originating from the empty cans or the salt used in packing should be considered as poor cleaning.

## Color

Color may be affected by workmanship in packing. Clams given an over-process will show discoloration. Improper preparation before canning will adversely affect the color of fish chowder or fish cakes and these products are easily discolored by processing at too high a pressure or for too long a period. "Stack-burning" may be a serious cause of discoloration, if canned fishery products are not promptly cooled. Some discoloration or "off" shades of color may be due to insufficient cleaning or washing.

## Cook

Sufficiency of cook or process may be determined accurately only on the basis of heat penetration studies and by the results of bacteriological examination. In some products, especially canned salmon, where the contents include relatively large vertebrae, friability of vertebrae is used in this determination. The vertebrae should crumble easily when rubbed between the finger tips, and rib bones should be brittle, snapping off readily without bending. According to Clark, *et al.* (1923), there does not seem to be any close relationship between sterility and softness of the bones. Cans containing soft bones have sometimes been found to contain living bacteria, while on the other hand, cans with hard bones are often sterile. Hardness of bone varies with species and size of fish. The gelatinous substance within the vertebrae should be well coagulated and opaque.

## Seasoning

Seasoning in the canned product depends on the individual taste and should not be subject to rigid standards. Salt or the various condiments should be added only in an amount to satisfy the average taste. The product should not be flat and insipid but neither should it be too heavily salted. Some products are salted more heavily than others to improve the texture, or in other cases to mask a "muddy" flavor. "Water-marked" salmon for example are salted more heavily than fish not in this condition. Increased salt seems to improve flavor in the more poorly flavored fish. It should also be remembered that salt and condiments, or sauces such as oil, require several months after canning to penetrate the flesh thoroughly. Tuna, salmon or sardines may seem insufficiently salted immediately after canning, yet will be found satisfactory a few months later.

Flavor, as distinguished from amount of salt, should be observed in some products, especially those containing a variety of ingredients. The salt or other seasoning may be sufficient, yet the combination may not give a pleasing flavor. Clam chowder, fish pastes, soups, fish cakes and most specialty products may be cited as belonging to this category. In simple products which have not undergone an elaborate process of manufacture, a caramelized or overcooked flavor may indicate over-processing. This applies especially to products such as minced razor clams or Norwegian style fish balls where the upper and lower limits of processing are close together.

## Net Weight

Each can must be correctly marked as to the net weight of the contents to comply with the requirements of the Food, Drug and Cosmetic Act. The examiner should weigh at least 12 unopened and 3 empty cans, calculating the net weight as an average. It is sometimes desirable to record also the net weight of each container. The net weight requirement is a minimum requirement. The average must at least equal, but may exceed it without violation of the law. It should be remembered that if there is a wide variation in net weight between individual containers of the sample, workmanship is poor, even though the average net weight meets the requirement. As a rule, the actual net weight of fishery products is in excess of that declared on the label.

It is also necessary to ascertain the drained weight in the examination of certain products such as clams, oysters and wet pack shrimp to which considerable brine or liquid is added in filling, or which are liable to shrinkage in processing.

## Chemical Examination

Chemical examination of a canned fishery product follows the same general method as the examination of any other protein food. As stated in the introductory paragraphs to this discussion, a chemical analysis is required in the regular examination of canned fishery products almost solely as confirmatory evidence where the question of spoilage is involved. As physical evidence obtained by the senses may be questioned in court, determination of freshness or the presence of decomposition products must be established by exact methods with the minimum possibility of mechanical error.

Several analytical procedures are or have been used to determine decomposition. A great deal of research has been conducted on this subject in the last twenty years, but a generally satisfactory method has not yet been developed. Volatile acid content in canned salmon may be useful as an index of decomposition, but this method has not yet been accepted for the examination of salmon and it is not known whether it can be adapted to the examination of other types of canned fishery products. Electrometric methods to determine the freshness of fish fillets have not yet been adapted for use with canned fishery products. The indole test, long used as the principal indicator of decomposition, is now discarded, as are methods based on the determination of ammonia or various nitrogen fractions.

## Records

It is necessary to keep systematic records of examinations for several reasons. In the first place the examiners' work may be readily questioned if this is not done. It is also essential for control purposes, especially in tracing responsibility for certain practices. Improvement of quality requires an immense amount of information on methods in present use. This is accurate and reliable only if detailed records have been kept of the examination of many sample cans of the products, packed at various times in the season, under varying conditions, and from as many locations as possible in the area where the product is canned.

## Cannery Inspection

Canneries packing fishery products are subject to inspection for regulatory purposes and for the purpose of checking on the methods in use, or to gather information which may enable canners to improve the quality of their pack, correct errors or increase efficiency in methods of production. The latter type is usually voluntary and is organized by trade associations for the benefit of their various members, or it may be a strictly private service operating as a part of the research department of a large canning firm.

Regulatory inspection may be under either State or Federal control. State control may vary from casual inspections by local public health or food and drug officials, to a constant control and check on every lot of canned fishery products processed. Federal control may vary from a general supervision which endeavors to remove adulterated or unfit goods from passing into interstate commerce, but does not set up standards of quality, to an inspection system devised for the supervision of packing individual products such as shrimp, and may control methods of production, packing and processing in individual canneries.

## Types of Inspection

For purpose of illustration, a good example of state cannery inspection is found in California. The canning of fishery products is controlled by state health authorities who prescribe equipment and processing methods. Automatic temperature control and recording devices are required. The pack must be coded and complete records must be kept of each lot canned. No product may be sold until the cannery inspection service is convinced that it is sufficiently processed and complies with all regulations. Packing and processing regulations are based on the result of research studies, data obtained from the packing industry and from the Federal government. Regulations are revised from time to time whenever the need of change is indicated.

In canning salmon, most important of the canned fishery products, dependence is placed on inspection of the pack rather than on inspection of production. The Northwest Branch, National Canners Association, inspects about 90 percent of the salmon packed in the United States and Alaska. Most of this work is performed at its laboratory in Seattle, Washington. This inspection is primarily intended for the detection and elimination of any lots of canned salmon which might be considered unfit for food, but if the packer so desires, a complete organoleptic examination is made, obtaining considerable information on workmanship in packing and similar factors, which has made it possible to improve methods of production and quality.

## Advantages and Disadvantages

Inspection at the cannery is regarded as more effective than inspection of the pack at the primary marketing point. The advantages of pack inspection are that it can be handled by a much smaller personnel and the cost of inspection is not as great. The disadvantages of pack inspection are that it is often made too late to change the current pack. Any alterations necessary must be instituted the following season, except that if improper sealing is noted it is possible to telegraph the cannery so that the necessary changes may be made. With pack inspection it is more difficult to pin down the exact cause of defects appearing in the pack or recommend changes necessary for improvement.

The disadvantages of cannery inspection are greater cost of operation and the difficulty of obtaining a properly trained personnel. Such work is apt to be seasonal with a comparatively brief period of employment. Properly qualified men with necessary technical or practical knowledge do not ordinarily care to accept employment under such conditions, or if they do, must be frequently replaced, which requires the training of new men with a loss of efficiency during the training period.

## Requirements For A Cannery Inspector

Men hired as fish cannery inspectors should meet certain requirements. They should have training in chemistry and bacteriology. Courses in sanitation and public health are desirable and also education in the fundamental principles of food preservation. Practical experience in the fish canning industry is also very useful. A man with academic training only will acquire a knowledge of methods in time, but as this is gained largely by observation is never as intimate or accurate as when gained by active participation. Over dependence on observation and on literature dealing with the industry is apt to lead to erroneous conclusions. On the other hand the man who has practical training only may not realize the importance of the fundamental principles of canning. A well balanced combination of practical experience with academic training should give the best results.

The personality of the inspector is as important as his training. He should be tactful and, while insisting on the enforcement of inspection regulations, should avoid an arbitrary attitude. Common sense is a necessary requisite. Lack of this factor is one of the principal causes of antagonism to inspection. For example, a regulation required that the raw material must not come in contact with wood during preparation and packing. A parcel of the *canned* product was packed in wooden shipping containers. The inspector refused to certify the parcel on the grounds that the product was in contact with wood. While his decision was ultimately reversed, the packer was caused loss and annoyance, which might have been avoided with application of common sense.

### Conduct Of A Cannery Inspection

Regulations issued by the inspection agency making the inspection usually give complete instructions adapted to the situation. To summarize very briefly, inspection begins with the raw material and ends with the shipping of the finished product, and to be effective, must control any factor within this period affecting the quality of the pack.

The cannery inspector should avoid issuing orders directly to the cannery crew whenever possible. If he wishes anything done, a request should be made to the cannery foreman which does not mean that the foreman should be frequently bothered by minor requests. However, the order should be issued by the cannery foreman if it is necessary to take a man from his regular tasks to perform some work in connection with the inspection.

The inspector should have his records up to date at all times and should take care to avoid possible errors. For this reason, records should be checked frequently. If the packer claims error even though the inspector feels him to be at fault, the claim should not be disputed at length but referred to the central inspection authority. Cordial relations should be maintained with the packer, but the inspector should not allow himself to perform duties outside of the field of inspection or fraternize to such an extent that he is forgetful of his obligations to the inspection service.

In a few instances inspectors are selected from among the employees of a cannery. They are paid by the inspection service, which is reimbursed by the packer. Inspections carried out under such conditions are apt to be ineffective. The inspector is too closely identified with the interest of the packer to insist on enforcement of regulations if the packer disobeys, or the plant is so familiar that he is unable to see clearly the errors he is supposed to eliminate. If it is necessary to choose inspectors from among cannery employees, they should not be stationed with the packer from whom they were obtained. ■ ■ ■

# Index

*Illustrated*

*Illustrated

*\*Illustrated*

*Illustrated

9

| DATE DUE | | | |
|---|---|---|---|
| | | | |
| | | | |
| | | | |
| | | | |
| | | | |
| | | | |
| | | | |
| | | | |
| | | | |
| | | | |
| | | | |
| | | | |